KEY THINKERS ON DEVELOPMENT

Since its publication in 2006 as *Fifty Key Thinkers on Development*, this invaluable reference has established itself as the leading biographical handbook in its field, providing a concise and accessible introduction to the lives and key contributions of development thinkers from across the ideological and disciplinary spectrum. This substantially expanded and fully updated second edition in the relaunched series without the numerical constraint includes an additional 24 essays, filling in many gaps in the original selection, greatly improving the gender balance and diversifying coverage to reflect the evolving landscape of development in theory, policy and practice.

It presents a unique guide to the lives, ideas and practices of leading contributors to the contested terrain of development studies and development policy and practice. Its thoughtful essays reflect the diversity of development in theory, policy and practice across time, space, disciplines and communities of practice. Accordingly, it challenges Western-centrism, Orientalism and the like, while also demonstrating the enduring appeal of "development" in different guises. David Simon has assembled a highly authoritative team of contributors from different backgrounds, regional settings and disciplines to reflect on the lives and contributions of leading authorities on development from around the world. These include:

- Modernisers like Kindleberger, Perroux and Rostow
- *Dependencistas* such as Frank, Furtado, Cardoso and Amin
- Progressives and critical modernists like Hirschman, Prebisch, Helleiner Sen, Streeten and Wang
- Political leaders enunciating radical alternative visions of development, such as Mao, Nkrumah and Nyerere
- Progenitors of religiously or spiritually inspired development, such as Gandhi, Ariyaratne and Vivekananda
- Development–environment thinkers like Agarwal, Blaikie, Brookfield, Ostrom and Sachs
- International institution builders like Singer, Hammarsköld, Kaul and Ul Haq
- Anti- and post-development thinkers and activists like Escobar, Ghosh, Quijano and Roy

Key Thinkers on Development is therefore the essential handbook on the world's most influential development thinkers and an invaluable guide for students of development and sustainability, policy-makers and practitioners seeking an accessible overview of this diverse field and its leading voices.

David Simon is Professor of Development Geography at Royal Holloway, University of London, UK, and Director of Mistra Urban Futures, Chalmers University of Technology, Gothenburg, Sweden. He is the author, editor and co-editor of many books and journal special issues on cities, development–environment challenges and climate change adaptation, most recently *Rethinking Sustainable Cities: Accessible, Green and Fair* (2016), *Urban Planet* (2018) and *Holocaust Escapees and Global Development: Hidden Histories* (2019).

KEY THINKERS ON DEVELOPMENT

Second Edition

Edited by David Simon

Routledge
Taylor & Francis Group
LONDON AND NEW YORK

Second edition published 2019
by Routledge
2 Park Square, Milton Park, Abingdon, Oxon, OX14 4RN

and by Routledge
52 Vanderbilt Avenue, New York, NY 10017

Routledge is an imprint of the Taylor & Francis Group, an informa business

First edition published by Routledge 2006

British Library Cataloguing-in-Publication Data
A catalogue record for this book is available from the British Library

Library of Congress Cataloging-in-Publication Data
Names: Simon, David, 1957- editor.
Title: Key thinkers on development / edited by David Simon.
Description: Second edition. | Oxon ; New York, NY : Routledge, 2019. | Originally published under the title: Fifty key thinkers on development | Includes bibliographical references and index.
Subjects: LCSH: Planners–Biography. | Economic development–Planning. | Economic development–Cross-cultural studies. | Economic policy–Cross-cultural studies.
Classification: LCC HD87.55 .F54 2019 | DDC 338.9–dc23
LC record available at https://lccn.loc.gov/2018056342

ISBN: 978-1-138-49428-2 (hbk)
ISBN: 978-1-138-49432-9 (pbk)
ISBN: 978-1-351-02630-7 (ebk)

Typeset in Bembo
by Integra Software Services Pvt. Ltd.

CONTENTS

CONTRIBUTORS

Rita Abrahamsen is Professor in the Graduate School of Public and International Affairs at the University of Ottawa. She is the author of *Disciplining Democracy: Development Discourse and Good Governance in Africa* (Zed Books, 2000) and *Security Beyond the State: Private Security in International Politics* (Cambridge University Press, 2011, with M.C. Williams).

V.N. Balasubramanyam (Baloo) holds the Chair in Development Economics (Emeritus), University of Lancaster. His publications include *Multinational Enterprises and the Third Word, The Economy of India, International Transfer of Technology to India, Meeting the Third World Challenge* (with A.I. MacBean) and *Conversations with Indian Economists.* He has edited several books on international trade and investment and published articles on trade, international investment and development issues.

Jo Beall is a Professorial Research Fellow at the London School of Economics and Political Science (LSE). Her expertise lies in international development policy and social development, particularly in the context of cities. She was formerly on the Executive Board of the British Council, Deputy Vice Chancellor at University of Cape Town and Professor in International Development at the LSE. Her specialist research interests are gender and development; urban poverty and urban governance; and cities in conflict-affected states. She has significant research experience in South Africa, Pakistan and Afghanistan and has also undertaken research in Colombia, in Bangladesh and India, and in Uganda and Zambia.

Anthony Bebbington is Australia Laureate Professor at the University of Melbourne, Higgins Professor of Environment and Society in the

Graduate School of Geography, Clark University and Professorial Research Fellow at the Global Development Institute, University of Manchester. Previously he was at the University of Colorado at Boulder and has also worked at Cambridge University, the World Bank, the Overseas Development Institute and the International Institute for Environment and Development. His research, primarily in Latin America, has addressed: extractive industries and socio-environmental conflict; NGOs and rural social movements; poverty and rural livelihoods; agricultural development; and the links between development interventions and political economy.

Jagdish Bhagwati is University Professor of Economics, Law, and International Affairs at Columbia University. He was recently Director of the Deepak and Neera Raj Center for Indian Economic Policies and Senior Fellow for International Economics at the Council on Foreign Relations. He has been economic policy advisor to Arthur Dunkel, Director-General of GATT (1991–1993), special advisor to the UN on globalization, and external advisor to the WTO. He has served on the expert group on the future of the WTO and the advisory committee to Secretary-General Kofi Annan on the NEPAD process in Africa, and was also a member of the Eminent Persons' Group under the chairmanship of President Fernando Henrique Cardozo of Brazil on the future of UNCTAD. He has co-chaired the Eminent Persons' Group on Developing Countries in the World Economy with former President of Finland Tarja Halonen.

Tony Binns has been Ron Lister Professor of Geography in the University of Otago, Dunedin, New Zealand since October 2004. Prior to this, he was based at the University of Sussex, Brighton, UK, where he is now Visiting Professorial Fellow. Tony has worked in the field of Development Studies for over 40 years, with particular experience in Africa, where he has had a longstanding interest in food production systems, food security, rural change and community-based development. He is co-author (with Alan Dixon and Etienne Nel) of *Africa: Diversity and Development* (Routledge, 2012), co-author (with Robert Potter, Jennifer Elliott, Etienne Nel and David Smith) of *Geographies of Development*, 4th Edition (Routledge, 2018), and co-editor (with Etienne Nel and Kenneth Lynch) of *The Routledge Handbook of African Development* (Routledge, 2018).

Roberto Saturnino Braga graduated in civil and economic engineering, with courses from the Economic Commission for Latin America and the Caribbean (ECLAC) and Superior Institute for Brazilian Studies (ISEB, in Portuguese). He was employee at the Brazilian Economic and Social Development Bank (BNDES, in Portuguese) since the first bank competition in 1956. He had a long political life, as a councillor, federal deputy and senator for three terms, and was the first Mayor of Rio de Janeiro elected by direct popular vote. He is also a writer with 15 published books on politics and fiction literature. He presides over the International Celso Furtado Center for Development Policy and the Casa Grande Cultural Institute, both in Rio de Janeiro.

John Brohman was Associate Professor of Geography and a member of the Latin American Studies Program at Simon Fraser University in Burnaby, BC, Canada, until his death in 2014. His general research interests concerned theories, strategies and practices of development, and the principal geographical area of his later research was Latin America, particularly Bolivia. That research focused on regional development and planning, rural development and agrarian reform, community participation and popular movements, and neoliberalism and alternative globalisations.

Sylvia Chant is Professor of Development Geography at the London School of Economics and Political Science, where she directs the MSc Urbanisation and Development. A specialist in gender, poverty and development with field experience in Asia, Africa and Latin America, her recent books include the edited volume, *The International Handbook of Gender and Poverty: Concepts, Research, Policy* (Elgar, 2010), and *Cities, Slums and Gender in the Global South: Towards a Feminised Urban Future* (with Cathy McIlwaine) (Routledge, 2016).

Reginald Cline-Cole, a native of Sierra Leone, human geographer by training and multidisciplinary researcher by choice, undertook his undergraduate and postgraduate studies at the Universities of Sierra Leone (Fourah Bay College) and Bordeaux (Centre d'Études de Géographie et *Écologie Tropicale)*, respectively. He is with the Department of African Studies and Anthropology (formerly Centre of West African Studies), University of Birmingham, UK. His publications include *West African Worlds: Paths through Socio-Economic Change, Livelihoods and Development* (Pearson, 2005, co-edited with Elsbeth

Robson) and *Contesting West African Forestry* (Ashgate, 2000, co-edited with Clare Madge).

John Connell is Professor of Geography of the University of Sydney. He has written numerous books and articles about development in the Pacific region, most recently *Change and Continuity in the Pacific* (with Helen Lee) (Routledge, 2018).

Stuart Corbridge is Vice Chancellor and Warden of Durham University. He is the author, with Sarah Jewitt and Sanjay Kumar, of *Jharkhand: Environment, Development and Ethnicity* (OUP, 2004), with Glyn Williams, Manoj Srivastava and Rene Veron of *Seeing the State: Governance and Governmentality in Rural India* (Cambridge University Press, 2005) and with John Harriss and Craig Jeffrey of *India Today: Economy, Politics and Society* (Polity, 2013).

Christopher Cramer (FAcSS) is Professor of the Political Economy of Development at SOAS, University of London. His research and teaching interests include the economics of Africa, rural labour markets, primary commodity processing, and the political economy of violent conflict. He is the author of *Civil War is Not a Stupid Thing: Accounting for Violence in Developing Countries* (Hurst, London, 2006) and led a DFID-funded research project on Fairtrade, Employment and Poverty Reduction in Ethiopia and Uganda. He is a Vice-Chair of the Royal African Society, a former Chair of the Centre of African Studies in the University of London, and is Chair of the International Scientific Committee of the African Programme on Rethinking Development Economics (APORDE). He is a co-editor of the *OUP Handbook on the Ethiopian Economy* (2019).

Vandana Desai is Senior Lecturer in Development Geography at Royal Holloway, University of London. Her research interests lie in the areas of community participation, informal housing, displacement, the role of non-governmental organisations, gender and ageing. Her recent books include the third edition of *The Companion to Development Studies* (2014) and *Doing Development Research* (2006), both co-edited with Robert Potter.

Richard S. Eckaus has been at MIT since 1962, as Ford Foundation Professor of International Economics and was Head of the Department from 1987 to 1990, retiring in 1996. He has specialised in problems of developing countries and researched in many of these countries. He received the Great Cross of the Oder of Prince Henry the Navigator in Portugal in 1995. In recent years his interests have extended to the economic problems associated with climate change.

Arturo Escobar is Professor of Anthropology at the University of North Carolina, Chapel Hill. His main interests are: political ecology, ontological design, pluriversal studies and the anthropology of development, social movements, and technoscience. Over the past 25 years, he has worked closely with several Afro-Colombian social movements in the Colombian southwest, particular the Process of Black Communities (PCN). His best-known book is *Encountering Development: The Making and Unmaking of the Third World* (1995, 2nd edn 2011). His most recent books are *Sentipensar con la Tierra. Nuevas lecturas sobre desarrollo, territorio y diferencia* (2014; French edn May 2018), *Otro posible es posible: Caminando hacia las transiciones desde Abya-Yala/Afro/Latino-América* (2018; English edn 2019) and *Designs for the Pluriverse: Radical Interdependence, Autonomy, and the Making of Worlds* (2018).

Ben Fine is Professor of Economics at the School of Oriental and African Studies, University of London. Recent books include (with A. Saad-Filho) *Marx's Capital* (6th edn, Pluto, 2016); *Macroeconomics: A Critical Companion*, with O. Dimakou (Pluto, 2016); *Microeconomics: A Critical Companion* (Pluto, 2016); *Developmental Politics in Transition: The Neoliberal Era and Beyond*, co-edited with Kyung-Sup Chang and Linda Weiss (Palgrave Macmillan, 2012); and *Beyond the Developmental State: Industrial Policy into the 21st Century*, co-edited with J. Saraswati and D. Tavasci (Pluto, 2013).

Tim Forsyth is Professor of Environment and Development at the London School of Economics and Political Science, where he has been based since 2000. He holds a PhD in Geography from the University of London (1993) and has published widely on various aspects of environmental policy and development, specialising in the governance of scientific knowledge, livelihoods and institutions, social movements, community-based adaptation and other local participation in climate change policy.

Jean-Jacques Gabas has been a researcher at CIRAD (Centre for International Cooperation in Agricultural Research for Development) since

2009. He was a Lecturer in Economics at the University Paris Sud/XI between 1982 and 2016 and teaches at Sciences Po Paris. He is Honorary President of GIS-GEMDEV (Group for the Study of Globalization and Development). After having been technical assistant in Ivory Coast at the Ministry of Planning from 1976 to 1978, he carried out numerous missions for the OECD/Sahel Club in Africa. He has published many books and articles on co-operation policies, including *Le temps de la Chine en Afrique* (Karthala, 2012) and co-edited several issues of journals, including "La mesure du développement" (*Revue Tiers Monde*, 2013). His current work focuses on the analysis of emerging and particularly Chinese and Brazilian actors in sub-Saharan Africa and the transformations in the financing of agricultural policies in sub-Saharan Africa.

Gary Gaile was, until his premature death on 13 February 2009, Professor of Geography at the University of Colorado, Boulder, and Executive Director of the Developing Areas Research and Teaching (DART) Programme. He was also Co-Chair of the Association of American Geographers' Developing Areas Specialty Group. He earned his PhD in Geography in 1976 from UCLA. His research interests focused on Africa and included food security, rural–urban linkages, and micro-enterprise credit.

Nandini Gooptu is Fellow of St Antony's College, Oxford. Educated in Calcutta and at Cambridge, and trained as a social historian, she is the author of *The Politics of the Urban Poor in Early-Twentieth Century India* (2001), editor of *Enterprise Culture in Neoliberal India* (2013), joint-editor of *India and the British Empire* (2012) and *Persistence of Poverty in India* (2014). While her past research has been on colonial India, her current research is concerned with social and political transformation and cultural change in contemporary India in the wake of economic liberalization and globalization. She has written on a variety of subjects, including caste, communal and religious politics, urban development, poverty and labour, work and employment, media and politics, and social movement of sex workers.

W.T.S. (Bill) Gould is an Emeritus Professor of Geography of the University of Liverpool, UK. His most recent book, *Population and Development* (2nd edn, 2015), synthesises his long-standing research and teaching interests on the interactions between demography, economy and culture in the contemporary world, and especially in the global South.

Robert Gwynne retired as Reader in Latin American Development at the School of Geography, Earth and Environmental Sciences, University of Birmingham. Prior to that, he was also Visiting Professor at the Catholic University of Chile. His research interests focused on industrialisation in the developing world and on the impacts of neoliberalism and globalisation on regional development in Latin America. He is the author of *Industrialisation and Urbanisation in Latin America* (Routledge, 1985); *New Horizons? Third World Industrialization in an International Framework* (Longman, 1990); co-author of *Alternative Capitalisms: Geographies of Emerging Regions* (Arnold, 2003); and co-editor of *Latin America Transformed: Globalization and Modernity* (Arnold, 2004).

John Harriss is Professor, and Founding Director of the School for International Studies at Simon Fraser University. He was previously Director of the Development Studies Institute at the London School of Economics, and earlier the Dean of the School of Development Studies at the University of East Anglia. His research interests are now in democracy, civil society and governance. He is the author of *Reinventing India* (with Stuart Corbridge, Polity Press, 2000); *Depoliticising Development: The World Bank and Social Capital* (Anthem Press, 2002); and *India Today* (with Stuart Corbridge and Craig Jeffrey, Polity Press, 2013). He served as a Managing Editor of the *Journal of Development Studies* (1998–2004) and as Editor-in-Chief of the *Canadian Journal of Development Studies* (2010–2014).

Philipp Horn is a Lecturer in the Department of Urban Studies and Planning at the University of Sheffield. His research interests include inclusive urban development, indigenous rights in cities, participatory planning, and critical urban theory. He has extensive research experience in Latin America, particularly in Bolivia and Ecuador.

Huang Ping (PhD Sociology, LSE London) is now Senior Research Fellow and Director of the Institute of European Studies at the Chinese Academy of Social Sciences (CASS). He is Vice-President of the International Institute of Sociology, Vice-President of the International Social Sciences Council and Vice-President of the Intergovernmental Council for Management of Social

Transformation, UNESCO. Before this position he was the Director of the Institute of American Studies at CASS. He has published books and papers on social change, labour, migration and globalisation.

Craig Johnson is Professor of Political Science at the University of Guelph, where he has been based since 2002. He holds a PhD in International Development from the London School of Economics and Political Science (2000), and has published widely in the field of development, focusing primarily on questions of urbanization, sustainable livelihoods and the environment. His latest book is *The Power of Cities in Global Climate Politics* (2018), within the Palgrave Macmillan Pivot series on "Cities and the Global Politics of the Environment".

Rainer Kattel is Professor and Deputy Director at the Institute for Innovation and Public Purpose, University College London. Previously he led the Ragnar Nurkse School of Innovation and Governance, Tallinn, Estonia, for ten years, building it into one of the leading innovation and governance schools in the region. He has served on various public policy commissions, including the Estonian Research Council and European Science Foundation. He has worked as an expert for the OECD, UNDP and the European Commission. Currently, he leads the Estonian Government's Gender Equality Council. He has published extensively on innovation policy, its governance and specific management issues. His research interests include also public-sector innovation, predictive governance and financialisation. His recent books include *The Elgar Handbook of Alternative Theories of Economic Development* (edited with Erik Reinert and Jayati Gosh, Elgar, 2016).

Cristóbal Kay is Emeritus Professor at the International Institute of Social Studies, The Hague, The Netherlands, and at FLACSO, Quito, Ecuador. He is also a Professorial Research Associate at the Department of Development Studies, SOAS, University of London. His main research interests are in the fields of rural development and theories of development and underdevelopment, mainly within the Latin American context. His most recent books include (all as co-editor) *Latin America Transformed: Globalization and Modernity*, *Transnational Agrarian Movements Confronting Globalization*, and *Peasants and*

Globalization: Political Economy, Rural Transformation and the Agrarian Question.

Kong Yuan is a researcher at the Institute of European Studies (IES), Chinese Academy of Social Science (CASS). He holds a PhD in Jurisprudence from Peking University Law School and was Postdoctoral Fellow, Tsinghua Institute for Advanced Study in Humanities and Social Sciences (TIAS). His research interests are development theory, legal theory and cultural studies. He is editing a book series on China, empire and history of international law.

Reinhart Kößler is the former Director of Arnold Bergstraesser Institute, Freiburg, and Professor in the Department of Political Science of Freiburg University. He is also a Visiting Professor and Research Associate at the Institute of Reconciliation and Social Justice, The University of the Free State, South Africa.

Kuntala Lahiri-Dutt is a Professor at the Crawford School of Public Policy, Australian National University. Trained as a human geographer, Kuntala has researched the social and ecological politics of resources and gender extensively, initially in India and subsequently in other Asian countries such as Indonesia, Lao PDR and Mongolia. In particular, she has written on the challenges of gendered livelihoods both in large, industrial mining and in informal, artisanal and small-scale mines and quarries, on the displacement of peasantry and indigenous peoples and on the transformations of land and livelihoods in mining areas, and has published widely on these thematic topics. A related field of research comprises the social and gender challenges in water resource management at different geographical scales.

John P. Lea, who was born in South India during the last years of the "Raj", was educated in England and has spent most of his academic career at the University of Sydney, Australia. His work on urbanisation and development in southern Africa, Australasia and the Pacific has focused on housing, urban planning and governance issues at the resource frontier.

David Lewis is Professor of Social Policy and Development at the London School of Economics and Political Science. A social anthropologist and rural sociologist by training, he has undertaken fieldwork mostly in South and South East Asia. He is the author of several books, including *Anthropology and Development: Challenges for the 21st Century* (with K. Gardner, 2015), *Non-Governmental Organizations, Management and Development* (2014), *Bangladesh: Politics, Economy and Civil Society* (2011), *Non-Governmental Organisations and Development* (2009, with Nazneen Kanji) and *Development Brokers and Translators: The Ethnography of Aid and Agencies* (2006, edited with David Mosse).

Cathy McIlwaine is Professor of Development Geography in the Department of Geography, King's College London. In addition to researching gender and development issues in the global South for many years, Cathy has also worked on international and transnational migration in London with a focus on the Latin American community in relation to transnational livelihoods, citizenship and political participation among migrants from a gendered perspective. She has published ten books including *Cities, Slums and Gender in the Global South* (with Sylvia Chant (Routledge, 2016), *Cross-Border Migration among Latin Americans* (edited, Palgrave Macmillan, 2011), and *Global Cities at Work: New Migrant Divisions of Labour* (with Jane Wills, Kavita Datta, Jo Herbert, Jon May and Yara Evans; Pluto, 2010).

Henning Melber holds a PhD in Political Science and a *venia legendi* in Development Studies at Bremen University. He has been a Senior Lecturer in International Politics at Kassel University (1982–1992), Director of the Namibian Economic Policy Research Unit (NEPRU) in Windhoek (1992–2000), Research Director at the Nordic Africa Institute (2000–2006) and Executive Director of the Dag Hammarskjöld Foundation (2006–2012), both in Uppsala, Sweden. He remains affiliated to both institutions, is Extraordinary Professor at the Department of Political Sciences/University of Pretoria and the Centre for Gender and Africa Studies at the University of the Free State in Bloemfontein, and a Senior Research Fellow at the Institute for Commonwealth Studies/School for Advanced Study of the University of London. Since 2017 he has been President of the European Association of Development Research and Training Institutes (EADI).

Ulrich Menzel, born 1947 in Düsseldorf, Germany, was until 2015 the Chair for International Relations and Comparative Politics at the Technical University of Braunschweig and Managing Director of its

Institute of Social Sciences. He has published more than 30 books and numerous articles on the subjects of International Relations, International Political Economy, Peace Studies and Development Theory. Further information about the author can be obtained at www.ulrich-menzel.de.

Marlène Monteiro, PhD, worked with Ignacy Sachs as a research assistant between 2007 and 2014, in particular on the issues of rural development and energetic transition.

Joshua Muldavin is Chair of Social Sciences and Professor of Geography and Asian Studies at Sarah Lawrence College, New York. His research and teaching address global political ecology and political economy of development, environment, and food and agriculture, with a regional focus on Asia, particularly China, Japan and the Himalayan Hindu Kush. He has undertaken nearly four decades of field-based research, primarily in rural China as well as the greater Himalayan Region. He recently concluded a research project with Professor **Piers Blaikie**, "Conservation, sustainability and poverty alleviation in the Himalayan Region: do participatory environmental policies work?," and he is currently completing a book, *The Environmental and Social Impacts of China's Reforms and Global Integration.*

Ronaldo Munck is Head of Civic Engagement at Dublin City University and Senior Researcher at the Interdisciplinary Institute for the Study and Research of Latin America (INDEAL), University of Buenos Aires. Professor Munck has written widely on the globalization problematic – including *Labour and Globalization: The New "Great Transformation"* (2002) – and on Latin America – *Rethinking Latin America: Development, Hegemony and Social Transformation* (2015). He is on the editorial boards of several international journals such as *Globalizations, Global Labour Journal, Global Discourse, Labour History* and *Review*. He was a lead author for Amartya Sen's project, "The International Panel on Social Progress on globalization and its discontents". He is currently co-ordinating an action research project on social movements and globalization in Latin America.

Dani W. Nabudere was a senior researcher and executive director of the Afrika Study Centre, Mbale, Uganda, until his death in November 2011. He studied at Lincoln's Inn, where he was called to the Bar in 1963. Later, he spent six years as Associate Professor of Law at the University of Dar es Salaam, Tanzania, where he was able to observe at close quarters the effects of the *Ujamaa* strategy between 1973 and 1979. Subsequently, he was a minister of Culture and Community Development in the post-Amin interim government in Uganda. During his final years, he promoted the establishment of the Marcus Garvey Pan-Afrikan Institute dedicated to research and mainstreaming of African indigenous knowledge and wisdom, and was a frequent visiting researcher at the University of South Africa in Pretoria/Tshwane.

Anders Närman was, until his sudden death on 15 November 2004, Associate Professor of Human and Economic Geography, Göteborg University. His research interests included various aspects of Development Geography. Educational matters were a central part of his research in Kenya and Tanzania. Until his death, he was involved in research on conflict in Uganda, the Greater Horn of Africa, and regional development in Sri Lanka. These were collaborative projects, with the universities of Dar es Salaam, Makerere and Kelaniya respectively.

Michael Parnwell retired as Reader in South East Asian Geography in the Department of East Asian Studies at the University of Leeds. He has specialised on the development process in South East Asia since the late 1970s. His PhD was on return-migration in North East Thailand, and he maintained research interests in Thailand during the 1980s and early 1990s, focusing on processes of regional development, rural industrialisation and extended metropolitanisation. More recently he worked on the human impact of deforestation in Sarawak, Malaysia, small-scale industries and urban sustainability in Sulawesi, Indonesia, sustainable development in South East Asia, and on coping mechanisms and sustainable livelihoods in Thailand, Vietnam and the Lao PDR.

Richard Peet is Professor of Geography at Clark University in Worcester, Massachusetts. He has been editor of *Antipode: A Radical Journal of Geography* and *Economic Geography* and is presently editor of *Human Geography*. His books include *Modern Geographical Thought* (Blackwell, 1998), *Theories of Development* (with Elaine Hartwick, Guilford, 1999, 2002, 2015) and *Unholy Trinity: The IMF, World Bank and WTO* (Zed Books, 2003). He is interested in issues of globalisation and development, global governance and economic policy, culture, consciousness and ideology, and the possibility of post-neoliberal societies.

Marcus Power is a Professor of Human Geography at Durham University. His research interests are focused on Southern Africa and the Lusophone world and are concerned with geopolitics and development, China–Africa relations and low carbon energy transitions.

Sarah Radcliffe teaches at the Department of Geography, University of Cambridge. Her research focuses on the dynamics between development and postcolonial intersectional hierarchies, especially in relation to rural indigenous women, and the politics of development and state formation in the Andes. Recent publications include *Dilemmas of Difference: Indigenous Women and the Limits of Postcolonial Development Policy* (Duke University Press, 2015).

Jordana Ramalho is a final-year PhD student in the Department of Geography and Environment at the London School of Economics and Political Science, where she graduated with distinction in the MSc Urbanisation and Development in 2013. In addition to her doctoral research on gender, risk and resilience in informal settlements in Metro Cebu, the Philippines, Jordana has also co-authored (with Sylvia Chant and Martina Klett-Davies) *Challenges and Potential Solutions for Adolescents in Urban Settings: A Rapid Evidence Review* (Overseas Development Institute, 2017).

Jonathan Rigg (FAcSS) is a geographer based at the University of Bristol. Until the end of 2018, he was at the National University of Singapore and the Director of NUS's Asia Research Institute. He has worked on agrarian change, the environment, development and disasters in Southeast and South Asia since the early 1980s. He is the author of *Challenging Southeast Asian Development: The Shadows of Success* (Routledge, 2015) and has recently completed a monograph on agrarian change in Thailand, *More Than Rural: Textures of Thailand's Agrarian Transformation*, to be published by Hawaii University Press.

Barbara Rugendyke is Professor of Geography and Dean of Arts and Social Sciences at Southern Cross University in Australia. Barbara's publications have encompassed local level development issues and the work of development NGOs.

M.A. Mohamed Salih is Emeritus Professor of Politics of Development at the International Institute of Social Studies, Erasmus University Rotterdam (EUR), in the Netherlands. Professor Salih is co-laureate, Nobel Peace Prize, awarded in 2007 to the Intergovernmental Panel on Climate Change (IPCC), for his contribution to the Fourth Assessment Report on Climate Change. His latest books include: *Economic Development and Political Action in the Arab Word* (Routledge, 2014); *Local Climate Change and Society* (edited) (Routledge, 2012); *Hermeneutics, Scriptural Politics and Human Rights* (co-edited) (Palgrave Macmillan, 2011); *Climate Change and Sustainable Development: Challenges to Poverty Reduction in the 21st Century* (edited) (Edward Elgar, 2009); and *Interpreting Islamic Political Parties* (edited) (Palgrave Macmillan, 2009).

Roberto A. Sánchez-Rodríguez was a Professor of Environmental Studies at the University of California, Riverside, and Director of the University of California Institute for Mexico and the United States (UCMEXUS) and is now at El Colegio de la Frontera Norte in Tijuana, Mexico. He is also co-chair of Working Group 2 of the Intergovernmental Panel on Climate Change's 6th Assessment Report. His research interests lie in development studies, sustainable development, trade and the environment, and the interactions between urbanisation and global environment change.

Tilman Schiel (born 1943) studied Cultural Anthropology at the University of Heidelberg (MA) and Sociology at the University of Bielefeld (Dr.rer.soc. and 'Habilitation'). Until his retirement in 2004 he worked as an itinerant scientist at several universities and research institutions.

Sucharita Sen is a Professor in the Centre for the Study of Regional Development, Jawaharlal Nehru University. She was a Nehru–Fulbright Fellow in 2009–2010, and was associated with Ohio State University, Columbus, Ohio, USA during this period. She is currently the Executive Director of South Asia Consortium for Interdisciplinary Water Resources Studies. She is an economic geographer and has published in areas of natural resources and rural livelihoods. Her recent engagement and publications are on labour, gender, water and societal aspects of climate. She is the managing Editor of *SAWAS* (*South Asian Water Journal*) and an editorial member of *Asia Pacific Journal of Regional Science* (Springer).

Amogh Dhar Sharma is a doctoral student at the Oxford Department of International Development. His research examines the "professionalization" of electoral and party politics in India. His research interests include political communication, the role of political professionals in electoral mobilization and the changing nature of citizen's participation in democratic politics.

The late **John Shaw** was associated with the UN World Food Programme for over 30 years, latterly as economic adviser and chief of its Policy Affairs Service before his retirement. He also served as consultant to the Commonwealth Secretariat, the FAO and the World Bank. He then served on the International Editorial Board of the journal *Food Policy* and continued to write on development and food security issues, as well as writing the definitive biography of Hans Singer.

Xiaojing Lynette Shi has been a translator and international conference interpreter for over 30 years. She has translated both fiction and non-fiction, and co-authored two works on translation between Chinese and English. She was Associate Professor at the Graduate School of Translation and Interpretation and Language Studies, Middlebury Institute of International Studies in Monterey, California, and is now retired in Beijing, China. She recently translated a collection of Fei Xiaotong's writings and speeches from the last dozen years of his life: *Globalization and Cultural Awareness*, compiled by Fang Lili, 2015, (Foreign Language and Teaching Press and Springer).

James D. Sidaway is Professor of Political Geography at the National University of Singapore. His research currently focuses on the history and philosophy of geography, urban geopolitics, and wider geographies of security and insecurity ("securityscapes"). Mostly in collaboration with other scholars, for over a quarter of a century he has worked on the interfaces of urbanisation, development and power in Mozambique, Persian Gulf cities and Southeast Asia.

David Simon (FAcSS) is Professor of Development Geography at Royal Holloway, University of London, and Director of Mistra Urban Futures, an international research centre on urban sustainability based at Chalmers University of Technology, Gothenburg, Sweden. His particular research interests include development theory and policy; the development–environment interface; urbanisation and urban–rural interaction; transport and geopolitics. He has extensive research experience in sub-Saharan Africa and tropical Asia. His most recent books are *Rethinking Sustainable Cities* (Policy Press, 2016), *Urban Planet* (co-editor) (Cambridge University Press, 2018) and *Holocaust Escapees and Global Development: Hidden Histories* (Zed Books, 2019).

Rana P.B. Singh has been Professor of Cultural Geography and Heritage Studies at Banaras Hindu University, Varanasi, India. He has been involved in mass awakening programmes and heritage planning in Varanasi region for the last 35 years as promoter, collaborator and organiser. On these topics he lectured at various centres in America, Europe, East Asia and Australia. His publications include over 295 papers and 40 books on these subjects. He is the founding Vice-President of Asian Cultural Landscape Association (Seoul, Korea), and also of the Big History Association of India.

Morris Szeftel was Senior Lecturer in Politics at the University of Leeds until his retirement. He has also worked and taught at universities in the United States, South Africa and Zambia. He has published on African politics, problems of democratisation, development and debt, and on corruption in Africa. He has served as Editor of the *Review of African Political Economy* and, until recently, was an editor of *The Journal of Southern African Studies*.

Pervez Tahir is Chairman, Bank of Punjab, Pakistan. He has been the Joan Robinson Memorial Lecturer, University of Cambridge; Chief Economist of Pakistan Planning Commission, Mahbub ul Haq Chair at the GC University, Lahore, Senior Fellow; Pakistan Institute of Development Economics and Member, Statistical Advisory Panel of the Human Development Report. His articles span human development, macro-economic policy, fiscal federalism, economic history and thought. He authored *Economic and Social Thinking of the Quaid-i-Azam* (Research Society of Pakistan, 1980) and co-edited *Social Sciences in Pakistan* (Council of Social Sciences, 2005). Besides being on the boards of the *Sustainable Development Review* and Social Policy and Development Centre, he contributes regularly to the *Express Tribune*.

Jan Toporowski is Professor of Economics and Finance at SOAS, University of London; Visiting Professor of Economics at the University of Bergamo, Italy; and Professor of Economics and Finance at the International University College, Turin, Italy. After studying economics at Birkbeck College, University of London, and the University of Birmingham, he worked in fund management and international banking. He has published widely on money, finance and economic development, and is the author of a two-volume biography of Michał Kalecki.

Juha I. Uitto is Director of the Independent Evaluation Office of the Global Environment Facility (GEF). He has worked for more than three decades on international development and environment issues in the United Nations and the World Bank, and in academia. A geographer by training, he received his MSc at the University of Helsinki in his native Finland and PhD from Lund University in Sweden. He is the author of about 25 articles in peer reviewed journals on sustainable development, hazards and evaluation. His recent books include *Evaluating Climate Change Action for Sustainable Development* (lead editor; Springer 2017), *Universities and the Sustainable Development Future: Evaluating Higher-Education Contributions to the 2030 Agenda* (co-author; Routledge 2017), *Sustainable Development and Disaster Risk Reduction* (lead editor; Springer 2016) and *Evaluating Environment in International Development* (editor; Routledge 2014).

Ton van Naerssen was Associate Professor in Development Geography at the Nijmegen School of Management (NSM), Radboud University of Nijmegen, the Netherlands. He is currently research associate of the Nijmegen Centre for Border Research (NCBR) at the same university. His fields of interest include development theories and international migration. He has carried out research mainly in Southeast Asia. His most recent books are (as co-editor) *Mobility and Migration Choices* (2015), *Women, Gender, Remittances and Development in the Global South* (2016) and *Trajectories and Imaginaries: The Migrant Actor in Transnational Space* (Routledge, forthcoming).

Krystyna Vinaver is a researcher in social sciences. She collaborated with Ignacy Sachs between 1970 and 1985 and co-founded the CIRED at the EHESS, where she worked between 1973 and 2013. She specialises in technology and growth; appropriate technology for development; ecodevelopment and habitat; environmental crisis in

Eastern Europe; rural and sustainable development; as well as regional sustainability in Poland.

Michael J. Watts is Class of 1963 Professor of Geography at the University of California, Berkeley, where he has taught geographical theory and development studies for 40 years. His 1983 book *Silent Violence* was reissued with a new Introduction in 2012; his recent books include *Grassroots Environmental Governance*, edited with Leah Horowitz, *Genealogies of Environmental Thought: The Lost Works of Clarence Glacken*, edited with Ravi Rajan, *Subterranean Estates: Life Worlds of Oil and Gas*, edited with Arthur Mason and Hannah Appel, and *West of Eden: Communes and Utopia in Northern California*, with Iain Boal, Cal Winslow and Janferie Stone. He is completing a book on petro-politics in Nigeria.

Katie Willis is Professor of Human Geography at Royal Holloway, University of London. Her main research areas are gender, development and migration, with a particular focus on Mexico, China and Singapore. She is Vice-President (Expeditions and Fieldwork) at the Royal Geographical Society (with IBG) and the author of *Theories and Practices of Development* (3rd edn, Routledge, 2019).

Francis Wilson was trained in Physics (at the University of Cape Town) and Economics (Cambridge). He has been working in the School of Economics at UCT for some 50 years. In 1975 he founded SALDRU, the Southern Africa Labour and Development Research Unit, which he directed until 2000, when he started Data First, a university resource unit gathering, cleaning and making available social science data from different countries in Africa. His main work has been in labour (gold mines; migrants; farms), South African history, data collection and poverty, on which he directed the second Carnegie Inquiry into Poverty and Development in Southern Africa during the 1980s. More recently he has been part of a national team working on strategies to overcome poverty and inequality.

Ben Wisner flew around the Caribbean in a small plane with Jim Blaut, learned about peasants and tropical soils from him and misses him sorely even though he has been gone for nearly two decades. Wisner is Visiting Professor in the Institute for Risk and Disaster Reduction, University College London.

Lakshman Yapa hails from Sri Lanka, where he took his first degree at Peradeniya University (then the University of Ceylon at Peradeniya). He holds a PhD in Geography from Syracuse University in New York. He is a Professor Emeritus of Geography at Pennsylvania State University and past Director of the service-learning project entitled "Rethinking Urban Poverty: Philadelphia Field Project". His publications have addressed critical perspectives on scarcity, poverty and poverty discourses in Sri Lanka and the USA.

Henry Wai-chung Yeung is Distinguished Professor at the Department of Geography, National University of Singapore. He specialises in the geography of transnational corporations, global production networks, East Asian firms and developmental states in the global economy. He has published six authored books (most recently *Strategic Coupling: East Asian Industrial Transformation in the New Global Economy*, Cornell University Press, 2016, and *Global Production Networks*, Oxford University Press, 2015), seven edited books, and over 95 journal papers and 50 book chapters. He is Editor of *Environment and Planning A* and *Economic Geography* and past Editor of *Review of International Political Economy*, and serves on the editorial boards of 20 other international journals.

Alfred Babatunde Zack-Williams is Professor of Sociology at the University of Central Lancashire, Preston, UK. He was formerly Senior Lecturer in Sociology at the University of Jos, Nigeria. Among his recent publications are: *Africa in Crisis: Challenges and New Possibilities* (ed. Pluto Press, 2002); *Africa Beyond the Post-colonial* (ed. with Ola Uduku; Ashgate, 2004); *The Politics of Transition: State, Democracy & Economic Development in Africa* (ed. with G. Mohan; James Currey, 2004). He is an editor for the *Review of African Political Economy*.

Aram Ziai is Professor of Development and Postcolonial Studies at the University of Kassel/Germany. His research areas include development theory and policy, postcolonial and post-development approaches, and global political economy. He is author of *Development Discourse and Global History: From Colonialism to the Sustainable Development Goals* (Routledge, 2016), editor of *The Development Dictionary @25: Post-Development and its Consequences* (Routledge, 2018), and one of the founders of the Alternatives to Development Network (A2DN).

David Zilberman is Professor and Robinson Chair, Department of Agricultural and Resource Economics, University of California at Berkeley. He works on agricultural and environmental policy, water, marketing, risk and innovation, and the bio-economy. He is Fellow and President-Elect (2018) of Agricultural and Applied Economics Association (AAEA) & Association of Environmental and Resource Economists (AERE). He won the AAEA Publication of Enduring Quality Award (2005, 2010) among other awards. He has published 350 refereed articles and has edited 25 books and loves to blog. He is the Co-Founder of the International Water Resource Economics Consortium and the International Consortium of Applied Bioeconomy Research. He established and is Co-Director of Berkeley's Beahrs Environmental Leadership Program and the Masters of Development Practice.

INTRODUCTION

This second edition of *Key Thinkers on Development* appears 13 years after its predecessor. The intervening period has witnessed profound global changes, from the prevailing gloomy *Zeitgeist* (literally "spirit of the time") almost everywhere and widespread conflict to the geopolitics of development and the environment, and understandings of and prospects for achieving more just and sustainable development at any scale. Some prominent ideas and approaches to development in the early to mid-2000s have faded from view or been eclipsed, while new priorities and perspectives have emerged. Inevitably too, many of the Key Thinkers featured in the first edition have died or retired from professional life. Others have continued their development engagements regardless of formal employment status. Remarkably, some remain professionally active well into their 80s.

Accordingly, this second edition updates their lives and work, while the relaunch by Routledge of this series without the constraint of 50 thinkers in each field has enabled new figures to be included. This has permitted me to fill many gaps inevitably left last time and also to diversify the pantheon further to reflect the emergence of new thought and practice leaders, or those whose work now is more prominent or important than previously. Readers will therefore notice yet greater representation of non-western thinkers and perspectives, particularly from Asia (including China) and Latin America, even though diversity was a key selection principle for the first edition (see below). Importantly, too, the gender balance of thinkers has been improved considerably, since the paucity of women in the first edition was marked.

The original invitation to edit the first edition of this book proved irresistible, despite several other simultaneous editorial commitments.

First, Routledge's confidence in a market for a development studies title in the already well-established and successful Fifty Key Thinkers series that spans many fields of study and research was pleasing. More particularly, though, I perceived the opportunity to address a long-felt gap in the development studies literature, namely a good-quality biographical reference work that brings together leading figures from the various constituent disciplines. Precisely because of its multi- (and to some extent but still inadequately inter- and trans-) disciplinary nature, as well as the very contested nature of both theory and policy, both of which continue to evolve rapidly in this relatively young field, development studies still has a remarkably underdeveloped centre or core. Consequently, there is less sense of shared heritage or of a widely agreed set of leading figures and personalities than in longer-established fields.

Moreover, most academic research and teaching still take place within traditional disciplines, a feature reinforced in the UK and a growing number of other countries worldwide as a result of the widespread implementation of discipline-based research assessment and audit cultures. This has accompanied sustained periods of cutbacks, rationalisation and demands for demonstrable market "relevance" in higher education as a whole. Multidisciplinary institutes or centres of development studies were established in various countries amid the optimism of the 1960s and early 1970s, but their subsequent fortunes have been mixed. In the USA, the primacy of conventional academic disciplines has never been seriously challenged. At another level, the development field was long beset by a veritable "town versus gown" divide between academics and practitioners. Michael Edwards's (1989) polemic both reflected and exacerbated the sense of division but did focus attention on the problem. Although he moderated his views within a few years (Edwards 1993; 1994), suggesting that rapid progress towards *rapprochement* had been made, the last decade or so has certainly seen more widespread interaction and collaboration, resulting in a considerable narrowing of the divide.

Attesting to the vitality of the field, numerous textbooks and compendia have sought to keep pace with the rapidity of change in the post-Cold War world and the profound debates over the very meaning and future of "development". Yet, despite harbingers of doom predicting its demise, and whether we prefer to denote contemporary perspectives by means of prefixes like anti- or post- or to refashion and reinvigorate the term "development" itself, there is certainly much ongoing theoretical debate, conceptually rigorous research and dynamism in policy-driven practice.

This volume therefore aims to make a substantial contribution through informed reflection on the lives and work of seminal thinkers and actors in the broadly defined field of development studies and practice. The 50th anniversary of President Truman's "Point Four" speech in 1999 was marked by several retrospective and prospective works. This book continued that trend, given additional impetus by the deaths in the late 1990s and early 2000s of several prominent, if in some cases highly controversial, figures like Walt Rostow, Charles Kindleberger and James Tobin. Many more have died since publication of the first edition, marking the now almost complete generational succession within the field. Indeed, Paul Streeten, who died in 2019 aged 101, was one of the very last survivors of the post-World War II flowering of "development" in theory, policy and practice, focused particularly on the demise of European imperialism and colonialism and intent to modernise the newly decolonised states as rapidly as possible. Streeten remained professionally active until his early 90s.

As already mentioned, the most difficult challenge in producing the first edition of this book came at the outset: how to narrow dow nto only 50 the long list of nearly 200 names that I jotted down in a very short time? It quickly became clear that no universal consensus would be reached, regardless of the final choice. Indeed, this was confirmed through a process of consultation with friends, colleagues, the publisher and the subsequent suggestions of referees of the proposal for this title. In reaching the final list, hard choices were inevitable and I undoubtedly made a few decisions that some people might have found surprising but which are now able to be remedied in this new edition. To wit, it was necessary to choose between Paul Rosenstein-Rodan and Aleksander Gerschenkron to represent the debate around un/balanced growth among early development economists. I opted for the latter, since Gerschenkron was less well known overall in Anglo-American development studies, and despite many people (contestably) regarding the former as the father of development economics. This time, it has been possible to include Rosenstein-Rodan, affectionately known as "Rosie" by those who knew him, as well. Moreover, that essay is written by two of his former students/mentees, giving it a more personal flavour. Several contributors to the first edition even admitted to not knowing of all the 50 thinkers themselves, perhaps thereby illustrating our individual disciplinary and regional biases. With the broader range of thinkers included in this edition, the same experience is likely. A bit of controversy in this diverse and contested field may thus be both good and necessary.

Let me explain the initial selection process. One helpful factor was that a number of leading lights whose contributions to development have been made as practitioners or activists rather than as "thinkers" could be excluded. This applies for instance to Chico Mendes, the champion of the indigenous Brazilian Amazonian rubber tappers, who was brutally murdered in 1988 by agents of the powerful ranchers and logging firms whose destruction was being challenged. As it happens, he is included in *Fifty Key Thinkers on the Environment* (Palmer 2001). Interestingly, five thinkers did appear in both the above title and the first edition of this book (Thomas Malthus, Karl Marx, Mohandas Gandhi, E.F. Schumacher and Vandana Shiva), while Marx and Gandhi also appeared in *Fifty Major Political Thinkers* (Adams and Dyson 2003). In such cases, the respective entries in this volume have been written by different authors and bring out different aspects or interpretations of their contributions, so readers could usefully consult more than one.

Overall, the choice of Thinkers was – and in this new and expanded edition still is – intended to enhance the relevance and significance of this book's contribution by including a group broadly representative of the diverse currents and movements in the development world. This has meant ensuring that the pervasive Anglo-American dominance of development theory and discourse (not least by economists) is challenged through the inclusion of people working in different disciplines, over different periods of time and – crucially – hailing from different walks of life and geographical regions. Both editions of this volume are, by definition, therefore very different from the self-reflective studies by 15 eminent Northern development economists (with commentaries by younger economists) commissioned by the World Bank some 30 years ago, when economic development and development were still commonly conflated (Meier and Seers 1987), although several of those economists are also featured here.

The sharp gender imbalance among the Thinkers featured in the first edition has already been mentioned; this reflected the strong male dominance in most areas of development thought. Economics is a partial exception, but efforts to balance the various disciplines and "Northern" versus "Southern" representation made for some difficult choices. This second edition has addressed the gender and generational imbalances considerably within the constraints of the available pool of contenders.

A few of the voices (both "Southern" and "Northern") included in this book are far less well-known "internationally" than they should be, precisely because of the linguistic, cultural, disciplinary and geographical limitations under which many of us labour and which – however unwittingly – sometimes serve to perpetuate the very simplifications and abstractions that we claim to challenge. Their inclusion is therefore deliberate, as a small contribution to overcoming insularity and promoting polyvocality in development, as befits a postcolonial approach. Importantly, it has been possible to include essays on two influential Chinese figures, Fei Xiaotong and Wang Hui, the latter being the intellectual leader of the New Left, written by Chinese and American co-authors. For this, I must immediately acknowledge my debt to Joshua Muldavan, who conceived the joint writing project and persevered with it to eventual successful completion, despite complex challenges within China. Conversely, one or two intended inclusions, such as the Iranian thinker Ali Shariati, often referred to as an intellectual inspiration of the Iranian revolution, had ultimately to be excluded because no willing author could be identified to write within the required timeline.

Several people like Norman Borlaug, progenitor of the Green Revolution, are included because of their vision and the profound impact of their work, even though this might arguably have been somewhat more "technical" than "conceptual". Inevitably, though, some difficult choices had to be made. For instance, despite the undoubted influence of his "world-systems theory", Immanuel Wallerstein has been omitted, in part because his ideas integrate less directly with development issues than the closely related work of Jim Blaut. That said, his early career was more directly in African development and political economy. Another important criterion for inclusion has been that, to allow "mature" reflection on the significance of the lives and contributions of the Thinkers, they should be relatively close to the end of their active careers, if not yet retired or dead. Given the relative youthfulness of Development Studies – as evidenced by Malthus and Marx being the only Thinkers included whose professional lives predated the twentieth century – this still leaves plenty of scope for topicality, while avoiding the temptation for authors to make premature judgements or risk incurring personal embarrassment.

Personally, the process of editing the essays and melding them into a hopefully coherent book was fascinating beyond my most optimistic expectations; even this second edition has been rewarding in terms of the comprehensive review over the last decade or so of endeavour by those whose essays have been updated, and also by virtue of the extensive consultations with existing and new authors. Indeed, it would be no exaggeration to say that this book has been the most rewarding editorial challenge I have undertaken to date. First and foremost, it has brought me into direct contact with two diverse sets of people, the authors and the Thinkers, some of whom I had never previously met or had contact with. Some of these new relationships from the first edition have continued, although in some cases curtailed by retirement from active work or death. Finding contributors willing to write about the identified Thinkers was in itself a challenge, particularly as I was soliciting only one entry per author for the first edition. In most cases, the task proved remarkably easy, perhaps because the idea of the book caught their imaginations too. In a few instances, it was difficult to recruit an author for a particular Thinker or to match a willing contributor to an appropriate Thinker not already "taken". Nevertheless, I must acknowledge the efficiency with which commitments were turned into well-crafted essays, and in most cases by the agreed deadlines, as a result of which it has proved possible to produce both editions of the book on schedule. Many of the authors also found their research and writing most illuminating. Some were actually able to communicate with, and occasionally even interview, the Thinker on whom they were working. The manuscript has benefited greatly as a result.

Inevitably, however, there have been challenges. Trying to capture the essence of full and fascinating lives, as well as assessing their lasting contributions, in 1,500–2,000 words has been a considerable challenge to authors. Those revising for the second edition have had an easier task, with only a small number of essays requiring substantive updating or revision. However, some of the authors are now in more senior professional or managerial positions and found it hard to allocate the necessary time. Conversely, many others are themselves now retired and have enjoyed the task, in a few cases even volunteering to write new essays on additional Thinkers they had proposed for inclusion. I owe them particular thanks. At the other extreme, no fewer than five of the original authors have died over the intervening years, while three others are in poor health and incommunicado. The task of updating all their essays has fallen to me. Fortunately I was familiar with the work of the respective thinkers, in the cases of John Friedmann, Albert Hirschman and Hans Singer, very well since they are included in my recent monograph on *Holocaust Escapees in Global Development* (Simon 2019).

My second source of inspiration as editor has been how much I have learned about the Thinkers and their lives and times, regardless of how much I knew of them and their work beforehand. This reflects the formula adopted for the book, namely to interweave their biographical "stories" and an appreciation of their work and legacies. One other distinctive feature of this second edition is that two Thinkers appear as both authors and subjects. Jagdish Baghwati, featured in both editions, has now kindly co-written the essay on Rosenstein-Rodan; conversely, Arturo Escobar, the author of the essay on Dudley Seers in both editions, is now also the subject of an essay by Aram Ziai,

Third, however, a fascinating set of insights emerged as the essays were edited and integrated into the manuscript of the first edition. Two, in particular, stand out and exemplify the "added value" of this compendium. The first was the tremendous impact of the Nazi regime on the subsequent evolution of Development Studies through the emigration or escape of many young European Jewish (and some non-Jewish) refugees to the UK and USA, where they later emerged through universities and political office as influential contributors to the emerging ideas and approaches. Holocaust survivors and escapees are surprisingly numerous among the Thinkers represented here. They number no fewer than 10 of the original 50 (Cernea, Frank, Friedmann, Gerschenkron, Helleiner, Hirschman, Schumacher, Singer, Streeten and Wolf), while Lipton was born in the UK shortly after his parents escaped Nazi Germany. Of the additional Thinkers included in the second edition, three (Adelman, Polanyi and Sachs) fit the category, while Kalecki and Rosenstein-Rodan left central Europe shortly before the Nazis took power. Moreover, Stewart is the daughter of Nicholas Kaldor, another key early economist keenly engaged in development and who came from Hungary to study in the UK shortly before the Nazi era. While many other implications of World War II for what was to become known as Development Studies or International Development, such as the stimulus to colonial liberation struggles and formulation of the Marshall Plan, for instance, are quite well known, this one has not previously been documented.

Fourth, and somewhat more broadly, the geopolitical shifts and ruptures of the "end of empire" represented by decolonisation opened up the possibility of interesting, indeed important, new interregional and experiential connections that had important influences on the subsequent thinking and work of the Thinkers concerned. This is exemplified by Arthur Lewis and Walter Rodney, two natives of the Caribbean who studied at the University of London (the former in the 1930s and the latter in the 1960s) and then worked for a time in Ghana and Tanzania respectively. Franz Fanon also hailed from the Caribbean and had a comparable path in the Francophone world, ultimately landing up in Algeria, where he made his name internationally. There are other widely known examples too. Decades earlier, Gandhi's experience fighting racism in South Africa had proved seminal to his subsequent strategy of non-violent direct action and resistance (*Satyagraha*) in India. Similarly, most of the "Northern" voices in this collection were profoundly influenced by growing up, travelling and/or working in parts of the "South" at an early age.

The fifth important insight was the often close and influential interconnections among some Thinkers whom most people today do not particularly associate with one another (e.g. Boserup, Lipton, Myrdal and Streeten with respect to the *Asian Drama* study in the 1960s) – something that may help explain the processes of interactions that spawned and popularised key ideas and theories at momentous times in the history of development thinking. The extent of such interconnections at various stages of Thinkers' lives – and by extension in different eras of development history – is remarkable. As an aid to readers, cross-references to other Thinkers are indicated throughout the text by means of their names appearing in bold type.

Hopefully readers will find this volume both informative as a good read and useful for reference purposes. Perhaps some will even be stimulated seek to explore the lives and contributions of other key figures in the field.

David Simon
Englefield Green, Surrey
October 2018

References

Adams, I. and Dyson, R.W. (eds) (2003) *Fifty Major Political Thinkers*, Abingdon: Routledge.

Edwards, M. (1989) "The Irrelevance of Development Theory", *Third World Quarterly* 11(1): 116–136.

—— (1993) "How Relevant is Development Studies?", in F. Schuurman (ed.), *Beyond the Impasse: New Directions in Development Theory*, London: Zed.

—— (1994) "Rethinking Social Development: The Search for Relevance", in D. Booth (ed.), *Rethinking Social Development*, Harlow: Longman.

Meier, G.M. and Seers, D. (eds) (1984) *Pioneers in Development*, Oxford: Oxford University Press.

Palmer, J.A. (2001) "Chico Mendes 1944–88", in J.A. Palmer (ed.), *Fifty Key Thinkers on the Environment*, London and New York: Routledge.

Simon, D. (2019) *Holocaust Escapees and Global Development: Hidden Histories*, London: Zed Books.

ADEBAYO ADEDEJI (1930–2018)

Born on 21 December 1930 in Ijebu-Ode in Ogun State, southwestern Nigeria, Adebayo Adedeji received his doctorate in economics from the University of London in 1967, following initial training in economics (BSc Hons, London, 1958) and Public Administration (Diploma, University College, Ibadan, 1954 and MPA, Harvard, 1961). After working initially as a civil servant, he joined the University of Ife (now Obafemi Awolowo University, Ile-Ife) in 1963, becoming Nigeria's first Professor of Public Administration in 1967 and, concurrently, Director of the Institute of Administration. Between 1971 and 1975 he served as Nigeria's (post-civil war) Minister of Economic Planning and Reconstruction, before joining the United Nations as Executive Secretary of the Economic Commission for Africa (UNECA) with the rank of Assistant Secretary-General in 1975. He was promoted to Under-Secretary-General in 1978. Adedeji resigned from the UNECA in 1991 to return to Ijebu-Ode in Nigeria, where he established an independent think-tank, the African Centre for Development and Strategic Studies (ACDESS), which he continued to head as Executive Director until his death in Lagos after a long illness on 25 April 2018, aged 87. He also played a key early role in the African Peer Review Mechanism, established by the African Union in 2003 to enhance African control of its governance agendas relative to the international financial institutions and donor agencies. Responding with sorrow to Adedeji's death, the Nigerian President, Muhammadu Buhari, described him as "a patriotic Nigerian, economist, administrator, scholar and international statesman" and noted that "Prof. Adedeji was one of the first people to realise, as far back as the early 1980s, that environment, gender, and governance, which have now become commonplace in development discourse, were important for sustainable development in Africa" (Adetayo 2018).

Adedeji was clearly influenced by his training in economics at a time when development thinking was dominated by notions of teleology and growth. However, this influence appears to have operated in a "malign" way, in the sense of, first, instilling and, subsequently, reinforcing a seemingly unshakeable belief in both the inability of inherited development policy and the inappropriateness of conventional development practice to respond adequately to the dynamic complexity of Africa's post-independence circumstances and realities. It is therefore probably in his articulation of a vision of a progressively "nationalist" but increasingly (and variously) integrated continent-of-regions, an Africa secure in itself and valued as an integral member of the global community, that Adedeji's most significant contribution to development thinking lies.

There is considerably more to his contribution, however, for in promoting the case for the elaboration of indigenous African development strategies, Adedeji also illustrated the vital demonstration effect that the skilful deployment of development administration can contribute to policy and planning, to the rethinking of practice and to the elaboration of strategy. Thus in the only comprehensive and coherent analysis of Adedeji's contribution to development thinking and practice, Asante (1991) acknowledges the contribution of like-minded colleagues at the UNECA to the formulation of the ideas about African development which came to be closely associated with the person of Adedeji. Asante also notes how Adedeji's tenure as UNECA Executive Secretary provided an indispensable platform for the dissemination of these ideas, pointing out that the start of Adedeji's sustained challenge to orthodox development thinking from an African perspective was concurrent with the emergence of the radical Latin American-based critiques by **Andre Gunder Frank** and Michael Todaro, while Adedeji's role at the UNECA paralleled that played by **Raúl Prebisch** at ECLAC/CEPAL (the Economic Commission for Latin America and the Caribbean).

Adedeji's thinking on African development had its origins in the ferment of decolonisation, the emergence of postwar development as planned socio-economic change and the growing lack of evidence for significant early post-independence trickle-down benefits (Adedeji 1977; 1981). It was also undoubtedly influenced by debates taking place in and around ECLAC, and drew inspiration from the work of both Raúl Prebisch and, to a lesser extent, **Arthur Lewis**, although Adedeji differed from both in important respects and almost certainly never considered himself a *dependentista* (Asante 1991).

Not surprisingly, he was an early questioner of the wisdom of conflating economic growth and material change (readily quantifiable indices – or *means* – of development) with the less easily measurable (because largely qualitative and continuously evolving) process – or *end* – of development which, for him, was always complex, holistic and people-centred, in addition to encompassing economic *as well as* social, cultural, environmental, political and other forms of non-economic change and transformation. "Development", he was quoted as saying, "is a collective responsibility in which all have to share in the labour as well as the fruits", for when "people become the end and the means of development, their interests, values and aspirations necessarily determine the content, strategies and modalities of [such] development" – and, in the process, serve to ensure that development remains anchored to its socio-cultural, political and historical bearings (Asante 1991: 6). For him, too, Africa's *actual* postwar experience of development diverged markedly from this ideal, partly because, their various descriptive labels notwithstanding, the policies and tactics adopted were inherited rather than home-grown, and thus overwhelmingly imitative rather than creative or locally responsive in nature. Second, development efforts failed significantly to alter the dependent status of, or indeed to stimulate wealth or opportunity redistribution within, African economies despite the varied but always heavy reliance of favoured strategies on external sources for inputs of all kinds and as markets for the continent's primary commodities (Adedeji 1981). Development also routinely failed to consider people as subjects rather than objects, he suggested, and in so doing neglected to engender a sense of ownership among the vast majority of Africans who were supposed to be the targets and beneficiaries of development intervention (Adedeji 1977).

By the mid-1970s, therefore, and in Adedeji's view, what was sorely needed was a route to "economic decolonisation", involving, among other things, the "indigenisation" of national and continental economic development, and the forging of an Africa which, "[h] aving inherited or borrowed development policy as well as political theory, [would subsequently be able] to revive its own economic assumptions and design its own orientations, just as it ha[d] come to reject much of its neo-colonial legal and organisational legacy" (Adedeji and Shaw 1985: 3).

It is this vision, driven in part by a firm belief that economic growth was merely a means to the more desirable end of economic restructuring, societal reform and national/continental transformation for increased self-reliance (Adedeji 1977), which informed his insistence on the need for an African-formulated alternative strategy for African development. Notable among the elements of such a strategy, as detailed in the 1981 Lagos Plan of Action (LPA), were its privileging of increased self-reliance (requiring capacity-building and the development of competence) and self-sustainment (involving both rural–urban, sectoral and national market integration); its questioning of the role of foreign trade as an engine of growth (as a prelude to the promotion of internal demand stimuli at the expense of external market demand); its attempt to separate, at least conceptually, internal socio-economic change from export market performance; and its repudiation of the assumption of a positive – rather than negative – correlation between the expansion of advanced and developing country economies (Adedeji 1985; OAU 1980).

Even though it did not advocate autarchy, the LPA was still tantamount to development "heresy" in its defiant foregrounding of its political-economic purpose. Not only did it represent "a political declaration, a development strategy, a set of priorities, sectoral programmes of action, and a blueprint for regional and subregional integration", but it also argued for "a complete departure from the past … substitut[ing] … an inward-looking development strategy for [an] inherited externally oriented one [and] put[ting] the development of the domestic market rather than dependence on foreign markets at the heart of … development" (Adedeji 1985: 15). Such defiance mattered little; a key goal of the LPA was, after all, for Africa and its peoples to recover, through the transformative power of a democratised/popularised development, a sense of self-confidence long undermined by slavery, colonialism and neocolonialism (Adedeji 1977). Furthermore, by striking a dynamic balance between autarchy and vulnerability, the LPA was thought to possess sufficient flexibility and potential to indicate ways out of economic crisis (Adedeji and Shaw 1985). The LPA stimulated widespread debate, generated critiques, and may even have provided a framework for policy formulation and the implementation of strategies; it does, however, need to be understood as part of a wider process of constructing a philosophy of development for Africa, one which also highlighted the UNECA's role "in the international marketplace of ideas about Africa in particular and development in general" (Asante 1991: 46).

Nonetheless, despite its capacity for addressing existing constraints and potential for contributing to long-term sustainability (Adedeji and Shaw 1985), implementation of the LPA was slow, hampered in part by the economic crisis of the 1980s but partly also by the gradual nature of the policy change/reform processes envisaged (Adedeji 1985). Confronted by the crises of the 1980s and 1990s, African policy-makers, economies and societies concentrated on immediate survival rather than long-term transformation, with key macro-economic policy being dictated by interpretations and recommendations contained in the Berg Report (World Bank 1981), the very antithesis of the LPA, with which it stood in fundamental contradiction, not least in its opposition to self-reliance and self-sustaining development and in its preference for export-orientation, the market principle and the minimal state (Adedeji 1985).

In response, Adedeji and his colleagues at the UNECA produced the African Alternative Framework to Structural Adjustment Programs for Socio-Economic Recovery and Transformation (AAF-SAP), incorporating diagnoses of the root causes of Africa's economic underdevelopment and combined socio-economic and political crises which had initially been set out in the LPA (UNECA 1989). Like the LPA, the AAF-SAP was considered crucial to Africa's development future; it was a "launching pad into the ne[w] decade and beyond" (Adedeji 1990: 112). Adedeji used the AAF-SAP as the basis for challenging the logic, wisdom and ethics of orthodox reform/adjustment, particularly its mantra that "There Is No Alternative" (TINA), by reiterating that the AAF-SAP *did* represent an alternative which was human-centred, consistent with the development objectives identified in the LPA, and advocated combining "the short-term objectives of stabilisation and structural adjustment [with] the requirements of long-term restructuring" (Adedeji 1990: 71). He was adamant that it would be inadvisable for the orthodox structural adjustment measures of the 1980s to be continued into the 1990s, and warned that this would plunge the continent into a downward spiral that would be extremely difficult to recover from. Taken together, the LPA and AAF-SAP can, by extension, be considered among Adedeji's most seminal contributions to development thinking; indeed, he would have approved (and probably did approve) their joint message: "no programme of adjustment or development makes sense if it makes people indefinitely more miserable." It is not surprising, then, that Adedeji (2002: 4) was less than sanguine that both the LPA and AAF-SAP were "opposed, undermined and jettisoned by the Bretton Woods institutions" which in this way impeded Africans "from exercising the basic and fundamental right to make decisions about their future".

Adedeji's departure from the UNECA in 1991 did not mark an end to his contribution to development thinking and administration in Africa. Certainly, in ACDESS, which, according to the late **Julius Nyerere**, founding president of its Board of Trustees, was established in response to a perceived "lack of opportunity to express, debate and test … ideas in an open environment, in an African context, and under African leadership … dedicated to thinking about and for Africa's future" (www.acdess.org), Adedeji has created a vehicle for the continuing pursuit of his interest in "prospective and strategic thinking in and about Africa" (Adedeji 1993). Just as significantly, perhaps, it is a vehicle which, like the UNECA, and via its research/training programme, consultancy/advisory operations and services, and its periodic national and international conferences, facilitates continuing interaction with (and between) politicians, policy-makers, bureaucrats, planners, technocrats, NGOs, (sub-)regional organisations and international institutions, universities and academic research institutions (www.acdess.org), thereby allowing Adedeji to continue to "combine theory with practical experience" (Asante 1991), and research with policy application (Adedeji 1999).

His reflections on the road travelled by Africa from the LPA to NEPAD (New Economic Partnership for Africa's Development) by the start of the new millennium are instructive in this regard (Adedeji 2002). Following a favourable representation of NEPAD as a renewed pan-Africanist attempt to reactivate intra-and inter-African integration as well as to rejuvenate Africa's partnerships with the international community, he cautioned against pursuing NEPAD's goals at the expense of the LPA's principles of structural transformation and socio-economic diversification. NEPAD, he suggested, should be about "the resources and policies of … international partners being devoted to achieving Africa-determined development goals" (Adedeji 2002: 17). In this as in so many of his earlier interventions he (re)focused attention on (collective) self-preservation and political will. But he also reiterated another of his enduring messages: that Africa's pursuit of sustainable development needs to begin with success in the long-running struggle to "indigenise" the paradigms, strategies and agendas which must, of necessity, guide the nature, pace, direction and dynamics of any such development. It is a struggle which he continued to wage through ACDESS and other channels until retiring from public life in 2010 when he turned 80.

Major works

Adedeji, A. (1977) *Africa: The Crisis of Development and the Challenge of a New International Economic Order*, Addis Ababa: Economic Commission for Africa.

—— (1989) *Towards a Dynamic African Economy: Selected Speeches and Lectures 1975–1986*, London: Frank Cass.

—— (1990) *Structural Adjustment for Socio-Economic Recovery and Transformation: The African Alternative; Selected Statements*, Addis Ababa: United Nations Economic Commission for Africa.

Economic Commission for Africa (ECA) (1989) *The African Alternative Framework to Structural Adjustment Programs for Socio-Economic Recovery and Transformation (AAF-SAP)*, Addis Ababa: ECA.

Organisation of African Unity (OAU) (1980) *Lagos Plan of Action for the Economic Development of Africa, 1980–2000*, Addis Ababa: OAU, www.uneca.org/itca/ariportal/docs/lagos_plan.pdf.

Further reading

Adedeji, A. (1981) "General Background to Indigenisation: The Economic Dependence of Africa", in A. Adedeji (ed.), *Indigenisation of African Economies*, London: Hutchinson.

—— (1985) "The Monrovia Strategy and the Lagos Plan of Action: Five Years After", in A. Adedeji and T.M. Shaw (eds), *Economic Crisis in Africa: African Perspectives on Development Problems and Potentials*, Boulder, CO: Lynne Rienner.

—— (1993) "Marginalisation and Marginality: Context, Issues and Viewpoints", in A. Adedeji (ed.), *Africa within the World: Beyond Dispossession and Dependence*, London: Zed Books.

—— (1999) "Comprehending African Conflicts", in A. Adedeji (ed.), *Comprehending and Mastering African Conflicts: The Search for Sustainable Peace and Good Governance*, London: Zed Books.

—— (2002) "From the Lagos Plan of Action to NEPAD and from the Final Act of Lagos to the Constitutive Act: Whither Africa?", keynote address to the African Forum for Envisioning Africa, Nairobi, 26–29 April, available at http://64.233.187.104/search?q=cache:AEoDAGQW71OJ:www.worldsummit_Hlt98338960_Hlt983389602002.org/texts/AdebayoAdedeji2.pdf+adebayo+ade_Hlt107369836d_Hlt107369836ej_Hlt107369839i_Hlt107369839&hl=en¡UTF-8.

Adedeji, A. and Shaw, T. (1985) "Introduction: Africa's Condition and Projections for the Future", in A. Adedeji and T.M. Shaw (eds), *Economic Crisis in Africa: African Perspectives on Development Problems and Potentials*, Boulder, CO: Lynne Rienner.

Adetayo, O. (2018) "Ex-UN chief, Adebayo Adedeji Dies at 87, Buhari Mourns", *Punch Nigeria* online, https://punchng.com/ex-un-chief-adebayo-adedeji-dies-at-87-buhari-mourns/, accessed 16 September 2018.

Asante, S.K.B. (1991) *African Development: Adebayo Adedeji's Alternative Strategies*, Borough Green, UK: Hans Zell.

Onimode, B. (ed.) (2004) *African Development and Governance Strategies into the 21st Century: Looking Back to Move Forward. Essays in Honour of Adebayo Adedeji at 70*, London: Zed Books.

Onimode, B. and Synge, R. (eds) (1995) *Issues in African Development: Essays in Honour of Adebayo Adedeji at 65*, Ibadan: Heinemann.

Sanmi-Ajiki, T. (2000) *Adebayo Adedeji: A Rainbow in the Sky of his Time. A Biography*, Lagos: Newswatch Books.

World Bank (1981) *Accelerated Development in Sub-Saharan Africa: An Agenda for Action*, Washington, DC: World Bank.

REGINALD CLINE-COLE WITH DAVID SIMON

IRMA ADELMAN (1930–2017)

Irma Adelman was one of the leading development economists of our time. She was a multidisciplinarian and creative thinker who emphasised the linkages within the economy and society. Her major contribution was the development of computable general equilibrium (CGE) models to capture the holistic impacts of development policies. The CGE models have become the workhorse of development planning, and Irma's applications of this approach to design a development strategy for Korea was a masterful application of this approach.

Irma Adelman (née Glicman) was born in Chernivtsi, Romania, in March 1930. The only child of a well-to-do socialist Jewish family, she was educated by French Catholic nuns. However, with the rise of anti-Semitism in Europe, her family's survival was threatened and Irma was given shelter by a family friend. After trials and tribulations, her immediate family were able to pay their way to arrive in Palestine in 1939. The rest of her family either perished in the Holocaust or drowned while trying to escape as refugees. Following this, Irma lost her belief in God and became an agnostic.

In Palestine, Irma attended high school and excelled in mathematics; that led her to become a Morse code transmission specialist for the Jewish underground and later the Israeli Army. Irma started her studies in French literature and social sciences at the Hebrew University, but her studies were cut short by the War of Independence in 1948. In 1949 she left to study at the University of California, Berkeley, where she stumbled upon economics. Irma enrolled in the Masters and PhD in economics at Berkeley and her dissertation was on the monetary theory of Walras. She augmented her mathematical skill studying under Robert Dorfman and George Kuznets with her peers Arnold Zellner, Zvi Griliches and Yair Mundlak. While in Berkeley, Irma married her husband, a physicist, and soon had their son, Alex.

Adelman completed her PhD in Economics in 1955 and, despite being the top of her class, was not offered a tenure-track position. She held temporary positions at UC Berkeley, Mills College, and Stanford University, facing the twin challenges of a heavy teaching load and child-rearing. Nevertheless, she was very productive. In 1959, she published her classic paper "The Dynamic Properties of the Klein–Goldberger Model" (Adelman and Adelman 1959). The paper included a pioneering simulation that confirmed Ragnar Frisch's theory of business cycles, where economies are driven through cycles by random shocks. In 1961, she developed with Griliches a methodology to compute price indices that take into account a product quality differential within product categories. This approach was applied to estimate automobile prices (Adelman and Griliches 1961). This was the first application of what would later become hedonic pricing, an essential tool in modern applied economics.

Despite Adelman's publications, she was not able to land a permanent position at the West Coast universities, which had no tenure-track female economics faculty at the time. In 1962 her husband got a job in Washington, DC, and Irma was able to secure an associate professorship at Johns Hopkins University. There, she established a collaborative relationship with Cynthia Taft Morris, an economic historian. They were among the pioneers of development economics, which did not exist as a distinct (sub)discipline at the time. Irma was attracted to the field because she had a background in international trade and an interest in what happened to the welfare and income of the poor. With Morris, she co-authored *Society, Politics and Economic Development: A Quantitative Approach* (Adelman and Morris 1967), where they applied joint principal-component models to analyse diverse quantitative and qualitative data on the properties of development processes. Their hypothesis was that different development processes were triggered by various socio-economic factors, and they aimed to identify these factors and quantify their effects. This pioneering application of principal-component models was very significant in its identification of the crucial role of income distribution in various institutions in the development process.

Irma moved from Johns Hopkins to a position at Northwestern University in 1966 and stayed until 1971. In 1973, she joined the World Bank in Washington while taking a professorial position at the University of Maryland. While in Washington, she also joined the United States Agency for International Development (USAID) at the urging of **Hollis Chenery**, who asked her to direct the research division. Her first assignment was South Korea, which was considered a lost cause and a "bottomless pit" for aid. Adelman dedicated some time to learn the situation at first hand. She changed the structure of Korea's development apparatus and ensured that the planning bureau, instead of reporting to the Finance Minister, who was concerned with short-term budgetary stability, reported directly to the Prime Minister. Based on her observations of the value placed on education and ambition in Korea and the scarcity of capital, she recommended a development strategy that was human-resource-intensive and export-oriented. She also encouraged the pursuit of a gradual industrialisation policy that would slowly transition Korea into an exporter of more advanced and capital-intensive products.

While working on the Korean project, Adelman created her most important methodological contribution, the large-scale computable general equilibrium (CGE) model. The distinguishing feature of the CGE model compared to the input/output model (used to assess impact of policy on prices at the time) was that it included a social accounting matrix. This allowed the model to distinguish between different classes of people, enabling policy-makers to assess the distributional impact of various policies. This design of the CGE model was motivated by Adelman's concern for the poor. Adding the distributional dimension enabled a better understanding of the forces that shaped the Korean development process. The result of the work on Korea was *Income Distribution Policy in the Developing Countries: A Case Study of Korea* (Adelman and Robinson 1978). It is an amazing study, computing 3,000 variables for a very disaggregated representation of the economy. It recognised both competitive and non-competitive behaviour, macroeconomic constraints, income distribution parameters and dynamic modelling of change over time of key parameters. This CGE model was the first general equilibrium model to solve the rate of change of key variables endogenously. The aggregate detail of the model allowed numerous policy experiments, and suggested that reduction of relative poverty is much more difficult than that of absolute poverty. It also suggested that in the case of Korea, the de facto development strategy was equity before growth. Adelman was awarded the Order of the Bronze Tower by President Park of Korea in 1972, commending her for "greatly contribut[ing] towards attaining the goals of economic self-sufficiency ... of the Republic of Korea".

She returned to UC Berkeley in 1979 and became a faculty member in the Department of Agricultural and Resource Economics for the rest of her career. In Berkeley, she inspired her colleagues to expand general equilibrium models and apply them to assess the impact of environmental and natural resource policies, and encouraged her students (including Edward Taylor) to take a holistic and quantitative approach to understand migration policies, especially from Mexico to the US. She was awarded the prestigious Berkeley citation for her scientific contributions, and for her service to the university as a member of the Budget Committee and as a role model. Adelman was an avid art lover: even as a poor student, she collected old masters, which became a foundation for her son's art dealerships in Oakland.

Adelman was among the first to realise the limitations of rigorous neoclassical economics and to develop a holistic, systematic approach expanding the toolbox of applied economics and of policy-makers. She blazed the trail that resulted in hedonic pricing, which allowed quantifying differentiation in quality when assessing products. Her pioneering CGE models incorporated multiple markets, institutions and qualitative elements, such as culture. She was the champion of multidisciplinary approaches, and incorporated considerations of factors like class and tools such as non-parametric factor analysis, which had previously been overlooked in quantitative development analysis by economists.

Irma Adelman's scientific contributions derived from her wisdom and insight. In her assignment by USAID in 1963, she observed that the tenurial arrangements in South Vietnam disincentivised the peasantry to support the military effort, and recommended reforming land policies but to no avail. On the other hand, her assignment to Korea led to structural reform in the planning process and an emphasis on labour-intensive, export-oriented policies that have contributed to the growth and success of Korea (Adelman, Zilberman and Kim 2014). Her colleagues and students benefited greatly from her keen eye, sharp mind and good artistic taste, and feel her loss.

Major works

Adelman, I. (1974) "On the State of Development Economics", *Journal of Development Economics* 1(1): 3–5.

—— (1975) "Development Economics: A Reassessment of Goals", *American Economic Review* 65(2): 302–309.

—— and Adelman, F.L. (1959) "The Dynamic Properties of the Klein–Goldberger Model", *Econometrica* 27(4): 596–625.

—— and Griliches, Z. (1961) "On an Index of Quality Change", *Journal of the American Statistical Association* 56(295): 535–548.

—— and Morris, C.T. (1967) *Society, Politics and Economic Development: A Quantitative Approach*, Baltimore, MD: Johns Hopkins University Press.

—— —— (1973) *Economic Growth and Social Equity in Developing Countries*, Stanford, CA: Stanford University Press.

—— and Robinson, S. (1978) *Income Distribution Policy in the Developing Countries: A Case Study of Korea*, Stanford, CA: Stanford University Press.

—— —— (1988) "Macroeconomic Adjustment and Income Distribution: Alternative Models Applied to Two Economies", *Journal of Development Economics* 29 (1): 23–44.

—— Zilberman, D. and Kim, E. (2014) "A Conversation with Irma Adelman", *Annual Review of Resource Economics* 6(1): 1–16.

Further reading

Johansen, L. (1982) *Econometric Models and Economic Planning and Policy: Some Trends and Problems*, Berlin: Springer.

Szenberg, M. and Ramrattan, L.B. (eds) (2004) *Reflections of Eminent Economists*, Cheltenham, UK: Edward Elgar.

DAVID ZILBERMAN

ANIL AGARWAL (1947–2002)

Activist, journalist and scholar, Anil Agarwal was a prominent Indian environmentalist who redefined environmental problems through the eyes of poor people, and who was not afraid to challenge powerful organisations and governments in order to do so. During the rise of global environmentalism in the 1960s and 1970s, it became common to blame poor people for environmental problems through acts such as population growth and deforestation. Agarwal was one of the first critics to challenge these generalisations, and to focus instead on questions of international justice in environmental politics and the choices and risks faced by poor people. Agarwal left various legacies. He founded the Indian think-tank, Centre for Science and Environment, in Delhi, which today remains one of the foremost centres of critical thinking about environment and development. More conceptually, however, Agarwal was a pioneer in debates that are today called political ecology and science and technology studies. Rather than accepting environmental explanations from large organisations as scientifically and politically neutral, Agarwal sought to expose the politics underlying each statement of causality, and to show how such science legitimised or delegitimised different policies. He demonstrated how justice, as a concept, could be integrated into environmental policy between North and South. Agarwal also brought his own style of influencing politics, through a combination of scholarly work, acerbic journalism and careful political campaigning.

Agarwal was born in Kanpur in Uttar Pradesh in 1947, the son of a local landowner. He attended the Indian Institute of Technology in Kampur, where he studied mechanical engineering, and learnt information about technology that was to characterise his later writings. In a change of career direction, in 1973 he became a science correspondent at the *Hindustan Times*. In 1974, he wrote about the Chipko movement in the Indian Himalayas, where local villagers opposed logging, and which has become an icon for local environmental struggles in the South in later years. His writing attracted international attention, and in 1979 he won the first A.H. Boerma Award given by the United Nations Food and Agricultural Organization in Rome.

In 1980, Agarwal founded the Centre for Science and Environment (CSE) in New Delhi. CSE was a non-governmental organisation that focused on environmental matters, and which sought to influence the Indian government and transnational corporations. At the time, mainstream environmental groups in India tended to focus on conservation, especially conservation of wilderness and wildlife, as their main concern. The CSE, however, highlighted environmental risks faced by poor people in India at a time when livelihoods were being challenged by the decline in traditional biomass-based rural economies and by the growth of industrialisation. Agarwal communicated these views widely by editing the CSE journal, *Down to Earth*, which included a supplement for children known as the *Gobar* (or Cowdung) *Times*. Much writing was translated into Hindi, Kannada and other Indian languages.

The approach adopted by Agarwal and CSE began to influence wider debates about the meaning of "sustainable development". His reports on *The State of India's Environment*, written with colleagues at CSE from 1982, challenged the elitist basis of environmentalism, and sought to portray environment as a political problem partly reflecting international and class-based divisions of power and wealth. Analysts have described this approach as "red–green environmentalism" – which acknowledges both resources and livelihoods – rather than just the "green" approach, which highlights conservation alone. Agarwal also believed that orthodox development thinking was wrong to place faith in rapid economic growth as the chief means of achieving social development. He proposed that a new concept of "gross nature product" should replace "gross national product" in order to express the impact of growth on environment and livelihoods. Agarwal was also sensitive to the roles of women in protecting resources, and to their vulnerability to environmental hazards. He argued that poverty and environment are interrelated, but that poor people were commonly more protective of resources than commonly thought, and that economic policy should be tailored more closely to address poverty.

Because of such writings, both Agarwal and CSE quickly developed international reputations. From 1983 to 1987, Agarwal chaired the Environmental Liaison Centre International (ELCI), a Nairobi-based network of environmentalists. His work was reported in the England-based *New Scientist* and *Economist* magazines, as well as the broadsheets

Le Monde (France) and *Asahi Shimbun* (Japan). In 1986, the then Indian Prime Minister, Rajiv Gandhi, invited him to address the Union Council of Ministers, and honoured him with the Padma Shri Award. Agarwal was later asked to address all 27 Parliamentary Consultative Committees in India to educate MPs about his concerns, and to initiate discussions to identify solutions. In 1987 he was elected to the Global 500 Roll of Honour of the United Nations Environment Programme.

Much of Agarwal's writing included a critical stance towards environmental science, and especially towards statements that blamed poor people for causing environmental degradation. Instead, he urged a more holistic appreciation of the social and political conditions that make environmental changes problematic, and described how proposed solutions may aggravate social injustice. Describing the oft-cited belief that upland deforestation causes lowland flooding in the Himalayas, for example, Agarwal argued that the phenomenon of floods was caused by various factors including lowland water demand, rather than simply deforestation in the uplands. Consequently, policies need to consider how resources (and access to resources) have changed, and for whom, rather than apply simple mechanistic controls on water flow or forest use. He wrote in *Down to Earth* in 1987,

> Floods and shifting of river courses is … inevitable. Deforestation can aggravate the problem but afforestation cannot get rid of it. Embankments and dams have become an important cause of floods. We need better flood plain management, rather than flood control.

This criticism of popular scientific statements, and concern about social justice, also affected Agarwal's work in international environmental politics. In one of his most famous works, *Global Warming in an Unequal World* (co-authored with Sunita Narain in 1991), Agarwal criticised the tendency for some analysts to assume that anthropogenic climate change should be addressed by controlling deforestation in developing countries. In particular, Agarwal and Narain condemned a report issued by the Washington, DC-based think-tank, World Resources Institute, which allocated national responsibilities for greenhouse gas emissions based on an index largely dependent on current rates of deforestation and methane emissions from wet rice and livestock. The report put the three developing countries of Brazil, India and China among the top six emitting countries.

Agarwal and Narain contested the report on various grounds. First, the report was based on total national emissions, rather than on per capita emissions, which of course were smaller in developing countries than in developed countries. Secondly, the index used highly simplistic estimates for both deforestation and methane emissions. For example, estimates of wet-rice methane emissions were extrapolated globally from Italian figures; deforestation was treated uniformly, with no distinction made between export-led logging and smallholder food production; and no account was taken of the impacts of vegetation that might replace forest. Thirdly, the index focused chiefly on current tropical deforestation, and did not consider historic deforestation in developed countries (which is important, as greenhouse gases can exist for many years). Fourthly, and perhaps most importantly, the index did not refer to questions of social justice in greenhouse gas emissions, such as acknowledging that much deforestation in developing countries may occur because of poverty and food production, whereas in developed countries burning fossil fuels may be linked to affluence. Agarwal's and Narain's criticisms of this index were a watershed in international environmental politics, and demonstrated that scientific reports about environmental problems should not be considered politically neutral, but contain deep political implications about which activities are considered damaging or not, and which countries or people may be considered responsible. Agarwal worked on this theme in the approach to the 1992 Rio Earth Summit (the United Nations Conference on Environment and Development), by advising both Indian Prime Minister P.V. Narasimha Rao and the former Tanzanian President **Julius Nyerere** at the South Centre in Geneva, and by joining India's official delegation to the Rio conference. The Rio Summit contained much discussion of sustainable development, and facilitated the signing of the United Nations Framework Convention on Climate Change, and Convention on Biological Diversity.

Agarwal's work after Rio involved a new attention to urban environmental problems, and to the justice of economic globalisation. In particular, he studied how trade and government policy can encourage the provision of clean technology to poor people in cities. In 1996, the CSE published a report on vehicular pollution in Indian cities, which blamed petroleum companies, car manufacturers and regulators and planners. The report was followed by a media campaign, and eventually by government action to phase out polluting cars. In a typically acerbic editorial in *Down to Earth*, Agarwal wrote (1996):

> The western economic dream is a toxic dream. And don't listen to the typical tripe from Indian scientists and officials that India's consumption

> and production of toxic substances per capita is zilch compared to Western countries. This is utter scientific nonsense trotted out to make you apathetic. It is the exposure levels that matter, which can be very high in India, because of among other causes, high pesticide residue in our food and low quality of drinking water.

Agarwal wrote a series of editorials and writings urging greater global democracy in how environmental problems were solved, and in the processes of globalisation. For Agarwal, it was unacceptable that trade should be used as a means to control environmental misbehaviour by richer countries when poorer countries who suffer pollution or rising sea levels because of these richer countries cannot impose trade sanctions. Yet globalisation – if conducted with attention to political conflicts and alliances between campaigners in North and South – could also bring opportunities for strengthening the political role of developing countries in international affairs.

Following some of his earlier writings, Agarwal and the CSE also continued to seek ways to demonstrate decentralised rural governance via village communities. Under a campaign entitled "Making Water Everyone's Business", the CSE supported experiments in water-harvesting and land management in Sukhomajri in Haryana, Ralegan Siddhi in Maharashtra and the Tarun Bharat Sangh in Rajasthan. But despite these actions, Agarwal was criticised by some for offering only muted support for the Narmada anti-dam movement in western India, and for allegedly abandoning some of his initial radical stances by becoming an adviser to the state, thus raising the question as to whether it is possible for a recognised environmentalist to remain radical. Many did not share these criticisms. In 2000, he was given an Environment Leadership Award from the Global Environment Facility – the multilateral funding agency for global environmental problems. In 2001, the Government of India bestowed on him the Padma Bhushan Award, an award reserved for people who have performed distinguished service to the nation.

Anil Agarwal died in 2002 at just 54. He had experienced a long battle with cancer, and had written about cancer care in India as another example of inadequate attention to social welfare. He left an important legacy through the creation of the CSE, and his personal writings pioneered current thinking about poverty and environment and the hidden politics of environmental scientific assessment. Agarwal made it clear that local questions of environment in developing countries were inherently linked to international political economy, and argued that creating knowledge about environmental problems should not be left to experts in developed countries. He also achieved these aims through establishing a system of campaigning and communication that both harnessed and educated many in poorer countries. Anil Agarwal was one of the most influential thinkers and writers on questions of environment and development because he fought to increase the representation of poor people in both the definition and solution of environmental problems.

Major works

Anil Agarwal wrote and edited more than twenty books on science and environment in India and numerous magazine articles and editorials, especially in *Down to Earth*.

Down to Earth, journal published by Centre for Science and Environment, www.downtoearth.org.in.

Floods, Flood Plains and Environmental Myths, edited with Sunita Narain, 1991, New Delhi: Centre for Science and Environment.

Global Warming in an Unequal World, with Sunita Narain, 1991, New Delhi: Centre for Science and Environment.

Homicide by Pesticides: What Pollution Does to Our Bodies, edited, 1997, New Delhi: Centre for Science and Environment, State of the Environment Series 4.

The State of India's Environment: A Citizens' Report, 1982, with Sunita Narain, New Delhi: Centre for Science and Environment.

Towards a Greener World: Should Global Environmental Management Be Built on Legal Convention or Human Rights?, 1992, New Delhi: Centre for Science and Environment.

Towards Green Villages: A Strategy for Environmentally Sound and Participatory Rural Development, 1989, with Sunita Narain, New Delhi: Centre for Science and Environment.

Further reading

Acharya, K. and Noronha, F. (eds) (2010) Section 10: "An Anil Agarwal Reader", in *The Green Pen: Environmental Journalism in India and South Asia*, New Delhi: Sage. DOI: http://dx.doi.org/10.4135/9788132107958.

Baviskar, A. (2002) "An Activist–Environmentalist: Anil Agarwal, 1947–2002", *Frontline* 19(2). www.flonnet.com/fl1902/19021210.htm.

CSE official biography and resource page: www.cseindia.org/node/216.

Guha, R. (2002) "Anil Agarwal and the Environmentalism of the Poor", *Capitalism, Nature, Socialism* 13(3): 147–160.

Kumar, S. (2002) "Anil Agarwal (1947–2002) from Rio to Rio +10: An Appreciation", *Local Environment* 7(3): 241–249.

TIM FORSYTH

ELMAR ALTVATER (1938–2018)

As a representative of the school of Critical Political Economy, Altvater analysed the limits of the dominant (capitalist) mode of production studiously and creatively. He stressed tirelessly that Fordist production structures and patterns of consumption could not be translated into a universally applicable avenue for the social development of all. Unusually among social scientists, he opened himself to basic natural science laws in searching for viable explanatory models for a future-oriented social development. He insisted that the entropy principle sets limits.

Born on 24 August 1938, Altvater studied economics and sociology in Munich and wrote a dissertation on "Environmental Problems in the Soviet Union". Teaching political economy (or rather the critique of political economy) from 1971 as professor in the Political Science Department of the Freie Universität in (West) Berlin, he influenced whole generations of students. Altvater's diverse interests did not sustain any damage from the turmoil of the later stages of a "student rebellion" in decline. As a representative of undogmatic Marxism, he filled his seminars to capacity. Declining profit rates, the internationalisation of capital and theories of the world market featured as much among his courses as introductions to *Das Kapital* by **Karl Marx** and seminars on state theories. The wide thematic range of his work was always focused on the creative application of Marxist theory.

Altvater's analyses of international financial capital and monetary and fiscal policy, as well as state interventionist questions regarding regulation of markets and societies, gained prominence from the early 1970s. During the 1980s, the intensified debt crisis in the countries of the periphery made him a powerful critic of development on loan (Altvater et al. 1991). Inspired by a sabbatical in the eastern Brazilian Amazon region, he devoted a rare case study to Brazilian issues but generalised them in a relevant way by presenting a pioneering study on "the world market as a force of circumstances" (*Sachzwang Weltmarkt*). In this groundbreaking analysis, he understands space as a real substratum of economic and social processes and their structuring, and probes the systemic and systematic connection between regional identity, state development policy and economic crisis tendencies in the world market. Given the latter's dominance over national development strategies and local conditions of ecologically adjusted production, a Fordist catch-up industrialisation strategy must come to nothing. In a chapter (in *Sachzwang Weltmarkt*) on "The Economy and Ecology", he explores the thermodynamic laws, for him an increasingly central argument for the natural limitation of social development as processes of material conversion, production and consumption are subject to increasing entropy. Hence the economic system and its tendencies cannot be imagined without limitation by natural laws.

Altvater's approach benefited from the fundamental works of Nicholas Georgescu-Roegen (1906–1994), a Romanian mathematician, who also studied at the Sorbonne and was apprenticed in economics at Harvard to Joseph Schumpeter, one of the leading names in economic theory and, in particular, the internationalisation of capitalism. Georgescu-Roegen, who finally settled at Vanderbilt University in the USA, gained increased influence in recent economic theory because of his ability to combine a new biological or evolutionary approach with economic theory. His *magnum opus* (Georgescu-Roegen 1971) emphasised the limits to growth in an economy on the basis of the Second Law of Thermodynamics (which can be translated as "useful energy gets dissipated") (see also Georgescu-Roegen 1976). Largely ignored by mainstream economics, Georgescu-Roegen's new approach to economic theory had lasting effects on the likes of Altvater, who discovered its relevance from the perspective of an environmentally sensitive school of thought, opening new discourses in the field of evolutionary economics.

Another pioneer inspiring a few social scientists – and even fewer economists – like Altvater in the application of economic theories to development studies was the Russian-born Ilya Prigogine (1917–2003), the Founding Director of the Centre for Statistical Mechanics and Thermodynamics at the University of Austin, Texas from 1967 (now the Ilya Prigogine Centre for Studies in Statistical Mechanics and Complex Systems). He was awarded the Nobel Prize in chemistry in 1977. Like Georgescu-Roegen, he contributed substantially to a new combination of natural and social sciences in theories of relevance for environmental and economic aspects in the reproduction of societies. His major works also highlighted the relevance of dissipative structures and contributed significantly to the understanding of irreversible processes, especially in systems far from equilibrium (see Kondepudi and Prigogine 1998; Prigogine and Stengers 1986; 1997).

Altvater derived environmental problems from the specific dynamic of current capitalism, which in his view was not "the end of history" in Fukuyama's sense of immanent future development but a demand for a continuing search for sound alternatives for the reproduction of social systems. In the course of this demand, Altvater harked back to the integration of the entropy laws into social science theory as documented in his two works *Die Zukunft des Marktes* (The Future of the Market) and *Der Preis des Wohlstands* (The Price of Prosperity). "Thermo-balance" was a key concept in his argument. Critics reproached him for a "thermo-dynamisation of the social sciences". Others scold him for utopian sandcastle games in which he now urged a solar instead of a socialist revolution without stressing the immanent contradictions of (post-)Fordist capitalist production or utilising the "transformist" renewal of the capitalist world order. These objections can be countered by pointing to the merit of his introduction of a new dimension into development discourse that considered the criterion of conscious metabolism in the sense of social conduct, and did not downgrade the environment and nature to mere objects of boundless exploitation (Köhn 1995). To some of the objections raised by a critic (Hein 1993), Altvater responded directly by pointing out that ten human generations have appeared since the "Promethean" industrial revolution of the world market. Nature is finite and has reached the limits of encumbrance and, as Altvater maintained, theoretical approaches that do not strive to integrate the ecological limit in their conceptual world are not up to date (Altvater 1994: 104). He challenged socio-economic theory to take into account the global range of the capitalist social formation and the (similarly global) ecological threat. He rejected the reproach for borrowing from thermodynamic physics the categories of "entropy" and "syntropy" that are out of place in socio-economic debate. Even if the cosmic process is understood as open and the earth as an energetic open system, a thermodynamic restriction remains. The existing material energy (fossil or nuclear) fuels industrial production cycles, and their emissions thereby contribute to global warming (p. 108). He argued further that a consideration of the categories of thermodynamic physics in socio-economic analysis was helpful in revealing the consequences for production. Hence he insisted on an expansion of socio-economic concepts, since conventional theories of the time were inadequate for comprehending the ecological problems, and he referred to the necessity for compatibility between the production and energy systems in humanity's history. Very presciently from today's vantage point, his underlying thesis was that the epoch of production based on fossil fuel will come to an end within a few decades, that the transition to a solar society awaits, and that the exchange of primary sources of energy cannot occur without radical social changes in the social systems of material and energy transformation (p. 110). He described the process of a necessary social, economic and political restructuring as a "solar revolution" without any substantive concretisation but vital as a future-oriented vision.

Together with his lifetime companion, Birgit Mahnkopf, Altvater illuminated the present *Grenzen der Globalisierung* (Limits to Globalisation) and took this further into the *Globalisierung der Unsicherheit* (Globalisation of Insecurity). He analysed the worldwide socio-economic processes that maintain competition between states and regions in an increasingly deregulated world market, and at the same time lead to a global "club society of the owners of financial assets", through the globalisation of financial markets and loss of state autonomy in the sense of its own regulatory possibilities.

His reference to the limited character of the Fordist model of development was also central to his argument. Neither the political nor the socio-economic qualities of industrialised countries can be generalised worldwide. He formulated the quintessence of this thematic approach in a concise essay (Altvater 1996) by warning that, in a fatal way, the world is finite, and that catch-up development by Third World countries by means of Fordist-type industrialisation of the sort undertaken by Western industrial countries is inconceivable. For Altvater, local social coherence is proven in global economic restrictions through the overlapping of functional spaces (see *Sachzwang Weltmarkt*, esp. chapters 2–5: 56–194). Development is determined twice. The observance of restrictions can succeed only when political institutions set limits. Development proposals must both obey general rules and be very specific. He recognised that the resulting political challenge consists in the lack of material prerequisites for local decisional freedom. Interdependence cannot arise as dependence.

As the editors of the *Festschrift* for his sixtieth birthday conceded in their introduction to the volume (Heinrich and Messner 1998), Altvater's holistic approach might have been controversial, but it accomplished orientation and integration in a very differentiated scientific world where we increasingly understand details but less and less the comprehensive connections. His ongoing work has so far been most influential in the debates concerning the reproduction of capitalism and the world market, ecological constraints, the theory of the state and more recently global public goods. Consequently, the contributions in honour of his sixty-fifth birthday centred on this topical issue (Brunnengräber 2003). Altvater pleaded for the regulation of energy consumption, working conditions and capital movements at the supranational level. In one of his few texts published in English, he summarised his position thus:

> In an era of globalisation, the conventional paradigm of economic policy is in need of radical rethinking. Such a paradigmatic shift, however, will necessarily have to be accompanied by practical efforts to re-embed the global economic system in qualitatively new social relations and forms of political regulation, on both local and global levels.
>
> (Altvater 2002: 88)

Altvater's contributions help us to keep sight of the essentials, including the political-moral ethics of a theory of development, and provide motivation to remain engaged. "At some time or other, we must begin," he said, in justification of his own beliefs:

> This sounds idealistic and the idealism reproach has been very harsh since Marx. Nevertheless, a process is initiated through which the fatal monetarisation and short-term materialism are revoked. Hopefully we will have this time. This is not certain. Unfortunately, catastrophe cannot be excluded. However, no leftist project for catastrophe is developed but rather a project for avoiding catastrophe.
>
> (Altvater 2004: unpaginated)

On 1 May 2018, perhaps fittingly the May Day workers' holiday in many countries, Elmar Altvater lost his final battle with cancer. The often emotional countrywide reactions from different generations in Germany provided another testimony to the lasting relevance of his thoughts.

Major works

Altvater, E. (1987) *Sachzwang Weltmarkt. Verschuldungskrise, blockierte Industrialisierung, ökologische Gefährdung – der Fall Brasilien*, Hamburg: VSA.

—— (1991) *Die Zukunft des Marktes. Ein Essay über die Regulation von Geld und Natur nach dem Scheitern des "real existierenden Sozialismus"*, Münster: Westfälisches Dampfboot.

—— (1992) *Der Preis des Wohlstands oder Umweltplünderung und neue Welt(un)ordnung*, Münster: Westfälisches Dampfboot.

—— (2002) "The Growth Obsession", in L. Panitch and C. Leys (eds), *Socialist Register 2002: A World of Contradictions*, London: Merlin Press, pp. 73–92.

—— and Mahnkopf, B. (1999) *Grenzen der Globalisierung. Ökonomie, Ökologie und Politik in der Weltgesellschaft*, Münster: Westfälisches Dampfboot (4th, revised and enlarged edn; originally 1996).

—— —— (2002) *Globalisierung der Unsicherheit. Arbeit im Schatten, schmutziges Geld und informelle Politik*, Münster: Westfälisches Dampfboot.

Further reading

Altvater, E. (1994) "Tchernobyl und Sonnenbrand oder. Vom Sinn physikalischer Kategorien in den Sozialwissenschaften. Replik auf die Kritik von Wolfgang Hein", *Peripherie* 14(54): 101–112.

—— (1996) "Von möglichen Wirklichkeiten. Hindernisse auf der Entwicklungsbahn", *Entwicklung und Zusammenarbeit* 37(2): 44–49.

—— (2004) "Time Needs Radicalism and Radicalism Needs Time", translated from the German original of 1998, posted to the indymedia website on 10 January 2004 and accessed on 6 February 2004 (http://sf.indymedia.org/news/2004/01/1671226.php).

—— Hübner, K., Lorentzen, J. and Rojas, R. (eds) (1991) *The Poverty of Nations: A Guide to the Debt Crisis – From Argentina to Zaire*, London: Zed Books (originally published in 1987 as *Die Armut der Nationen. Handbuch zur Schuldenkrise von Argentinien bis Zaire*, West Berlin: Rotbuch).

Brunnengräber, A. (ed.) (2003) *Globale Öffentliche Güter unter Privatisierungsdruck. Festschrift für Elmar Altvater*, Münster: Westfälisches Dampfboot.

Georgescu-Roegen, N. (1971) *The Entropy Law and the Economic Process*, Cambridge, MA: Harvard University Press.

—— (1976) *Energy and Economic Myths: Institutional and Analytical Economic Essays*, New York: Pergamon Press.

Hein, W. (1993) "Elmar Altvater: Entropie, Syntropie und die Grenzen der Metaphorik", *Peripherie* 13(51/52): 155–170.

Heinrich, M. and Messner, D. (eds) (1998) *Globalisierung und Perspektiven linker Politik. Festschrift für Elmar Altvater zum 60. Geburtstag*, Münster: Westfälisches Dampfboot.

Köhn, R. (1995) "Gesellschaftliche Grenzen der Entropie. Wider die Thermodynamisierung der Sozialwissenschaften", *Peripherie* 15(59/60): 180–193.

Kondepudi, D. and Prigogine, I. (1998) *Modern Thermodynamics: From Heat Engines to Dissipative Structures*, New York: Wiley.

Prigogine, I. and Stengers, I. (1986) *Order Out of Chaos*, New York: Bantam.

—— ——(1997) *The End of Certainty: Time, Chaos, and the New Laws of Nature*, New York: Free Press.

HENNING MELBER

SAMIR AMIN (1931–2018)

Samir Amin was the best-known Egyptian/Arab thinker in the field of "**Marx**ist" development theory, formidable critic of capitalism, radical political economist and one of the ferocious champions of antiglobalisation activism. His contribution to political theory was comparable only to those of contemporary Marxist critics of capitalism, including Paul Baran, **Andre Gunder Frank** and Immanuel Wallerstein. Commonly, they applied Marxist development theories in an effort to explain the consequences of capitalist economic, political, cultural and military development and expansion to the developing countries.

Amin's earlier academic writings preceded the emergence of what is now commonly known as development studies. As an economist, Amin began his research (1957–1970) with the study of the economies of individual countries: Mali, Congo Brazzaville, Egypt, Senegal, Ghana, Côte d'Ivoire and the Maghreb countries. Since then, his work has been informed by these early theoretical developments, which originated in historical Marxism.

Influenced by the Cold War economic and ideological rivalry and a disappointing first decade of development, Amin's four empirically based works, *The Maghreb in the Modern World* (1970), *Neo-colonialism in West Africa* (1973a), *Unequal Development* (1976b) and *The Arab Nation: Nationalism and Class Struggle* (1976a), echoed the frustrations of a generation of intellectual leftists in developing countries who witnessed the obliteration of the optimism of the decolonisation euphoria. Describing "independence" as neocolonialism, the relationship between the newly independent Maghreb and West African states and the economically dominant former colonial powers, Amin was able to foresee the overriding patterns of future economic development of these countries and the developing world in general. This realisation influenced his intellectual development for decades, as reflected in his development theorising. His doctoral dissertation (Institute of Economics, University of Paris, 1957) was entitled "The Structural Effects of the International Integration of Precapitalist Economies: A Theoretical Study of the Mechanism Which Has Engendered the So-Called 'Under-developed' Economies".

Amin's first major contribution to development theory was so powerful that it continues to surface in most of his current intellectual works. The two-volume work, *Accumulation on a World-Scale* (1974b), was instantly recognised as a major contribution, probably not for its originality, but for its Marxist orientation that found resonance in the Cold War ideological schism. In this study, Amin argues that one of the anomalies of capitalist development theory is that it confuses under-development with poverty (pp. 261–262). Because it describes bare economic characteristics of developing countries, capitalist development theory artificially separates economics from the sphere of social and political organisation, and by doing so it ignores the fact that under-development consists of more than the outward appearances of poverty. Underdevelopment is described as a cumulative sum of the whole history of capitalist expansion structurally constructed as a world system with centre and periphery. Amin called this his "theory of capitalist social formation", which exhibits certain characteristics of underdevelopment (pp. 15–20).

Amin explained underdevelopment as an outcome of three factors. (1) Unevenness of productivity between spheres, or "sectoral unevenness of productivity" between centre and periphery. (2) The disarticulation, astructuration or distortion of the underdeveloped economies of the periphery made up of non-integrated sectors, with less flow of internal exchanges in the bid to satisfy the external demands imposed on them by the economies of the centre. (3) The centre that dominates economically, socially and politically over the periphery.

If dependence and integration into the world capitalist system engenders underdevelopment, Amin's alternative to disarticulation was delinking. In essence, Amin's theorising of development is closely associated with dependency theory and its intellectual heritage.

One of Amin's major contributions to development theory was his ability to reinvent the insights of Marxism–Leninism and **Mao**ism (1974b: 63, 112–116) on the national question in order to explain the origins of underdevelopment. Methodologically, he extricated Marxism from its Western antecedents and redeployed it to explain underdevelopment as an ultimate result of past colonial experiences or the hegemony of the ex-colonial powers, the centre over their ex-colonies – which constitute the periphery. As a historical process, underdevelopment is the cumulative result of unequal exchange, unequal development and imperialism (*Unequal Exchange*, 1973b; *Unequal Development*, 1976b; *Imperialism and Unequal Development*, 1977). The central premise is that while economic growth characterises the economic sectors at the centre, it contributes to the development of underdevelopment and the disarticulation of the social formations of the periphery.

> Delinking has nothing to do with exclusion or autarkic withdrawal. It is a matter of subjecting the mutual relations between the various nations and regions of the whole world of the planet to the varying imperatives of their own internal development and not the reverse.
>
> (Amin 1990a: xii)

Delinking is therefore a manifesto for change, a rejection of the idea that the expansion of capitalism, and subsequently the currently dominant neoliberal paradigm, is inescapable and therefore leaves no possibility for national autonomy. Amin offers polycentricity (i.e. one planet with several competing systems) (Amin 1990a: xii) as an alternative to neoliberal exclusionary and polarising development, also referred to as "maldevelopment" (pp. 80, 94–97, 129–136).

Amin's theoretical strand, ideological bent and activism are best described as a search for praxis or a theory of practice. Much of his writings during the closing decade of the 20th century and until his death on 12 August 2018 were devoted to the critique of globalisation, which he described as a peculiar new form of managing the international economy. The implications of globalisation for development and development theorising are horrendous. In *Capitalism in the Age of Globalisation* (1997), Amin revisits the idea of polycentric globalisation and argues the need for "renewing the perspective of global socialism" (p. 6) in order to usher in an era of an "alternative humanist project of globalisation" (p. 10). Such an agenda cannot be realised and the crisis of development cannot be resolved until popular, democratic forces capable of dominating society get together again (p. 135). In this scheme, the intelligentsia is assigned a vanguard role in establishing bonds between its own productive thinking and the aspirations and actions of the popular classes, making them social partners for change.

Obsolescent Capitalism: Contemporary Politics and Global Disorder (2003) is Amin's ultimate commentary on the expansion of capitalism from the Thirty Years' War (1914–1945) to the current age of neoliberal globalisation. Here, Amin draws much from his earlier writings, including *Maldevelopment: Anatomy of a Global Failure* (1990b); *Transforming the Revolution: Social Movements and the World System* (1990c); *Empire of Chaos* (1992b) and *Re-reading the Post-war Period* (1994). The future of capitalism is "prophesied" with the main aim of answering one important question, "Are the current developments of world capitalist systems permanent or transient [or] rather, signs of obsolescence of a system that must be overcome if human civilisation is to survive?" (Amin 2003: 1). Theoretically, the volume offers confirmation of Amin's Marxist interpretation of the state of the world, claiming that "Marx's law of pauperisation resulting from capitalist accumulation has been ever more strikingly confirmed on a world scale over the past two centuries" (p. 3).

The realisation that former communist countries such as Russia and China succumbed to the economic power of global capitalism (*Empire of Chaos*, 1992b) created the necessity for Amin to rethink development theorising. In his latest writings, he defines development as a concept that does not connote "catching up", but instead "involves the project of a very different (alternative) society", with a twofold aim: (a) to free humanity from economistic alienation, and (b) to end the legacy of polarisation on a world scale. These objectives, according to Amin, cannot be realised without the active participation of the whole world population because the problems facing humanity have an ever-deeper global dimension (1992b: 3–4).

Capitalism is obsolete because it is an ageing system that has entered a state of permanent disorder, leading to a long transition to socialism or catastrophe and the suicide of humanity. Capitalism is obsolete because: (1) the current scientific and technological revolution shows that capitalism has exhausted its reliance on labour, thus hindering the possibility of continual accumulation; (2) the collective Triad[1] of imperialism operating on a world scale no longer allows the pursuit of dependent capitalist development in the peripheries (Amin 2003).

Because the US hegemony and its new right neoliberal project thrive on obsolescence capitalism, its alternative is a non-American 21st century whose basic requirements include: (1) the dislocation of the current unipolar world system with a multipolar (democratic and regionalised) one, implying that obsolescence capitalism corresponds to "the dictatorship of transnational capital, attacks any idea of 'self-reliance,' and treats 'delinking' and 'national construction' as regressive protectionism" (2003: 30). (2) The 21st century should be more radical than the twentieth; instead of looking back to historical Marxism, historical Keynesianism and national populism, neo-Marxism, neo-Keynesianism and postcapitalism became the counter responses to globalised liberal capitalism (pp. 136–138). (3) A non-American 21st century requires the building of convergences of social and political movements that give expression to the victims of global neoliberal capitalism, a task which in turn demands respect for diversity. Unlike the 20th century, the 21st century requires building a Left with alternative strategies and tactics for a united front in support of social and international justice (pp. 140–147).

In recent years Amin consolidated his critique of global capitalism and liberalism in *The Liberal Virus: Permanent War and the Americanization of the World* (2004), *Global History: A View from the South* (2011b), *Ending the Crisis of Capitalism or Ending Capitalism* (2011a), *The Implosion of Contemporary Capitalism* (2013a), *From Capitalism to Civilization: Reconstructing the Socialist Perspective* (2013b), *Modern Imperialism, Monopoly Finance Capital, and Marx's Law of Value* (2018). In all these, his message is unrelenting: the developing countries' efforts to develop their economies, societies and cultures are undercut by a ruthless hegemonic system of accumulation at a world scale. Modern capitalism is a replica and testimony to Marx's major economic contributions in relation to capital and the theory of value.

Amin's other major recent contributions targeted the USA in books such as *The Liberal Virus: Permanent War and the Americanization of the World* (2004), *Beyond US Hegemony: Assessing the Prospects for a Multipolar World* (2013c) and what he referred to as Russia's return to socialism in *Russia and the Long Transition from Capitalism to Socialism* (2016a). He also engaged the Arab Spring in a string of articles echoing George Antoniu's (1939) book *The Arab Awakening*, which captured the early rise of Arab nationalism and struggle for decolonisation and Arab unity. Amin's book, *The Reawakening of the Arab World: Challenge and Change in the Aftermath of the Arab Spring* (2016b), explains the origins of the uprisings against the backdrop of decades of slow and non-consequential economic and political reforms and corruption. These negative attributes of the Arab world could not prevent the eventual unravelling of Arab social orders and the making and unmaking of Arab regimes. The failure of the Arab Spring is due to political and economic structural fault-lines that go deeper than many analysts' expectations for a quick fix.

Samir Amin was able to maintain a level of theoretical coherence and persistence, probably unmatched by his contemporaries. He consciously and systematically employed his talent and innovative thinking in the search for alternative development paths to global capitalism. Evidently, after decades of research and writing on development theory and as an intellectual activist, Samir Amin was still an "optimist Marxist" until his death, arguing that development's ultimate aim is to displace obsolescent capitalism and realise the long-awaited transition to world socialism.

Note

1 "Triad" refers to the Marxist dogma that human history is predetermined, evolving in three basic stages (triads), with the two final stages being socialism and communism.

Major works

Amin, S. (1970) *The Maghreb in the Modern World*, Harmondsworth: Penguin.

—— (ed.) (1972) *Modern Migrations in Western Africa*, Studies presented and discussed at the Eleventh International African Seminar, Dakar, April.

—— (1973a) *Neo-colonialism in West Africa*, Harmondsworth: Penguin.

—— (1973b) *Unequal Exchange, Imperialism and Underdevelopment: An Essay on the Political Economy of World Capitalism*, Calcutta: Ranjit Sau, and Oxford University Press.

—— (1974a) "Accumulation and Development: A Theoretical Model", *Review of African Political Economy* 1: 19–26.

—— (1974b) *Accumulation on a World-Scale: A Critique of the Theory of Underdevelopment*, New York: Monthly Review Press.
—— (1976a) *The Arab Nation: Nationalism and Class Struggle*, London: Zed Press.
—— (1976b) *Unequal Development: An Essay on the Social Formations of Peripheral Capitalism*, Hassocks, UK: Harvester Press.
—— (1977) *Imperialism and Unequal Development*, Brighton, UK: Harvester Press.
—— (1978) *The Law of Value and Historical Materialism*, New York: Monthly Review Press.
—— (1980) *The Arab Economy Today*, London: Zed Press.
—— (1981) *The Future Maoism*, New York: New Left Review.
—— (1987) "Democracy and National Strategy in the Periphery", *Third World Quarterly* 9: 1129–1156.
—— (1990a) *Delinking: Towards a Polycentric World*, London: Zed Books.
—— (1990b) *Maldevelopment: Anatomy of a Global Failure*, London: Zed Books.
—— (1990c) *Transforming the Revolution: Social Movements and the World System*, New York: Monthly Review Press.
—— (1991) "The Ancient World System *versus* Modern Capitalist World System", *New Left Review* 14(3): 349–385.
—— (1992a) "Thirty Years of Critique of the Soviet System", *Monthly Review* 44 (1): 43–50.
—— (1992b) *Empire of Chaos*, New York: Monthly Review Press.
—— (1993) "Historical and Ethical Materialism", *Monthly Review* 45(1): 44–56.
—— (1994) *Re-reading the Postwar Period: An Intellectual Itinerary*, New York: Monthly Review Press.
—— (1997) *Capitalism in the Age of Globalisation*, London: Zed Books.
—— (1998) *Specters of Capitalism: A Critique of Current Intellectual Fashions*, New York: Monthly Review Press.
—— (2003) *Obsolescent Capitalism: Contemporary Politics and Global Disorder*, London: Zed Books.
—— (2004) *The Liberal Virus: Permanent War and the Americanization of the World*, New York: Monthly Review Press.
—— (2011a) *Ending the Crisis of Capitalism or Ending Capitalism*, New York: Monthly Review Press.
—— (2011b) *Global History: A View from the South*, Oxford: Pambzuka Press.
—— (2013a) *The Implosion of Contemporary Capitalism*, New York: Monthly Review Press.
—— (2013b) *From Capitalism to Civilization: Reconstructing the Socialist Perspective*, New Delhi: Tulika Books.
—— (2013c) *Beyond US Hegemony: Assessing the Prospects for a Multipolar World*, London: Zed Books. (First published in 2006, Beirut: World Book Publishing.)
—— (2016a) *Russia and the Long Transition from Capitalism to Socialism*, New York: Monthly Review Press.
—— (2016b) *The Reawakening of the Arab World: Challenge and Change in the Aftermath of the Arab Spring*. New York: Monthly Review Press. (Parts of this

book were published in 2011 under the title: *The People's Spring: The Future of the Arab World*, Oxford: Pambazuka Press.)

—— (2018) *Modern Imperialism, Monopoly Finance Capital, and Marx's Law of Value*. New York: Monthly Review Press. (Most of the main ideas informing this book were published in 1978 under the title *Law of Worldwide Value*, New York: Monthly Review Press.)

Antonius, G. (1939) *The Arab Awakening: The Story of the Arab National Movement*, Philadelphia: J. B. Lippincott. https://ia801209.us.archive.org/17/items/McGillLibrary-rbsc_isl_arab-awakening_DS636A461939-16015/rbsc_isl_arab-awakening_DS636A461939.pdf.

M.A. MOHAMED SALIH

ALICE AMSDEN (1943–2012)

Alice Amsden was a heterodox political economist who profoundly influenced contemporary thinking in development theory, policy, and practice. Conceptually rich and empirically "thick", her path-breaking work on late industrialisation and state-led development in East Asia has fundamentally challenged market-oriented neoclassical explanations and Marxian-inspired dependency theories in development thinking over the past four decades. Her 1989 masterpiece, *Asia's Next Giant*, remains the most cited work on the political economy of East Asian development. To some observers, she stands in company with J.K Galbraith, Thorstein Veblen, J.R. Commons, Adolf Berle and **Karl Polanyi**, the most original and influential political economists of the earlier period (Clifton et al. 2017a: 7).

Born in New York's Brooklyn on 27 June 1943, Amsden went to Midwood High School, one of New York City's public schools in their heyday of excellence, and graduated with honours. While her parents were not economists, she grew up discussing major economic issues over the dinner table passionately with them and her sister, Myra Strober, who went on to become a fellow economist at Stanford. She went on to complete her undergraduate degree at Cornell University and PhD at the London School of Economics. She started her professional career in 1971 as an economist with the OECD at a time when only 7 per cent of the economics doctorates in the United States were held by women.

Determined to excel in this gendered world of academic economists, Amsden took an unusual route to outstanding intellectual work by engaging in extensive fieldwork (beginning with studying industrial policy in Poland and Hungary) and taking an inductive approach to theory development. This careful and grounded experience would later give prominence to her major works that define her as one of the most influential political economists in the later 20th century. After teaching in several major universities such as UCLA, Columbia and Harvard, she joined MIT as a professor in the Department of Urban Studies and Planning in 1994. Until 1999, she held the Ellen Swallow Richards Institute Chair, and from 2000 until her untimely death on 14 March 2012, she was Barton L. Weller (1940) Professor of Political Economy.

While Amsden had published her PhD thesis as her first book, *International Firms and Labour in Kenya: 1945–70* in 1971, it was not until her next book, *Asia's Next Giant*, published in 1989, that her enormous influence on our understanding of economic development began to be truly felt across the social sciences. By the early 1990s, even radical development scholars had come to acknowledge that Amsden and other proponents of the developmental state "now constitute an impressive intellectual force, giving legitimacy to what I call the new conventional wisdom, a position that advocates state capitalism along South Korean lines as the most promising strategy for third world development" (Hart-Landsberg 1993: 97). Starting from her inductive and detailed empirical work on Kenya in the 1960s and later Taiwan in the 1970s and South Korea in the 1980s (see Chu 2017), Amsden moved from conducting micro-level studies of national economies to developing macro-level theory of late industrialisation in the 1990s and beyond. During these five decades of sustained contribution to the economic analysis of diverse development trajectories in late industrialising economies, she identified three critical facets of the development process: (1) national firms; (2) the state; and (3) global institutions.

First of all, though most readers tend to associate Amsden's best work with championing the role of the state in late industrialisation, it is important to point out that the *firm* was always central in her analysis of industrial transformation in developing economies.

> The translation of high growth rates of output into high growth rates of productivity depends on what happens *inside the unit of production*. Closing the loop between growth and productivity, therefore, involves an analytical shift, a change in the center of gravity from the state to the other key institution of industrialization, the firm.
>
> (Amsden 1989: 112; original italics)

In fact, her 1989 treatise on South Korea's emergence as Asia's next giant (after Japan) offers a very detailed study of two major national champions – Hyundai Heavy Industries and Pohang Iron and Steel Company (POSCO). Several chapters in that book were devoted to firm-specific management practices, such as corporate power, professional management and human resources. Building on a critique of the absence of the firm in conventional growth theories, industrialisation theories, and market-conforming paradigms, she focused on firm-level practices as "the agent of expansion in all late-industrializing countries" below the level of the state (Amsden 1989: 8–9). To her, late industrialisation is possible if national firms can learn from first movers in advanced countries and acquire the necessary capabilities in project execution and organisational know-how for industrial development (Amsden and Hikino 1994). The firm thus serves as an important unit of production and the productivity gain of its workers from learning-by-doing on the shop floor can be translated into high growth and rapid catching-up at the national level.

Once the centrality of national firms in late industrialisation had been established, Amsden proceeded to explain the political-economic conditions that favoured their rapid emergence and competitiveness. This was the occasion for her to bring in *the state* as the key domestic institution for steering developmentalism in latecomer economies. Indeed, Amsden has often been described as a pioneer in the genre of development theory associated with what is commonly known as "developmental state" theory. The other major contributors are Chalmers Johnson, Robert Wade, Gordon White, Peter Evans, and Stephan Haggard (see Haggard 2015; Yeung 2017; Bishop et al. 2018). And yet, Amsden's approach to the developmental state was less theoretical and more practical. Unlike some developmental state theorists (e.g. Haggard and Evans) whose works focused much more on the politics of state institutions, Amsden was more interested in state instruments and policies and in reciprocal state–firm relationships, i.e. what the state did to make firms work and therefore development happen.

By focusing on state action and its efficacy on firm-level dynamics, Amsden developed some of the most original insights on the role of the state in late industrialisation. Her now widely cited notions of the developmental state "getting relative prices wrong" and deploying "reciprocal control mechanisms" have informed much of the subsequent work by generations of development scholars. Arguing against neoclassical explanations of price-based market competition and dependency theories of foreign domination, she believed in the possibilities of an interventionist state that deliberately distorts relative prices to stimulate economic growth. As a disciplined institution advancing capital rather than enriching itself, the developmental state can take on the entrepreneurial function of industrial planning and the co-ordinating function insofar as it controls financial capital and long-term credit.

In particular, the state can intervene in the market economy by "picking winners" and offering deliberate subsidies and licences to these favoured national firms. Together with other policy instruments such as tariffs, price controls, and capital flow restrictions, the state intentionally distorts relative prices by lowering capital costs and creating profitable investment opportunities in order to push these firms to enter into export-oriented industries and markets they otherwise would not venture into. She argued that "the government established multiple prices for loans, only one of which could possibly have been 'right' according to the law of supply and demand. Moreover, the most critical price – that for long-term credit – was wildly 'wrong' in a capital-scarce country, its real price, due to inflation, being negative" (Amsden 1989: 144). In exchange for these subsidies, cheap credits and limited licences, the state disciplines these favoured firms by imposing stringent performance standards and monitoring in the areas of exports, R&D, or new product introduction. Taken together, Amsden's work on the firm and the state has provided a solid micro-economic foundation to the developmental state theory of late industrialisation. It showcases how national firms evolve in response to state-led industrial policies and financial distortions in the capitalist market economy.

Beginning with her 2001 work, *The Rise of "The Rest"*, Amsden "up-scaled" her shrewd analysis of national economies to the *global institutions* of development by examining how late industrialisers and developing economies could indeed challenge the domination of the "West" in the global economy (also Amsden 2007). Here, she argued for the rise of "the rest" – defined as latecomers in industrial transformation – through pure learning, a process through which these latecomers depend completely on the advanced technological capabilities of first movers in the global economy. This total initial dependence on the commercialised technology of advanced countries to establish modern industries "was one of the phenomenal changes in the last half of the twentieth century. For the first time in history, backward countries industrialized *without proprietary innovations*" (Amsden 2001: 2; original italics). National firms from these latecomer economies have managed to benefit from globalisation through learning from their foreign ventures and/or alliance partners and developing "manufacturing experience" via global outsourcing arrangements. Pre-empting the critical importance of globalisation as an institutionalised platform for latecomer catching-up, she recognised divergent development pathways since the 1980s in relation to the differential industrial targeting by home states across developing economies (see also Amsden and Chu 2003; Yeung 2016).

Incorporating this global dimension of development more explicitly in her *Escape from Empire*, Amsden (2007) pointed further to the critical importance of *geopolitics* in shaping the developmental trajectories of latecomer economies (see also Yeung 2017). She distinguished two major phases of such geopolitical conditions in relation to America's two Empires: 1950–1980 and 1980–the late 2000s. In this trenchant critique of the laissez-faire approach to economic development, she argued that latecomers in East Asia chose to deviate from free market ideology by using their manufacturing experience and state interventions to march into mid-tech industries, such as automobiles, electronics, petrochemicals, shipbuilding and steel. Under the First American Empire, these developing economies could grow rapidly and catch up in these industries "because Washington chose *not* to implement the principle of reciprocity, hoping that 'redistributive trade' would win support for capitalism among the poor and sustain an attitude of indifference among the rich" (Amsden 2007: 45; original italics). However, the rise of the so-called "Washington Consensus" at the beginning of the Second American Empire greatly damaged the prospect for latecomer catching-up due to the financialisation of the real economy and the rapid cross-border flows of "hot" money. The developing world was therefore "doomed to Hell" because "the Second American Empire let opening markets and 'getting the prices right' crowd out acquiring technology and building institutions to exploit it" (Amsden 2007: 11). This geopolitical imperative supported by financialisation "killed" the developmental state that flourished during the First American Empire.

To sum up, there is no doubt that Amsden's work has profoundly reshaped the ways in which we research into and think of economic development and late industrialisation (see recent reflections in Clifton et al. 2017b). While highly argumentative in the usual academic sense, her work is characterised by deep experience from decades of field-based research. In her unpublished manuscript written shortly before her death in 2012, Amsden clearly affirmed that "learning to develop requires a shift from relying on pure deductive theories to relying on inductive live experiences in real time" (quoted in Clifton et al. 2017a: 3). Justin Lin (2017: 78), Chief Economist of the World Bank between

2008 and 2012, paid tribute to Amsden for her significant influence on his own thinking in relation to promoting structural transformation and the state's role in late industrialisation. In 2002, together with Harvard's Dani Rodrik, Amsden was named a co-winner of the Leontief Prize for advancing the frontiers of economic thought and broadening the field of economics to better comprehend urgent contemporary issues. In the same year, she was also featured in the December issue of *Scientific American* as one of fifty "visionaries from the worlds of research, industry and politics whose recent accomplishments point toward a brighter technological future for everyone". Amsden was honoured for identifying strategies for late industrialisation that could be of enormous value to developing countries on the rise. On 14 March 2012, development studies prematurely lost one of its intellectual giants. We can only hope that Amsden's non-conformist spirit and heterodox legacy will continue to endure and inspire future generations of development scholars and practitioners.

Major works

Amsden, A.H. (1971) *International Firms and Labour in Kenya: 1945–70*, London: Cass.

—— (1977) "The Division of Labor Is Limited by the Type of Market: The Case of the Taiwanese Machine Tool Undustry", *World Development* 5(3): 217–233.

—— (1979) "Taiwan's Economic History: A Case of Etatisme and a Challenge to Dependency Theory", *Modern China* 5(3): 341–379.

—— (1985) "The State and Taiwan's Economic Development", in T. Skocpol, P. B. Evans, and D. Rueschemeyer (eds), *Bringing the State Back In*, Cambridge: Cambridge University Press, 78–106.

—— (1989) *Asia's Next Giant: South Korea and Late Industrialization*, New York: Oxford University Press.

—— (1997) "Bringing Production Back In: Understanding Government's Economic Role in Late Industrialization", *World Development* 25(4): 469–480.

—— (2001) *The Rise of "The Rest": Challenges to the West From Late-Industrializing Economies*, New York: Oxford University Press.

—— (2007) *Escape from Empire: The Developing World's Journey through Heaven and Hell*, Cambridge, MA: MIT Press.

—— (2012) "The Rational Revolution: Developing from Role Models, Deserting Deductive Theory", unpublished MS.

—— (2014) "Securing the Home Market: A New Approach to Korean Development", in T. Mkandawire and I. Yi (eds), *Learning from the South Korean Developmental Success*, New York: Palgrave, 54–90.

—— and Chu, W.-W. (2003) *Beyond Late Development: Taiwan's Upgrading Policies*, Cambridge, MA: MIT Press.

—, DiCaprio, A. and Robinson, J.A. (eds) (2012) *The Role of Elites in Economic Development*, New York: Oxford University Press.

— and Hikino, T. (1994) "Project Execution Capability, Organizational Know-How and Conglomerate Corporate Growth in Late Industrialization", *Industrial and Corporate Change* 3: 111–147.

— Kochanowicz, J. and Taylor, L. (1994) *The Market Meets its Match: Restructuring the Economies of Eastern Europe*, Cambridge, MA: Harvard University Press.

Further reading

Bishop, M.L., Payne, A., Sen, K., Breslin, S., Öniş, Z., Muzaka, V., Booth, D., Lindsay, C. and Yeung, H.W.-C. (2018) "Revisiting the Developmental State", SPERI Paper no.43, Sheffield Political Economy Research Institute, University of Sheffield.

Chu, W.-W. (2017) "Inductive Method and Development Perspective: Alice Amsden on Taiwan and Beyond", *Cambridge Journal of Regions, Economy and Society* 10(1): 15–34.

Clifton, J., Glasmeier, A. and Sheth, A. (2017a) "Revisiting Development Theory: Alice H. Amsden's Impact on the Field", *Cambridge Journal of Regions, Economy and Society* 10(1): 3–14.

— — — (2017b) *Revisiting Development Theory: A Celebration of Alice Amsden*, special issue of *Cambridge Journal of Regions, Economy and Society* 10(1): 1–205.

Haggard, S. (2015) "The Developmental State Is Dead: Long Live the Developmental State!", in J. Mahoney and K. Thelen (eds), *Comparative Historical Analysis in Contemporary Political Science*, Cambridge: Cambridge University Press, 39–66.

Hart-Landsberg, M. (1993) *The Rush to Development: Economic Change and Political Struggle in South Korea*, New York: Monthly Review Press.

Lin, J.Y. (2017) "Alice H. Amsden's Contributions to Development Economics", *Cambridge Journal of Regions, Economy and Society* 10(1): 77–81.

MIT News Office (2012) "Development Economist Alice Amsden Dies at 68", *MIT News*, 16 March (also at news.mit.edu/2012/alice-amsden-0316, accessed on 24 April 2018).

Rasiah, R., Singh, A. and Ernst, D. (2015) "Alice Hoffenberg Amsden: A Consummate Dirigiste on Latecomer Economic Catch-Up", *Institutions and Economies* 7(1): 1–8.

Van der Hoeven, R. (2008) "Alice Amsden", *Development and Change* 39(6): 1091–1099.

Yeung, H.W.-C. (2016) *Strategic Coupling: East Asian Industrial Transformation in the New Global Economy*, Ithaca, NY: Cornell University Press.

— (2017) "Rethinking the East Asian Developmental State in its Historical Context: Finance, Geopolitics, and Bureaucracy", *Area Development and Policy* 2 (1): 1–23.

HENRY WAI-CHUNG YEUNG

A.T. ARIYARATNE (1931–)

A.T. Ariyaratne's distinctive contribution as a key thinker lies in his lifelong efforts to follow a new path to development, independent of both capitalism and socialism. He is the leader of the Sarvodaya Shramadana movement, the largest non-governmental organisation (NGO) engaged in development and poverty alleviation in Sri Lanka today. Sarvodaya's vision for a new society with "no poverty" and "no affluence" was based on **Gandhi**'s philosophy of "truth", "non-violence" and "self-sacrifice". The term *Sarvodaya* comes from two Sanskrit words, *sarva* (universal) and *udaya* (awakening). Ariyaratne uses the word *sarvodaya* in two ways, to mean the awakening of all people and the awakening of individuals in all spheres – psychological, moral and spiritual, as well as social, economic and political. The term *shramadana* is also derived from two Sanskrit words, *shrama* (labour) and *dana* (gift), that is, the gift of labour.

In the Sinhalese language the two words *sarvodaya shramadana* have come to mean "the sharing of one's time, thought, and energy for the awakening of all" (*Dana*, Feb. 1987: 15). Sarvodaya believes that development involves more than material growth. It involves psychological, moral and spiritual dimensions as well as social, economic and political ones. Shramadana or "gift of labour" implies both physical and mental labour. Ariyaratne's *shramadana* draws on social networks, donated labour, skills and co-operation, showing how social capital can create material wealth.

Dr Ariyaratne was born in Unawatuna village, Galle District, and after graduation worked as a teacher in Galle town before attending a teachers' training college. In 1958, as a newly arrived science teacher at Nalanda College, a Buddhist high school in Colombo (where he served till 1972), along with his students he organised the first of several voluntary workshops (a *shramadana*) in one of the poorest villages on the island. The workshop consisted of digging wells, building pit latrines, planting gardens and opening up rural roads using the co-operative labour of students and villagers. Performing manual labour alongside poor villagers was a transformative experience for students who came from upper- and middle-class urban homes. This was an early example of what we now call "service learning".

Ariyaratne received a Bachelor of Arts general degree from the Vidyodaya University of Sri Lanka, graduating in economics, Sinhalese language and education. Later he received an honorary doctorate from the same university, and a doctor of humanities from Amelio Aguinaldo

University in the Philippines. As the founder and leader of the Sarvodaya Shramadana movement, Ariyaratne has received several international awards, including the Raman Magsaysay Award for Community Leadership from the Philippines (1969), the Feinstein World Hunger Award from Brown University in Rhode Island (1986), the Niwano Peace Prize from Japan (1992) and the Mahatma Gandhi Peace Prize from India (1996).

Although there are many publications under his name (Ariyaratne 1988; 1999) and many more about him (Bond 2004; Macy 1985), Ariyaratne is not widely regarded as a development theorist. According to his own homepage (www.sarvodaya.org), "he was not guided by theory. He wanted to practice first and enunciate theory later. And the practice should be meaningful; the theory should only follow it." After remaining active in community development throughout his life, in the 1990s he increasingly turned his attention to one of the central problems of Sri Lanka: political violence and the long-drawn-out military conflict with the Liberation Tigers of Tamil Eelam (LTTE). The Liberation Tigers had been fighting since 1983 for a separate state for minority Tamils on the grounds that they suffered discrimination at the hands of the Sinhalese majority. In 2009 the Liberation Tiger movement collapsed with the killing of its leader, Velupillai Prabhakaran, by the Sri Lankan army.

According to Bond (2004: 7–42), since its inception in 1958, Sarvodaya seems to have evolved through four phases. However, it is best to add a fifth phase to take account of recent changes with the gradual withdrawal of Dr Ariyaratne from administrative leadership of the organisation. During the period 1958–1967, Sarvodaya was primarily a volunteer work camp movement. The work camp begun by Ariyaratne and his students in 1958 was very successful, and launched a larger social movement that quickly spread to other high schools and villages. At that time resources came entirely from local donations and voluntary labour (*sharmadana*). During the second period (1967–1983), Sarvodaya became a formal NGO. In 1972 the movement was recognised by an Act of Parliament and incorporated as a legal body. It began to attract generous foreign funding, became a fully-fledged NGO with a large portfolio of projects for village economic development, and adopted methods of cost accounting, monitoring and evaluation. As it grew, Sarvodaya moved away from the ideology of social revolution, co-operated more closely with the government and acted in the capacity of an extension agency. Apart from local village schemes, Sarvodaya also undertook several well-funded national projects for enterprise development, alternative technology (see also **Fritz Schumacher** and **Frances Stewart**) and child care. During the third period (1983–1997), the conflict between the government and the Liberation Tigers intensified and spread; civilian life for both Sinhalese and Tamils was severely disrupted. With funding from overseas, Sarvodaya operated a large programme offering rehabilitation and relief work in villages most affected by the Tiger insurgency.

During this period the close co-operation with the government came to an end because Dr Ariyaratne opposed the policy of seeking a military solution to the LTTE problem. Drawing on **Gandhi**an and Buddhist principles of non-violence, Sarvodaya laid out a plan for the peaceful resolution of the conflict through spiritual means, but the larger conflict continued. In the period (1997–2009), we see clear evidence of Sarvodaya taking an even stronger stand against a military solution. Sarvodaya publicly declared that neither the government nor the LTTE could "win" the war; all they could do was to draw out the conflict (www.sarvodaya.org). On the other hand, Sarvodaya claimed it knew how to help transcend the war. To that end it organised a large-scale peace movement, announced an alternative framework of power for conflict resolution and held several well-attended public peace meditations. It also started a programme of "sister villages", with villagers from the south travelling to the war-ravaged villages of the north to do rehabilitation work repairing houses, wells, tanks, schools, toilets and places of worship. Even though Ariyaratne was convinced of the potential of his philosophy to achieve both personal empowerment and national reconciliation through non-violence, spirituality, compassion and Buddhist principles of "right livelihood", he could not prevent the brutal and bloody civil war which finally ended in May 2009. It is difficult to fix an exact date for the final phase of the history of Sarvodaya, because the changes that have taken place happened slowly with the gradual withdrawal of Dr Ariyartne from the central administration of the organisation. I shall return to an assessment of this contemporary period at the end of the essay.

The Sarvodaya philosophy has a strong moral and spiritual foundation. Ariyaratne was well aware that religion, in its institutional form, historically has not played a progressive role in the material transformation of society. And yet he chose this vehicle for his programme of social advancement, a path not unlike that taken by **Gandhi** in India, and by the worker priests of the Catholic-based communities in Latin America. Sarvodaya philosophy is founded on Buddhist teaching; Ariyaratne's contribution has been to show that Buddhism can be used to address two principal problems facing contemporary Sri Lanka – poverty and violence. The Sarvodaya philosophy is vast, but for the purpose of this exercise I shall present it under three headings.

Personal agency in structural change: Sarvodaya holds that structural change in society must begin with personal change. According to Buddhism, the principles of virtuous living include loving kindness, respect for all living beings, compassion for others, sharing joy in the completion of projects intended for someone else's benefit and even composure in the face of both joy and sadness. All Sarvodaya workers are urged to practise these in their routine everyday engagements. Beginning with personal awakening (*puroshodaya*), Sarvodaya expands outwards to community or village awakening (*gramodaya*), and beyond that to the awakening of the country (*deshodaya*), and finally to the awakening of the world (*vishovodaya*). All social change must radiate out from the agency of the individual living and practising a righteous life.

Buddhist theory for overcoming suffering: central to Sarvodaya philosophy is the Buddhist notion of four "noble truths". The first states that the normal condition of existence is suffering (*Dukkha*). Second, the root cause of suffering is greed, craving and desire (*Thanha*). Third is the claim that suffering can be overcome (*Niradha*). Finally is *Marga*, the Buddhist path to overcome suffering in the world. *Marga* contains eight elements: right understanding, right thought, right speech, right action, right livelihood, right effort, right mindfulness and right concentration. The "Eightfold Path" points to the centrality of personal agency in both Buddhist and Sarvodaya philosophy. Dr Ariyaratne uses the first of the "Four Noble Truths" (there is suffering in the world) by starting from an actual village with great poverty. This becomes the basis for analysis and reflection on the realities of the village. The second principle (craving as a cause of suffering) is presented to villages by describing how distrust, competition, enmity and egocentricity sap the energy of villagers. The third principle (suffering can cease) is translated into a discourse on affection, compassion, kind words, sharing and mutual self-help. Finally, the eightfold path of *Marga* recommends the avoidance of two extremes – the pleasure of sensual indulgence and the pain of self-mortification. Ariyaratne adapted this Buddhist concept of the "Middle Way" by recommending a development path of "no poverty, no affluence". Sarvodaya does not believe that poverty can be eradicated through creation of wealth and economic development. Sarvodaya begins this task by stating a manifesto of ten basic human needs which includes: a clean and healthy physical environment, water, clothing, food, housing, health care, communication, fuel, education, and spiritual and cultural needs (www.sarvodaya.org) (see also **Paul Streeten**). These needs can be satisfied by using local resources, self-reliance and *shramadana*. Sarvodaya believes that the dominant models of economic development constitute obstacles to meeting basic needs because such models emphasise increased production, expansion of desire, unlimited growth and open markets.

Non-violence in conflict resolution: a core principle of Buddhism and Hinduism is non-violence, which advocates respect for all sentient beings. Following the examples of **Gandhi** and Vinoba Bhave in India, Ariyaratne argued that fundamental social change can be achieved through non-violence. The adoption of the eightfold path of right livelihood automatically excludes the use of physical violence as a means of achieving objectives. Sarvodaya reached out to the Tamils through membership of the organisation and by doing relief and rehabilitation work in villages of the North. It organised peace marches and mass meditations, and offered to mediate between the government and the LTTE.

Ariyaratne attributes Sri Lanka's current economic problems to British imperialism which destroyed self-reliant village economies. In that regard, his analysis is very similar to that of dependency theorists like **Andre Gunder Frank** (Frank 1966). He also believes that unequal global structures and giant transnational organisations are responsible for increasing poverty, armed conflict and ecological destruction in Sri Lanka and throughout the world. Despite that structural analysis, his proposals do not focus on changing global structures but on personal agency and right living. Nowhere in its literature is a discussion of how "righteous living" can overcome the power of global capital. Indeed, what the "Open Economy" has created in the last forty years is a very far cry from the Sarvodaya vision of a just, moral, righteous and non-violent society. In 2018 Ariyaratne was 87 years old. He has an active, inquiring mind and is still knowledgeable about politics in Sri Lanka, but he has no serious unified analysis of the political economy of global capital, consumerism, environmental destruction, political corruption and commercial fraud.

Sarvodaya's material critique of development arises from its explicit rejection of affluence. The rejection of capital-intensive, high-technology, open economies is clearly at odds with the dominant models of development. Economic growth requires the constant expansion of consumption and desire, which according to the Buddhist theory of craving is what leads to suffering. Economists measure living standards principally by how much people consume. Sarvodaya rejects this path of development.

Even though the concept of no affluence has the potential to yield a serious critique of development, Sarvodaya has weakened the argument by framing it entirely within the context of Buddhist advocacy of controlling one's craving. Craving is presented in Sarvodaya philosophy as an intrinsic human condition that must be overcome through reflection and moral advancement. But it is not enough to advocate the voluntary limiting of consumption without also addressing the myriad forces of the social construction of craving through advertising and the presentation of the "middle class" as the accepted norm of success. It is indeed ironic that Buddhist monks from the more prestigious Temples of the island usually travel in very expensive, luxury motor cars despite clerical vows of modesty, piety and frugality. Monks, professionals and really anyone who can afford to finance a car, all choose to travel in private vehicles, while public transport remains crowded and time-consuming. I have not seen any Sarvodaya literature on the social construction of craving.

Sarvodaya's advocacy of a non-violent solution to the ethnic problems of Sri Lanka is very courageous. A number of Tamil Sarvodaya workers were killed by the Liberation Tigers, and Ariyaratne himself lived under constant threats to his life. But there are some ironies to Sarvodaya's offering non-violence as a solution to the ethnic problem. Despite its ecumenical outlook, Sarvodaya in Sri Lanka is perceived primarily as a Buddhist organisation. The conflict in Sri Lanka arises from Tamil reaction to what they perceive as the hegemony of Sinhala Buddhist nationalism. The identity of the Sinhalese and their perception of self are both intricately tied to Buddhism, its social history and its myths. Conversely, the Sinhalese perception of the Tamil is that of "the other as non-Buddhist" – a dynamic which the Tamils in turn have incorporated into their perception of the Sinhalese and of themselves. Given the signification of Buddhism to the two main parties in the conflict, it is important to reflect on the limits of a social movement that uses Buddhist theology to bring about a non-violent resolution of the protracted ethnic conflict.

ADDENDUM

The original essay on Ariyaratne was written in 2005. The following is based on notes compiled by the author over six months in 2017 and 2018 during field visits to Sarvodaya and meetings with Dr Ariyaratne, administrative officers of the organisation and a few rural beneficiaries of Sarvodaya services.

I make two sets of comments. The first concerns a theoretical observation on Sarvodaya philosophy. The word *Sarvodaya* means universal awakening of the human being in all its senses – spiritual, social, cultural, economic and political. It can also mean the awakening of all human beings. Sarvodaya also encompasses the concept of *Deshaodaya* – awakening of the region moving through the geographic hierarchy of a neighbourhood, village, district and country. Changing the world through Buddhist-inspired values of kindness, generosity, sharing and simplicity, "no poverty and no affluence" is the core of Sarvodaya thinking.

Using the ideas of post-structural analysis (Foucault 1980), I argue that this is a "totalising theory" which has little chance of bringing about lasting social change. Here is a summary of this argument. The sociopolitical economy in Sri Lanka, as in every other nation, is a very complex system of interdependent variables. It is simply impossible to enter such "overdetermined" systems armed with the moral authority of the philosophy of righteous living. It would make far more sense for Sarvodaya to have taken a substantive approach to social change by giving leadership or by actively co-operating with other organisations in Sri Lanka to work on topics such as preventive health, nutrition, public transport and the environment. I would argue that a totalising philosophy of universal awakening can provide an overarching framework of good behaviour, but that it cannot offer practical guides to social change. Within a larger moral framework of kindness and justice, we need to address substantive topics related to (quotidian) routine life issues such as nutrition, preventive health and pressure to conform to middle-class models of success. Such substantive discourses and action are missing from the Sarvodaya literature, office conversations I heard, administrative priorities and projects I witnessed.

Again, based on my field observations of 2018, I wish to make a second comment on Sarvodaya as a practical organisation which I also present as the contemporary period of its history. With Dr A.T. Ariyaratne's diminished role due to age and poor health, Sarvodaya has begun to resemble an institution of development administration much like a government department. Dr Ariyaratne's son, Dr Vinya Ariyaratne, became the Executive Director of Sarvodaya in 2000 and gradually took over policy planning and day-to-day administration of the organisation. Dr Vinya is a highly qualified health professional with degrees from very prestigious universities. He has a modest demeanour, is a good listener at meetings, and appears to be a competent administrator of a complex organisation with thousands of employees. He lacks the charisma of his father, does not have access to foreign funds at levels his father received during the heyday of the organisation, has the burden of a huge payroll and is unable to rekindle the spirit of voluntary labour (*shramadana*) so characteristic of Sarvodaya's past.

One example serves to illustrate what I saw as the changed character of Sarvodaya. During a presentation I made to its administrators in September 2018, I raised the following points: (1) that Sarvodaya lacked a coherent social theory that could help counter the detrimental effects of global capitalism in Sri Lanka; and (2) that Sarvodaya should pay attention to Foucault's (1980) critique of totalising social theory and focus on substantive sites of social change related to daily needs of people, with programmes requiring only modest capital inputs and employing locally available technical knowledge and inputs. As an example of such a substantive project, I suggested creating an urban garden on Sarvodaya land in its headquarters in Moratuwa. Diet-related health problems such as high blood pressure, diabetes, cardiac disorders and obesity are very common in urban Sri Lanka. The Sarvodaya garden could provide a model of intensive vegetable cultivation on a small area of land. It would have educative and instructional value to other Sarvodaya centres throughout Sri Lanka. The proposal was to create a garden growing commonly consumed greens, using principles of organic cultivation, vertical gardening, and recycling greywater from the kitchen and bathrooms. Despite the obvious advantages of such a project, modest costs and an offer to share costs, I found it difficult to move the proposal forward. To clear the ground for a small initial plot we had to hire labourers, as no one in the organisation was available to donate labour – this from an organisation that contains the term *shramadana* (donation of labour) in its name. Even though Dr Ariyaratne senior wanted the project, and Dr Vinya Ariyaratne was in favour, a period of fully six months was not enough to launch the project.

As for the contemporary period of Sarvodaya history, it is my contention that volunteerism is essentially dead; talk of the spirit of universal awakening seems distant and quaint, and Sarvodaya seems more interested in working with large donors. The focus on funds seems totally understandable, as Sarvodaya has a very large number of paid employees.

In conclusion, I believe that A.T. Ariyaratne is the most important development thinker that Sri Lanka has produced. Even before Ivan Illich's (1978) critique of economic development, Dr Ariyaratne had arrived at the conclusion that development cannot eradicate poverty. Even though Sarvodaya has changed much, I believe that a philosophy of "no poverty and no affluence" and personal agency should be an integral part of a serious post-structural approach to the political economy of Sri Lanka.

Major works

Ariyaratne, A.T. (1988) *The Power Pyramid and the Dharmic Cycle*, Ratmalana, Sri Lanka: Sarvodaya Vishva Lekha Press.

—— (1999) *Collected Works*, 7 vols, ed. N. Ratnapala, Ratmalana, Sri Lanka: Sarvodaya Vishva Lekha Press.

Further reading

Bond, G.D. (2004) *Buddhism at Work: Community Development, Social Empowerment and the Sarvodaya Movement*, West Hartford, CT: Kumarian Press.

Foucault, M. (1980) *Power/Knowledge*, New York: Pantheon Books.

Frank, A.G. (1966) "The Development of Underdevelopment", *Monthly Review* 18: 17–31.

Giddens, A. (1984) *The Constitution of Society*, Berkeley, CA: University of California Press.

Illich I. (1978) *Towards a History of Needs*, New York: Pantheon Books.

Macy, J. (1985) *Dharma and Self Development: Religion as Resource in the Sarvodaya Self-Help Movement*, West Hartford, CT: Kumarian Press.

Sarvodaya (1987) *Dana: Journal of the International Sarvodaya Shramadana.*

LAKSHMAN YAPA

JAGDISH BHAGWATI (1934–)

Jagdish Bhagwati is currently a Professor at Columbia University and Senior Fellow at the Council of Foreign Relations. He was born and brought up in Bombay (Mumbai) and was educated at Cambridge, Oxford and MIT. Bhagwati has been an adviser to a number of international organisations, including the WTO. His vast published output extends to more than 300 articles and 50 volumes. His work has so dominated thinking on a number of economic policy issues in our own time that Paul Samuelson has referred to it as "the Age of Bhagwati". This essay provides a brief overview of Bhagwati's work centred on his advocacy of free, non-discriminatory trade, his assault on misguided protectionist policies, his analysis of the so-called phenomenon of the brain drain and his latest contribution on the primacy of growth.

Bhagwati's relentless advocacy of free trade is often referred to as militant and aggressive. This may be so, but his perspective is grounded in economic theory and his analysis and observation of the welfare costs of India's protectionist trade and investment regime for more than four decades since the 1950s.

Students of international economics are familiar with the gains from free trade based on the comparative advantage of trading nations. There is, though, a long list of exceptions to the case for free trade, including the infant industry argument, distortions in the labour and product markets in the domestic economy, and externalities. In a seminal article (1963), co-authored with the late V.K. Ramaswami, using the conventional tools of the trade theorist, Bhagwati demonstrated that the best way to address distortions or domestic market failures was through domestic policies while maintaining free trade. The best policy in the presence of various distortions is an appropriate subsidy which rectifies the distortion at its source but leaves the international price unchanged and avoids the loss of consumer surplus associated with a tariff. If the distortion is in the trade sphere, such as monopoly/monopsony power in international trade, it should be rectified with appropriate trade policy instruments such as an export tax or an import tariff. This message relating to trade carries over to policies regarding foreign capital inflows. It is that capital inflows in the presence of a tariff can be immiserising.

Bhagwati enunciated the concept of immiserisation in an article (1958) written whilst he was an undergraduate student reading for the Economics Tripos at Cambridge. The message of the article is that economic growth may so worsen a country's terms of trade that it may actually reduce economic welfare. Extensions of the idea by Bhagwati himself (1968; 1973) and others demonstrate that capital inflows in the presence of a tariff on an importable good in a two-good two-country mode can be immiserising. Whilst the capital inflow induces growth at a constant tariff inclusive of domestic prices, the tariff imposes the familiar production and consumption costs, and the rewards accruing to foreign capital must also be reckoned as a cost.

Yet another of Bhagwati's policy prescriptions relating to trade and capital flows is that countries pursuing an export promotion (EP) policy are likely both to attract a higher volume of foreign direct investment (FDI) and to experience substantially higher rewards from it than countries pursuing import substitution (IS) policies (1978). An EP policy here is defined as one which is neutral in terms of the protection it provides for importables and the incentives it affords exportables. In such an environment, investment decisions of foreign firms would be based on comparative advantage dictated by market forces and not by artificial and uncertain policy inducements. Econometric tests suggest that Bhagwati's twin propositions are robust (Balasubramanyam, Salisu and Sapsford 1996). It is worth noting that Bhagwati's interpretation of the EP policy does not rule out tariffs, but merely requires that investment decisions should be based on comparative advantage and should not be influenced by artificial policy incentives which distort labour and product markets.

Bhagwati is known for identifying a variety of policies which reduce welfare and labelling them in memorable phrases. Immiserising growth, in relation to FDI and Directly Unproductive Profit Seeking (DUP, pronounced DUPE), is now a part of the lexicon of economics. DUP, as defined by Bhagwati (1982), refers to ways of making profits by undertaking activities which do not produce goods and services. A whole array of activities, including tariff-seeking, lobbying, tariff evasion, procuring import licences and seeking premium prices for such licences and smuggling, constitute DUP. All such activities, with rare exceptions, decrease welfare by drawing resources away from production-oriented, utility-enhancing activities into non-production-oriented, profit-seeking enterprise. Most DUP activities arise because of various sorts of distortions induced by policy interventions in trade.

An arresting feature of Bhagwati's work is its enduring quality: his insights into trade policy are applicable to the ever-changing scenarios in the world economy. One such scenario, which Bhagwati refers to as "ironic role reversal", is the growing fear of globalisation in the developed countries, which mirrors the fears expressed by developing countries. The fears of developing countries concerning trade and foreign investment are well known; integration with the global economy would exacerbate income inequalities, perpetuate backwardness and result in dependency. The opposition of developed countries to free trade with the developing countries is based on the relatively low wages and poor standards relating to protection of the environment in the developing countries. Bhagwati has relentlessly questioned the wisdom of these arguments for protection and restated the proposition, which he enunciated in the 1960s, that market failures should be fixed with appropriate domestic policies and free trade should be maintained externally. His provocative book on globalisation (2004) denounces many popular fallacies associated with globalisation current in both the developed and developing countries.

During the late 1970s and the early 1980s, Bhagwati produced a series of articles and books analysing the welfare implications of emigration of talented people from developing countries to the developed world (1974; 1977). Here again his analysis questions received wisdom encapsulated in the so-called "cosmopolitan model" of brain drain, which argues that the emigration of talented people improves welfare all round – of the emigrants, the countries from which they depart and those which receive them. Bhagwati detects a variety of situations in which the emigration of talented people could impose welfare costs on the countries from which they depart. These include situations in which the wage paid to the prospective emigrants is less than their social marginal product and in some cases less than their private marginal product also, and cases where those who leave have to be replaced by others in the labour market with state-funded subsidies for training and education.

These and other welfare-reducing effects of the brain drain underlie Bhagwati's advocacy of a brains tax or what is now known as the "Bhagwati tax", which requires skilled migrants to pay a special tax in addition to the taxes they pay in the countries to which they emigrate. The proceeds of the tax are to be transferred to their home countries. The economic rationale for a brains tax is that immigration of skilled people benefits developed countries, there are capital flows embedded in the flow of skilled people from developing to developed countries, and the former deserve to be compensated for this transfer of capital received by the latter. It is worth noting that Bhagwati's advocacy of a tax on skilled emigrants is based on considerations of equity and efficiency and not presented as a device for restricting emigration. The advocate of free trade is also an advocate of free flows of labour, be they of the skilled or the unskilled variety. He has frequently championed a liberal humanitarian approach to the regulation of illegal immigrants into the USA.

Bhagwati's contribution to development economics is manifold, but what has evoked widespread debate and, in fact, controversy is his questioning of that which most economists regard as received wisdom. In a debate organised by *The Economist* magazine (2011), Bhagwati questioned the belief that manufacturing rather than services is the leading sector for growth and development. His thesis that services were complementary to manufacturing, and that there was no reason to believe that services could not be the leading sector, was not only grounded in theory and facts but also presented with as much wit as wisdom.

Yet another controversy followed Bhagwati's oration to the Lok Sabha, the lower house of India's Parliament (2010), that endorsed the primacy of investment in growth as opposed to investments in social infrastructure such as health and education. In other words, stage one of the development process should be centred on growth that would provide the resources for stage two of the reforms that target heath and education. This proposition, though endorsed by most economists, triggered a widespread debate in India's press and academia. Bhagwati's continued endorsement of his growth-first proposition in India's leading newspapers gave publicity to an issue of importance for policy that had hitherto been discussed mostly in academic publications (Bhagwati and Panagariya 2013; Drèze and Sen 2013).

This brief overview of Bhagwati's contribution to the literature on international economics hardly does justice to the breadth of his scholarship, his many insights and his economic philosophy developed over decades of research and teaching. His crusade for free trade, grounded in the belief that free trade is an important moral force for good, has evolved over long years of theoretical work and analysis of trade policy of both the developed and developing countries. These ideas have influenced economic policy both in India and abroad.

Major works

Bhagwati, J.N. (1958) "Immiserising Growth: A Geometrical Note", *Review of Economic Studies* 25 (June): 201–205.

—— (1968) "Distortions and Immiserising Growth: A Generalisation", *Review of Economic Studies* 35(104): 481–485.

—— (1971) "The Generalised Theory of Distortions and Welfare", in J.N. Bhagwati et al. (eds), *Trade, Payments and Growth*, Amsterdam: North-Holland, 69–90.

—— (1973) "The Theory of Immiserising Growth: Further Applications", in A. Swoboda and M. Connolly (eds), *International Trade and Money*, London: George Allen and Unwin, 45–54.

—— (1977) "The Brain Drain: International Resource Flow Accounting, Compensation, Taxation and Related Policy Proposals", prepared for the Division of Transfer of Technology, UNCTAD.

—— (1978) "Anatomy and Consequences of Exchange Control Regimes", *Studies in International Economic Relations* 1(10): 232.

—— (1982) "Directly Unproductive Profit Seeking (DUP) Activities", *Journal of Political Economy* 90(5): 988–1022.

—— (2010) "Indian Reforms Yesterday and Today", Hiren Mukherji lecture to the Lok Sabha.

—— and Hamada, K. (1974) "The Brain Drain, International Integration of Markets for Professionals and Unemployment: A Theoretical Analysis", *Journal of Development Economics* 1(1): 19–24.

—— and Ramaswami, V.K. (1963) "Domestic Distortions, Tariffs and the Theory of Optimum Subsidy", *Journal of Political Economy* 71(1): 44–50.

—— and Panagariya, A. (2013) *Why Growth Matters: How Economic Growth in India Reduced Poverty and the Lessons for Other Developing Countries*. New York: Public Affairs.

Further reading

Balasubramanyam, V.N. (ed.) (1997) *Jagdish Bhagwati: Writings on International Economics*, Oxford: Oxford University Press.

Balasubramanyam, V.N., Salisu, M.A. and Sapsford, D. (1996) "Foreign Direct Investment and Growth in EP and IS Countries", *Economic Journal* 106(434): 92–105.

Bhagwati, J.N. (1998) *A Stream of Windows: Unsettling Reflections on Trade, Immigration and Democracy*, Cambridge, MA: MIT Press.

—— (2004) *In Defense of Globalisation*, Oxford: Oxford University Press.

Drèze, J. and Sen, A. (2013) *An Uncertain Glory: India and its Contradictions*. London: Allen Lane.

The Economist (2011) The Economist.com debate: "Manufacturing".

V.N. BALASUBRAMANYAM

PIERS BLAIKIE (1942–)

Piers Blaikie has the reputation of an iconoclast. In a tribute when Blaikie retired from the School of Development Studies at the University of East Anglia (UEA) in 2003, Frank Ellis wrote of a "totally over-the-top character, given to extravagant gestures and even more extravagant juxtapositions of ideas".[1] A second characteristic of Blaikie's academic career has been his willingness, even delight, in up-ending and challenging his own assumptions and writings to push the research frontier on to new ground.

Blaikie took a first degree in Geography at the University of Cambridge in 1963. In 1964, funded by a Hayter Studentship, he began a PhD at Cambridge under the supervision of Professor Benny Farmer, spending a year in India and undertaking fieldwork in the Punjab and Rajasthan. He successfully defended his thesis on the "spatial organisation of some villages in Northern India" in 1967 (see Blaikie 1971). He then went on to join the Department of Geography at the University of Reading as a Lecturer, where he stayed until 1972. From 1972 through to 2003 Blaikie was successively Lecturer, Senior Lecturer, Reader and, from 1990, Professor in the School of Development Studies at the

University of East Anglia. It was here that he made his mark. He has also held visiting positions at the Norwegian University of Science and Technology in Trondheim; the Department of Geography at the University of California, Los Angeles; the Department of Geography at Berkeley; the East–West Center in Hawaii; and the Department of Geography at the University of Hawaii at Manoa. Since he published his first book in 1975, Blaikie has authored or co-authored nine major volumes. His journal output has been, by contrast, relatively modest: thirty mainstream academic papers since 1970. It would seem that Blaikie successfully resisted the publication demands of the UK's Research Assessment Exercise (RAE). Indeed, in a paper in 2008 he lamented how, notwithstanding "rhetorical gestures", the RAE cannot "reward time away from publication ... however worthy and effective engagement outside the academy may be" (Blaikie 2008: 768).

Blaikie's contributions to scholarship and policy have also been reflected in a number of significant honours and awards: an honorary doctorate from the Norwegian University of Science and Technology (2009); the Netting Award from the Association of American Geographers (2009) for his work bridging the gap between geography and anthropology; and the Edward Heath Award from the Royal Geographical Society (2004) for his work on Asia and Africa.

Blaikie's contributions to development thinking have been in three key areas: environment, environmental processes and environmental change; agrarian change, particularly in Nepal; and AIDS and family planning. While, at first glance, these may seem to be quite disparate threads of scholarship, the underpinning approach to each reveals evident resonances: first, a broad interest in the relationships between people and their environment (broadly construed) and, second, a wish to let empirics speak to theory.

Of all Blaikie's books, the one that has had the greatest impact on the most people is his *Political Economy of Soil Erosion in Developing Countries* (1985). Like many influential books, the core idea on which the wider argument rests seems, with hindsight, to be so obvious and unremarkable that it is difficult to believe that nobody had thought of it earlier. It is that soil erosion is not due to mismanagement, overpopulation or environmental context, but due to the tendency for surplus to be extracted from farming households by the nature of the political and economic system, requiring that farmers, in their turn, extract surplus from the land, leading to erosion and degradation. In this way, class relations become centrally implicated in the process. "A principal conclusion of this book", Blaikie writes, "is that soil erosion in lesser developed countries will not be substantially reduced unless it seriously threatens the accumulation possibilities of the dominant classes" (p. 147).

Another surprising aspect of the book is that it was marketed as an introductory text for "colleges and universities", is written in clear and accessible prose, and is just 188 pages in length. The argument that physical processes like soil erosion and land degradation require more than technical solutions and are products and reflections of the political, social and economic domains may now be common currency in social science, but the message still needs to be communicated to policy-makers in many places, and to scientists in some fields. This is all the more pressing given the sprectre of climate change. At the time of its publication, however, the book was seminal in bringing such views into the mainstream, helping to spawn a whole new subset of intellectual endeavour under the mantle of political ecology. Prominent scholars like Simon Batterbury, Tony Bebbington, Raymond Bryant, James Fairhead, Tim Forsyth, Melissa Leach, Nancy Peluso and Jules Pretty all owe an intellectual debt to this low-key and simply expressed undergraduate textbook. The book has been criticised for its rather essentialist and non-dynamic approach to politics (see Watts in *PIHG* 1997). But there can be few seminal works – in 1997 it was labelled a "classic" in human geography (see Watts 1997) – that have been so modestly attired.

Blaikie further developed these ideas, with **Harold Brookfield**, in their part-jointly authored, part-edited *Land Degradation and Society* (1987). In this book the argument that land degradation is an inter-disciplinary issue *par excellence* is promoted with vigour, and the authors also propose that the task of explanation lies, in major part, with social scientists, for the reason that natural scientists have extracted the process of land degradation from its political, social and economic context. The first sentence of the opening chapter sets out the book's central claim: "Land degradation should by definition be a social problem" (p. 1).

In 1994, Blaikie co-authored with Terry Cannon, Ian Davis and Ben Wisner a third foundational text, in this instance in the field of Disaster Risk Reduction (DRR): *At Risk: Natural Hazards, People's Vulnerability and Disasters* (1994, and Wisner et al. 2003). Like the earlier two volumes, *At Risk* arose out of shared dissatisfaction with prevailing views that disasters "were "natural" in a straightforward way" (Wisner et al. 2003: xiii). In the book the authors connect risk to vulnerability and vulnerability to livelihoods and normal living: "The crucial point about understanding why disasters occur is that it is not only natural events that cause them. They are also products of the social, political, and economic environment because of the way it structures the lives of different groups of people" (p. 3).

Together, these three books have been instrumental in forging Blaikie's reputation as a leading, indeed founding, political ecologist, even though he has never really identified himself as a mainstream political ecologist (see Blaikie 2008; Muldavin 2008: 689). After retirement he wrote: "By about 1983, I found myself on a research course which seemed to be aligned with others who would also be labelled political ecologists at a later date" (2008: 766). All the more surprising, then, that in 2008 *Geoforum* published a special issue honouring the life-work of Piers Blaikie in the field of political ecology. Raymond Bryant and Michael Goodman (2008: 713) conclude their piece by commenting that it is "inconceivable" to talk of political ecology without mentioning Blaikie's foundational contribution to the field.

Less well known, but just as influential among a smaller group of scholars and development practitioners, is Blaikie's work (with collaborators) on Nepal, which stretches over more than three decades.

In 1973 the Overseas Development Group at UEA were awarded funding from the Economic and Social Committee for Overseas Research (ESCOR) of the Ministry of Overseas Development (MOD) to evaluate the impact of roads in West-Central Nepal. The funders got more than they bargained for. The three project directors, Blaikie among them, wrote a long, three-volume report which, as well as detailing the impacts of roads, also placed this squarely within a broader politico-economic context. The reports were later condensed into a summary report, *The Effects of Roads in West-Central Nepal* (1977). Three years on from this, *Nepal in Crisis: Growth and Stagnation at the Periphery* (1980) was published. Neo-Marxist in vision and using the language of dependency, this book was a profoundly pessimistic view of Nepal's present and likely future. Blaikie and his co-authors believed that road-induced and market-led integration would "not deliver the benefits of increased agricultural production, increased commercialisation, and trade as forecast in the economic appraisal documents" (Blaikie et al. 2002: 1256). Rather, the outcome would be a deepening dependency and growing underdevelopment. The authorities in Nepal banned it for a time and the country's leftists scrambled to get hold of a copy. The UK's ambassador in Kathmandu, apparently, wished it had never been written. It became, as the authors put it in a 2001 reprint, "somewhat of a political 'hot potato'" (Blaikie et al. 2001: 276).

Blaikie and his collaborators returned to the hills of Nepal in 1998 to review and update their conclusions (Blaikie et al. 2002). While not saying that they were wrong in their 1980 diagnosis (they were correct in predicting that agriculture would not be invigorated and that dependency would deepen), they do admit that events did not evolve in quite the way they had anticipated. In particular, they failed to appreciate how the development of non-farm opportunities would make stagnation in agriculture less critical in livelihood terms. They conclude: "The original model underestimated the capacity of the global labour market to provide work and remittances to sustain rural life and to stave off a more generalised crisis" (2002: 1268–1269). This admission also, and interestingly, provides a valuable lens for looking at *The Political Economy of Soil Erosion*. Just as *Nepal in Crisis* failed to gauge the trajectory of global and national change and the degree to which individuals would be able to benefit rather than classes lose out, so too with the soil erosion volume. The dependency perspective, the general pessimism and the view that individuals as members of classes would be squeezed and controlled looks rather dated in the light of what has actually happened over the intervening years.

The final area where Blaikie has made a contribution to development thinking is in family planning and AIDS research. Blaikie's interest in family planning can be traced back to 1970, when he took leave of absence to work on the diffusion of family planning (and agricultural) technologies in Bihar (India). Following a paper in *Population Studies* in 1970, he published a research monograph on *Family Planning in India: Diffusion and Policy* (1975). With its strong locational, model-building approach, it was very much a product of its time. There are resonances, once again, with Blaikie's other work, in particular in his pessimistic take on the future and his view that people do not plan for families in the poor, rural Indian context largely for structural reasons. This research also provided the inspiration for a paper in *Progress in Human Geography* on innovation diffusion (Blaikie 1978). Blaikie's main contribution to research on AIDS was a book, co-authored with Tony Barnett, *AIDS in Africa: Its Present and Future Impact* (1992). This book was one of the first to draw on thorough fieldwork (in this instance in Uganda) to map out the likely dynamic of the disease.

Stepping back from the minutiae, it is possible to make three general statements about Blaikie's work and his contribution to scholarship in the field of international development studies. First, his publications show a clarity of argument and expression which is becoming increasingly rare. Second, Blaikie's work is invariably a careful blend of the empirical and the theoretical. He gets into the field, collects data and is involved at the sharp end of development thinking. But, as an academic and not a consultant, he also made it his mission to theorise up (as well as down) and to embed his results in wider considerations and perspectives. The third feature of his work is a common thread that runs through and links his best publications, whether on land degradation or agrarian change: a desire to challenge those who might apply simple solutions of a technical or managerial nature to "problems" that have complex politico-economic roots. Time and again his work emphasises that "space is what the political makes it" (1978: 289).

Note

1 Frank Ellis, "Farewell to Piers Blaikie and Stephen Biggs", www.uea.ac.uk/der/newsletter/ellis.htm.

Major works

Blaikie, P. (1975) *Family Planning in India: Diffusion and Policy*, London: Edward Arnold.

—— (1985) *The Political Economy of Soil Erosion in Developing Countries*, Harlow, UK: Longman.

Blaikie, P. and Barnett, T. (1992) *AIDS in Africa: Its Present and Future Impact*, London: Belhaven Press.

Blaikie, P. and Brookfield, H. (1987) *Land Degradation and Society*, London: Methuen.

Blaikie, P., Cameron, J. and Seddon, D. (1980) *Nepal in Crisis: Growth and Stagnation at the Periphery*, Oxford: Clarendon Press.

—— (2002) "Understanding 20 Years of Change in West-Central Nepal: Continuity and Change in Lives and Ideas", *World Development* 30(7): 1255–1270.

Blaikie, P., Cannon, T., Davis, I. and Wisner, B. (1994) *At Risk: Natural Hazards, People's Vulnerability and Disasters*, London: Routledge.

Wisner, B., Blaikie, P., Cannon, T. and Davis, I. (2003) *At Risk: Natural Hazards, People's Vulnerability and Disasters*, 2nd edn, London: Routledge.

Further reading

Blaikie, P. (1970) "Implications of Selective Feedback in Aspects of Family Planning Research for Policy-Makers in India", *Population Studies* 26(3): 437–444.

—— (1971) "Spatial Organisation of Agriculture in Some North Indian Villages", *Transactions of the Institute of British Geographers* 53: 15–30.

—— (1978) "The Theory of the Spatial Diffusion of Innovations: A Spacious Cul-de-Sac", *Progress in Human Geography* 2: 268–295.

—— (2000) "Development, Post-, Anti-, and Populist: A Critical Review", *Environment and Planning* A32: 1033–1050.

—— (2008) "Epilogue: Towards a Future for Political Ecology that Works", *Geoforum* 39(2): 765–772.

Blaikie, P., Cameron, J. and Seddon, D. (1977) *The Effects of Roads in West-Central Nepal*, Norwich: University of East Anglia.

—— —— —— (2001) *Nepal in Crisis: Growth and Stagnation at the Periphery*, 2nd edn, New Delhi: Adroit Publishers.

Bryant, R. and Goodman, M.K. (2008) "A Pioneering Reputation: Assessing Piers Blaikie's Contributions to Political Ecology", *Geoforum* 39(2): 708–715.

Muldavin, J. (2008) "The Time and Place for Political Ecology: An Introduction to the Articles Honoring the Life-Work of Piers Blaikie", *Geoforum* 39(2): 687–697.

Watts, Michael (1997) "Classics in Human Geography Revisited: *The Political Economy of Soil Erosion in Developing Countries*", *Progress in Human Geography* 21 (1): 75–80.

JONATHAN RIGG

JAMES M. (JIM) BLAUT (1927–2000)

Jim Blaut was an early and rigorous critic of modernisation theory and development policy. In common with another geographer, David Harvey, Blaut's work has influence far beyond that discipline. However, Blaut's critique of Eurocentrism and diffusionism as essential pillars of Western imperialist ideology and world view is quintessentially geographic. The works he produced during the last ten years of his life (Blaut et al. 1992; Blaut 1993; 2000; but see also his earlier embryonic works, 1969; 1970a; 1976; 1987a) were detailed, scholarly yet polemical treatments of the origins and trajectory of a world system we know today as neoliberal globalisation. In this he was a pioneer, with such authors as Immanuel Wallerstein, Edward Said, Franz Fanon, **Andre Gunder Frank** and **Samir Amin**, in conceptualising the world from the point of view of the subaltern, the colonised and the oppressed (Sheppard 2011; Denemark and Gills 2012).

Blaut argues that diffusionists believe that "independent invention is rather uncommon, and therefore not very important in culture change in the short run and cultural evolution in the long run" (1993: 11). In fact, he finds that many historians and proponents of modernisation-as-development believe that "only certain select communities are inventive" (p. 12). Diffusionism's basic model assumes that Greater Europe has been, and is, the "Inside" (source of modernity and innovation) and that non-Europe is the "Outside". This division and the supposedly innate superiority of Europe are based on the assumption that Europe enjoys a better mode of thinking (rationality) as well as better climate and better soils (Blaut 1994). Diffusionism then makes seven claims (Blaut 1993: 14–17; see also Blaut 2000: 3–12):

- Europe naturally progresses and modernises.
- Non-Europe naturally remains stagnant, unchanging, traditional and backward.
- The basic cause of European progress is an intellectual or spiritual factor, a system of values.
- Likewise, the backwardness of non-Europe is attributable to the absence of these values, in particular the lack of rationality.
- Non-Europe progresses by the diffusion of innovation from Europe.
- The historical reverse flow of material wealth from non-Europe to Europe is justifiable as a partial repayment for the priceless gift of rationality, innovation and modernity.
- Non-material reverse flow from non-Europe to Europe is entirely made up of ideas that are ancient, savage, atavistic, uncivilised or even evil.

This model of the world began to take shape in the fifteenth and sixteenth centuries, flowering fully in the eighteenth and nineteenth centuries. Although these propositions may seem extreme and unlikely to influence popular understanding or professional statecraft, a look at bestsellers such as Ben Barber's (1996) *Jihad vs. McWorld*, Thomas Barnett's (2004) *The Pentagon's New Map* and Jared Diamond's *Collapse* (2005) reveal that what Blaut exposed as "inside" and "outside" ("core" and "gap" in Barnett's phrase) is alive and well; just as other popular books revive environmental determinism, for example, Jared Diamond's *Guns, Germs and Steel* (1997), Robert Kaplan's *Revenge of Geography* (2012) and Tim Marshall's *Prisoners of Geography* (2015).

The year before he died, Blaut presented his "Null Hypothesis" (Blaut 1999) at a conference at University of California, Davis. The sweeping and absolute ambition of his project of de-construction is evident:

> Most present-day historians, I am certain, are opposed to Eurocentrism. Many are engaged in the task of disproving the old Eurocentric arguments […] about the superiority of pre-modern Europe. But most historians seem to believe that, at the end of the day, when all the fallacies have been rooted out and discarded,

> *something* will be left: some one incontestable argument that proves the superiority or priority of pre-modern Europe over non-Europe, or some essential historical force for change toward modernity that was present in Europe and nowhere else, or was more effective in Europe than elsewhere. I suggest that *nothing* will be left: that is my null hypothesis.

His uncompromising position earned his work praise (Frank 1995; Hughill 1995) as well as condemnation (Kennedy 1996). Blaut's stridency should not be confused with dogmatism. The brilliance and lasting impact of his work is to insist on the insidious, deeply ingrained assumptions of Eurocentrism that can be seen even to influence views, decisions and policies of intellectuals and politicians in the global South (Allen 2002). Resistance to Blaut's revelations of Eurocentrism was similar to pushback by scholars against arguments asserting the thoroughgoing, suble influence of white privilege (Litowitz 1997). Blaut referred to such a reaction as "circling the wagons" (Blaut 1996).

Blaut's writings are characterised by an exceptional level of detail and meticulous historical research. Although only two of the planned three volumes in his project to demolish Eurocentrism (Blaut 1993; 2000) were finished by the time of his death, he went a very long way toward that goal. For students of development, these two volumes provide essential vaccine against a new, virulent Crusader Virus that began to sweep the world from the US White House and Pentagon shortly after 11 September 2001. Although the notion of "spatial modernisation" was unique to geography, and therefore a specialist and obscure concept, assumptions about diffusion and innovation run like fat throughout the marbled meat of development theory and practice. Probably the best-known canon of that school is **Walt Rostow**'s (1960) *The Stages of Economic Growth: A Non-Communist Manifesto*. Blaut's critique, therefore, provides as important a corrective lens as, for example, Wolfgang Sachs's archaeology of the concept of "development" (Sachs 1999) or of Said's deconstruction of the idea of "orientalism" (Said 1988).

Another special contribution by Blaut to such critical re-assessment of development thinking is his argument that contingent accumulation of wealth from the so-called New World both enriched Europe and enabled further conquest and accumulation while it acted to self-justify notions of European superiority (Hughill 1995: 259). In 2018, that is precisely the vicious circle that US President Trump articulated as he denigrated African countries (Davis et al. 2018), banned Muslims and attempted to wall off Latin America. Trump's "America first" rhetoric is built on underlying assumptions of white supremacy and that, in turn, is reinforced by Trump's assumption that wealth accumulated by the US empire came by virtue of America's "greatness".

What also sets Blaut's contribution apart is its range and methodological complexity, as well as its unity over nearly fifty years. Beginning his career in the 1950s and 1960s as a cultural geographer concerned with the practicalities of tropical agricultural production, he quickly began to criticise top-down attempts to "modernise" peasant agriculture by "tropicalising" the techniques of European and North American farming (Blaut 1967; 1970b). He also appreciated before many others the importance of what has become known as indigenous technical knowledge – and he employed, as did a few other geographers and anthropologists of his generation, techniques that would later become standard for researchers and NGOs alike under its contemporary name, participatory research. A good example is Blaut's study of farmer perceptions of the causes of soil erosion in Jamaica's Blue Mountains (Blaut 1959). Blaut had a lifelong preoccupation with methodological and theoretical questions that balanced and informed his "mud on the boots" fieldwork and approach to the practice of development. Therefore he also theorised his early work with peasant farmers (Blaut 1979).

His experiences with the rich knowledge possessed by farmers in Singapore, Jamaica, St Croix, Costa Rica and Venezuela led Blaut to more general questions. How do children learn, informally, to become farmers? How do children learn to map and to navigate on the surface of the planet? These spawned a series of empirical studies and theoretical debates that began in the 1970s and were still ongoing at his death.

The significance for Blaut's contribution to development studies of this work on place learning is twofold. First, he and his close colleague, David Stea, discovered that mapping skills learned spontaneously very early in childhood are actually "unlearned" during formal schooling (Blaut and Stea 1974; Blaut 1997; Sowden et al. 1997). Second, Blaut built on a series of cross-cultural studies of children's place learning (Blades et al. 1998) to develop a theory of mapping as a human universal (Blaut 1991; Stea et al. 1996). Eurocentrism and diffusionism – bedrocks of Western imperialism to the present day – necessarily privilege one group's knowledge and skill over another's. Just as those assumptions denigrated and ignored peasant farmers' local knowledge, so they also tend to ignore and, indeed, to destroy early childhood mapping skill. Some of the fiercest academic polemics to involve Jim Blaut centred not on his overt political ideas or anti-imperialism, but on whether (following the Swiss psychologist Piaget) children have built-in stages and limitations in their ability to understand spatial relations. Seeing this as yet another Western, rationalist idea, the roots of which go back to the 18th-century philosopher Kant, Blaut coined the phrase "Can'tianism" – employing a rather esoteric pun (thus, the belief that "children can't …") and attempted to refute it, asserting, "Children Can!" (Sowden et al. 1997).

Ultimately, Eurocentrism must deny human universalities such as mapping and spontaneous, local innovation by peasants. Contemporary imperialism may give lip service to the universality of human rights, but its Eurocentric assumptions demand that the West name, enumerate and protect them. African philosophers – whose work Blaut knew – have also countered the claim by Western academic philosophy that African cultures have produced no "philosophy" in the strict sense, only "folk wisdom". Thus one sees a unity in Blaut's contribution that embraces his studies of tropical agriculture, of place learning, and his critique of Eurocentrism.

Blaut integrated these themes nicely in a conference paper (Blaut 1994), noting that Eurocentric diffusionism leads to certain typical claims about tropical soils and family farmers in the tropics, namely:

- Soils are difficult to manage, and small-scale farmers destroy them when they try.
- European guidance is necessary to bring these soils into sustainable production.

From the 1930s onwards, tropical soil science was producing data that contradicted these claims, yet the dominant assumptions remain and still influence policy. In one striking example, he shows how "irrational" refusal to drain hillside farms on the contour is based on local knowledge that such drainage is likely to lead to landslides.

Blaut's contribution to development studies is also noteworthy for the way in which he put his ideas into practice. He was a strong advocate of Puerto Rican independence from the United States and wrote extensively on the national question, publishing in both Spanish and English, in aid of that struggle (Blaut 1987b; Blaut and Figueroa 1988). He also wrote a good deal about ghettos as internal colonies, his intellectual efforts again guided by his solidarity with the cause of Puerto Ricans (and other minority, oppressed groups) (e.g. Blaut 1974). Blaut was also active promoting the rights of the Palestinian people and in support of the anti-apartheid struggle in South Africa.

Jim Blaut was born in New York City in 1927 and attended the University of Chicago at the young age of 16 (Matthewson and

Stea 2003). He had the unique experience of studying at the Imperial College of Tropical Agriculture in Trinidad during the twilight of the British Empire. He then completed his PhD at Louisiana State University – writing at great length on a one-acre family farm in Singapore (Blaut 1953). While undertaking this fieldwork, he was an instructor at the University of Malaysia (1951–1953).

Blaut later held academic positions at Yale (1956–1961), Cornell (1960), Clark (1967–1971) and the University of Illinois – Chicago Circle Campus (1972–2000). He maintained a strong interest in the Caribbean, often teaching and consulting there. Blaut served as agricultural consultant to the Venezuelan government (1963–4) and as UNESCO adviser to the Dominican Republic's Planning Board in 1964. He taught at the University of Puerto Rico from 1961 to 1963 and again in 1971–1974. He directed the Caribbean Research Institute based in the Virgin Islands from 1964 to 1966 and acted as a consultant to the institute from 1966 to 1967. As late as 1982, he was back in the field, climbing on foot into the mountains of Grenada to interview young *rastamen* who produced charcoal for sale in Georgetown, once again using the gifts he had been developing for three decades for conversational-style interviews.

Blaut says of himself that he was an "activist as a teenager in the Old Progressive Party (of Henry Wallace), working in Georgia" (Blaut 2005). He was later an activist in the Puerto Rican Socialist Party and was arrested (with a number of his graduate students) for blocking a draft board entrance in Worcester, Massachusetts as a Vietnam War protest. Blaut was an enormously gifted and effective teacher who has left behind several generations of left-wing intellectual workers, spread throughout the world (Matthewson and Stea 2003; Wisner and Matthewson 2005).

The rewards for scholar-activism in bourgeois society include academic censure. Blaut was denied promotion and tenure at Clark University despite arriving at the decisive faculty meeting literally with a wheelbarrow filled with his publications. The Graduate Student Association at Clark University responded by voting no confidence in the then director of the Graduate School of Geography, Saul Cohen. Later in life Blaut did receive recognition, receiving the Association of American Geography's "Distinguished Scholar of the Year" award in 1997. His importance as an architect of world-systems history and its mission of dismantling Eurocentism continues (Stuenkel 2014; Denemark and Gills 2012).

Major works

Blaut, J.M. (1953) "The Economic Geography of a One-Acre Farm in Singapore: A Study in Applied Microgeography", *Journal of Tropical Geography* 1(1): 37–48.

—— (1983) "Nationalism as an Autonomous Force", *Science and Society* 46: 1–23.

—— (1987a) "Diffusionism: A Uniformitarian Critique", *Annals of the Association of American Geographers* 77: 30–47.

—— (1987b) *The National Question: Decolonising the Theory of Nationalism*, Atlantic Highlands, NJ: Zed Books.

—— (1993) *The Colonizer's Model of the World: Geographical Diffusionism and Eurocentric History*, New York: Guilford Press.

—— (1994) "Eurocentrism, Diffusionism, and the Assessment of Tropical Soil Productivity", paper read at the International Conference on Advances in Tropical Agriculture in the Twentieth Century and Prospects for the Twenty-First Century, University of the West Indies, Trinidad, 4–9 September.

—— (1997) "Piagetian Pessimism and the Mapping Abilities of Young Children", *Annals of the Association of American Geographers* 87(1): 168–177

—— (1999) "Null Hypothesis", paper presented at Conference on the Origins of the Modern World: Comparative Perspectives from the Edge of the Millennium, University of California at Davis, 15–17 October. www.columbia.edu/~lnp3/mydocs/Blaut/a_null_hypothesis.htm.

—— (2000) *Eight Eurocentric Historians*, New York: Guilford.

—— (2005) "My Legacy (Assessed by Me, Jim Blaut, on 16 September, 2000. Incomplete and Hasty with Some Redundancy)", in B. Wisner and K. Matthewson (eds), *The Work and Legacy of J.M. Blaut*, special issue of *Antipode*, 37 (5), November.

—— with contributions by S. Amin, R. Dodgshon, A.G. Frank and R. Palan, and with an introduction by P.J. Taylor (1992) *Fourteen Ninety-Two: The Debate on Colonialism, Eurocentrism, and History*, Trenton, NJ: Africa World Press.

Further reading

Allen, T. (2002) "Book Review, *The Colonizer's Model of the World & Eight Eurocentric Authors*", *Journal of Black Studies* 32(6): 735–738.

Barber, B. (1996) *Jihad vs. McWorld: How Globalism and Tribalism Are Reshaping the World*, New York: Ballantine.

Barnett, T. (2004) *The Pentagon's New Map: War and Peace in the 21st Century*, New York: Putnam's.

Blades, M., Blaut, J.M., Darvizeh, Z., Elguea, S., Sowden, S., Soni, D., Spencer, C., Stea, D., Surajpaul, R. and Uttal, D. (1998) "A Cross-Cultural Study of Young Children's Mapping Abilities", *Transactions of the Institute of British Geographers* n.s. 23: 269–277.

Blaut, J.M. (1959) "A Study of Cultural Determinants of Soil Erosion and Conservation in the Blue Mountains of Jamaica", *Social and Economic Studies* 8: 403–420.

—— (1967) "Geography and the Development of Peasant Agriculture", in S.B. Cohen (ed.), *Problems and Trends in American Geography*, New York: Basic Books, 200–220.
—— (1969) "Jingo Geography: Part I", *Antipode* 1(1): 10–13.
—— (1970a) "Geographic Models of Imperialism", *Antipode* 2(1): 65–85.
—— (1970b) "Realistic Models of Peasant Agriculture", in A.J. Field (ed.), *Town and Country in the Third World*, Boston: Schenkman, 213–224.
—— (1974) "The Ghetto as an Internal Neocolony", in R. Morrill and J. Eichenbaum (eds), *New Perspectives in Urban Location Theory*, special issue of *Antipode* 6(1): 37–42.
—— (1976) "Where Was Capitalism Born?", *Antipode* 8(2): 1–11.
—— (1979) "Some Principles of Ethnogeography", in S. Gale and G. Olsson (eds), *Philosophy in Geography*, Dordrecht: Reidel, 1–7.
—— (1991) "Natural Mapping", *Transactions of the Institute of British Geographers* n.s. 16(1): 55–74.
—— (1996) "To the Editor", *American Historical Review* 103 (October): 1352–1353.
—— and Figueroa, L. (1988) *Aspectos de la cuestión nacional en Puerto Rico*, San Juan: Editorial Claridad.
—— and Stea, D. (1974) "Mapping at the Age of Three", *Journal of Geography* 73 (7): 5–9.
Davis, J.H., Stolberg, S.G. and Kaplan, T. (2018) "Trump Alarms Lawmakers with Disparaging Worlds about Haiti and Africa", *New York Times*, 12 January: www.nytimes.com/2018/01/11/us/politics/trump-shithole-countries.html?emc=edit_th_180112&nl=todaysheadlines&nlid=46661045.
Denemark, R.A. and Gills, B.K. (2012) "World-System History: Challenging Eurocentric Knowledge", in S.J. Bobones and C. Chase-Dunn (eds), *The Routledge Handbook of World-Systems Analysis*, New York: Routledge, 163–171.
Diamond, J. (1997) *Guns, Germs and Steel*, New York: W.W. Norton.
Diamond, J. (2005) *Collapse: How Societies Choose to Fail or Succeed*, New York: Penguin.
Frank, A.G. (1995) "Book Review, *The Colonizer's Model of the World*", *Annals of the Association of American Geographers* 85(3): 589–591.
Hughill, P.J. (1995) "Book Review, *The Colonizer's Model of the World: Geographical Diffusionism and Eurocentric History* by J.M. Blaut", *Geographical Review* 85 (2): 259–261.
Kaplan, R. (2012) *Revenge of Geography*, New York: Random House.
Kennedy, D. (1996) "Book Review, *The Colonizer's Model of the World*", *American Historical Review* 102 (July): 148–149.
Litowitz, D.L. (1997) "Some Critical Thoughts on Critical Race Theory", *Notre Dame Law Review* 72(2):503–529: http://scholarship.law.nd.edu/ndlr/vol72/iss2/5.
Marshall, T. (2015) *Prisoners of Geography*, New York: Scribner's.
Matthewson, K. and Stea, D. (2003) "James M. Blaut (1927–2000)", *Annals of the Association of American Geographers* 93(1)(March): 214–222.

Rostow, W.W. (1960) *The Stages of Economic Growth: A Non-Communist Manifesto*, Cambridge: Cambridge University Press.
Sachs, W. (1999) "Archeology of the Development Idea", in W. Sachs (ed.), *Planet Dialectics: Explorations in Environment and Development*, London: Zed Books, 3–22.
Said, E. (1988) *Orientalism*, New York: Vintage.
Sheppard, E. (2011) "Geography, Nature and the Question of Development", *Dialogues in Human Geography* 1(1): 46–75.
Sowden, S., Blaut, J., Stea, D., Blades, M. and Spencer, C. (1997) "Children Can", *Annals of the Association of American Geographers* 87(1): 152–158.
Stea, D., Blaut, J.M. and Stephens, J. (1996) "Mapping as a Cultural Universal", in J. Portugali (ed.), *The Construction of Cognitive Maps*, Boston: Kluwer.
Stuenkel, O. (2014) "Book Review: *The Colonizer's Model of the World* by J.M. Blaut", *Postwestern World*: www.postwesternworld.com/2014/12/06/geographical-diffusionism-eurocentric/.
Wisner, B. and Matthewson, K. (eds) (2005) *The Work and Legacy of J. M. Blaut*, special issue of *Antipode* 37 (5), November.

BEN WISNER

NORMAN BORLAUG (1914–2009)

During the 1960s, rapid increases in agricultural yields, particularly in wheat and rice, in parts of Asia and Latin America were heralded as a solution to the problems of hunger facing millions of the world's poorest people. The use of high-yielding varieties (HYVs), chemical fertiliser, irrigation and machinery was termed "the Green Revolution". Debates about the relative successes and failures of these changes in agricultural production continue, but Norman Borlaug, sometimes termed "the Father of the Green Revolution", remains almost unknown.

Borlaug was born in Cresco, Iowa, USA on 25 March 1914, and grew up on a farm. His future interests were strongly influenced by this environment. During the 1930s he studied at the University of Minnesota, where he gained a BSc in Forest Management in 1937, followed by a Masters in Plant Pathology in 1939. His doctoral thesis, completed in 1942, was on a common fungus, rust, which attacks a wide variety of crops. He then worked as a microbiologist for the du Pont de Nemours Foundation in Delaware, where his research concentrated on agricultural chemical products such as fungicides and bactericides.

Borlaug's scientific research focused on how technology could improve agricultural yields. His convictions regarding the role that science could play in agriculture were based not only on this work but also on his observations growing up in the US Midwest. The "Dust Bowl" of the 1930s is often used as an example of how farming methods inappropriate for a particular physical environment can cause long-term environmental as well as social damage. As the economic depression of the 1930s worsened, farmers sought to increase yields by intensifying production. This left large areas of land without appropriate vegetation cover, leading to high levels of soil erosion. Borlaug observed that it was not the use of scientific agricultural techniques which caused these problems, but their misuse or lack of use. He argued that those farmers who adopted appropriate scientifically informed practices did not suffer the same losses.

The 1940s saw the beginning of large-scale "development assistance" from the global North to global South. In 1943, the Rockefeller Foundation, in conjunction with the Mexican government, set up the Comparative Wheat Research and Production Program in Mexico. Borlaug went to Mexico in 1944 to become head of this Program. This provided him with the opportunity to put his ideas into practice.

The Program was meant to concentrate on teaching Mexican farmers how to improve their agricultural techniques, but under Borlaug's leadership it also developed a very strong focus on innovation. Borlaug was determined to breed new forms of wheat that would help increase yields and reduce risk for poor farmers. These innovations included developing a strain of wheat (*ceredo*) which was insensitive to the number of hours of sunlight in a day and, most famously, varieties of dwarf wheat. Borlaug and his fellow researchers argued that traditional wheat's long stalks limited yields, because of the energy expended on growing the inedible long stalks rather than the ears of wheat, and also because tall stalks often got damaged in wind or rain, so making harvesting more difficult. Dwarf strains had much higher yields if grown with appropriate fertiliser and irrigation. Experiments on dwarf rice strains were being conducted at the same time at the International Rice Research Institute in the Philippines and at China's Human Rice Research Institute.

In 1963 the Mexican Program was transformed into a new institution known as the International Centre for the Improvement of Maize and Wheat, often referred to by its Spanish acronym CIMMYT (Centro Internacional de Mejoramiento de Maíz y Trigo). Given the success of the dwarf wheat programme in Mexico, Borlaug was impatient to transfer the technology and practices to other parts of the world where starvation and hunger were much more widespread. This led him to focus on India and Pakistan. At that time, seed distribution was controlled by state companies, so he focused his attention on these organisations. Borlaug's focus on wheat, rather than indigenous crops of the subcontinent, was driven not because he felt that wheat was intrinsically better than lentils or other local staples, but rather because high-yielding varieties of indigenous crops had not been developed, wheat can grow in a wide variety of physical environments, and it provides significant calories. This latter point was key as Borlaug's mission was to address the perceived mismatch between population size and food supply.

Unsurprisingly, there were significant obstacles to his attempts to introduce HYVs to India and Pakistan as it represented a mammoth shift in cultural acceptance and understandings of the role of particular foodstuffs, as well as farming methods. Borlaug would not give up, and eventually the governments agreed to limited adoption of HYV wheat because of the widespread famine in parts of their countries. Borlaug often argued that the military hostilities between India and Pakistan in 1965 meant that he was able to introduce his ideas with little government interference once approval had been given. As well as using HYVs, irrigation was important, as was the use of inorganic fertiliser.

Yields increased very rapidly: by 1968 Pakistan was self-sufficient in wheat and by 1974 India was self-sufficient in all cereals. These figures were particularly timely, as neo-Malthusian ideas about the "population timebomb" were becoming increasingly widespread in the global North. India was often used as an example of how rapid population growth was outstripping food supply and would lead to widespread famine, disease and war. Borlaug argued that **Malthus**'s predictions did not take into account scientific advances in food production such as those he promoted. However, he was, and continued to be until his death, concerned about population growth and its implications for food security.

In 1970, Borlaug was awarded the Nobel Peace Prize. This reflected the awareness that precarious food supply can lead to major tensions and violence between individuals, communities and countries. In his acceptance speech, Borlaug stated: "Food is the moral right of all who are born into this world." He acknowledged that people had other rights as well, but said that "without it [food] all other components of social justice are meaningless". The acceptance speech demonstrates his passion for practical and effective research, and his drive to address problems of food supply throughout the poorest regions of the world. His commitment to developing research capacity in the countries of the South is also apparent. For example, in his overview of the early work of the Mexico Program, he stressed the provision of training and fellowship programmes, emphasising that "researchers in pursuit of irrelevant academic butterflies were discouraged" (Borlaug 1970). Borlaug's own career focus on the practical implementation of scientific developments, rather than the writing of academic papers, perhaps partially explains his low profile among the academic community working on "development", particularly in the global North.

Borlaug retired in 1979, but this did not represent an end to his work. He continued to be involved with research and projects to promote improved agricultural practices in sub-Saharan Africa. Despite earlier financial support from charitable organisations such as the Rockefeller and Ford Foundations and multilateral agencies including the World Bank, a significant amount of Borlaug's later work was funded by a Japanese foundation. In 1986 he helped launch the Sasakawa-Global 2000 Program. The work of this Association involves Green Revolution-style projects in numerous countries in sub-Saharan Africa.

Borlaug and supporters of his methods claimed that the increasing reluctance of US-based organisations to fund his work reflected the growing power of environmental movements. The use of non-indigenous crops and HYVs, as well as the increased use of irrigation, pesticides and inorganic fertiliser has been blamed for decreasing soil fertility, water pollution, soil erosion and other environmental problems, particularly in marginal environments (Shiva 1991; 2000). In addition, while the Green Revolution may have made dramatic increases in yields at the start, these increases are impossible to sustain. Finally, the social impacts of the Green Revolution have been highlighted. In the vast majority of cases Borlaug was not criticised directly, with the target being the changes in agricultural practices which he promoted. For example, given the costs of HYVs, pesticides and inorganic fertilisers, the Green Revolution, it is argued, has often exacerbated existing class and caste divisions. While richer farmers can afford these new inputs and benefit from increased yields, poorer farmers are left behind and may have to sell their land and become landless labourers or urban migrants (Das 2002; Lawrence 1988).

Hunger and starvation remain daily realities for millions of people in the global South today. New agricultural technologies and the possibilities opened up by genetic modification are, some argue, the answers to these problems. Borlaug supported the use of genetic modification, arguing that opponents who claim such processes are "unnatural" fail to understand the genetic mixing that happens in nature without human interference. He was also passionate in his criticisms of many groups and individuals who, he claimed, lobby against the use of pesticides, fertilisers and GM crops from the "comfort" of Europe or the USA.

These views, however, do not necessarily reflect the criticisms that certain forms of agricultural practice and the use of GM seeds, in particular, have received from people living in the global South. In India and Mexico, for example, despite past crop yield increases, there have been widespread protests against GM crops. Borlaug admitted that at times scientists had not been successful in presenting their work effectively, implying that if people were presented with the "facts" then there would be no protests.

When discussing his later work in sub-Saharan Africa, as in an interview in 2000 (Bailey 2000), Borlaug mentioned continued obstacles to meeting individuals' food needs. While agricultural technology may improve yields, there are issues of distribution to consider. This involves not only physical distribution in terms of infrastructure but also social distribution, encompassed in **Amartya Sen**'s ideas of entitlement. Sufficient food may be produced, but if you do not have enough money to buy it then you will still starve. For some development theorists and practitioners, it is these issues of distribution which should now get far greater attention than the focus on increasing global food supply.

Norman Borlaug's commitment and passion in addressing a key development debate of how to feed an increasing population with a finite amount of land was admirable. His focus on implementing projects, influencing governments and working with local people demonstrated not only his scientific credentials but also his ability to adapt and work within specific social and cultural contexts. While the enthusiasm that greeted the "Green Revolution" in the 1960s and 1970s may have been overly optimistic, the Green Revolution's contribution to saving millions of people from starvation cannot be underestimated. Borlaug's Nobel Prize acceptance speech stressed that he was just one part of a large team which deserved recognition. This is certainly true, but without Borlaug's vision and determination the team's results might have been less successful and much less widely implemented.

Major works

Borlaug, N. (1970) "The Green Revolution: Peace and Humanity", speech on the occasion of the awarding of the 1970 Nobel Peace Prize: www.nobelprize.org/nobel_prizes/peace/laureates/1970/borlaug-lecture.html.

—— (2002) "Agriculture and Peace: The Role of Science and Technology in the 21st Century", U Thant Distinguished Lecture Series, United Nations University Tokyo, October: http://archive.unu.edu/hq/library/Collection/PDF_files/UNU/E00034.pdf.

Further reading
Bailey, P. (2000) "Billions Served", *Reason*, April: https://reason.com/archives/2000/04/01/billions-served-norman-borlaug.
CIMMYT (Centro Internacional de Mejoramiento de Maíz y Trigo): www.cimmyt.org.
Das, R. (2002) "The Green Revolution and Poverty: A Theoretical and Empirical Examination of the Relation between Technology and Society", *Geoforum* 33(1): 55–72.
Easterbrook, G. (1997) "Forgotten Benefactor of Humanity", *The Atlantic*, January: www.theatlantic.com/magazine/archive/1997/01/forgotten-benefactor-of-humanity/306101/.
Hesser, L. (2006) *The Man Who Fed the World: Nobel Prize Laureate Norman Borlaug and His Battle to End World Hunger. An Authorized Biography*. Dallas, TX: Durban House Publishing.
Lawrence, P. (1988) "The Political Economy of the Green Revolution in Africa", *Review of African Political Economy* 42: 59–75.
Lerner, A.B. (2018) "Political Neo-Malthusianism and the Progression of India's Green Revolution", *Journal of Contemporary Asia* 48(3): 485–507.
Lipton, M. and Longhurst, R. (1989) *New Seeds and Poor People*, London: Unwin Hyman.
Mann, C. (2018) *The Wizard and the Prophet: Two Groundbreaking Scientists and Their Conflicting Visions of the Future of our Planet*, London: Picador.
Sasakawa Africa Association: www.saa-tokyo.org.
Shiva, V. (1991) *The Violence of the Green Revolution: Third World Agriculture, Ecology and Politics*, Lexington: University of Kentucky Press.
Shiva, V. (2000) *Stolen Harvest: The Hijacking of the Global Food Supply*, London: Zed Books.

KATIE WILLIS

ESTER BOSERUP (1910–1999)

Ester Boserup was a development economist, who worked as a civil servant for two decades and later as an independent researcher and consultant for the United Nations and its agencies concerning issues and problems in developing countries.

Boserup was born into an upper-class family in Copenhagen on 18 May 1910. She joined the University of Copenhagen in 1929 and in 1931 married Mogens Boserup, whom she had met when she joined the Danish Marxist high school movement. She graduated in 1935 as "candidatus politices". The main emphasis of her studies during the Great Depression was theoretical economics but she also attended lectures in sociology and agricultural policy. Part of her degree work led to a paper comparing **Marx** and Keynes, a shorter version of which was published in the Danish economic journal *Nationaløkonomisk tidsskrift* (Boserup 1936). She involved herself for a time with a small group of independent socialist intellectuals and later continued to participate in the socialist Danish *Clarté* movement.

After graduation she worked for twenty years as a civil servant, initially for over a decade in the Danish economic administration[1] (1935–1947), during which time she gave birth to three children. From 1947 until 1965, she was employed by the Research and Planning Division of the Economic Commission for Europe in Geneva, focusing on agricultural trade policy. Her work contributed greatly to the early success of the annual Economic Surveys.

In 1957 she and her husband moved to New Delhi, accepting a proposal from **Gunnar Myrdal** to engage in a joint study of South and Southeast Asian agriculture. She travelled extensively within India, discussing with Indian agricultural experts and advisers and experiencing unfamiliar agricultural systems. She soon came to the opinion that "the generally accepted theory of zero marginal productivity and agrarian surplus population in densely populated developing countries was an unrealistic theoretical construction" (Boserup 1999). Her and her husband's developing scepticism towards prevailing development theory put them at odds with Myrdal's evaluation of the Asian situation. Unable to continue in the joint study, they returned to Denmark when their contract expired in May 1960, after delivering the agreed chapter to *Asian Drama* (Myrdal 1969). She continued to work closely together with her husband until his death in 1978.

Key contributions

Population and agricultural growth

Thereafter, she accepted short-term consultancies but continued to reflect on her experiences from Asia and consequently wrote the controversial book *The Conditions of Agricultural Growth* (1965), which outlined the relationship between population and environmental resources. In this book she argued that population growth is the major cause of agricultural change, and that the principal mechanism of change is the intensification of land use through increases in the frequency of cropping. While such intensification would not occur without population growth, which reduces labour productivity – a contentious claim that remains unproven (Grigg 1979) – it is, she argues, a necessary collective response to population pressure. Furthermore, such intensification leads to technological advances, such as the adoption of new fallowing systems, which require new tools and techniques and which shape institutions, land tenure systems and settlement forms (Grigg 1979). This theory overturns the direction of causation implicit in **Malthus**ian and neo-Malthusian approaches, which see technological change as an autonomous process inducing rather than proceeding from population growth. Economists were generally unenthusiastic about Boserup's model, but it refocused many modern debates (see Roumasset 2008) and stimulated further thought on the issues (Giovanni 2001). Other responses, such as expansion at the extensive margin, adoption of new crops or out-migration, can also relieve population pressure, and other influences such as urban growth or development of trade can stimulate agrarian change, yet the thesis remains "a fruitful interpretation of agrarian change" (Grigg 1979). In spite of all its shortcomings, *The Conditions of Agricultural Growth*, which Boserup was invited to discuss at the International Geographical Union in 1967, remains a small masterpiece of enduring importance.

Women in development

In 1964–1965 she spent a year in Dakar, where her husband was director of the United Nations Institut de Développement Economique et de Planification (IDEP). During this period she also became a consultant to the Center for Industrial Development (later UNIDO) and started focusing on the linkages between industry and agriculture and both rural–urban and rural–rural migration. Her earlier work had denied the Malthusian claim that marginal productivity would be reduced to zero on agricultural land, and she now emphasised the role of desire for money incomes rather than land shortage in generating the large labour migration from poor subsistence villages to urban, mining and export-producing centres. She was intrigued by the importance of the gender division of labour to the pattern of African migration. Predominantly male migrants could leave wives and children to take care of themselves for long periods, since in traditional villages it was the women and children who supplied nearly all the subsistence food and fetched water and fuel for cooking. Her interest led her to gain funds to gather data on the hours worked by men and women in both paid and unpaid work, including domestic services and the gathering of fuel and water, in addition to agricultural work and non-agricultural production and services.

Many students of gender studies will have first come across Ester Boserup's work through her well-known book *Woman's Role in Economic Development* (1970),[2] which formed the output of this project and provided intellectual inspiration for awareness of this important topic. Her work sought to make the "invisible woman" visible by drawing attention to the economic contribution made, for example, by women in agricultural families, virtually all of whom, despite their economic productiveness, were formally classified as outside the labour force, as housewives or "non-active" persons. She highlighted the role of what she called the "bazaar and service sector", stressing that what many considered as largely concealed unemployment was in fact production of necessary goods and services with pre-industrial technology catering largely for the market of low-income groups (1970: 175–86). Such activities were later dubbed the "informal sector" by the United Nations' International Labour Organization (ILO). This highlights how effortlessly Boserup integrated the roles of women as both producers and entrepreneurs and as consumers and reproducers into theory, analysis and policy prescription, thereby demonstrating her ability to bring the theoretical rigour of a critical economist to bear on case study material.

In 1970 she received the Rosenkjaer Pris of the Danish Broadcasting Company, and took the opportunity to popularise her ideas on economic development and the sexual division of labour in six radio broadcasts. She also suggested that improvement of women's education and status might lead to a reduction of family size (Boserup 1970: 224–225). Such ideas have been gradually moving into the mainstream among agencies concerned with development and the empowerment of women and the poor. This influenced her appointment in 1972 as a vice-chair of the UN Symposium on Population and Development in Cairo, in preparation for the UN World Population Conference in Bucharest in 1974, and later inspired the UN's International Women's Year (1975) and the UN Decade for Women (1976–1985).

In 1981 she published the book *Population and Technological Change: A Study of Long-Term Trends*, in which she extended the scope of her earlier theories on population growth and adoption of new tools and techniques in agriculture to include non-agricultural activity. Throughout history, both the invention of new technologies and their spread have been demand-induced. Even the most "primitive" people speculate and experiment with new methods, materials and weapons, when the motivation for change becomes strong. Here population growth plays a significant role. She suggested that many technologies can be properly exploited only if the population is dense enough. Population growth makes urban civilisation possible. Population size enables labour specialisation and thus allows economies of scale to be realised in industrial production.

Through the following years she continued to research the interrelationships between population, family structure, status of women, national culture and economic development, both in rural and in urban areas, always arguing for recognition of the complexity of the issues taking *all* the relevant determinants into consideration, instead of using simplified demographic models. The role of fertility remained a central concern, and she argued that high fertility may result in both higher agricultural production and higher investments and savings in kind. Her strong analytical standards for engaging the multifaceted realities at the grassroots provided the foundation for the Women in Development (WID) perspective, and highlighted that women were discriminated against at all levels of the development process.

Her contribution was recognised through the award of three honorary doctorates: in Agricultural Sciences by the Agricultural University of Wageningen, The Netherlands, in 1978; in economics by the University of Copenhagen in 1979; and in Human Sciences by Brown University, Providence, USA in 1983. In 1989 she was elected Foreign Associate of the National Academy of Sciences in Washington on an interdisciplinary vote. She spent the rest of her career as a consultant and independent researcher.

It is important to put her work into the context of the period when she worked. She was in a small minority both as a female civil servant and as a development economist.[3] Much of her work appeared as discussion papers which informed UN publications and lectures at symposia and invited conferences. She challenged conventional wisdom first with regard to the relations between agrarian systems and population, and later with regard to women in development and technological change. Her models have stood the test of time. Her ideas and arguments were mainly exposed to the academic community through her much later work in the 1970s and 1980s (aged in her 60s and 70s) mainly through her three books (1965; 1970; 1981) and some journal articles. All this was managed without the benefit of a university post. Shortly before her death on 24 September 1999, she published an unpretentious and objective historical account of all her published work (Boserup 1999). Ester Boserup's work provoked multidisciplinary debates and has influenced various disciplines, particularly anthropology, demography, agricultural economics, geography and sociology.

Though she was an economics graduate, she saw the limitations of narrow economic analysis. She emphasised this especially for least-developed countries, for which the use of monetised transactions for production, consumption, investment and income (both personal and national income) as proxies for totals including non-monetised transactions led to false conclusions (Boserup 1996). She was at times very critical of classical economics and its assumption of fixed capacities of land, labour and capital, all subject to variable capacity utilisation. She argued for long-term analysis focusing on changes in the capacities themselves, and emphasised the need for this to address changes in those structures which economists usually leave to be studied by other scientific disciplines, for instance national cultures. Such models therefore need – like her own – to be based on interdisciplinary understanding, and in later papers she attempted to lay out an overarching framework for long-term interdisciplinary modelling (Boserup 1996). Such a framework involves delineation of the flows of pressure between six structures: environment; population; technology level; occupational structure; family structure; and culture.

Her work concentrated on examining human society and its processes of social change, especially in the context of dynamic relationships between natural, economic, cultural and political structures, while evaluating both economic and non-economic factors. Her analyses of rapid technological change are even relevant today in the context of the processes of globalisation and social change in developing countries. Reflecting on technological changes in the 1970s and 1980s, she concluded that rapid technological change has created conflicts with national culture through its radical influence on the way of life: cultural attitudes and behaviour, which may have been rational before, are no longer so. But many groups and governments attempt to bring about economic development and technological change without cultural change, and this leads to serious conflicts within and between countries. The importance of these problems for economic development is overlooked by economists, when they make the assumption that rational behaviour is the rule whatever the circumstances (Boserup 1995). She elaborated on the positive and negative effects of technological change in both the short and long run, and on the need for government control of the private sector, both nationally and internationally, to ensure that it uses sustainable technology.

On 15–17 November 2010 a centennial tribute was held in Vienna, Austria, entitled the "Ester Boserup Conference", and sponsored by the Austrian Ministry of Science, the City of Vienna, the Austrian Academy of Science and UniCredit Group.

Notes

1 As a chief of the planning office, she was responsible for estimating future export earnings and suggesting the distribution of import licences by commodity and country, discriminating against luxuries and items that could be obtained by Danish production and against countries which did not buy much from Denmark.
2 Translated in Swedish in 1971, Danish 1972, Italian 1982, German and French 1983, Indonesian 1984 and Spanish 1993.
3 Women have probably made greater inroads into rhetoric, but there are still concerns regarding the inclusion of women in the material as well as the personnel of economics. See website of the Royal Economic Society, Women in Economics www.res.org.uk. At the 2002 World Bank ABCDE seminar, Ravi Kanbur pointed out that most of the economics profession still adheres to and teaches the unitary "black box" model of the household, which effectively makes gender allocations a private matter.

Major works

Boserup, E. (1936) "Nogle centrale økonomiske spørgsmål I lys af den marxistiske teori" (Some Central Economic Issues in Light of Marxian Theory), *Nationaløkonomisk Tidsskrift* 75: 421–435.

—— (1965) *The Conditions of Agricultural Growth: The Economics of Agriculture under Population Pressure*, London: George Allen and Unwin. (Reprinted London: Earthscan, 1993.)

—— (1970) *Woman's Role in Economic Development*, London: Earthscan.

—— (1981) *Population and Technological Change: A Study of Long-Term Trends*, Chicago: University of Chicago Press.

—— (1995) "Obstacles to Advancement of Women during Development", in T. Paul Schultz (ed.), *Investment in Women's Human Capital*, Chicago: University of Chicago Press, 51–60.

—— (1996) "Development Theory: An Analytical Framework and Selected Applications", *Population and Development Review* 22(3): 505–515.

—— (1999) *My Professional Life and Publications 1929–1998*, Copenhagen: Museum Tusculanum Press, University of Copenhagen.

Further reading

Giovanni, F. (2001) "Review of Ester Boserup, *The Conditions of Agricultural Growth: The Economics of Agrarian Change under Population Pressure*", Economic History Services, 16 April: https://eh.net/book_reviews/the-conditions-of-agricultural-growth-the-economics-of-agrarian-change-under-population-pressure/.

Grigg, D. (1979) "Ester Boserup's Theory of Agrarian Change: A Critical Review", *Progress in Human Geography* 3(1): 64–83.

Kergard, N. and Henriksen, I. (2000) "Ester Boserup", in R.W. Dimand, M.A. Dimand and E.L. Forget (eds), *A Biographical Dictionary of Women Economists*, Cheltenham, UK: Edward Elgar, 65–72.

Myrdal, G. (1969) *Asian Drama: An Enquiry into the Poverty of Nations*, 3 vols, New York: Twentieth Century Fund (chapters particularly influenced by Ester Boserup's research are 22 and 26).

Roumasset, J. (2008) "Population and Agricultural Growth", in S.N. Durlauf and L.E. Blume, *The New Palgrave Dictionary of Economics*, 2nd edn, London: Palgrave Macmillan.

Schultz, T.P. (1990) *Economic and Demographic Relationships in Development*, Baltimore, MD: Johns Hopkins University Press. (Articles selected, edited and introduced by T. Paul Schultz.)

Turner, B.L. and Fischer-Kowalski, M. (2010) "Ester Boserup: An Interdisciplinary Visionary Relevant for Sustainability", *Proceedings of the National Academies of Science*, 21 December, 107(51): 21963–21965.

VANDANA DESAI

HAROLD BROOKFIELD (1926–)

It is rare for geographers to be seen as critical thinkers on development issues. Indeed, even this collection, edited by a geographer, includes very few. Of these, Harold Brookfield is assuredly the doyen. One of the others, **Terry McGee**, described Brookfield as "the acknowledged 'guru' of development geography" (1978: 71). A much-lauded book, *Interdependent Development* (1975a), which emphasised the necessity for development to be seen in a diversity of interlocking ways, together with a longstanding focus on local studies (tied to the belief that indigenous knowledge is a key to development) and on the environment and sustainability, were just some of his contributions to development studies. Such diverse interests were shaped around a search for pragmatic solutions and the eschewing of arid theorising, dogma, ever-changing semantics and political correctness. Brookfield preferred the language of everyday life, of ordinary people and their problems, and was unafraid to use words such as "poverty".

Harold Brookfield grew up in southern England, completing his BA and PhD degrees at the London School of Economics. Completion of his PhD thesis in 1950, on post-18th-century urban development in coastal Sussex, was the precursor to a peripatetic life. After briefly being Lecturer in Geography at Birkbeck College, London, Brookfield became Lecturer in charge of the nascent Department of Geography at the University of Natal. There he first engaged with the development and social justice issues concerning South Africa and Mauritius. After nearly three years he left for the University of New England (UNE) in the small city of Armidale in Australia, the first of many moves to and within Australia. He spent over half his life in Australia where he became one of the most distinguished of Australian geographers. From Armidale he went on to the Australian National University (ANU), eventually his longstanding academic home, though Chairs soon followed at Pennsylvania State University, McGill University and Melbourne University, with a two-year period as a Fellow at the Institute of Development Studies, University of Sussex. By 1982 Brookfield returned to ANU, where he remained, publishing substantial works well into his eighties.

Early work on the historical geography of southern England and Ireland was soon followed by detailed work in South Africa, where he wrote an early critique of apartheid (Brookfield 1957), and in Mauritius, where his interest in the evolution of plural societies was further fuelled. The shift to ANU in 1957 brought the start of a lifelong fascination with Papua New Guinea and detailed fieldwork in the highlands of PNG, where contact had been relatively recent. Collaboration with the anthropologist Paula Brown brought his first major book on the region (Brookfield and Brown 1963). Other books on Pacific marketplaces, and the extraordinarily innovative and successful text *Melanesia* (Brookfield with Hart 1971), marked his association with the region, and emphasised his own commitment to both small-scale local studies and the comparative method, "characterised by the careful collection of data and intimate knowledge of a region built up through many years" (McGee 1978: 70). In some respects, that commitment became the hallmark of work on the Pacific region, celebrated in his edited collection, *The Pacific in Transition* (1973) – mainly the work of his own students and colleagues. The focus on PNG was so strong that the prominent American geographer, Marvin Mikesell, identified a "New Guinea syndrome" where "this once remote and mysterious island has played a role in the recent history of cultural geography comparable to the influence of Sauer and his students from Mexico" (1978: 8).

As his colleague, Oskar Spate, observed, this work's value lay in "the realisation that other people also had ways of life which were not based on classical economics, but were not irrational and were tied up with their culture and environment and hence the Brookfield style with its close links to anthropology" (quoted in Rugendyke 2005: 308). A focus on problems, the comparative method and indigenous authority was innovative. Brookfield believed that a key contribution was the "bringing of geographers and geographical work into contact with that of other disciplines which were concerned with people and land, anthropology and pre-history, in particular" (p. 309). Critical of the study of patterns and processes of development derived from statistical methods, and arguing for "a process-oriented discipline in which the question 'How?' should be at least as important as the question 'Why?'", Brookfield largely avoided quantitative research methodologies. In addressing the former question, he "found that it was far easier to seek cause and effect at microscale, where individual decision makers could be seen in action. I therefore advocated this scale of work to my colleagues" (Brookfield 1984: 35). Remarkably, in this sense, his empirical focus at the local scale, on the interactions between people, their cultures and their environments, rather than on quantitative methodologies and spatial models, pre-empted postmodern reflections that "there is no universal or unique understanding of development or the environment" (Simon 2003: 35).

Brookfield's work in Papua New Guinea, especially, was with a thoroughly multidisciplinary group of prehistorians, economists and anthropologists, and, while at the research school in Canberra, "transdisciplinary exchange was as important as interdisciplinary discussion" (Brookfield 1984: 27). He said, "I became very much associated with anthropologists, so much that at one stage Anthony Forge said to me that 'you are an anthropologist who earns his pay masquerading as a geographer'" (Rugendyke 2005). Indeed, it is not surprising that Brookfield chose to title his recollections of that period of geography *Experiences of an Outside Man*, as an iconoclast tangling with anthropology and other disciplines and aghast at the irrelevance of the quantitative revolution. Culture is written on the land, and Brookfield wished to be remembered most for his work in cultural ecology.

Later, his work extended to the smaller islands of the Caribbean (though he wrote little directly on this) and later to the smaller eastern islands of Fiji. Perhaps surprisingly for a geographer, issues of scale were never directly addressed, other than in one book chapter (Brookfield 1975b). Simultaneously the wide-ranging reflections on development evident in *Melanesia* were developed further in the companion volume, *Colonialism, Development and Independence: The Case of the Melanesian Islands in the South Pacific* (1972). They drew together strands of culture, history and geography, in a broad political, social and economic context, at the time of the Melanesian nations' first faltering steps towards independence. These diverse issues came together at a world scale in his magisterial *Interdependent Development* (1975a), a *tour de force* later to become a "classic in human geography revisited" (Corbridge 1996; O'Connor 1996), and republished by Routledge as a classic as recently as 2011. That book began a theoretical backlash against the irrelevance of several geographical traditions, arguing the case for a more historical and environmental perspective, allied, however, to an empirical position that was to become stronger in the future.

Brookfield always emphasised that his own principal long-term interest was the "adaptation of the use and management of land to variability and change in society, economy and natural environment": the "soul" of geography (2004: 40). That was evident, and two further books established his reputation in this area. The first, largely neglected because of its seemingly peripheral focus, was an intricate account of the lives of villagers in the eastern islands of Fiji which brought together a number of human and physical geographers, while drawing on the preliminary work of economists and others (Bayliss-Smith et al. 1988). In some respects, that was a local version of the jointly written *Land Degradation and Society* (Blaikie and Brookfield 1987) that had appeared in the previous year. There and elsewhere, Brookfield stressed the interdisciplinary analysis of environmental issues, arguing that processes of land degradation cannot be understood if divorced from their political, social or economic contexts. In its turn it too became a classic, reprinted five times, as recently as 2016 in the Routledge Revival series (see **Blaikie** in this volume). By then the focus on the environment, the leitmotif of Brookfield's work, had been drawn into wider debates over development. *Interdependent Development* was described as "a striking anticipation of green development thinking" (Corbridge 1996: 86) and what sounded "remarkably like a call for what would now be called 'sustainable development'" (O'Connor 1996: 88).

While Brookfield never lost interest in the Pacific, and especially PNG, his interests began to shift towards Asia. A long period of research in Malaysia brought a renewed focus on environmental issues. It also heralded work on both urbanisation and industrialisation (Brookfield et al. 1991; Brookfield 1994) in a nation that was changing much faster than the island states of the Pacific and elsewhere. Renewed focus on the environment resulted in a book on environmental change in Borneo and Malaysia (Brookfield et al. 1995) and a growing emphasis on agrodiversity. That led to Brookfield's commitment to the more than decade-long United Nations University's project on People, Land Management and Environmental Change (PLEC), which preoccupied him and became an abiding theme of his publications.

The PLEC project, an acronym very close to the Tok Pisin (PNG) word for village, *ples*, began in 1992 – the year of the Earth Summit in Rio de Janeiro – as an initiative to examine relationships between population and environmental change in rural areas. The original focus was on research, centred around five "clusters" – PNG, Amazonia, West and East Africa and montane South-East Asia – but it evolved into a practical concern for agrodiversity, defined in 1994 as "the many ways in which farmers use the natural diversity of the environment for production, including not only their choice of crops but also their management of land, water and biota as a whole" (Brookfield and Padoch 1994: 9). Four years later, PLEC joined the Global Environment Facility (GEF) and became a large demonstration and capacity-building project, involving the participation of small farmers as experts in conservation and sustainable development, alongside indigenous researchers and institutions. Again this project invoked the comparative method, to "bring together 'bottom-up' research in different regions" (Brookfield and Padoch 1994), and enable hypothesis generation, with an emphasis on indigenous management practices. At a time when others might have retired, or merely gazed from distant offices, Brookfield continued to emphasise and demonstrate the value of fieldwork and commitment. Amidst many papers and reports, three more books emerged from the PLEC project (Brookfield 2001; Brookfield et al. 2002; 2003), but these were not the last words. Publications arising from the project, and within which agrarian change and indigenous management practices remained the central theme, emerged until Brookfield was 89, with geographical foci as diverse as China, Ghana and India (Fu et al. 2009; Brookfield 2011; Brookfield and Cairns 2011; Brookfield 2015).

The family farm became the focus of some of Brookfield's latest works. The outcome was an ambitious volume, global in focus, written jointly with Helen Parsons, which focused on the adaptibility and efficiency of the family farmer, positing the survival of the family farm ino the future (Brookfield and Parsons 2007; Brookfield 2008). In this, Brookfield reaffirmed "the intelligent optimism that the world's rural people have shown in all the time since the first cultivation began" (Brookfield 2015: 60).

Remarkably, for over six decades, Brookfield worked in the people–environment tradition; one reviewer noted his "many original and insightful contributions to issues of tropical human ecology and development" which "furnishes significant conceptual gunpowder for the long-awaited resurrection of the human-environment tradition in geography" (Airriess 1996: 3). As many geographers uneasily shifted their focus from troubling development issues, Brookfield remained true to the essence of the discipline and allied disciplines, ever seeking to encourage continued focus on those complexities. In emphasising the necessity for theory to be closely linked to practice, he triumphed in the ethnographic method that gave voice to the participants in development, cherished interdisciplinary approaches, and forged the way in recognising the relationship between cultural ecology, the environment and development.

Major works

Bayliss-Smith, T.P., Bedford, R.D., Brookfield, H.C. and Latham, M. (1988) *Islands, Islanders and the World: The Colonial and Post-colonial Experience of Eastern Fiji*, Cambridge: Cambridge University Press.

Blaikie, P. and Brookfield, H. (eds) (1987) *Land Degradation and Society*, London: Methuen.

Brookfield, H.C. (1957) "Some Geographical Applications of the Apartheid and Partnership Policies in Southern Africa", *Transactions of the Institute of British Geographers* 23: 225–247.

—— (1972) *Colonialism, Development and Independence: The Case of the Melanesian Islands in the South Pacific*, Cambridge: Cambridge University Press.

—— (1973) *The Pacific in Transition*, London: Arnold.

—— (1975a) *Interdependent Development*, London: Methuen.

—— (1975b) "Multum in Parvo: Some Questions about Diversification in Small Countries", in P. Selwyn (ed.), *Development Policy in Small Countries*, London: Croom Helm, 54–76.

—— (1984) "Experiences of an Outside Man", in M. Billinge, D. Gregory and R. Martin, *Recollections of a Revolution: Geography as Spatial Science*, London: Macmillan, 27–38.

—— (1994) *Transformation with Industrialisation in Peninsular Malaysia*, Kuala Lumpur: Oxford University Press.

—— (2001) *Exploring Agrodiversity*, New York: Columbia University Press.

—— (2004) "American Geography and One Non-American Geographer", *GeoJournal* 59: 39–41.

—— (2008) "Family Farms Are Still Around: Time to Invert the Old Agrarian Question", *Geography Compass* 2(1): 108–126.

—— (2011) "Scott and Others on History in the Southeast Asian Uplands", *Asia Pacific Journal of Anthropology* 12(5): 489–494.

—— (2015) "Shifting Cultivators and the Landscape", in M.F. Cairns (ed.), *Shifting Cultivation and Environmental Change: Indigenous People, Agriculture and Forest Conservation*, Abingdon, UK: Earthscan, 25–61.

—— and Brown, P. (1963) *Struggle for Land: Agriculture and Group Territories among the Chimbu of the New Guinea Highlands*, Melbourne: Oxford University Press.

—— and Cairns, M. (2011) "Composite Farming Systems in an Era of Change: Nagaland, Northeast India", *Asia Pacific Viewpoint* 52(1): 56–84.

—— and Gyasi, E.A. (2009) "Academics Among Farmers: Linking Interventions to Research", *Geoforum* 40(2): 217–227.

—— Hadi, A.S. and Mahmud, Z. (1991) *The City in the Village*, Kuala Lumpur: Oxford University Press.

—— with Hart, D. (1971) *Melanesia: A Geographical Interpretation of an Island World*, London: Methuen.

—— and Padoch, C. (1994) "Appreciating Agrodiversity: A Look at the Dynamism and Diversity of Indigenous Farming Practices", *Environment* 36(5): 6–45.

—— Padoch, C., Parsons, H. and Stocking, M. (2002) *Cultivating Biodiversity*, London: Intermediate Technology Publishing.

—— and Parsons, H. (2007) *Family Farms, Survival and Prospect: A World Wide Analysis*, London: Routledge.

—— Parsons, H. and Brookfield, M. (2003) *Agrodiversity: Learning from Farmers Across the World*, Tokyo: United Nations University Press.

—— Potter, L. and Byron, Y. (1995) *In Place of the Forest: Environmental and Social Transformation in Borneo and the Eastern Malay Peninsula*, Tokyo: United Nations University Press.

Fu, Y., Brookfield, H., Guo, H., Chen, J., Chen, A. and Cui, J. (2009) "Smallholder Rubber Plantation Expansion and its Impact on Local Livelihoods, Land Use and Agrobiodiveristy: A Case Study from Daka, Xishuangbanna, Southwestern China", *International Journal of Sustainable Development and World Ecology* 16(1): 22–29.

Further reading

Airriess, C. (1996) Review of Brookfield, H.C., Potter, L. and Byron, Y. (1995), *In Place of the Forest: Environmental and Socio-Economic Transformation in Borneo and the Eastern Malay Peninsula*, *Geographical Review* 86: 611–613.

Connell, J. (1998) "Development Studies in Australian Geography", *Australian Geographical Studies* 26: 157–171.

—— and Waddell, E. (2005) "Introduction: That Most Remarkable of Outside Men – Harold Brookfield's Intellectual Legacy", *Asia Pacific Viewpoint* 46(2): 192–233.

Corbridge, S. (1996) "Classics in Human Geography Revisited, on Brookfield, H., 1975, *Interdependent Development*, Commentary 1", *Progress in Human Geography* 20: 85–86.

McGee, T. (1978) "The Geography of Development: Some Thoughts for the Future", *Australian Geographer* 14: 69–71.

McTaggart, D. (1974) "Structuralism and Universalism in Geography: Reflections on Contributions by H.C. Brookfield", *Australian Geographer* 12: 510–516.

Mikesell, M. (1978) "Tradition and Innovation in Cultural Geography", *Annals of the Association of American Geographers* 68: 1–16.

O'Connor, A. (1996) "Classics in Human Geography Revisited, on Brookfield, H., 1975, *Interdependent Development*, Commentary 2", *Progress in Human Geography* 20: 87–88.

Rugendyke, B. (2005) "W(h)ither Development Geography in Australia?", *Geographical Research* 43(3): 306–318.

Simon, D. (2003) "Dilemmas of Development and the Environment in a Globalising World: Theory, Policy and Praxis", *Progress in Development Studies* 3(1): 5–41.

JOHN CONNELL AND BARBARA RUGENDYKE

FERNANDO HENRIQUE CARDOSO (1931–)

Fernando Henrique Cardoso was born in Rio de Janeiro, Brazil, in 1931 and trained as a sociologist at the University of São Paulo. He became a prolific author and one of the most distinguished social scientists in the world, but also a successful political leader who served as President of Brazil. Having started his academic career at his alma mater, he subsequently spent a considerable period in France, including as Associate Director of Studies at the École des Hautes Études en Sciences Sociales in Paris. His principal contributions to development thinking are often referred to as his study of dependency and development in Latin America during the late 1960s and early 1970s, although initially he worked on slavery. Less attention has been devoted to the evolution of his ideas around development during his transition from scholar to political leader since the 1980s. Many of those ideas appeared in speeches, interviews, and in the column he wrote regularly in the newspaper *Fohla de São Paulo* for many years. They can also be found in the policies he implemented during his term as president of Brazil. His academic and political careers have generated considerable scholarly commentary. Cardoso's work includes more than 200 articles, books and book reviews, as well as hundreds of speeches, interviews and journalistic notes. His most important speeches are available through the internet, and many publications focusing on his work are available in English. A large database of his life and work is available at the Fundação Fernando Henrique Cardoso (http://fundacaofhc.org.br/acervo_i.htm).

His most recent academic appointment was as professor-at-large at Brown University, Providence, Rhode Island for the five years following his presidency (2003–2008), while he has been a longstanding member of the Institute of Advanced Study at Princeton University, both in the USA. He also has remained politically active in his party and in his US-style presidential foundation in Brazil, while he is a member of The Elders, the eminent group of former heads of state working to settle conflicts and human rights abuse around the world. His many international honours include the Prince of Asturias Award in 2000 for international co-operation, the Kluge Prize from the US Library of Congress in 2012, and several honorary doctorates, most recently from Harvard University in 2016.

His best-known book, *Dependency and Development in Latin America*, written with Enzo Faletto in 1969, addresses the problems of economic development through an interpretation emphasising the political character of the process of economic transformation. Cardoso and Faletto (1969) incorporated in their analysis the historical situation in which the economic transformations occurred in order to better understand these changes and their structural limitations. Cardoso argued that the situation of dependency was neither stable nor permanent. Although he recognised that the dynamic elements of international capitalism lie outside the periphery and its control, he argued that dependent development should not be interpreted as a lack of dynamism in parts of the periphery (Resende-Santos 1997). Cardoso recognised the changing nature of international capitalism, and that the reconfiguration of the international distribution of labour opened opportunities for the periphery (Cardoso 1973; Goertzel 1999). For Cardoso, dependency was about position and function within the international economy, about how Latin American countries participated in the global economy (Cardoso 1982). He rejected the assertion that dependency produced underdevelopment. For him, dependency was fundamentally a question of power and domination, internationally and locally (Cardoso and Faletto 1969; Cardoso 1973).

The dependency and development thesis has had a significant impact on generations of scholars concerned with development, but as mentioned above, it would be a mistake to base Cardoso's contribution to the study of development only on that thesis. Cardoso developed further elements of his thesis as central elements in his understanding and practice of development. His attention to social as well as institutional analysis allowed him to identify that politically, dependent development in Brazil produced institutional deformities that constituted important obstacles to democracy and development (Font 2001). Brazil's political system was characterised by its clientelism, corporatism and corruption, and Cardoso was an outspoken advocate of fundamental institutional restructuring in Brazil. He was convinced of the importance of strengthening democratic practices in order to assure a stable transition to the next stage in Brazil's development. For him, many of the solutions to the defects in Brazil's political and economic system were bound up in his notion of social democracy (Resende-Santos 1997).

Cardoso became progressively drawn to an active role in the pro-democracy movement in Brazil during the 1970s. His commitment to open opportunities for development in Brazil led him to take an active role in the restructuring of political institutions. This approach was conducive to the political career he initiated in 1978 when he was elected an alternate senator, and later a senator in 1983. Cardoso's involvement in understanding and shaping the process of political and structural reform in Brazil led him to become a founder member of the Brazilian Social Democratic Party (PSDB) in 1982 and to a successful career as a senator. Following a short spell as Foreign Minister in 1992/1993, his successful performance as Brazil's finance minister in controlling hyperinflation in the early 1990s positioned him as a popular presidential candidate. Cardoso was elected and served two terms as President of Brazil (1995–2003). His political philosophy stressed change and the need continuously to redefine ideas and political agendas as circumstances change. That philosophy allowed him to maintain an efficient relationship between ideas and policy choices.

It is precisely the bridge between ideas and the practice of development that targeted criticism at Cardoso. For many of his critics, Cardoso abandoned his **Marx**ist roots to become a defender of liberal democracy and capitalist development. One set of criticisms accused him of embracing political institutions he once criticised. This is direct reference to his early work in the 1960s and 1970s, when he allegedly regarded political parties and other political institutions as formal expressions of class domination; he is criticised for becoming involved in those same political institutions since the late 1970s (Packenham 1992).

A second set of criticisms focuses on Cardoso's term as president of Brazil, labelling him as a neoliberal serving the interests of the multinational business elite. These criticisms focused on Cardoso's macroeconomic policies, including the privatisation of state-owned enterprises (Lesbaupin 1999). Scholars studying Cardoso's work and life have suggested these criticisms are based on misconceptions of Cardoso's ideas. For Goertzel, Cardoso was interested in Marxism as social theory, not as a political dogma (1999). The analysis of Cardoso's presidential administrations in Brazil documents his achievements in critical areas for national, regional and local development (Amann and Baer 2000; Dantas 2003; Goertzel 2003).

Resende-Santos (1997) suggests that Cardoso's critics had a difficult time reconciling the progressive social philosophy and reformist idealism that in their view defined Cardoso's early years as a scholar with the neoliberal political leader. For Resende-Santos, this is due to a misunderstanding of Cardoso's political philosophy that stresses change and the need for constant redefinition of ideas and political agendas as circumstances change:

> Cardoso has always rejected dogmatic thinking, especially as displayed in the political projects of both the left and the right in Latin America. He has always maintained that ideas must change as circumstances change … Cardoso is, and has always been, much more a pragmatist than a revolutionary. His is a moderate, gradualist agenda that stresses what is viable, rather than one committed to absolutes.
>
> (Resende-Santos 1997: 146)

Cardoso's scholarly work and political career illustrate his contributions to the study of development, and invite reflection on our own understanding of this concept. He conceptualises development as a dynamic multidimensional process that does not propose a finished model but a road towards the transformation of society. He emphasises three central dimensions of that process: social equity, institutional rebuilding leading to democracy, and alternatives leading to sustainable economic growth. Cardoso's attention to the dynamic interactions among these dimensions illustrates his understanding of the complex reality that development seeks to address.

For Cardoso, the deficiencies in Brazil's political and economic system are central to understanding the problem of social equity. Social democracy makes it possible to correct those deficiencies and improve social equity. But social democracy has only an instrumental value in the sense that it creates a path towards the transformation of society (Resende-Santos 1997). A central element in addressing Brazil's economic and political deficiencies is the role of the state. For Cardoso, the main reason for state reform is not neoliberal ideology, as many critics have suggested, but rather a profound general crisis of the state. The state remains an important actor in development, but it needs to redefine its mission, refocus, and increase its effectiveness according to the changing conditions in the world economy and within Brazilian society (Font 2001). Cardoso realised that in order to compete successfully in the global economy, Brazil required internal reforms to enhance the process of international insertion. This analysis continues to have great pertinence for understanding the current turmoil in Brazilian politics.

Cardoso considers that the periphery is more in danger of being left behind and excluded from participating in the global economy than of remaining dependent as a supplier of raw materials. For him, Brazil's best chances of a successful future lie in a strong market economy, an efficient regulatory state and effective social programmes. The development policies that he implemented during his presidency of Brazil reflect those ideas. Cardoso accepted the leading economic role of the market while maintaining that it does not address all the needs of society, creates problems of its own, and tends to dissolve human solidarity. For him, the state plays a fundamental role in reducing inequalities, poverty and other social problems. But Cardoso understood that development also needed a strong civil society:

> One has to overcome the simplistic notion that what is in the best interest of the citizenry has to originate necessarily in the State. The mechanism for regulating "privatised public" activities must be guided by the needs of the people, and to that end, the direct participation by representatives of civil society in the bodies concerned is fundamental. The State must be porous and permeable to the needs of the citizens. The identification of the State with national interest is a political construction work that requires major efforts to reach consensus ... In sum, politics is less "the art of the possible" and more the "art of making possible that which is necessary".
>
> (Cardoso 2001: 286)

In summary, Cardoso has made two significant contributions to our understanding and practice of development that have not yet been adequately assessed. In terms of ideas, Cardoso has brought remarkable clarity to conceptualising development as a dynamic multidimensional process, clearly identifying its critical dimensions and the interactions among them. There are two principal reasons why scholars have not devoted enough attention to Cardoso's perspective on development. First, he has not taken the time to present his conceptual framework explicitly in an integrated form. Second, the evolution of his critical thinking around development has not appeared in traditional academic publications. The most interesting part of that evolution is when he became a practitioner of development and had the opportunity to confront, first-hand, his ideas with reality. Of course, the best way he had to express his thoughts on development as President of Brazil were his speeches in domestic and international events.

The second contribution is his experience and performance as a development practitioner. The practice of development in the context of crises deserves better attention, to enhance our understanding of opportunities and obstacles for social change. Poverty, inequality, deficiencies in the political system, financial crises and the pressure from a challenging range of domestic and international interests created obstacles to putting his ideas on development into practice. Clearly his administration was not perfect, but it was encouraging to see positive results conducive to social change in Brazil. The lack of attention to Cardoso's legacy in the practice of development is in part due to the focus on the short-term performance of his administration's policies rather than on their contribution to a comprehensive process of development. Time is proving Cardoso right. Despite criticisms of his social and economic policies, many of these were maintained by Luiz Inacio Lula da Silva, Cardoso's successor as President and leader from the political left, but who subsequently became embroiled in corruption allegations and at the time of writing was in prison and withdrew from the 2019 presidential election.

A final issue worth noting about Cardoso's legacy is his remarkable balance in resisting the corrosive effect of power and pursuing good governance. He regarded the presidency as a political institution with the obligation to account for its acts. This contrasts with many other Latin American Presidents. In times when development is contested as a concept guiding transformations of society, Cardoso's legacy should be considered an inspiration to strengthen the ideas and practice of development as a tool for social change. Cardoso himself insists on having his performance judged on its ability to set Brazil on a new course.

Major works

Cardoso, F.H. (1973) *Estado y sociedad en América Latina*, Buenos Aires: Ediciones Nueva Vision.

—— (1982) *Politicas sociales para la decada en América Latina*, Montevideo: Centro Latinoamericano de Economía Humana.

—— (2001) "Radicalizing Democracy", in M. Font (ed.), *Charting a New Course: The Politics of Globalization and Social Transformation: Fernando Henrique Cardoso*, Lanham, MD: Rowman & Littlefield, 284–288.

—— and Faletto, E. (1969) *Dependencia y democracia en América Latina*, Mexico: Siglo XXI Editores. (A Portuguese version, *Dependencia e desenvolvimiento na América Latina.* appeared in 1970, while the University of California Press published an updated English version, *Dependency and Development in Latin America*, in 1979. The book has also been translated into French, German and Italian.)

Further reading

Amann, E. and Baer, W. (2000) "The Illusion of Stability: The Brazilian Economy under Cardoso", *World Development* 28(10): 1805–1819.

Dantas, F. (2003) "After Eight Years, Lula Inherits a Better Country. Brazil Advanced during Fernando Cardoso's Two Terms as President But It Is Still Fragile and Unequal", www.estadoa.com/eleicoes/governolula/noticias/2003/jan/01/24.htm. Last accessed 20 September 2004.

Font, M. (2001) "Introduction. The Craft of a New Era: The Intellectual Trajectory of Fernando Henrique Cardoso", in M. Font (ed.), *Charting a New Course: The Politics of Globalisation and Social Transformation: Fernando Henrique Cardoso*, Lanham, MD: Rowman & Littlefield, 1–34.

Goertzel, T. (1999) *Fernando Henrique Cardoso: Reinventing Democracy in Brazil*, Boulder, CO: Lynne Rienner.

—— (2003) "Eight Years of Pragmatic Leadership in Brazil", a supplement to *Fernando Henrique Cardoso: Reinventing Democracy in Brazil*, http://crab.rutgers.edu/~goertzel/fhc.htm. Last accessed 10 September 2004.

Lesbaupin, I. (organizador) (1999) *O desmonte da nacao. Balanco do Governo FHC*, Petropolis: Editora Vozes.

Packenham, R. (1992) *The Dependency Movement: Scholarship and Politics in Development Studies*, Cambridge, MA: Harvard University Press.

Resende-Santos, J. (1997) "Fernando Henrique Cardoso: Social and Institutional Rebuilding in Brazil", in J. Domínguez (ed.), *Technopols: Freeing Politics and Markets in Latin America in the 1990s*, University Park, PA: Pennsylvania State University Press, 145–94.

Wikipedia (2018) Fernando Henrique Cardoso https://en.wikipedia.org/wiki/Fernando_Henrique_Cardoso, accessed 14 September (contains many details of his academic and political careers and awards).

ROBERTO SÁNCHEZ-RODRÍGUEZ WITH DAVID SIMON

MICHAEL CERNEA (1934–)

"Development anthropology is a contact sport," Michael Cernea tells his students. His career, from researcher in the Romanian Academy of Sciences in the late 1950s and early 1960s to Senior Social Advisor for Sociology and Social Policy in the World Bank in the 1990s, is testament to this observation. Cernea's professional life has been characterised by constant high-stakes struggles over social development ideas within different bureaucratic and political settings. His has been a career where the thinking through of ideas, and the acting upon them, have been one and the same process – as a thinker in development, Cernea has to be understood as much in terms of his relationships to particular institutions as to development anthropology and sociology. Indeed, many might argue that Michael Cernea's most critical contribution has not been just the body of intellectual work that has elaborated on culture and development, participation and local knowledge, or the inherence of risk in the development process, but rather the embedding of those ideas in the World Bank. From the 1970s to the mid-1990s, Cernea contributed more than anybody in pushing the "social envelope" at the Bank. He had the ear of Presidents and Vice-Presidents in the process, and from time to time was unrepentant in giving those ears a good chewing. He has been one of those quintessential reformists (and at times small-"r" revolutionaries) inside the Bank recognised by outside commentators and analysts as critical to any form of progressive change in the institution (Fox and Brown 1998).

How far personal history determines subsequent careers is always a matter of interpretation, but in Michael Cernea's case there are at least apparent continuities. He grew up as part of the Jewish community in Iaşi (Jassy), a Romanian town close to the Russian border. As a child, he watched the violent unravelling of that community in the face of anti-Semitism and fascism (Simon 2019). Along with all other Jewish children in the community, he was expelled from elementary school. Shortly before that, his father had lost his job as an engineer for being a Jew. During the pogrom of Iaşi in 1941, "one of the worst in Romania's history" (Simon 2019), he witnessed mass killings and his father being badly beaten. While Michael insists that he and his family did not suffer nearly as much as did very many others, the experience of persecution and resettlement was nonetheless visceral – he was forced out of his home twice, first as a result of the pogroms (when the family went into hiding) and then in the face of the advancing German army. The experiences left him politically active and sensitised, and after the war he became a member of the Socialist Youth Movement and a young journalist. Though initially a supporter of the Romanian left's rise to government power in the postwar years, he grew more concerned with time at the directions taken by the regime. His PhD thesis, "About Dialectics in the Socialist Society" (University of Bucharest, 1962), in which he tried to legitimise theoretically the persistence of contradictions, conflicts and tensions within Romania's allegedly homogeneous society and structures, reflected these growing concerns. That he chose such a topic also presaged strategies of enquiry later in life at the World Bank – strategies in which he aimed to change an institution by challenging its foundational ideas from within that same institution.

As would later be the case at the Bank, this strategy generated resistance, and so – coupled with the effects of anti-Semitism in the Romania of the late 1950s – he had to wait four years simply to schedule the defence of his PhD. Still, by the time he made his defence, the effects of the Khrushchev thaw had changed the political context in Romania again, and Cernea was finally able to address the topic that really moved him, peasant economic rationality, although not without political fallout. This early work (Cernea 1970; 1971) operated at the boundaries of anthropology, sociology and rural economics, exploring cultural dimensions of productive strategies, and the relationships between peasant rationalities and co-operative agriculture. This concern for the sociocultural foundations of the economy would reappear later in a different guise in his sustained questioning of what he referred to as the "economic reductionism" of the World Bank (see Cernea 1994 and below).

As the thaw continued, foreign scholars visited Romania in increasing numbers, a process that ultimately came to define a turning point both in Cernea's career (see his Malinowski Award lecture, Cernea 1996) as well as in the future that the World Bank was yet to have. Cernea was nominated by one such visitor for a fellowship at the Center for Advanced Studies in the Behavioral Sciences (CASBS) at Stanford, one of the US's most prestigious academic fellowships. The politics behind Cernea being able to assume the fellowship were convoluted, but shortly following the meeting of Ceauşescu and Nixon (a meeting triggered by Ceauşescu's resistance to the Russian invasion of Czechoslovakia), Cernea finally received the visa allowing him to travel to California, albeit only in time for the latter part of the fellowship. His time there (1970–1971) led to the distillation in English of his ideas on peasant society, but more importantly, to a friendship with US sociologist Robert Merton, and a growing reputation in the USA.

The links between the year at Stanford, his initial contacts with the World Bank and his being interviewed for a position in Washington are neither direct nor without intrigue, but led to his being offered a job there in 1974 – following a talk given at Bank headquarters on the role of the family plot in collective farms. His position was as the first sociologist to work in McNamara's newly created central Rural Development Division, a division that in many respects occupied a pivotal piloting role in McNamara's rural-development-led approach to poverty reduction, in the taming of rural radicalisation (albeit more implicitly) and in changing the Bank. Cernea approached the Bank as simultaneously an institution he worked (and largely lived) in, as an object of analysis and as something he wanted to change. He believed that such change could come from the injection of sociological knowledge into the very foundations of the institution's way of interpreting and acting on the world.

In his recounting of the experience, it took him three or four years to understand properly how the institution operated, and the different ways in which social dimensions were excluded from the World Bank's way of intervening. Among other things, he concluded that operational guidance and rules had to be revised if the institution was going to change. An early effort was to develop, with colleagues, guidelines on the Bank's lending for dams – which, interestingly, received endorsement from McNamara. This paved the way for work that Cernea, together with the team of social scientists that he helped build over the years, would do to develop other operational directives and policies on a range of issues related to the social, environmental and cultural impacts of the Bank. These "safeguards" became a new set of rules that would constrain project design from within, and at the same time give civil society activists instruments through which to hold the Bank accountable against its own standards. With time, these safeguards, or variants of them, were adopted by other development banks and bilateral agencies.

Underlying this work on safeguards was the belief that if social science analyses were to have any teeth in the Bank, they had to be turned into bureaucratic instruments that made such knowledge a requirement of normal practice in the institution. One part of this process was to generate the ideas and write the operational policies and directives that derived from them. The harder challenge was to get these proposals through the Bank's approval process and then, once they had been approved, keep them alive in the face of the constant pressure from other parts of the Bank to get rid of them: a process of constant lobbying and insistence involving not just Cernea but also a broader network of social and environmental scientists within and outside the Bank.

At a more operational level, Cernea also studied the Bank's project cycle in depth with a view to exploring the entry points for sociological knowledge in the processes through which the Bank designed and implemented its operations. As part of this initiative, he began inviting outside social scientists to speak on the different ways in which sociological knowledge and an awareness of the social dimensions of development could be brought into the project cycle. This cycle of seminars culminated in the book, *Putting People First* (Cernea 1985), which – notwithstanding its focus on the World Bank – became one of the early foundational texts on participatory rural development. Cernea's introduction to the collection argued that the World Bank was in the business of financially induced development, but that for any such induced change to be successful it was imperative that the Bank understand the social structures that existed among the populations being affected by the project and that could play constructive roles in this process of induced development – echoes of Merton's influence (Cernea 1985). Social science knowledge – and social scientists – were thus essential for the Bank (Cernea 1994).

A recurrent theme in his work has been that social science should be conducted not only for its own sake but also for development's sake, and that one of the key purposes of such social science is to challenge the ideas underlying economists' and others' models of development. While this argument would be made intellectually – and Cernea has been a prolific writer – most strategically for him, it had to be made bureaucratically. The culmination of that effort to turn social science ideas into bureaucratic form began in an event in late 1995 on the occasion of the Bank's formal recognition of Cernea's 20 years working for the institution. At the event, he caught the new Bank President, James Wolfensohn, and managed to persuade him why it was that social development had to be central to the poverty orientation that Wolfensohn was promoting. This led to a series of exchanges in which Cernea convinced Wolfensohn to create a social development task force that would report on the state of social development work in the institution. Negotiated and contested ("development anthropology is a contact sport"), this was a process that the Development Economics Vice-Presidency and two Managing Directors were able to contain and to some degree capture. But not entirely, for in 1997, and against the advice of the Bank's Chief Economist, Wolfensohn approved the creation of a Social Development Department in the Bank.

Cernea's work has also pushed thinking on the links between culture and development, on cultural patrimony and on the environment, on the concept of social policy and on articulating various social policies. However, of most significance perhaps – both intellectually and also for human welfare – has been his work on involuntary resettlement, risk and vulnerability. Cernea joined the Bank during the period in which the large-scale rural development and infrastructure programmes that would later bring the institution into such criticism were being hatched. As early as 1978 he began elaborating guidelines for resettlement, which were subsequently issued as Bank policy (1980). Not long after came a sharp fight within the Bank over the provisions for resettlement in India's infamous Narmada Dam project. This became the seed for a review of the Bank's overall performance on resettlement (Cernea 1986) and two subsequent documents giving policy guidelines for involuntary resettlement (Cernea 1986; 1988). This latter piece has been translated into Bahasa, Turkish, Chinese, Italian, Spanish and French and has had over a dozen print runs. If the 1986 review was critical, a later 1993–1994 review (Cernea et al. 1994) pulled even fewer punches. It systematically drew attention to the failures of the Bank to follow its own policy. It also constituted one of Cernea's key – empirically sustained – statements on the way in which development inherently brings risk to people at the same time as bringing opportunity.

The resettlement reviews were not anti-development statements, but they did make clear that development processes handled irresponsibly, without adequate social science insight, without adequate consultation and involvement of poor people, are likely to increase vulnerabilities, and very often for the poorest. The intellectual core of this argument was later captured in his "Impoverishment Risks and Reconstruction Model" (Cernea 1997; 2001; 2002; 2004; Cernea and McDowell 2000) of the eight risks that he came to identify as inherent in the process of displacement: landlessness, joblessness, homelessness, marginalisation, food insecurity, loss of access to common property resources, increased morbidity and community disarticulation (Cernea 1997; also Mahapatra 1999). The model typifies Cernea's intellectual and professional project in that it offers a framework intended not only to help understand risk but also – through identifying and making explicit such risks – to trigger a response to that risk such that it might be mitigated even before its effects fully manifest themselves. Its purpose is to be analytical, predictive and methodologically useful. It has become a leading model used internationally in research and policies about development triggered involuntary resettlement (Koenig 2002).

Cernea's contributions to development are many, reflected in a long list of publications, advisory roles and academic honours. Perhaps the most important among them, however, was to have been part of a process that changed an institution through the sustained and forceful insistence on a few ideas: that social science knowledge is critical to development, that induced development will fail in any meaningful sense if ordinary people are not involved in shaping the forms it takes, and that the cherished disembodied concepts of so much development theory cannot be considered separately from the social structures in which they are embedded. These ideas became embodied in the World Bank in different ways: the recruiting and crafting of a strong, mutually supportive, occasionally contentious network of social scientists who carried this agenda forward in their work in different parts of the institution; the conversion of these ideas into operational directives and policies, rules to govern the behaviour of the institution as a whole; the translation of these ideas into project design and in some cases fully fledged loans; and the elaboration of a sort of institutional public sphere in which these ideas became part of the ether such that, even if many task managers, country directors and others found them annoying, they could not easily contest them openly or brazenly.

Thought of slightly more analytically, what Cernea did was to lead the building of the ideational, relational and normative bases for institutional change. This process was, of course, only possible because of the sustained hard work of the broader social science and environmentalist community within the World Bank – none of this is the work of Cernea alone. But, along with figures such as Gloria Davis, he was one of the very early players in these processes of change, and played an instrumental role in devising strategy and in facilitating the creation of this social science community in the first place.

What did not change, however – and this obviously exceeded the capacities of any network of social scientists – was the political economy of the World Bank as an institution. There are three dimensions to this observation. First, the Bank will always be an institution dominated by economists and economic and financial logic. The social and environmental science community will always be a very small minority, and so to sustain some of the achievements noted above is a Sisyphean task. Indeed, once the strength of continuing insistence on these achievements diminishes, progress is undone and the boulder rolls back downhill again. Second, like all institutions, the Bank will always be in the business of self-protection, attempting to deflect and reduce critique. Third, the Bank is ultimately dependent for its survival on its chief financial contributors (for its concessional lending), its chief borrowers (without whom there is no demand for the Bank) and the approval and repayment of loans. These (and other) pressures constantly push back against the social development agenda. Consequently in 2018, much more so than in 2005 when the first version of this essay was written, the long-term survival of Cernea's impacts on the Bank is less certain. The Social Development Department no longer exists, the network of social scientists that carried Cernea's early ideas forward has now largely left the institution, and social science at the Bank has once again been rendered less powerful than it was in the 1990s and 2000s. World Bank Senior Management, other epistemic communities within the Bank and financially and politically powerful member countries together push for the substantial watering down of the policies and directives that derived from Cernea's and colleagues' work. The incentive is to get more loans out of the door, and to reduce criticism that the Bank has not followed its own policies. Indeed, in 2016 the Bank eliminated its Safeguard Policies, replacing them with the far less demanding *Environmental and Social Framework* and *Environment and Social Standards* documents. Cernea worries greatly about these reversals and the need to find new ways to pressure the institution to protect the livelihoods and dignity of vulnerable people (Cernea and Maldonado 2018).

Acknowledgements

I am very grateful to Michael Cernea for the interviews, materials and time he gave me during the preparation of this entry, as well as for our conversations over the years. David Simon also graciously shared insights and information.

Major works

Cernea, M. (1970) *Two Villages, Social Structures and Technical Progress* (in Romanian, senior author, research co-ordinator), Bucharest: Edit. Politica.

—— (1971) *Changing Society and Family Change: The Impact of the Cooperative Farm on the Traditional Peasant Family*, Stanford, CA: Center for Advanced Studies in Behavioral Sciences.

—— (ed.) (1985) *Putting People First: Sociological Variables in Rural Development Projects*, New York: Oxford University Press.

—— (1986) *Involuntary Resettlement in Bank-Assisted Projects: A Review of the Application of Bank Policies and Procedures in FY79-95 Projects*, AGR, Operations Policy Staff, World Bank, February.

—— (1988) *Involuntary Resettlement in Development Projects: Policy Guidelines for Bank-Financed Projects*, Washington, DC: World Bank.

—— (1994) "A Sociologist's View on Sustainable Development", in I. Serageldin and A. Steer (eds), *Making Development Sustainable: From Concepts to Action*, Environmentally Sustainable Development Occasional Paper Series no. 2. Washington, DC: World Bank.

—— (1996) *Social Organization and Development Anthropology: The 1995 Malinowski Award Lecture*, ESD Studies and Monographs Series no. 6. Washington, DC: World Bank.

—— (1997) "The Risks and Reconstruction Model for Resettling Displaced Populations", *World Development* 25(10): 1569–1588.

—— (2001) "Eight Main Risks: Preventing Impoverishment During Population Resettlement", in C. de Wet and R. Fox (eds), *Transforming Settlement in Southern Africa*, Edinburgh: Edinburgh University Press for the International African Institute, 237–252.

—— (2002) *Cultural Heritage and Development: A Framework for Action in the Middle East and North Africa*, Washington, DC: World Bank.

—— (2004) "The Typology of Development-Induced Displacements: Field of Research, Concepts, Gaps and Bridges", paper presented at the US National Academy of Sciences Conference on the Study of Forced Migration, 22–23 September, Washington, DC.

—— and Freidenberg, J. (2007) "'Development Anthropology Is a Contact Sport': An Oral History Interview with Michael M. Cernea by Judith Freidenberg", *Human Organization* 66(4): 339–353.

—— and Guggenheim, S. (eds) (1993) *Anthropological Approaches to Resettlement: Policy, Practice, and Theory*, Boulder, CO: Westview Press.

—— —— and Associates (1994) *Resettlement and Development: The Bankwide Review of Projects Involving Involuntary Resettlement 1986–1993*, Washington, DC: World Bank.

—— and Maldonado, J. (eds) (2018) *Challenging the Prevailing Paradigm of Displacement and Resettlement: Risks, Impoverishment, Legacies, Solutions*, London and New York: Routledge.

—— and Mathur, H.M. (eds) (2008) *Can Compensation Prevent Impoverishment? Reforming Resettlement Through Investments and Benefit-Sharing*, New York: Oxford University Press.

—— and McDowell, C. (eds) (2000) *Risks and Reconstruction: Experiences of Resettlers and Refugees*, Washington, DC: World Bank.

Further reading

Fox, J. and Brown, D. (eds) (1998) *The Struggle for Accountability: The World Bank*, Cambridge, MA: MIT Press.

Koenig, D. (2002) *Toward Mitigating Impoverishment in Development-Induced Displacement and Resettlement*, Oxford: Refugee Studies Centre, University of Oxford.

Mahapatra, L.K. (1999) *Resettlement, Impoverishment and Reconstruction in India*, New Delhi: Vikas.

Simon, D. (2019) *Holocaust Escapees and Global Development: Hidden Histories*, London: Zed Books.

ANTHONY BEBBINGTON

ROBERT CHAMBERS (1932–)

Robert Chambers has been at the forefront of attempts since the early 1980s to place the poor, the destitute, the marginalised, the excluded and the powerless at the heart of the processes of development problem identification, decision-making, policy formulation and project implementation. He popularised the notions of "farmer first" and "putting the last first", and has been highly influential in challenging professionals in all echelons of the development sector to foster critical self-awareness in their approach to their role and activities. In so doing, he has played a catalytic role in the introduction, refinement and dissemination of the participatory development ideology and associated methodologies. His "new paradigm" has rapidly become the new orthodoxy.

Robert John Haylock Chambers was born on 1 May 1932. He received an Open Scholarship to Cambridge in the natural sciences in 1949, but this was interrupted by National Service from 1950 to 1952, when he was commissioned in the Somerset Light Infantry. Studying at St John's College, he graduated with a Double First in the BA History Tripos in 1955, and was awarded an MA in 1959. His lifelong appetite for travel and discovery was whetted in 1955 as a member, and short-term leader, of the first scientific expedition to Gough Island in the South Atlantic – then one of the last places in the world about which little was known. He still lists running, mountaineering and rock-climbing as his principal recreational pursuits.

Chambers was appointed a District Officer in Kenya between 1958 and 1962, in the process gaining valuable insight into both colonial administration and local rural development. He then shifted his emphasis towards training and research, becoming a Lecturer in Public Administration at the Kenya Institute of Administration, and then a Research Officer at the East African Staff College. After receiving his PhD from the University of Manchester in 1967 for a thesis on settlement schemes in tropical Africa, and after short teaching appointments at the Universities of Manchester and Glasgow, he returned to Kenya to become a Senior Research Fellow in the Institute for Development Studies at the University of Nairobi from 1969 to 1971. From 1972 until 1997 he was a Fellow at the Institute of Development Studies (IDS), University of Sussex, where he established an international reputation as a leading advocate for and promoter of appropriate and just forms of development. During the 1970s, Robert Chambers continued to work on the management of rural development in eastern Africa, later extending his areal expertise to include India and Sri Lanka. Also, while at the IDS, he spent a period in the mid-1970s with the UNHCR, working with rural refugees in eastern Africa and southeast Asia.

He continues to work at the IDS in the role of Research Associate. He was awarded the OBE in 1995, and has received Honorary Doctorates from the University of East Anglia (1995) and the University of Edinburgh (1998). The two principal influences on his way of looking at the development world were **E.F. Schumacher** and Ivan Illich, although he professes not to have been in total agreement with their ideas.

Although Chambers' journey on the path towards appropriate development probably started in the late 1950s, it was a paper published in the journal *World Development*, in 1981, entitled "Rural Poverty Unperceived: Problems and Remedies", that, in his words, "started all this off". A number of themes and issues that recur in his later work were first introduced in that article. Its central argument, and that of much of Chambers' work, is that history has presented us with particular ways of looking at and approaching the poor and the disadvantaged, and "development" more generally, which are often far removed from frequently unseen and unperceived realities. "Doing development" typically falls to outsiders – people who, however well-intentioned, are neither from the target areas nor have directly experienced the problems of poverty, deprivation, insecurity, vulnerability, exploitation and so on. Their visions, interpretations and actions are thus built upon preconceptions and misperceptions about the rural poor, and are often influenced by stereotypical views of their circumstances and capabilities. Their interventions are also frequently constrained by a number of institutionalised biases that are influenced by practical realities (time, access, resources), attitudes (planning arrogance, political correctness, timidity), inequalities (hierarchies, differentials of power), segmentation (sectoralisation, project focus) and so on which determine that those who are in greatest need, or who are less visible and vociferous, remain unseen, unheard and unhelped. The result is that mountains of cash, eons of effort and one of the world's largest "industries" have collectively failed to make much more than a dent in deprivation and destitution.

The paper sets the challenge to the principal actors in development to retrain and reprogramme themselves. Chambers emphasises the need for a series of "reversals" where practitioners, academics, administrators, scientists and the like step away from what he pointedly calls "rural development tourism" and start to bring about a revolutionary change in forms of cognition and behaviour. One of the principal reversals – now a mantra for the burgeoning non-governmental sector – is in the learning process, with the outside expert becoming the unimportant pupil who learns from and with those in need of development assistance, as opposed to the hitherto pervasive "I talk, you listen" approach and attitude. The challenge is to see rural poverty as it really is, rather than as it is perceived to be, and to tailor actions accordingly in order truly to make a difference. The need is to confront global cores and peripheries of knowledge by looking for, being receptive to and finding means of incorporating local forms of knowledge that, time and again, have been shown to be tried and tested, appropriate but woefully overlooked by outside "experts" and seriously compromised by the predilection for modern science and technology. Imagination and inventiveness are needed to find ways of reaching the poorest of the poor, the socially excluded, the politically voiceless, the geographically peripheral and the environmentally marginal.

The ideas in the *World Development* article were developed, honed and expanded in a series of accessible books produced during the 1980s and early 1990s, most of which were published via the Intermediate Technology Group (now Practical Action) founded by **E.F. Schumacher**. *Rural Development: Putting the Last First* (1983) emphasised the need for new professionals and a new professionalism with which to rise to the challenge of an inclusive and emancipatory development. The emphasis was placed squarely on "the last": the hundreds of millions of largely unseen people in rural areas who are poor, weak, isolated, vulnerable and powerless. The book also exposed the gulf between the two principal cultures involved in the development project – the academic and the practical – and the tension between scientific/technical and local/indigenous knowledge. It looked at tactical means of confronting the "superiority complex" of the development profession, of bringing about reversals in learning and of underpinning effective practical action.

Farmer First (Chambers et al. 1989) fleshed out a "new paradigm" centred on localised, particularistic, contextualised, sympathetic, democratic and participatory means of supporting people in marginal contexts. The idea was not to come up with "another development" of the kind advocated by post-development writers such as **Manfred Max-Neef**, **Arturo Escobar** and Wolfgang Sachs, but to suggest forms of intervention involving a "new professionalism" that complemented and enhanced existing science and technology-based approaches. The solutions to farmers' problems lay, in part, within their own capacities and capabilities. The challenge was to find ways of unlocking their development potential by creating a suitable enabling environment. The targets of development were to assume increasing control of the development process. Outsiders were to become the facilitators, not the paternalistic controllers, of local development. New institutions and institutional practices were to become established in order to allow the new paradigm to flower and flourish.

Farmer First adopted a holistic view of rural people's development challenges, introducing for the first time a concept – "sustainable livelihoods" – that has become the leitmotif of the development industry in the early part of the new century. The principles of the sustainable livelihoods approach were first put forward in an IDS Discussion Paper (1992) jointly authored by Robert Chambers and Gordon Conway, and centred on ideas of capability, equity and sustainability. The livelihoods approach to rural development has subsequently been adopted by the UK's DFID (Department for International Development), Sweden's Sida (Swedish International Development Co-operation Agency), the United Nations Development Programme (UNDP) and several other bilateral and multilateral development agencies. Chambers' work on the hidden facets of poverty and deprivation, and the need to inform our understanding of these problems based on the experiences of those who are beset by them, inspired a large-scale body of work, leading up to the production of the World Bank's *World Development Report 2000/01: Attacking Poverty*, entitled "Consultations with the Poor" and "Voices of the Poor", where, for perhaps the first time, the Bretton Woods institutions' understanding of the poverty phenomenon became to some extent informed by the experiences and opinions of the poor themselves.

Robert Chambers became increasingly involved in rising to the methodological challenge that had arisen from the need to understand poverty and deprivation from the perspective of the poor and deprived. The first challenge was to find a cost-effective means of probing and penetrating beneath the apparent surface within target rural communities. Up until the late 1970s, constraints of time, resources and willingness had contributed to the prevalence of what he called "quick and dirty" local survey methods, where outside experts or support teams engaged in superficial and uncritical rural appraisal as a precursor to project development, often using questionnaires ("survey slavery") that reflected their preconceived ideas about rural areas and the challenges of development. Chambers helped pioneer the tool of Rapid Rural Appraisal (RRA), which was neat and nimble but, most importantly, incorporated a range of devices to "include the excluded" and to obtain a broad, flexible, insightful and locally contextualised understanding of poverty and the poor, or of other local development challenges. RRA came to constitute not just a new research methodology but also a new way of understanding development. It centred on developing and using a range of imaginative techniques – key informants, group appraisal, triangulation, progressive learning, identification and incorporation of indigenous knowledge. These both yielded an insightful understanding of local conditions as revealed by a representative cross-section of the communities themselves and freed up time for learning with and from local people, and for searching out the poor, the marginal and the excluded.

RRA steadily evolved into Participatory Rural Appraisal (PRA), and subsequently into a wider set of participatory tools that now constitutes the dominant development methodology and ideology. PRA was presented in *Whose Reality Counts? Putting the First Last* (1997) (the sequel to *Rural Development: Putting the Last First*) as a methodological solution to the challenge of building development efforts around local knowledge, experiences and capabilities. It is based on the observed ability of local communities across the developing (and developed) world to express and analyse their own complex realities in a way which has often been shown to be at variance with the perceived realities held by professionals operating from the top down. PRA centres on a raft of recipes for including the excluded and enabling local communities to express their development realities in ways which were suitable to their own cultures, literacy levels and so on. Much PRA, Participatory Action Research and Participatory Learning and Action in recent years has been driven by innovative development workers in and from the South. A prerequisite for practitioner/facilitators is critical self-awareness: a reflexive sense of the baggage of values and behaviour that they bring to the table of development courtesy of their background and identity. A key word in PRA is "empowerment". Robert Chambers sees participatory development as an "exciting revolution in learning and action", and claims that the turn of the 21st century has been one of the most stimulating and challenging periods to be involved in the development profession.

Long after formal retirement, he has remained very active in research, advocacy and publication from his base at IDS. Integrated water and sanitation have been an important focus, particularly in community-led total sanitation (CLTS) and the links between (mal)nutrition, water and sanitation, and community health (e.g. Cavill et al. 2015; Kar and Chambers 2008).

Robert Chambers would, we're sure, be the first to emphasise that his contribution to the reversal of development directionalities and the construction of appropriate methodological tools has been a collaborative and participatory endeavour, be that with colleagues and passers-through at the IDS, practitioners in the field and in big offices, or the poor and the peripheral themselves. Nonetheless, his thumbprint is everywhere one looks in the modern development field: bottom-up development, participatory development, sustainable livelihoods, the redefinition of poverty, the strengthening of civil society, and the entire ethos of appropriate development. In both the developing and development worlds, Robert Chambers has become one of the best-known, most influential and most widely followed persons in contemporary development over the last five decades.

It is therefore both fitting and valuable that in recent years he has published what amounts to a trilogy of reflective volumes addressing big-picture challenges of and in development as a whole. The first two, *Revolutions in Development Inquiry* (Chambers 2012) and *Ideas for Development* (Chambers 2013), focused on methodological progress and concepts and principles respectively. Finally, *Can We Know Better? Reflections for Development* (Chambers 2017) draws together key insights and lessons from his experience on the ground in what is simultaneously a retrospective and forward-looking book. Written in his characteristically modest and accessible style for multiple audiences, he reflects on the inescapable complexity of development issues, on why so many well-intentioned interventions have failed, and on how we might seek to do better in future. Inevitably, participatory methods and tools, in fields such as information and communications technologies (ICTs) and statistics, feature prominently, along with the Reality Check Approach. Underpinning these and, indeed, all appropriate approaches, he argues, reflexivity, facilitation, humility and ground-truthing remain essential attributes of mind and practice.

So many of us who, as outsiders, work on development issues genuinely but forlornly hope that our efforts will make a difference to the problems with which we are concerned; Robert Chambers is one among very few who can confidently claim to have made a profound difference to the ideology, paradigm and practice of development.

Major works

Chambers, R. (1981) "Rural Poverty Unperceived: Problems and Remedies", *World Development* 9(1): 1–19.

—— (1983) *Rural Development: Putting the Last First*, Harlow: Longman (translated into Arabic, French, Indonesian, Italian, Japanese, Portuguese and Vietnamese).

—— (1993) *Challenging the Professions: Frontiers for Rural Development*, London: Intermediate Technology Publications.

—— (1994a) "The Origins and Practice of Participatory Rural Appraisal", *World Development* 22(8): 953–969.

—— (1994b) "Participatory Rural Appraisal (PRA): Analysis of Experience", *World Development* 22(9): 1253–1268.

—— (1994c) "Participatory Rural Appraisal (PRA): Challenges, Potentials and Paradigms", *World Development* 22(10): 1437–1454.

—— (1997) *Whose Reality Counts? Putting the First Last*, London: Intermediate Technology Publications.

—— (2012) *Revolutions in Development Inquiry*. London and New York: Routledge.

—— (2013) *Ideas for Development*. London and New York: Routledge.

—— (2017) *Can We Know Better? Reflections for Development*. Rugby, UK: Practical Action Publishing.

—— and Gordon Conway (1992) "Sustainable Rural Livelihoods: Practical Concepts for the 21st Century", IDS Discussion Paper 296, Brighton, UK: Institute of Development Studies.

—— A. Pacey and L.A. Thrupp (eds) (1989) *Farmer First: Farmer Innovation and Agricultural Research*, London: Intermediate Technology Publications.

Further reading

Cavill, S., with Chambers, R. and Vernon, N. (2015) "Sustainability and CLTS: Taking Stock", Frontiers of CLTS: Innovations and Insights 4, Brighton, UK: Institute of Development Studies, CLTS Knowledge Hub: www.communityledtotalsanitation.org.

Kar, K. and Chambers, R. (2008) *Handbook on Community-Led Total Sanitation*. Brighton, UK: Institute of Development Studies.

MICHAEL PARNWELL WITH DAVID SIMON

HOLLIS B. CHENERY (1918–1994)

Hollis Chenery's importance as a development economist reaches beyond his contributions to the academic debate, which were considerable. As Vice-President of the world's premier multilateral development institution, the World Bank, Chenery's ideas and worldview widely influenced development thinking and practice during the 1970s and 1980s.

Hollis Chenery was born into a wealthy family in Richmond, Virginia, USA, in 1918. His father was an oil and gas magnate. He grew up in Virginia as well as Pelham Manor in New York state. His upbringing in affluent suburban settings there would leave a permanent mark on him. A private passion of his was racehorses, and one of those he owned would once win the Triple Crown.

Chenery was educated both as an engineer and as an economist. Before World War II, Chenery attended the universities of Arizona and Oklahoma, earning undergraduate degrees from both. During the war, he served as an Army Air Officer in the US military. He then went on to earn two master's degrees from the California Institute of Technology and the University of Virginia respectively, before gaining a doctorate in economics from Harvard University in 1950.

From the outset, Chenery's career combined academia with positions in international development. After World War II, he served as a Marshall Plan economist in Europe, throwing himself into the European reconstruction programme. This job, and being part of the postwar reconstruction process, greatly influenced his thinking regarding industrial development, economic growth and societal change. It was decisive in focusing his intellectual energies on development economics, the field which he would make his life's work.

Chenery's initial academic position after returning to the United States and receiving his PhD was as Professor of Economics at Stanford, a post which he held from 1952 until his appointment as a Guggenheim Fellow in 1961. That same year he joined the newly established United States Agency for International Development (USAID), which was created by decree in President John F. Kennedy's Foreign Assistance Act. Chenery quickly rose through the Agency's ranks to become assistant administrator. Yet he stayed with USAID only until 1965, when he returned to academia. He was appointed Professor of Economics at the university where he earned his doctorate fifteen years earlier. Harvard would remain Chenery's academic home for the rest of his life, to which he would return after his tenure at the World Bank.

Chenery's academic work prior to joining the international development institutions was focused on empirical studies regarding factors of growth and development. His work was concerned with the role of investments and industrialisation in the development process, and the design of development projects and programmes (e.g. 1953; 1955; 1958; 1959). Chenery's book with Paul Clark on *Interindustry Economics* (1959) combined the qualities of a textbook with a systematic reader for professionals. His 1961 article, "Comparative Advantage and Development Policy", was received as a significant examination of the debate on resource allocation in less-developed economies and the classical principle of comparative advantage. The paper considered development policy from an economic theoretical standpoint, yet based on an analysis of actual government policies. He concluded that much of the confusion resulted from the failure to distinguish theoretical and operations research.

Hollis Chenery joined the World Bank in Washington, DC, in 1970 to serve as Economic Adviser to the Bank's then President, Robert McNamara, the former US Secretary of State. In a World Bank oral history (Asher 1983), Chenery speculates why McNamara chose him for the position. He recalls meeting McNamara a few times before: very casually when they both worked in the government in the 1960s (when Chenery was with the newly established USAID) and at a conference early in 1970. Their most extensive contact had been at the Harvard Kennedy School, where Chenery briefed McNamara on economics research when the latter first became the World Bank President. He further recalls: "At our first encounter, it was quite clear that the reason he (McNamara) was interested in me was to establish not only a research capability in the Bank, but a quantitative research capability."

Two years later, McNamara handpicked Chenery as Vice-President for Development Policy at the Bank, a position he would hold for more than a decade until 1983. In this position, Chenery played a key role influencing both the World Bank's internal operations as well as its lending policy towards developing countries.

Chenery was one of the first development economists to suggest that there was no linear path to development, thus rejecting the views of earlier development theorists such as economic historians **Alexander Gerschenkron** and **Walt W. Rostow**. Their "stage theory" had stipulated that all countries passed through the same historical stages of economic development that the now industrialised had experienced earlier. According to this view, the developing countries were only at an earlier stage of development, i.e. primitive versions of developed countries. Furthermore, this view tended to equate development simply with industrialisation and economic growth, which would lead to a trickle-down effect that would increase the overall wealth of the population. Based on his broad empirical work, Chenery recognised that there were differences in the experiences of individual countries, while at the same time acknowledging that there were similar patterns that could be identified in different countries' paths to development.

Chenery understood that it was not inevitable that countries would all pass through the same linear phases. For that reason, he saw it as a major task for development economists to suggest ways in which developing countries could leapfrog over certain stages to catch up with industrialised countries. These shortcuts would be based on a systematic learning from the development histories and experiences of developed and more advanced underdeveloped countries.

Hollis Chenery brought these nuanced views of the development process with him to the World Bank. Citing, in particular, the experiences of Brazil and Sri Lanka with their contrasting development policies, Chenery in the oral history states, "Internally, these discussions of individual countries were probably as important as any method of sharpening our analysis and, in some cases, changing our policies." His worldview coincided well with that of McNamara, reflecting the liberal developmentalism of the era, including the notion that poverty was the key factor in the spread of communism and it was, therefore, essential to combat poverty in Asia, Latin America and Africa to prevent them from falling into the communist bloc. Poverty reduction and income distribution, not only growth at an aggregate level, thus became important to the World Bank. An influential book, *Redistribution with Growth*, edited by Chenery with others (1974), epitomised this change in the Bank policy. The Bank, jointly with the International Labour Organization (ILO), also worked on the "basic needs" concept and its implications for lending operations – which at the time was controversial within the Bank, where many economists saw it involving excessive government intervention.

As the Vice-President for Development Policy in the mid-1970s, Chenery initiated a study on the role of the Bank in promoting economic development since 1950. The study, which was carried out by an outside consultant, David Morawetz, found that economic growth around the world had been swift and spectacular, but that its results were generally poorly distributed. At the same time, time-series data showed that despite impressive economic growth in the developing countries, its benefits have only reached the poor to a very limited degree (Ahluwalia et al. 1979). As a consequence, governments and the World Bank as an international financial institution needed to refocus their efforts on poverty eradication, even at the cost of increasing indebtedness. During McNamara's tenure, supported by the empirical research and theoretical thinking under Chenery's leadership, the World Bank's lending expanded considerably and an explicit poverty-alleviation focus was brought to its operations. Chenery chaired an in-house task force established in 1981 that articulated the Bank's poverty focus.

When the first oil crisis hit in 1973, Chenery was concerned with its impacts on the world economy. In the spirit of the era, as exemplified by the studies of the Club of Rome and writings of Paul and Anne Ehrlich, he had adopted a view that the rapid growth in the world economy was pushing against the productive capacity and the potential rate of expansion of world supplies of many raw materials. He wrote an article in the influential *Foreign Affairs* (1975b) journal in which he argued, with hindsight over-optimistically, that the need for the adjustment was bringing forth an energy transition that would reduce the need for formal international agreements that had proven to be elusive.

Overall, Chenery had a fundamental impact on the World Bank's research programme and its influence on the Bank's policy-making. According to him, the Bank's best research was in the field of empirically based comparative studies. He saw that the World Bank's research programme could optimally combine micro-oriented studies based on its own lending operations in various countries with more theoretical, speculative macro-economic research. He pursued this policy successfully, getting the Bank's Board to accept the approach as a legitimate way of influencing policy. Chenery took it upon himself to build up the research capacities within the Bank by recruiting a number of highly qualified economists (including **Mahbub ul Haq**), many from academia. He bridged the gap between research and operations by establishing a rotational system for staff. The in-house research capacity was complemented by bringing in outside expertise through consultants and academics participating in research work. A Research Committee was established, and research findings were widely disseminated internally through "management seminars" and internal discussions based on the cases of individual borrower countries. While the management seminars were chaired by Chenery, McNamara frequently participated, usually sitting in the back row absorbing the debates. These were all tools intended to improve the Bank's analysis and its consequent lending policies.

During Hollis Chenery's time, the World Bank started producing a wide variety of policy papers that became authoritative statements on development issues. These included the "Prospects for the Developing Countries" series that Chenery initiated in response to the rise in oil prices in the autumn of 1973. Recognising the drastic impact the oil crisis had on developing countries, Chenery felt that this was the pre-eminent problem on which the Bank as a development institution had to focus. The series then led to the preparation of the first annual *World Development Report* (WDR), the Bank's flagship publication to date. Chenery credits strong backing from McNamara enabling him to pursue the WDR. It was one of Chenery's other recruits to the Bank, Ernie Stern – later to become Senior Vice-President – who recruited the team to produce what became the first WDR (Simon 2019).

After McNamara's retirement from the World Bank Presidency in 1981, the new management under Alden Winship (Tom) Clausen changed course. This coincided with the emergence of neoliberal ideology as personified by the rise to power of Ronald Reagan and Margaret Thatcher in the USA and UK respectively. At the World Bank, Anne Krueger was hired to replace Hollis Chenery, with the result that neoliberal dogma replaced the poverty-alleviation focus. However, while the official policy of the Bank changed, at the operational levels many staff economists still maintained their pragmatic approaches to development policy.

After leaving the Bank, Chenery returned to Harvard, where he continued to put his empirical experiences with international development to good use in writing and teaching. He collaborated with T.N. Srinivasan in editing a *Handbook of Development Economics* (1988; 1989), which became the most comprehensive sourcebook on development economics available.

International development theory has undergone considerable changes since the 1970s and 1980s, when Chenery was an active participant in its making. Some of his work has inevitably been overtaken by events and advances in our understanding of development processes. Nevertheless, he made a lasting impact on international development theory, and in particular on policy-making and praxis. Despite significant ups and downs since his move from the World Bank, his legacy lives on, and is today perhaps more evident in the institutional policies of the Bank than immediately after his departure, because of the sharpened poverty focus of its current activities. In 1983, Hollis Chenery expressed concern that income distribution within countries and in the world, including in the United States, has become considerably worse. Since then, the problem has become, if anything, even more acute.

Major works

Adelman, I. and Chenery, H.B. (1966) "Foreign Aid and Economic Development: The Case of Greece", *Review of Economics and Statistics* 48(1): 1–19.

Arrow, K.J., Chenery, H.B., Minhas, B.S. and Solow, R.M. (1961) "Capital–Labor Substitution and Economic Efficiency", *Review of Economics and Statistics* 43(3): 225–250.

Chenery, H.B. (1952) "Overcapacity and the Acceleration Principle", *Econometrica* 20(1): 1–28.

—— (1960) "Patterns of Industrial Growth", *American Economic Review* 50(4): 624–654.

—— (1961) "Comparative Advantage and Development Policy", *American Economic Review* 51(1): 18–51.

—— (ed.) (1971) *Studies in Development Planning*, Cambridge, MA: Harvard University Press.

—— (1975a) "A Structuralist Approach to Development Policy", *American Economic Review* 65(2): 310–316.

—— (1975b) "Restructuring the World Economy", *Foreign Affairs* 53(2): 242–263.

—— (ed.) (1979) *Structural Change and Development Policy*, New York: Oxford University Press/World Bank.

—— (1983) "Interaction Between Theory and Observation in Development", *World Development* 11(10): 851–903.

—— and Clark, P.G. (1959) *Interindustry Economics*, New York: John Wiley and Sons.

—— Duloy, J.H. and Jolly, R. (eds) (1974) *Redistribution with Growth: Policies to Improve Income Distribution in Developing Countries in the Context of Economic Growth*, London: Oxford University Press.

—— and Strout A. (1966) "Foreign Assistance and Economic Development", *American Economic Review* 56(4): 679–733.

Syrquin, M. and Chenery, H. B. (1989) *Patterns of Development*, 1950–1983. Washington, DC: World Bank.

Further reading

Ahluwalia, M., Carter, N. and Chenery, H.B. (1979) "Growth and Poverty in Developing Countries", in H.B. Chenery (ed.), *Structural Change and Development Policy*, New York: Oxford University Press.

Asher, R. (1983) "Hollis Chenery", Oral History Program, transcript of interview, 27 January 1983, World Bank/IFC Archives.

Chenery, H.B. (1953) "The Application of Investment Criteria", *Quarterly Journal of Economics* 67(1): 76–96.

—— (1955) "The Role of Industrialization in Development Program", *American Economic Review* 45(2): 40–57.

—— (1958) "Development Policies and Programmes", *Economic Bulletin for Latin America* 3(1): 51–77.

—— (1959) "The Interdependence of Investment Decisions", in M. Abramovitz et al. (eds), *The Allocation of Economic Resources: Essays in Honor of Bernard Francis Haley*, Stanford, CA: Stanford University Press.
—— (1981) "Restructuring the World Economy: Round II", *Foreign Affairs* 59(5): 1102–1120.
—— (1982) "Industrialization and Growth: The Experience of Large Countries", World Bank Staff Working Paper no. 539, Washington, DC: World Bank.
—— and Srinivasan, T.N. (eds) (1988 and 1989) *Handbook of Development Economics*, vols I and II, Amsterdam: North-Holland.
Easterly, W. (2001) *The Elusive Quest for Growth: Economists' Adventures and Misadventures in the Tropics*, Cambridge, MA: MIT Press.
Pace, E. (1994) "Hollis B. Chenery Dies at 77; Economist for the World Bank", *New York Times*, 5 September.
Simon, D. (2019) *Holocaust Escapees and Global Development: Hidden Histories*. London: Zed Books.

JUHA I. UITTO

DIANE ELSON (1946–)

Since the late 1970s, and throughout numerous institutional and disciplinary incarnations, Diane Elson has been at the forefront of gender and development (GAD) scholarship at a world scale. Under the broad umbrella of analysing socio-economic processes from a gender perspective, Elson's work has spanned a truly diverse range of themes. Until the turn of the 21st century, these included women's position in the international division of labour; gender dimensions of structural adjustment; the significance of gender inequality and unpaid labour within the household or "reproductive economy" (sometimes called the care economy, or sphere of social reproduction) for the operation of national economies and international trade and finance; gender budget initiatives (GBIs); and gender and globalisation. Building on this trajectory, Elson's work since the early 2000s has maintained a focus on GBIs, exploring the links between poverty reduction, macro-economic policy and women's rights. More recently, and alongside her research into gender-equitable inclusive economic growth, she has been active in developing a new body of scholarship concerning human rights and economic policy (Balakrishnan and Elson 2011; Balakrishnan et al. 2016). Elson has also published seminal pieces on the impacts of the 2008 financial crisis in both developed and developing countries (Elson 2013).

As one of a still relatively small number of feminist economists able to draw from an impressive multidisciplinary expertise encompassing sociology and politics as well as economics, Elson's work has been distinguished, *inter alia*, by its ability to challenge and make accessible to non-economists the concepts underlying thinking and modelling in the predominantly male field of macro-economics. Macro-economic models often masquerade under the guise of gender-neutrality, but in reality emanate from gender-blindness, in turn resulting in gender bias. As Elson has so eloquently argued, the need to "talk to the boys" (Elson 1998a), in the form of making gender heard in an arena where there are few feminist voices conversant with the language and concepts deployed within international development economics has been an abiding principle in much of her work over nearly five decades. The call to "engender" macro-economics and international economics has featured not only in Elson's numerous academic publications and policy reports but in major curriculum innovations in the UK and overseas.

Elson was born on 20 April 1946. She grew up in the English Midlands, attending Nuneaton High School for Girls. She was the first member of her family to attend university, and graduated from Oxford with a BA (Hons) in Philosophy, Politics and Economics in 1968. She then worked for three years as a Research Assistant at the Institute of Commonwealth Studies at Queen Elizabeth House, Oxford, and was a member of St Antony's College. In 1985 she was appointed as Lecturer in Development Economics at the University of Manchester (where she later attained her PhD by publication and became Professor of Development Studies and Research and Graduate Dean). In the meantime, she held a range of teaching and research positions at the Universities of Sussex, York and Oxford, as well as fitting in consultancy and lecturing at the Open University and Manchester with the birth and pre-school care of her son, Paul. After 15 years of full-time employment at Manchester, Elson moved to the University of Essex, where she was appointed to a Chair in Sociology, and taught on gender, the sociology of development and human rights. After her retirement in 2011, she continued to carry out research as an Emeritus Professor. She is also a Visiting Professor at the Centre for Research on Women in Scotland's Economy at Glasgow Caledonian University, a Research Associate at the Center for Women's Global Leadership at Rutgers University and an Honorary Associate of the Institute of Development Studies, University of Sussex.

Elson stands out as an agenda-setting individual in her own right, particularly in opening up and establishing new lines of enquiry and conceptual development; but she has also conscientiously and productively engaged in numerous collaborations with fellow academics and with writers, researchers and practitioners from other disciplines and backgrounds, both in the UK and internationally. Some of her collaborative ventures have resulted from her consultancy, advisory and research work for major development institutions, including the Swedish International Development Cooperation Agency (Sida), the United Nations Development Programme (UNDP), UNIFEM (The United Nations Development Fund for Women), where between 1998 and 2000 she was a special advisor to the Executive Director, Noeleen Heyzer, and the UN Committee for Development Policy, to which she was appointed by the UN Secretary General in 2013. Elson has also maintained substantial involvement in women's groups and networks, and in several forms of public service. The latter includes membership of the Management Committee of the UK Women's Budget Group and the UN Millennium Project Task Force (2002–2005); service as Vice-President of the International Association for Feminist Economics (2004–2006); membership of the Advisory Committee for UN Women for *Progress of the World's Women 2018*; and her position on the G7 Gender Equality Advisory Council commencing in 2018. In the interests of advancing gender equality through changing the ways in which public revenue is raised and expended, Elson has also helped to co-found a global network of gender budget initiatives comprising civil society groups and government personnel. In recognition of her significant contributions to this field of study, she was awarded the Leontief Prize for Advancing Frontiers of Economic Thought in 2016, and in 2017 she received an Honorary Doctorate from the Open University.

Elson's longstanding commitment to movements and initiatives for change outside as well as within academia has strengthened the applied as well as the theoretical aspects of her work. Her writings are distinguished, for example, by a remarkable accessibility, given the complexity of some of the issues she has dealt with, especially in the realm of macro-economics, trade and financial governance (Elson et al. 2000; Van Staveren et al. 2007; Young et al. 2011). In turn, a keen commitment to policy relevance, to judicious and critical interrogation of empirical data, and to linking the macro with the micro, has lent itself to an enviable capacity to bridge the often wide divide between analysis and action. This has helped to earn Elson credibility across several different sets of stakeholders in the development field as a whole, including global policymakers, grassroots activists and countless PhD students who, under her tutelage, have between them worked on gender issues in most parts of the world. Among GAD scholars and practitioners more generally, she is recognised as having been one of the movement's most valuable assets in respect of establishing gender as a legitimate and enduring issue, both conceptually and pragmatically.

Elson is the author and editor of numerous books, including her groundbreaking *Male Bias in the Development Process* (1991), as well as special journal issues, papers and policy documents. Many of her writings have rapidly assumed the status of classics in the GAD field, as evidenced *inter alia* by their extensive reprinting and translation into other languages. In light of the prolific nature of her work, it is clearly hard to do justice to the vast repertoire of ideas which she has injected into different debates, and which indeed have spawned debate in their own right. One of her earliest and most significant contributions was a co-authored article (Elson and Pearson 1981), which not only provided one of the first comprehensively researched yet concise accounts of women's involvement in export-oriented industrialisation but also introduced the idea which – contrary to conventional WID (Women in Development) orthodoxy that women had been "left out" of the development process – drew attention to the need to focus on the frequently exploitative ways in which women were actually *integrated*. More specifically, Elson and Pearson identified the need to distinguish between three, non-mutually exclusive tendencies in the relation between the disproportionate concentration of women in un- and semi-skilled jobs in export-oriented factories and their subordination as a gender. These were, respectively: "a tendency to *intensify* the existing form of gender subordination; a tendency to *decompose* existing forms of gender subordination; and a tendency to *recompose* new forms of gender subordination" (Elson and Pearson 1981: 110, emphasis in original). Various possibilities for how women workers could challenge these different forms of subordination were drawn out. Even if women made few instrumental gains in their positions (for example, improvements in pay and in working conditions), evaluation of women's struggles as a gender should be made in relation to the way that "the struggle itself develops capacities for self-determination" (p. 105). In the rich and ever-growing literature on gender and export manufacturing to which Elson herself has continued to contribute (Van Staveren et al. 2007), the 1981 paper remains a central reference point.

Emphasis on the potential for women's resistance, despite considerable barriers to their mobility and empowerment, also appears in Elson's pioneering work on gender and neoliberal restructuring, which became a major new genre of enquiry in the late 1980s and early 1990s. Highlighting how women frequently absorbed the costs of structural adjustment policies by increasing their loads of paid and, in particular, unpaid labour, Elson not only argued for the need for macro-reforms such as the democratisation of states and the restructuring of the international financial system, but emphasised how seeds of transformation also lay at the level of women's involvement in self-help groups at the community level. Yet, Elson's contribution to feminist critiques of structural adjustment went far beyond this: she synthesised the observations drawn from localised case studies, linked the micro with the meso and macro, made the most convincing case for economies being gendered structures and demonstrated the extreme gender bias inherent in structural adjustment programmes (along with macro-economic models more generally) (Elson 1993). More particularly, she drew attention to the failure to incorporate gender in the macro-economic modelling of restructuring, marked above all by complete disregard for women's extensive unpaid labour in the "reproductive economy". This exclusion, she argued, led not only to pernicious effects on women's own quality of life but, given their role in (re)producing human resources (and *ipso facto*, the labour force), to economically inefficient outcomes (Çağatay et al. 1995; Elson 1995). Out of this came the conclusion that any macro-economic model which neglected gender and the reproductive economy was not only fundamentally flawed on conceptual grounds but could exacerbate social inequality and render possibilities for sustainable economic growth untenable. As articulated in her widely cited *World Development* article:

> Though maintaining gender inequality may sometimes lead to higher profits in the short run, it also tends to generate negative feedbacks which hamper the process of restructuring to a development path which is sustainable in the long run. The outcome of macroeconomic policies depends upon the gendered social matrix in which they are introduced. But that gendered social matrix could in principle be changed in ways that promote both more effective macro-economic policies and greater gender equality.
>
> (Elson 1995: 1865)

Similar arguments on the importance of recognising, valuing and redistributing unpaid care and domestic labour remain central to her more recent assertions that incorporating these vital contributions into gross domestic product valuations is crucial if we are ever to address the gender pay gap:

> The gender wage gap will never be closed by measures that aim to make women's working lives more like men's. Now we need more radical measures, those that will transform men's working lives to make them more like those of women … As well as raising the rate of women's participation in paid work, we need to raise the rate of men's participation in unpaid care and domestic work.
>
> (Elson 2017a: 58)

Elson's continuing enthusiasm for taking on big ventures, and for transforming policy, includes her lead authorship on the 2000 and 2002 UNIFEM reports on *The Progress of the World's Women* (Elson 2000; Elson and Keklik 2002), and her work on GBIs (Elson 1998b; 2004; Elson et al. 2002). In the impressively detailed yet accessibly presented 2002 UNIFEM report, on the third UN Millennium Development Goal (MDG) to "promote gender equality and empower women", Elson criticised the four indicators selected to represent the target (the ratio of boys to girls in primary, secondary and tertiary education; the ratio of literate women to men in the 15–24 age group; the share of women in non-agricultural employment; and the proportion of parliamentary seats held by women) for their inherent weaknesses as well as continued data deficiencies, and also argued that more targets should be included, such as gender equality in the labour market. This constituted an important contribution to a larger and growing body of feminist critique of the gap between rhetoric and genuine commitment to addressing gender inequality in global development frameworks, and one which appears to have been partially conceded as indicated, *inter alia*, by Sustainable Development Goal 8, which aspirationally seeks to "promote sustained, inclusive and sustainable economic growth, full and productive employment and decent work for all".

With regard to her work on GBIs, Elson's contributions not only include key justifications for integrating a gender perspective into national budgetary processes but also provide important guidelines for "engendering" budgets in accordance with principles of equality and non-discrimination central to human rights and social justice (Elson 2016). She has also documented numerous case study examples from among the 50-plus countries in which GBIs have thus far taken place. This includes a co-authored book chapter with Rhonda Sharp (2010) exploring gender-responsive budgeting as a means of facilitating poverty alleviation, where they persuasively show notable improvements in female development outcomes achieved through including women in project planning and budgeting processes. Drawing on evidence from Tanzania, Uganda, India, Indonesia, South Africa, the UK and Australia, the authors highlight that when women are actively sought out and given a space to advocate for their needs, priorities and interests, and to hold governments accountable, women grow in confidence and develop new skills and knowledge while helping to direct funds to services and infrastructure that are meaningful to them. Although Elson and Sharp caution against viewing GBIs as a panacea for poverty alleviation, their findings reinforce the unique and exciting potential of GBIs to deliver impacts at the grassroots through improved taxation and expenditure. Similar ideas extend to Elson's subsequent publications (Elson 2013; 2017b), showcasing the utility of employing a human rights perspective in broader economic analyses, including deficit reduction strategies and the evaluation and sequencing of public policies and procedures. Especially pertinent are her critical examinations of the impacts of global finanical crises on women and the reproductive sphere (Elson 2013; Pearson and Elson 2015). These revealed the limited extent to which macro-economic analysis and policy have taken account of gender, despite the increase in the availability of data, not to mention the reality that women around the globe bear the bulk of costs of austerity measures.

Elson's writings have also embraced more general development issues outside GAD, including development theory (Elson 1998b), the socialisation of markets (Elson 1992), and the UN Global Compact. Often richly informed by Elson's feminist and socialist political perspectives, such contributions have made her a major name in the development field in general, as well as within GAD circles. It is within GAD, however, that her pioneering and enduring contribution to scholarship is most apparent. No student of gender and development today completes a course at undergraduate or postgraduate level without some exposure to Elson. Given her ongoing ventures into cutting-edge issues in the GAD field – currently comprising fiscal policy in relation to women's economic and social rights, globalisation and gender equality, and the importance of gender-inclusive economic growth which acknowledges the primacy of carework– and their hallmark rigour and vision, this situation is unlikely to change for a long time to come. Elson will also be present as one of GAD's outstanding and trailblazing scholars, activists, and ambassadors in any retrospective review of gender and development that may be compiled in the future.

Major works

Balakrishnan, R. and Elson, D. (eds) (2011) *Economic Policy and Human Rights Obligations*, London: Zed Books. (Introduction by Balakrishnan and Elson (1–27); Chapters 4 (99–130) and 8 (199–221) by Aydinar-Avsar and Elson.)

—— Heintz, J. and Elson, D. (eds) (2016) *Rethinking Economics for Social Justice: The Radical Potential of Human Rights*, London: Routledge.

Çağatay, N., Elson, D. and Grown, C. (eds) (1995) *Gender, Adjustment and Macroeconomics*, special issue, *World Development* 23(11): 1825–2017.

Elson, D. (ed.) (1991) *Male Bias in the Development Process*, Manchester: Manchester University Press. (Chapters 1 (1–28), 7 and 8 (164–210) by Elson; 2nd edn 1995, with two additional chapters by Elson (211–79).)

—— (ed.) (2000), *Progress of the World's Women 2000*, New York: UNIFEM.

—— Çağatay, N. and Grown, C. (eds) (2000) *Growth, Trade, Finance and Gender Inequality*, special issue, *World Development* 28(7): 1145–1390.

Van Staveren, I., Elson, D., Grown C. and Çağatay N. (eds) (2007) *Feminist Economics of Trade*, London: Routledge. (Introduction by Elson, Grown and Van Staveren (1–10); Chapter 3 (33–52) by Elson, Grown and Çağatay.)

Young, B., Bakker, I. and Elson, D. (eds) (2011) *Questioning Financial Governance From a Feminist Perspective*, London: Routledge. (Introduction by Young, Bakker and Elson (1–10); Chapter 6 by Elson and Warneke (110–131).)

Further reading

Elson, D. (ed.) (1992) "The Economics of a Socialised Market", in R. Blackburn (ed.), *After the Fall: The Failure of Communism and the Future of Socialism*, London: Verso.

—— (1993) "Gender-Aware Analysis and Development Economics", *Journal of International Development* 5(2): 237–247.

—— (1995) "Gender Awareness in Modelling Structural Adjustment", *World Development* 23(11): 1851–1868.

—— (1998a) "Talking to the Boys: Gender and Economic Growth Models", in C. Jackson and R. Pearson (eds), *Feminist Visions of Development*, London: Routledge, 155–170.

—— (1998b) "Integrating Gender Issues into National Budgetary Policies and Procedures: Some Policy Options", *Journal of International Development* 10: 929–941.

—— (2002) "Gender Justice, Human Rights, and Neo-liberal Economic Policies", in M. Molyneux and S. Razavi (eds), *Gender Justice, Development and Rights*, Oxford: Oxford University Press, 78–114.

—— (2004) "Engendering Government Budgets in the Context of Globalisation (s)", *International Feminist Journal of Politics* 6(4): 623–642.

—— (2013) "Economic Crises From the 1980s to the 2010s: A Gender Analysis", in G. Waylen and S. Rai (eds) *New Frontiers in Feminist Political Economy*, London: Routledge, 189–212.

—— (2016) "Gender Budgeting and Macroeconomic Policy", in J. Campbell and M. Gillespie (eds), *Feminist Economics and Public Policy: Reflections on the Work and Impact of Ailsa McKay*, London: Routledge, 27–37.

—— (2017a) "Recognize, Reduce, and Redistribute Unpaid Care Work", *New Labor Forum* 26(2): 52–61.

—— (2017b) "Financing for Gender Equality: How to Budget in Compliance with Human Rights Standards", in Z. Khan and N. Burn (eds), *Financing for Gender Equality: Gender, Development and Social Change*, Basingstoke: Palgrave Macmillan, 25–48.

—— Budlender, D., Hewitt, G. and Mukhopadhyay, T. (2002) *Gender Budgets Make Cents*, London: Commonwealth Secretariat.

—— Faune, M., Gideon, J., Gutierrez, M., Lopez de Mazier, A. and Sacayon, E. (1997) *Crecer con la Mujer*, San José, Costa Rica: Embajada Real de los Paises Bajos.

—— and Keklik, H. (2002) *Progress of the World's Women 2002*, New York: UNIFEM.

—— and Pearson, R. (1981) "Nimble Fingers Make Cheap Workers: An Analysis of Women's Employment in Third World Export Manufacturing", *Feminist Review* 7 (Spring): 87–107.

—— and Sharp, R. (2010) "Gender Responsive Budgeting and Women's Poverty", in S. Chant (ed.), *The International Handbook of Gender and Poverty: Concepts, Research, Policy*, Cheltenham: Edward Elgar, 522–527.

Pearson, R. and Elson, D. (2015) "Transcending the Impact of the Financial Crisis in the United Kingdom: Towards Plan F – A Feminist Economic Strategy", *Feminist Review* 109: 8–30.

SYLVIA CHANT AND JORDANA RAMALHO

ARTURO ESCOBAR (1952–)

The Colombian-American anthropologist Arturo Escobar is one of the leading figures of the Post-Development debate in development theory. His analyses of the discourse of development from a post-structuralist, Foucauldian perspective, inspired by work with activists of the network of Afro-Colombian organisations *Proceso de Comunidades Negras* (PCN), have been immensely influential in critical development studies. He has also contributed works on political ecology, social movements, globalisation, science and technology studies and pluriversal and transition studies. His seminal monograph, *Encountering Development* (1995a), has been cited over 11,000 times according to Google Scholar, indicating that Escobar is one of the most prominent writers in development studies today. He is a co-ordinating member of the World Anthropologies

Network (WAN) and part of the Latin American Modernity/Coloniality group. Further, he has been active in the anti-globalisation movement and the World Social Forum.

Arturo Escobar was born in Manizales, Colombia, in 1952. After studying chemical engineering at the Universidad de Valle in Cali from 1969 to 1975 and completing a BSc (and one year of graduate work in the same subject), he went to the US for an MA in food science and international nutrition at Cornell University in 1978. From 1978 onwards, he pursued a interdisciplinary PhD in the philosophy, policy and planning of development at the University of California at Berkeley. During 1981–1982 he worked in Colombia's Department of National Planning in Bogotá. In 1987, he received his PhD for a thesis entitled "Power and Visibility: The Invention and Management of Development in the Third World". He taught Latin American Studies at the University of California in Santa Cruz (1985–1989) and anthropology at Smith College (1989–1994), at the Universidad del Valle in Cali (1993), and at the University of Massachusetts at Amherst (1994–1999). Since 2000, he has taught anthropology at the University of North Carolina at Chapel Hill, as a Kenan Distinguished Teaching Professor until retirement in 2018, when he received the title of Distinguished Professor of Anthropology Emeritus. Apart from that, he has taught at many other institutions in the USA, Colombia, Argentina, Finland, Spain, the Netherlands and Great Britain.

Escobar's best-known work is concerned with the critique of development discourse. Following Foucault in investigating the nexus between discourse and power, Escobar argues that the way we talk and write about the "Third World" or "developing countries" is imbued with relations of power and thus has a political dimension. He analyses development as a formation constituted by forms of knowledge (development theories and strategies), forms of subjectivity fostered by it (a perception of the self as "developed" or "underdeveloped") and a system of power that regulates its practice (in specific projects and programmes) (1995a: 10). This discourse was articulated in the post-World-War-II era, around the construct "underdevelopment" and on the basis of certain material conditions baptised as such (p. 53). Its function was the establishment and reproduction of rule over the Third World through an apparatus for producing knowledge about and exercising power over it (p. 9). It created a space in which only certain things could be said and done (p. 39), in which the perception of social problems in Africa, Asia and Latin America ("underdevelopment") and possible solutions ("development") was structured in a specific manner. The conditions of possibility of this discourse include the economic and geopolitical interests of the USA during the Cold War, the anticolonial struggles in Africa and Asia, rising nationalism in Latin America, the perception of a "population problem" in the South, faith in science and technology and increasing experience with economic planning and interventions (pp. 31ff.). Escobar claims that the discourse became so influential that even

> those who opposed the prevailing capitalist strategies were obliged to couch the critique in terms of the need for development, through concepts such as "another development", "participatory development", "socialist development", and the like.... one could criticize a given approach and propose modifications or improvements accordingly, but the fact of development itself, and the need for it, could not be doubted. Development had achieved the status of a certainty in the social imaginary.
>
> (Escobar 1995a: 5)

Thus Escobar is concerned with a process of systematically conceiving non-European areas in Eurocentric categories and transforming them accordingly. Going beyond the traditional leftist dependency critique, he endeavours to analyse capitalism not only as a system of production but also as a system of representations and disciplinary mechanisms (1988: 438). In analogy to Foucault's gaze of surveillance characterising the modern prison system, Escobar emphasises the crucial role of visibility. In the era of development, the "poor, underdeveloped masses" of the Third World entered the horizon of Northern elites. It was only on the basis of this visibility that new academic fields and disciplines like development economics could rise, which generated knowledge about the Third World legitimising interventions. In due course, new aspects of the Third World became visible (small farmers, women, the environment ...) which expanded the reach of the interventions based on expert knowledge. The basic pattern employed by the discourse was the construction of "abnormalities" (the underdeveloped, the illiterate, the malnourished) which it would later treat and reform (1985: 387). Its aim was "not simply to discipline individuals but to transform the conditions under which they live into a productive, normalized social environment: in short, to create modernity" (1995a: 156). This is why, for Escobar, the analysis of development discourse was part of an anthropology of modernity.

The development discourse became powerful through the mechanisms of professionalisation and institutionalisation. "Professionalisation" refers to practices organising and controlling the generation and diffusion of knowledge, creating and maintaining a "politics of truth" (1988: 430). "Institutionalisation" refers to the production of techniques which organise visibilities, put this knowledge into operation and enable the exercise of power (p. 431). Through these mechanisms, interventions appear as rational, scientific solutions to objective problems, neglecting the underlying political conflicts and different cosmovisions. "Target groups" are reduced to one aspect of their livelihood (e.g. inability to read or access to land) and persons are turned into individual cases – poverty is thus linked more to their characteristics, not to structural factors. In order to gain benefits from the interaction with the donor (e. g. access to resources), people adjust to their categorisation: the discourse thus shapes subjectivities and practices. As the discourse of "development" describes material want as the central quality of the non-Western world, the objective of economic growth and the discipline of economics are awarded special status. This, however, neglects the fact that this discipline is, in Escobar's view, "above all a cultural production" (1995a: 59), based upon specific notions of rationality, market, production and labour which are historically contingent, such as the notion that we are rational maximisers of utility and it is rational to produce goods and services primarily for individual gain.

Beyond this Eurocentric and technocratic discourse, says Escobar, we need a different conception of non-Western societies which allow for a different politics and a "post-development era". He sees this conception in the works of writers like Gustavo Esteva, Ashis Nandy, Vandana Shiva, Majid Rahnema, Shiv Visvanathan and Orlando Fals Borda. They share an interest not in alternative development but in alternatives to development, rejecting the entire paradigm, an interest in local culture and knowledge, a critical stance towards established scientific discourses and the defence and promotion of localised, pluralistic grassroots movements (1995a: 212ff.).

Of course this begs the question what these alternatives to development look like. Escobar's second major English book – *Territories of Difference: Place, Movements, Life, Redes* (2008) – could be seen as a lengthy answer to this question. Here, he explores how place, capital, nature, development, identity and networks are conceived of by the Afro-Colombian activists of the Proceso de Comunidades Negras (PCN, Process of Black Communities) in Colombia. He shows that, despite making use of some projects supporting organising processes and food autonomy, for them, "development" is not compatible with their cultural values concerning the use of surplus for festivities, reciprocity in the community and the relation to nature. Consequently, in the face of the current economic crises, "overcoming the model of modern liberal capitalist society has become a must for survival" (Escobar 2008: 303) to the PCN. Thus alternatives to "development" are also alternatives to capitalist modernity which is inextricably intertwined with colonial and patriarchal ways of perceiving the world. This means dispensing with the idea of a single modernity according to Western standards, and "provincializing Europe" in the sense of displacing it from the centre of epistemic and historical imagination (p. 305): there are numerous ways of imagining progress or positive social change or a good life beyond the lineal scale of "development" which provides only one particular Euro-centric way of doing so. It also means abandoning the idea of a "radical discontinuity between humans and other modes of being and the utilitarian attitude toward the earth" (p. 308).

This shift towards political ecology and a different relation to nature is reflected in his more recent works as well. In the Preface to the new edition of *Encountering Development* (2012), Escobar maps the historical transformations (the rise of China, September 11, and the end of the Washington Consensus), assesses Post-Development theory (which paved the way for more nuanced critiques of "Development" theory and practice), but above all sketches the emergence of "transition discourses" (p. xix). These discourses share ecological (decentralised organic food and energy systems) as well as epistemological concerns (place-based, relational and non-universal knowledge); the Latin American debate about *Buen Vivir* is but one example. What they also have in common is an orientation towards the pluriverse, which is in line not only with the World Social Forum's postulate that "Another world is possible" but also with the desire of the Zapatistas in Chiapas to create "a world where many worlds fit" (Escobar 2018: xvi, 15ff.) – contrary to the "Development" universe, where there is only one scale to fit all, be it the GDP or the HDI. *Pluriverse* is therefore also the title of the *Post-Development Dictionary*, edited by Escobar and others (Kothari et al. 2018), which outlines the myriads of alternatives to the present models of the economy, politics and knowledge. This transition to the pluriverse involves a turn away from patriarchal values such as competition, domination, appropriation of resources and a rational justification of all that in the name of truth, and towards inclusion, participation, collaboration, respect, sacredness and the renovation of life (p. 13), requiring a reconstruction of civilisation based on empathy and love. Thus Escobar has progressed from the Foucauldian critique of power in "development" to a spiritual call for a more holistic world view in order to overcome not only capitalism but colonial and patriarchal modernity altogether.

Major works

Alvarez, S., Dagnino, E. and Escobar, A. (eds) (1998) *Culture of Politics, Politics of Culture: Revisioning Latin American Social Movements*, Boulder, CO: Westview Press.

Escobar, A. (1985) "Michel Foucault and the Relevance of His Work to the Third World", *Alternatives* 10: 377–400.

—— (1988) "Power and Visibility: Development and the Invention and Management of the Third World", *Cultural Anthropology* 3(4): 428–443.

—— (1991) "Anthropology and the Development Encounter: The Making and Marketing of Development Anthropology", *American Ethnologist* 18(4): 658–682.

—— (1992) "Planning", in W. Sachs (ed.), *The Development Dictionary*, London: Zed Books, 132–145.

—— (1995a) *Encountering Development: The Making and Unmaking of the Third World*, Princeton, NJ: Princeton University Press.

—— (1995b) "Imagining a Post-Development Era", in J. Crush (ed.), *Power of Development*, London: Routledge, 211–227.

—— (1996) "Constructing Nature: Elements for a Poststructural Political Ecology", in R. Peet and M. Watts (eds), *Liberation Ecologies: Environment, Development, Social Movements*, London: Routledge, 46–68.

—— (2000) "Beyond the Search for a Paradigm: Post-Development and Beyond", *Development* 43(4): 11–14.

—— (2004) "Beyond the Third World: Imperial Globality, Global Coloniality, and Anti-Globalisation Social Movements", *Third World Quarterly* 25(1): 207–230.

—— (2007) "Post-development as Concept and Social Practice", in A. Ziai (ed.), *Exploring Post-Development*, London: Routledge, 18–31.

—— (2008) *Territories of Difference: Place, Movements, Life, Redes*, Durham, NC: Duke University Press.

—— (2012) *Encountering Development: The Making and Unmaking of the Third World*, 2nd edn, Princeton, NJ: Princeton University Press.

—— (2018) *Designs for the Pluriverse: Radical Interdependence, Autonomy, and the Making of Worlds*, Durham, NC: Duke University Press.

—— and Alvarez, S. (eds) (1992) *The Making of Social Movements in Latin America: Identity, Strategy and Democracy*, Boulder, CO: Westview Press.

Esteva, G. and Escobar, A. (2017) "Post-Development @ 25: On "Being Stuck" and Moving Forward, Sideways, Backward and Otherwise", *Third World Quarterly* 38(12): 2559–2572.

Harcourt, W. and Escobar, A. (eds) (2005) *Women and the Politics of Place*, Bloomfield, CT: Kumarian Press.

Kothari, A., Salleh, A., Escobar, A., Demaria, F. and Alberto Acosta, A. (eds) (2018) *Pluriverse: A Post-Development Dictionary*, Delhi: Authors Up Front and Tulika.

Mignolo, W. and Escobar, A. (2010) *Globalization and the Decolonial Option*, Abingdon: Routledge.
Sen, J., Anand, A., Escobar, A. and Waterman, P. (eds) (2004) *The World Social Forum: Challenging Empires*, Delhi: Viveka.

Further reading

Batterbury, S. and Fernando, J.L. (2004) "Arturo Escobar", in P. Hubbard et al. (eds), *Key Thinkers on Space and Place*, London: Sage, 113–120.
Lehmann, D. (1997) "An Opportunity Lost: Escobar's Deconstruction of Development", *Journal of Development* 33(4): 568–578.
Sachs, W. (ed.) (1992) *The Development Dictionary: A Guide to Knowledge as Power*, London: Zed Books.
Rahnema, M. and Bawtree, V. (eds) (1997) *The Post-Development Reader*, London: Zed Books.
Ziai, A. (ed.) (2007) *Exploring Post-Development: Theory and Practice, Problems and Perspectives*, London: Routledge.
—— (ed.) (2018) *The Development Dictionary @25: Post-Development and its Consequences*, London: Routledge.

ARAM ZIAI

FRANTZ FANON (1925–1961)

Of all of the Third World public intellectuals, activists and postcolonial leaders who strode onto the world stage during the rise of the Afro-Asian Conferences of the 1950s and 1960s, it is Frantz Fanon and his writing on race, nationalism, violence and liberation who seems to have left the most compelling legacy of the "Bandung generation". Twenty-nine countries participated at the first Bandung Conference in 1955, representing nearly one-quarter of the Earth's land surface and a total population of 1.5 billion people including leaders such as Zhou Enlai, Jawaharlal Nehru and Gamal Abdul Nasser. Fanon's radicalism challenged the comfortable consensus of Banding built around nationalist aspirations and non-alignment. Edward Said (1993: 328) noted that Fanon was something of a prophetic genius, sensing in his foundational text, *The Wretched of the Earth* (1965b), a radical divide between the nationalist bourgeoisie in Algeria and the Front de Libération Nationale's (FLN's) liberationist tendencies, which stood awkwardly with respect to much of the nationalist and non-alignment rhetoric of Bandung. Fanon's standing as a theorist of anti-colonial struggle, of the psychology of racial oppression, of the emancipatory possibilities in the revolutionary violent confrontation of imperial oppression, and not least of the threats imposed by nascent post-colonial elites is simply without peer.

Fanon was the first major theorist of anti-imperialism to appreciate that nationalism followed the same track as imperialism. Conceding power to an emergent nationalist bourgeoisie, it was in fact really extending the hegemony of colonial subjection. As Fanon put it bluntly: "History teaches clearly that the battle against colonialism does not run straight away along the lines of nationalism" (Fanon 1965b: 119). The pitfalls of national consciousness and its complicity with bourgeois interests were profound; a national bourgeoisie might dispense with popular legitimacy, turning its back more and more on the oppressed classes, on true political emancipation, and on the realities of uneven development. Instead, argued Fanon, bourgeois nationalism resembled crony capitalism, endemic pillaging of the public purse, one-party states, tribalism and exporting enormous profits to foreign countries in the name of development. The task, said Fanon, is to move – and to move quickly – from national consciousness to political and social consciousness. The challenge was to capture the emancipatory and revolutionary potential of decolonisation:

> Decolonisation is quite simply the replacing of a certain "species" of men by another "species" of men. [W]e have chosen to speak of that kind of tabula rasa which characterizes at the outset all decolonisation. Its unusual importance is that it constitutes, from the very first day, the minimum demands of the colonised. To tell the truth, the proof of success lies in a whole social structure being changed from the bottom up.
>
> (1965b: 35)

In his understanding, drawing from the "new nations" he witnessed arising from anti-colonial struggle, a sterile and unproductive bourgeoisie collectively constituted a comprador elite who represented a radical threat to the sovereignty of the nation. As Fanon (1965b: 141) eloquently put it, the elites are "incapable of great ideas and inventiveness and do not even succeed in extracting spectacular concessions from the West, such as investments which would be of value for the country's economy". A champion of the masses and of revolutionary change, Fanon's merciless and ferocious assault on postcolonial elites – the lethargy of parties and masses alike, the rise of ethnic chauvinism and tribal dictatorship, the weakness of "under-developed" middle classes – proved to be prescient, if not prophetic, particularly for Africa and the

Arab world which concerned him so deeply. What was required, said Fanon (1965b: 188–189), was an attentiveness to a "national culture" of the masses – "it is around the people's struggles that African-Negro culture takes on substance and not around songs, poems or folklore" – and its harnessing to revolutionary impulses, a transformative process that would create "new souls". Fanon closes his 1959 book *L'An V de la Révolution Algérienne* (*A Dying Colonialism*, 1965a), penned during his time in Algeria during the anti-colonial struggles, with the words: "The revolution in depth, the true one, precisely because it changes man and renews society, has reached an advanced stage. This oxygen which creates and shapes a new humanity – this, too, is the Algerian revolution" (Fanon 1965a: 181).

If Fanon's thought and practice emerged as a sort of counterweight to the relative safety of the Bandung consensus during the 1960s and 1970s, his influence travelled far and wide, quickly being drawn into the discourses of revolutionary struggle worldwide, influencing black militancy in the US while entering (one might even say colonising) the Anglo-American academy since the mid-1980s, particularly in psychoanalytic and cultural theory (Kipfer 2007). A revived and repurposed Fanonism in academia was centred around postcolonial theory in general and in Homi Bhabha's (1994) influential rereading of Fanon in particular. In the Anglo-American context, Fanon's elevated fashionable status reflected "the convergence of the problematic of colonialism with that of subject-formation. As a psychoanalyst of culture, as a champion of the Damned of the Earth, he is an almost irresistible figure for a criticism that sees itself as both oppositional and postmodern" (Gates 1999: 252).

Ironically, at the same time that Fanon was hailed as a key precursor to postcolonial theory and forms of literary deconstruction within the English-speaking world, he was rendered almost invisible in the Francophone world, a silencing propelled by debates within and around structural Marxism, his putative promotion of violence and what Kipfer (2007) refers to as a general amnesia about France's colonial past. Fanon's central position within the postcolonial canon has been contentious at many levels in part because of what is sometimes seen as his overly Manichean view of race and liberation and his much-vaunted new humanity. For some, a new "heterodox" wave of Fanon scholarship has carefully reconstructed his complex engagements of arguments for and against Negritude, his critical reworking of Sartre's view of anti-Semitism and his engagement with a more bodily grounded notion of lived experience. Others have investigated his adaptation of left-wing social psychiatry to colonial situations and his purportedly proto-theoretical feminism; some have returned to his historical materialist and strategic aspects of his writings on decolonisation in Algeria and Africa, and to his sensitivity to space and spatial analysis (see Kipfer 2007; Verges 1996; Macey 2000). What emerges from these re-interpretations is not only a lost radical Fanon compatible with certain Hegelian, Marxist, phenomenological and radical Black modernist traditions (Sekyi-Out 1996; Gibson 2003) but, as Kipfer (2007) says, also "a novel set of interpretative strategies that insist on the continuities between Fanon's early phenomenological analyses of everyday racism and his more explicitly political and historical writing on national liberation" (p. 706).

A product of the interwar Caribbean, Frantz Fanon was born in the French colony of Martinique on 20 July 1925 into a family that could plausibly qualify as part of the black bourgeoisie. His father, Casimir Fanon, was a customs inspector and his mother, Eléanore Médélice, owned a hardware store in downtown Fort-de-France, the capital of Martinique; both were the descendants of slaves deployed on the sugar plantations. If the conventional aspirations for members of Fanon's social class were assimilationist and an identification with French culture, Fanon himself, while schooled in both traditions, was radicalised by the philosophy of negritude and racial identity taught to him by Aimé Césaire, Martinique's other great critic of empire. In 1943, at the age of 17, Fanon left his home and sailed to the Caribbean island of Dominica; from there he travelled to France to join the Free French Forces who were resisting the Nazi forces of Germany. While fighting for the French, he spent time in the French colony of Algeria, and for his military service he was awarded the Croix de Guerre for bravery (Nicholls n.d.). As a result of his military service Fanon was able to secure support for education in psychiatry in Lyon. During his studies, he attended Maurice Merleau-Ponty's lectures in philosophy and read the works of Mauss, Heidegger, Lévi-Strauss, Marx, Hegel, Lenin and Trotsky, and in this sense came to be exposed to dialectical forms of thinking. He finally received his psychiatric licence through training with the famous Spanish humanist psychiatrist, Francois Tosquelles. During this period Fanon also completed two plays, *L'oeil se noie* and *Les Mains parallels*, and met and married Marie Josephe Duble, a young white French woman of Corsican and Gypsy heritage, with whom he had a son.

While studying at University in Lyons, he ran straight into a sort of crude French racism which seemed, in his view, to stand in sharp contrast to the complex, class-permeated distinctions of shades of lightness and darkness he was familiar with in the Caribbean. Drawing from these experiences, he was to draft an essay entitled "An Essay for the Disalienation of Blacks", which in turn laid the foundations for what would eventually become *Peau noire, masques blancs* (1952) (*Black Skin, White Masks*, 1967a). Fanon was never a Marxist in any simple sense despite his anti-imperialist writing and political practice, but his time in France exposed him to Marxist and existentialist ideas that would inform the radical departure from the assimilation-negritude dichotomy that *Black Skins, White Masks* sought to inform.

Fanon returned to the Caribbean following his studies but quickly left again and, after a period of residence in Paris, realised he wanted to serve as director of any psychiatric ward in the French-speaking world. Applying to numerous institutions, he was finally considered at the Blida-Joinville hospital in Algiers in Algeria. Shortly after his arrival in Algeria, the war of independence against France was unleashed by the FLN and brutally repressed by French armed forces. Working in a French regional hospital in and around the front lines of the struggle, Fanon was increasingly responsible for treating both the psychological distress of the soldiers and officers of the French army who carried out torture in order to suppress anti-colonial resistance, and the trauma suffered by the Algerian torture victims.

By 1956, Fanon's deepening radicalisation led him to reject in any fashion the French efforts to repress a decolonisation movement that commanded his political sympathies. As a result, he resigned his position at the hospital and began secretly helping the FLN, soon becoming a full-time organiser and writer for the movement. In early 1957, he was exiled to Tunisia by the French colonial government, a move which escalated his interest in political activism; as a consequence he emerged as a mouthpiece and promoter for various African independence movements, and travelled around the continent preaching the gospel of anti-colonial struggle. Fanon established a magazine called *Moudjahid*, served as the ambassador for the Algerian rebel movement's provisional government, trained FLN members in techniques for resisting torture and combat techniques (which he had learned from his wartime and psychiatric experiences) and served as strategist during his visits to insurgent camps and training grounds. In Tunisia, he also contributed articles about the movement to sympathetic publications, including *Présence africaine* and Jean-Paul Sartre's journal *Les Temps modernes*. Some of Fanon's writings from this period were published posthumously in 1964 as *Pour la Révolution Africaine* (*Toward the African*

Revolution, 1967b). In 1959 Fanon published a series of essays, *L'An V de la Révolution Algérienne* (published in English as *A Dying Colonialism*, 1965a), which detail how the oppressed natives of Algeria organised themselves into a revolutionary fighting force.

While in Ghana as part of his ceaseless promotion of anti-colonial movements across the continent, Fanon was diagnosed with the leukaemia that would ultimately result in his untimely death. Notwithstanding his failing health, Fanon nevertheless spent ten months of his last year of life writing the book for which he would be most remembered, *Les Damnés de la terre* (1961) (*The Wretched of the Earth*, 1965b). In it, Fanon offers a devastating indictment of the violence and savagery of colonialism, a warning for the immediate dangers of the postcolonial project, and finally a passionate call for a new history of humanity to be initiated by a decolonised Third World. As his condition deteriorated, Fanon decided to leave the continent for the Soviet Union for treatment. Fanon was informed that the best care available was in the United States of America, and he travelled there using a false identity under the name of Ibrahim Fanon. Upon arrival, the Central Intelligence Agency (CIA) detained him for ten days without treatment. Fanon subsequently contracted pneumonia, which led to his death on 6 December 1961. Taken to Algeria, the country where Fanon obtained his citizenship, his body received a heroic welcome as a revolutionary *avant la lettre*. He was finally buried with military honours in an FLN veterans' graveyard.

There are three core theoretical and intellectual arenas in Fanon's *oeuvre*. The first is race, sometimes referred to as Fanon's contribution to phenomenology – that is to say, an analysis of the pre-conscious forces shaping the self that organise themselves around race as a foundational category (Silverman 2010). Drawing from his own life and work experiences, Fanon explores the existential challenges faced by black human beings in a white social world: the central metaphor of *Black Skin, White Masks* is of course masking – black people wear "white masks" to operate and survive in a white world. There is an echo here of W.E.B. Du Bois's double consciousness living under a white power structure. Fanon emphasises how race deforms the lives of both men and women in the French Caribbean, in France and in colonial conflicts in Africa, and how traumatic the confrontation is between the self and categorisation by others as inferior due to an imposed racial identity. For Fanon, "negrification" produces a phenomenological violence and trauma, promoting negative attitudes toward other blacks and Africa, and normalising attitudes of debasement; race offers an all-encompassing way of being in the world without alternatives. Fanon identifies, nevertheless, how the self might recuperate a sense of identity and a cultural affiliation independent of violent imperialist racism (Nicholls n. d). Only through an oppositional stance to, and independence of, racist white norms and values might black men and women shake off the psychological colonisation that a racist phenomenology imposes. While *Black Skins* is a compelling analysis, feminist critiques raise questions Fanon failed to ask about gender and gender oppression, not least within the FLN. For Fanon, exploitation is a process by which a person is transformed into an object; however, most concerned with the form of exploitation that arises out of colonialism, "he found a similar structure in anti-Semitism and racism, and other forms of exploitation: the key characteristic of exploitation is that it denies another's humanity" (Tronto 2004: 249).

Fanon came to see language as one of the most pervasive agents of racism, and argues that it is impossible to learn and speak this (colonial) language without subconsciously accepting the cultural meanings embedded, for example, in equations of purity as white and venality as black. *Black Skins* focuses on the potential colonising aspects of the French language, but in *A Dying Colonialism* he offers an alternative account in which a decision is made by the revolutionary forces in Algeria in 1956 to give up their previous boycott of French and instead start using it as the lingua franca that could unite diverse communities of resistance (Nicholls n.d). Adopting French was subversive, extending beyond the convenience of a common language to cast doubt on the belief of French colonialists that all French speakers in Algeria were loyal to the colonial government.

The political dimensions of negrification that call for decolonisation receive fuller treatment in *The Wretched of the Earth*. Here Fanon situates his diagnosis within an unambiguous ethical commitment to a sort of universalism, to the equal right of every human being to have his or her human dignity recognised by others. Everyone is entitled to moral consideration; the notion that no one is dispensable is the principled core of his decolonisation theory. *A Dying Colonialism* is Fanon's first-hand account of how the Algerian people mobilised themselves into a revolutionary fighting force and repelled the French colonial government, tracing the interplay of ideological and pragmatic choices including the creation of a nationalist radio station ("The Voice of Fighting Algeria") that challenged colonial propaganda. As Fanon noted, the provisional government also offered the radically inclusive notion that all people living in Algeria would be considered citizens of the new nation – a rejection of European imperialism on the model of Haiti's first constitution (1805), which dissolved hierarchies of social privilege predicated on skin colour by declaring that all Haitian citizens would be considered black.

The second and third themes explored in *The Wretched of the Earth* are violence, and revolution and the decolonisation project, of which the book provides a more theoretically inclined analysis of colonialism and revolution. Sometimes described as a handbook for black revolution, the book ranges over the necessary role Fanon thinks violence must play in decolonisation struggles: "violence alone … makes it possible for the masses to understand social truths" (1965a: 118). He charts the false paths that decolonising nations take when "natives" permit a bourgeois elite class and the former colonisers to negotiate their freedom rather than mobilising the masses – and the revolutionary lumpenproletariat in particular – as a popular fighting force. Controversially, in the preface that Jean-Paul Sartre wrote for the book ("violence can heal the wound it has inflicted"), some have found fault with the broader prescription of *The Wretched of the Earth* for what they see as its endorsement of necessary and cleansing violent insurgency. But much of this sort of analysis misreads Fanon, and ignores his argument made in the chapter entitled "Concerning Violence" to the effect that violence is *the* fundamental element of colonisation and colonial oppression. The unequivocal choice for the wretched of the earth is to accept violence passively – absorbing and displacing it upon themselves or upon other members of the oppressed – or to reject it and return it to those who impose it. Fanon displays a consistent and compelling existentialist commitment to human praxis which means that decolonisation can only happen when the oppressed colonial subject refuses to occupy the position of violence-absorbing passive victim. At this point Fanon said the Third World was

> starting a new history of Man, a history which will have regard to the sometimes prodigious theses which Europe has put forward, but which will also not forget Europe's crimes, of which the most horrible was committed in the heart of man, and consisted in the pathological tearing apart of his functions and the crumbling away of his unity.
>
> (1965b: 31)

Fanonite thought was subject to a multiplicity of influences. He was introduced to the philosophy of negritude by Aimé Césaire, but more generative were his studies in France of Hegel, Marx and Husserl. In particular, his use and development of dialectical analysis offered the prospect of how the alienated self could respond to racist trauma in a positive fashion, and an understanding of the emancipatory politics and challenges of independence. Fanon offered an account of the central challenge of the immediate postwar period: decolonisation. Only when all human beings have dignity, he asserted, will we have ended exploitation and be ready to move forward. As Tronto (2004) put it, describing his worldwide significance, it was not reparations, not recriminations, but action that was imperative. As Fanon himself put it, the task was to "educate man to be actional, preserving in all his relations his respect for the basic values that constitute a human world, is the prime task of him who, have taken thought, prepares to act" (1967a: 222).

Major works

Fanon, F. (1952) *Peau noire, masques blancs*. Paris: Editions du Seuil.
—— (1959) *L"An V de la Révolution Algérienne*, Paris: François Maspero.
—— (1961) *Les Damnés de la terre*. Paris: François Maspero.
—— (1964) *Pour la Révolution Africaine*, Paris: François Maspero.
—— (1965a) *A Dying Colonialism*, trans. Haakon Chevalier from *L"An V de la Révolution Algérienne*, New York: Grove Press.
—— (1965b) *The Wretched of the Earth*, trans. Constance Farrington from *Les Damnés de la terre*, New York: Grove Press.
—— (1967a) *Black Skin, White Masks*, trans. Charles Lam Markmann from *Peau noire, masques blancs*, New York: Grove Press
—— (1967b) *Toward the African Revolution*, trans. Haakon Chevalier from *Pour la Révolution Africaine*, New York: Grove Press.

Other references cited

Bhabha, H. (1994) *The Location of Culture*, Abingdon: Routledge.
Cherki, A. (2006) *Frantz Fanon: A Portrait*, trans. Nadia Benabid, Ithaca, NY: Cornell University Press.
Gates, H.L. Jr (1999) "Critical Fanonism", in N.C. Gibson (ed.) *Rethinking Fanon: The Continuing Dialogue*, Amherst, MA: Humanity Books.
Gibson, N. (2003) *Fanon: The Postcolonial Imagination*, Cambridge: Polity.
Gordon, L., Sharpley-Whiting, T.D. and White, R.T. (eds) (1996) *Fanon: A Critical Reader*, Oxford: Blackwell.
Kipfer, S. (2007) "Fanon and Space: Colonization, Urbanization, and Liberation from the Colonial to the Global City", *Environment and Planning D: Society and Space 25*(4): 701–726.
Macey, D. (2000) *Frantz Fanon: A Life*, Oxford: Granta.

Nicholls, T. (n.d.) Frantz Fanon, *Internet Encyclopedia of Philosophy*. www.iep.utm.edu/fanon/, accessed 1 September 2018.

Said E. (1993) *Culture and Imperialism*, New York: Vintage.

Sekyi-Out, A. (1996) *Fanon's Dialectic of Experience*, Cambridge, MA: Harvard University Press.

Silverman, M. (2010) *Frantz Fanon's* Black Skin, White Masks*: New Interdisciplinary Essays*, Manchester: Manchester University Press.

Tronto, J. (2004) "Frantz Fanon", *Contemporary Political Theory* 3: 245–252.

Verges, F. (1996) "To Cure and to Free: The Fanonian Project of 'Decolonized Psychiatry'", in L. Gordon, T.D. Sharpley-Whiting and R.T. White (eds), *Fanon: A Critical Reader*, Oxford: Blackwell, 85–99.

Secondary work

Adam, H.M. (1993) "Frantz Fanon as a Democratic Theorist", *African Affairs* 92 (369): 499–518.

Etherington, B. (2016) "An Answer to the Question: What Is Decolonization? Frantz Fanon's *The Wretched of the Earth* and Jean-Paul Sartre's *Critique of Dialectical Reason*", *Modern Intellectual History* 13(1): 151–178.

Majumdar, M. (2016) "*The Wretched of the Earth*: Then and Now", *International Journal of Francophone Studies* 19(1): 95–103.

Mbembe, A. (2012) "Metamorphic Thought: The Works of Frantz Fanon", *African Studies* 71(1): 19–28.

Nagy-Zekmi, S. (2007) "Frantz Fanon in New Light: Recycling in Postcolonial Theory", *Journal of Caribbean Literatures* 4(3): 129–139.

Salem, S. (2018) "Reading Egypt's Postcolonial State Through Frantz Fanon: Hegemony, Dependency and Development", *Interventions: International Journal of Postcolonial Studies* 20(3): 428–445.

Trojan, C. (2016) "Revolution as Restoration or Foundation? Frantz Fanon's Politics of World Building", *Contemporary Political Theory* 15: 399–416.

MICHAEL J. WATTS

FEI XIAOTONG (1910–2005)

Fei Xiaotong was born in southern Jiangsu Province on 2 November 1910. As "scholar officials", Fei's family was typical of the Chinese landed gentry. His initial education was at Yanjing University in Beijing in Physical Anthropology (BSc), though he shifted to Cultural Anthropology at Tsinghua University for his MA. Tsinghua, established in 1911 by the United States, was the first such Western-style university in China, and had both US and European faculty. These connections allowed Fei Xiaotong to go abroad on a Boxer Indemnity Scholarship to study and receive his PhD from the London School of Economics in 1936 under the supervision of pioneering social anthropologist Bronisław Malinowski, who promoted the practical uses of anthropological research. Thus, Fei was one of the earliest "modern" anthropologists and sociologists in China. His PhD thesis was on peasant life in China and focused on rural industry along the Yangtze river near Shanghai in the early 20th century. Fei worked in academia before and after 1949, when the People's Republic of China was founded, and after the Cultural Revolution (1966–1976) he held both political and academic posts. He remained active in research and involved in national-level policy discussions until his death in 2005 at the age of 94.

Later in life, Fei was widely recognised, but earlier in his career he was under very intense critique. One of the key points of contention concerning his thought and contributions is on the question of whether or not China was a class-based society. Fei saw the gaps between rich and poor, and therefore focused on rural–urban connections to bring about development. While his family was of the landed gentry, he was well aware of the problems of gaping social inequality. This, however, did not protect him from critique for his class background, evolutionary approach, and idealistic conceptual framework. Yet ultimately, thanks to his strong belief that social science must propose pragmatic solutions to practical problems, and that scientists have a duty to make their research relevant, Fei triumphed.

Fei began his studies in China in the midst of a long period of civil conflict and competition between different political and economic forces. Well before he went to university, China was perceived internationally as a weak nation encountering the West. During his studies, China was in an era of social revolution: the republican revolution led by Dr Sun Yat-sen, seen as the "father of modern China", as well as the early industrial revolution which had greatest impact in major cities such as Shanghai, and spurred the creation of a new working class as well as new cities. This emerging working class was led by Chinese Marxists under Soviet influence through the Third International, an international communist organisation (1919–1943) that advocated world communism, and posited a workers' revolution in China. Initially, it co-operated with the bourgeois revolution promoted by Dr Sun Yat-sen, yet was later in competition and conflict with the movement led by Chiang Kai-shek. Interestingly, given the focus of these revolutions on the industrial working class, both revolutionary theory developed by **Mao** during the 1930–1940s, and the anthropological studies by Fei during the same period emphasised the significance of peasants for China's future, either for the revolution or for rural development. In the midst of this era of deep social ferment and rapid transformation—primarily in urban China—Fei carried out groundbreaking work on rural industrialisation and its various effects in transforming peasant life and the rural economy.

Fei's first book, *Peasant Life in China* (1939), is widely considered a milestone in the discipline of anthropology because it shifted the norms for anthropological studies of society and development. Fei's focus was on rural industry in China. Its three key contributions were:

1. Until then, anthropologists had typically looked at "primitive" societies, and yet China was not considered a primitive society, but rather a society with a long history and rich culture.
2. Anthropologists did not usually study their own societies, instead focusing on "the other".
3. Anthropologists did not focus on industry, and yet Fei's focus was on rural industry in Jiangsu, where he did his fieldwork.

Fei spent a year in the US in 1943–1944, by invitation from the US government, and was affiliated with the University of Chicago's Department of Anthropology and Sociology (Redfield n.d.). During his stay, he set out many of his key ideas for what would later be published in China as *From the Soil: The Foundations of Chinese Society* (1948a). The book was also in part based on the lectures he gave during the 1940s in Kunming, in China's southwest, at the "combined universities" (Tsinghua, Peking and Nankai Universities) that had moved operations there during this period. *From the Soil* was based on Fei's prior field studies in rural Guangxi, Yunnan, and his home villages in Jiangsu. It provided the first original interpretation of rural China to a Western anthropological audience. As Hamilton and Wang (1992: 4) state in their introduction to the English-language version published in 1992, Fei's was the "first and only effort to construct a non-Western theoretical foundation for a sociology of Chinese society", and "represents one of the few and certainly one of the most insightful efforts to build a sociology of a non-Western society". In *From the Soil*, Fei developed the concept of *chaxugeju*, an original framework for understanding societal change. Though difficult to translate (versions include: "differential mode of association", "gradual differential mode of relationship in community", and "differential pattern"), it poetically described the process of societal formation in China as resembling a stone thrown into a lake which causes ripples from the centre to the shores. It was not a class-based theory, but one of regional diffusion, and of social stratification based on cultural distance from the centre. Nor was it a linear or binary concept of change, but rather a gradualist, incremental and interactive view of change through all societal institutions – from family to village to larger social institutions of religion and the state. Thus, his idea was not rural versus urban, but the complex interactions and ties between them.[1]

During China's revolutionary era of competition between the Communists and Nationalists (1927–1949), Fei ran into trouble. While he was not a Communist revolutionary, he joined the China Democratic League, one of the main political parties in coalition with the Communist Party from 1945 to 1949. The Nationalists' assassination of two leading intellectuals in 1946 – Li Gongpu and Wen Yiduo – which he strongly condemned in public – identified him with the anti-Nationalist camp. He escaped to England via the US Consulate in Kunming, which allowed him to evade political persecution and assassination by the Nationalists. He returned to China in 1947, joining the Chinese Peasants' and Workers' Democratic Party in coalition with the Communist Party of China from 1947 to 1949. During this period, Margaret Park Redfield (wife of Robert Redfield, Chair of Anthropology and Sociology and former Dean of Social Sciences at the University of Chicago) worked with him on the translation of his papers while the Redfields were at Tsinghua University in Beijing in 1948. This effort resulted in his next major contribution in English to Western development discourse, and the first analysis of the gentry in China, entitled *China's Gentry: Essays in Rural–Urban Relations*, published in 1953 by the University of Chicago Press. He also established the first Department of Rural Sociology in Kunming, the centre of scholarship in China at the time. This laid the basis for the later founding of the Sociology Department at Nankai University in the early 1980s.

In the early 1950s, Fei was Professor and Vice-President of the National Minorities University, and a member of the Chinese Academy of Sciences. He led the study of minority groups in China organised by the National People's Congress. From 1950 to 1956 he undertook the official identification and categorisation of 56 ethnic peoples that continues to be utilised in China today. As a leader of this classification, he was not only a professor but was also a senior official of the National People's Congress.

In 1956, Mao launched the movement "Let a Hundred Flowers Bloom" to encourage relatively free political expression, particularly amongst artists and intellectuals. In response to this opening, Fei visited different universities encouraging scholars to speak openly, and wrote a piece for the national newspaper, *People's Daily*, "Early Spring for Intellectuals". In 1957, he also organised a small roundtable of professors at his home, later referred to as the "six-professor group". In the late summer of 1957, despite his role as an official, Fei and the other five scholars were judged as having gone too far in challenging the system. This was the beginning of the Anti-Rightist Movement (1957–1959), which targeted critics, some of whom were severely persecuted, and re-imposed a more limited context for dissent.

Over a period of twenty years from 1957 to 1977, leading up to and during the Cultural Revolution, Fei did not publish any original scholarly articles or books. As a "rightist", he did not undertake field studies in the countryside for many years, but was comparatively fortunate as he was not sent to do manual labour in the countryside. Fei returned to the University of National Minorities as a Professor of Ethnic Studies in 1972. His primary work at the time was translating Western histories into Chinese. In 1979 he became president of the Chinese Sociology Society. In 1980 he was awarded the Malinowski Award by the Society for Applied Anthropology, and became a member of the Society. In 1982 he was named a Professor of Sociology at Peking University.

Fei's early scholarly publications had impacts both in the West and in China. During the 1930s and 1940s there were two peasant movements in China. One was the revolutionary peasant movement led by **Mao Zedong**. The other was an evolutionary peasant reform movement led by James Yen (Yan Yangchu) and others, focused on building rural community, for which Fei's research served as a foundation. Along with the ideas of Yen and Liang Suming (who debated with Mao in the caves in Yannan over the issue of class society), this non-class, non-conflict-based movement developed experimental villages in rural China in the 1930s and 1940s. Fei provided scholarly contributions to these experiments in rural reform. From an empirical basis, his focus on rural industry was novel. From a theoretical position, his contribution was the gradual transformation of society brought about without major conflict, using the above-mentioned concept of *chaxugeju*.

From 1978 onwards, following Deng Xiaoping's launching of reforms, Fei (as a former rightist now "rehabilitated") was asked to re-establish Anthropology and Sociology in China, (disciplines abolished in 1952 under Russian influence); he was the first Director of the Sociology Institute in the Chinese Academy of Social Sciences (CASS), and the first chair of the Department of Sociology at Nankai University. Fei once again took up field research. His next major work was *Small Towns in China* (1986), written with a small group of graduate students. The major contribution of this work was its argument for the importance of developing small towns and rural industries (called Township and Village Enterprises, or TVEs) as part of China's rapid rural reforms. The focus on small towns and TVEs was recognised as a major contributor to rapid rural growth, subsequently becoming national development policy.

The three most important steps of Deng's post-1978 reforms were arguably not top-down dictates, but rather the initiatives of villagers in different parts of the country. These included re-allocation of the land of the former collectives to individual households, development of TVEs, and allowing rural to urban migration. Fei was one of the first scholars to propose these as research topics and analyse the impacts and possibilities, grounded in his much earlier rural industry studies. He argued that China could not develop just with large cities, on a Western model, but must develop small towns and townships (out of the former communes) as a means to encourage widespread industrial development in rural China. This would be less expensive, consume fewer resources and cause less pollution than urban-based industry, as well as being easier to manage for both the state and communities.

In the 1990s, he developed the idea of "Go West" based on his study of national minority ethnic peoples in southwest and northwest China. He argued that the state should shift its development focus from the eastern seaboard to the inland western areas of the country. Thus, the national policies of small towns, rural industrialisation and Go West came together to help overcome the geographic and economic inequalities of the country, all influenced by Fei's theories of social and economic development.

As a professional social anthropologist, Fei is thus unique in his major contribution to national development policy formation and implementation. Policies developed based on his recommendations were implemented over a 20-year period. The current much-discussed Belt and Road Initiative of infrastructure-building proposed by the Chinese state, which extends from China to Africa and through all of Asia, is in many ways an extension of Fei's ideas on cohesion and harmony through geographical integration.

Fei's final contribution was his effort to lay out China's pathway for development, and how the world could similarly develop to avoid clashes and conflicts amongst diverse peoples. Fei did not agree with Samuel Huntington's "clashes of civilisations". Instead, he proposed the idea of *duoyuan yiti* ("diversity within integration"), arguing that a multiplicity of cultures and approaches to daily life can exist within a commonly accepted whole, and be integrated into a national development strategy without oppressing or destroying the unique aspects of diverse communities. The logic of this idea now forms the basis of an important thread in the approach of China's current leaders in national and foreign affairs, and thus Fei's star has risen again. Today he is widely quoted by China's leadership in explaining China's national and global strategy: the concept of China pursuing integration on a path that is inclusive, mutually beneficial and peaceful. Fei's major impact on contemporary China's development pathway and policies continues to reverberate in the ongoing debates over the relevance of China's path in global discourse on development.

Note

1 This is what Huang Ping has referred to as "Rurbanization" (1996).

Major works

Fei, X. (1939) *Peasant Life in China: A Field Study of Country Life in the Yangtze Valley*, Preface by Bronislaw Malinowski, London: Routledge and New York: Dutton. (Various reprints and a Japanese translation.)

—— (1946) *Neidi de nongcun* 《内地的農村》 (*Villages of the Interior)*, Shanghai: Shenghuo.

—— (1948a) *From the Soil* (*Xiangtu Zhongguo*, 《鄉土中國), Shanghai: Guancha. (Translated in 1992 as *From the Soil: The Foundations of Chinese Society*, Berkeley: University of California Press.)

—— (1948b) *Xiangtu chongjian* 《鄉土重建 (*Rural Recovery)*, Shanghai: Guancha.

—— (1953) *China's Gentry: Essays in Rural–Urban Relations*, Chicago: University of Chicago Press.

—— (1986). *Small Towns in China: Functions, Problems and Prospects*, Beijing: New World Press.

—— (1999) 《费孝通文集》 (*Collected Works of Fei Xiaotong)*, 15 vols, Beijing: Qunyan chubanshe.

—— (2015) *Globalization and Cultural Self-Awareness*, trans. Lynette Shi, Heidelberg: Springer.

—— and Chang Chih-I (1945) *Zhang Zhiyi* 张之毅], *Earthbound China: A Study of Rural Economy in Yunnan*, Chicago: University of Chicago Press.

—— and Luo, H.X. (1998). *Rural Economy Comparison Model*, Chongqing: Chongqing Press (in Chinese).

Further reading

Arkush, R.D. (1981) *Fei Xiaotong and Sociology in Revolutionary China*, Harvard East Asian Monographs vol. 98, Cambridge, MA: Harvard University Asia Center, DOI: 10.2307/j.ctt1tfjcs4, https://www.jstor.org/stable/j.ctt1tfjcs4.

Bruckermann, C., and Feuchtwang, S. (2016) *The Anthropology of China: China as Ethnographic and Theoretical Critique*, Singapore: World Scientific: https://doi.org/10.1142/p1089.

Feuchtwang, S (2015) "Social Egoism and Individualism Surprises and Questions for a Western Anthropologist of China: Reading Professor Fei Xiaotong's Contrast between China and the West", *Journal of China in Comparative Perspective* 《中国比较研究》 1(1): 128–145.

Hamilton, G. and Wang, Z. (1992) Introduction, in *From the Soil: The Foundations of Chinese Society*, Berkeley: University of California Press.

Huang, P. (1996) *Experiences from Western China*, (2006) Beijing: Social Science Document Press.

Huang P. (1996) *Seeking Survival: A Study of Rural Urban Migration in Southern Jiangsu*, Kunming: Yunnan People's Press.

Huang P. and Wang X. (eds) (2011) *Reconstructing Community in Rural China*, Beijing: Social Science Document Press.

Redfield, R. (n.d.) Papers, Special Collections Research Center, University of Chicago Library.

Sun, A. (2017) "Fei Xiaotong's Humanism Infuses 'From the Soil'", *Harvard Divinity Bulletin* 45(1,2).

Tan, T. (2010) "Socialism qua Civilisation Encounters the Confucian Tradition: Morality, Power and Social Structure in Bridge Village, Hunan Province, China", Working Paper no. 122, Halle: Max Planck Institute for Social Anthropology.

Zhao, X. 2015. "Toward a Chinese Phase of Anthropology" [迈向人类学的中国时代], *Journal of Social Sciences* [社会科学] (Shanghai) 4: 74–85.

HUANG PING, JOSHUA MULDAVIN AND XIAOJING LYNETTE SHI

ANDRE GUNDER FRANK (1929–2005)

One of the world's most prominent and productive radical political economists, Andre Gunder Frank left a massive corpus of work[1] – 44 books in 140 different editions, 169 chapters in 145 edited books covering 13 languages, and 400 articles translated into 20 languages. This makes him by some measures the most-cited economist in the world. Indisputably, he has left an indelible mark on the critical study of development and the world system. Frank led an extraordinarily peripatetic life, rarely living in one place for more than four years until he took up a professorship at the University of Amsterdam in 1983. He described his nomadic existence – as a refugee from Nazi Germany (see Simon 2019) to a teenage gardener in Santa Monica (California) and on, via Chicago, to Mexico, Brazil, Chile, Germany, the UK and Holland – in vivid terms:

> I was born in Berlin in 1929 and at the age of 4 I left there with my parents, who went to Switzerland as political exiles[2] when Hitler came to power in Germany in 1933. I would return to the place of my birth 40 years later, but by then as a political exile myself and my family from the military coup in Chile in 1973. In the meantime and indeed also since then, my usual stay in any one place around this world lasted one to a couple of years or less … After I arrived in Switzerland, I spent the next 8 years there, but going from each of its three principal language regions to another … In 1941, during the second year of World War II in Europe and in my 12th year of age, I left Europe and went to the United States, where I remained until the age of 31, when in 1961 I began my further Odyssey around the world. The only exceptions to my usual 1–2 year stays in one place were 4 years in the same college in the United States and 5 years each in Chile from 1968 to 1973 and then at from 1978 to 1983 in England – until from 1983 to 1993 I lived in Amsterdam for twice my previous 5 year maximum.
>
> (www.rrojasdatabank.info/agfrank/personal.html)

In this remarkable odyssey of scholarship, activism and politics, Frank brushed shoulders with everyone from Milton Friedman to Che Guevara. His life outside of academia was as colourful and varied as it was within. In Washington State Frank worked in a saw mill and then as a logger and gandy-dancer; in Michigan, he built automobiles at Willow Run; in New Orleans he offered his services as a private eye, as well as in the French Quarter tourist industry as a waiter on Bourbon Street; in San Francisco, he served in a removals company, as an usher at the San Francisco Opera House and, most implausibly, in a swanky Union Square store wrapping Christmas presents (see www.rrojasdatabank.info/agfrank/personal.html).

Frank's name – a story in itself, since he was named Andreas at birth but in his wanderings changed his name to Andrew and Andre, and added Gunder as a high-school nickname invoking a famous Swedish athlete! – will always be sutured to the body of radical scholarship that emerged from Latin America in the 1960s operating under the sign of dependency theory, but his most enduring contributions reside elsewhere (see below). A combative and iconoclastic intellectual, Frank's work has always been controversial and provocative, drawing as much fire as it has acclaim. It is a measure of his stature and influence that in a collection of essays published upon his retirement, his interlocutors included Immanuel Wallerstein, **Samir Amin** and **Eric Wolf**. His stature as a theorist of the world system – certainly the equal of Fernand Braudel – is irrefutable, yet at the same time he vastly expanded the circumference of world systemic thinking by pushing back the longue durée of the system by 5,000 years, and seeing its early dynamics in terms of the hegemony of Asia (Chew and Denemark 1996).

Frank's early exile laid the foundation for what became his multilingual fluency (he had control over seven languages). His nomadism was to some extent a product of his own restless, inquisitive nature; but in equal measure he was in a sort of permanent exile, at once too radical, too ornery and too unconventional for most universities on both sides of the Atlantic. His life was intimately shaped by the Cold War and the rise of Third World nationalism and anti-imperialist movements, and by his critical engagement with development policy and practice. His reach and influence surged during the 1960s and 1970s, when his ideas drew strength from, and indeed animated to a certain extent, the worldwide student radicalism of the period.[3]

In 1941 Frank moved with his parents to the USA, but as a teenager his life was equally mobile – California, Idaho, Michigan, New York. He finished school at Ann Arbor High in Michigan and entered Swarthmore College, where he became – by his own admission – a "Keynesian economist". It is one of the enduring paradoxes of radical social science that Frank was educated in Economics at the University of Chicago, where he studied with Milton Friedman. He was awarded his PhD in 1957 for a dissertation on Soviet agriculture in the Ukraine. Not surprisingly, his career at Chicago was unorthodox. Bert Hoselitz in Anthropology introduced him to development and modernisation theory, on which he was to unleash a ferocious critique directed at the likes of Talcott Parsons, Wilbert Moore, Everett Hagen, Daniel Lerner and David McClelland. **Walt Rostow**, whom he later met at MIT in 1958, was his *bête noire*. Frank published little of his dissertation during the period 1957 to 1962, when he taught at Michigan State University, Wayne State and the University of Iowa. He visited the Soviet Union in 1960 but, as he put it, "all official doors in Moscow and Kiev were closed to me". He later returned to the question of socialism, but this time to its vulnerabilities in relation to the world system and the unevenness of its capitalism's cyclical development. Frank was sometimes credited with having predicted the collapse of the Soviet Union in 1989.

In 1962 Frank resigned from his position at Michigan State, disillusioned both by the impress of the Cold War on academia and by the realisation that most social science research was "part of the problem". His radicalisation had begun earlier, largely through self-discovery and what he calls "years of wandering through the woods". His resignation marked the beginning of a worldwide trek across many countries, universities and disciplines. In 1962 he taught at the University of Brasilia (unable to teach at Leipzig or Havana as he intended), and thereafter successively at UNAM in Mexico City, George William University in Montreal and the International Labor Organisation (ILO) in Chile (from where he was fired). In 1968, at the instigation of Salvador Allende, he was appointed to the Faculties of Sociology and Economics in Santiago. Any return to the US was foreclosed by a 1965 decision to deny him entry on the grounds of his writing for *Monthly Review* and "his identification with the Chinese Communist position" as the Immigration and Naturalisation Department put it.

In Chile, Frank settled at the Center for Socio-Economic Studies (CESO) at the University of Chile, a hotbed of dependency theory and Marxism in Latin America. The Center attracted prominent Latin American exiles such as Theotonio dos Santos, Ruy Mauro Marini, Vania Bambirra, Toma Vasconi, Emir Sader and Marco Aurelio García, as well as Chilean scholars like Marta Harnecker, the liberation theologist Gonzalo Arroyo SJ, Jose Bengoa and many others. Marta Fuentes, his wife-to-be, was CESO's librarian. Prominent left-wing scholars and activists, such as Eric Hobsbawm, Alain Touraine, Paul Sweezy, **Fernando Henrique Cardoso** and **Aníbal Quijano** all passed through CESO. But with the military *coup d'état* which overthrew the socialist government of Allende on 11 September 1973, practically all staff, including Frank, were dismissed, most seeking asylum in embassies for fear of their lives (Kay 2005).

Frank's Chilean sojourn was foundational for his work on dependency theory, but it also marked the beginning of a new concern with the need for global resistance to the crisis of capitalist accumulation that, in his view, had begun in 1970. For three unhappy years he moved between Berlin, Munich and Frankfurt, before being appointed Professor of Development Studies at the University of East Anglia in 1978. From there he moved to Amsterdam in 1983. Over this period Frank claims to have applied unsuccessfully for 80 positions in the USA. Frank taught in at least 16 universities and across ten different disciplines. In 1994 he went into mandatory retirement from the Faculty of Economics at the University of Amsterdam. At the time of his death, he was a member of the Graduate Faculty in Sociology at the University of Toronto, but between 2000 and 2004 he held visiting appointments at the Universities of Miami, Nebraska, Northeastern and Calabria. For almost half a century – a stretch of time that included the death of his first wife, Marta Fuentes, in 1993, the divorce from his second, Nancy Howell, and four major medical operations – his productivity and intellectual vitality did not wane.

Frank's large body of work was all directly shaped by **Marx** (though he has rarely engaged with the larger body of modern Marxist theory from Luxemburg to Lenin to Trotsky to Gramsci to Horkheimer). The centrality of accumulation, and particularly accumulation on a world scale, encompasses the five major themes in his corpus. The first addresses the socialist bloc, beginning with his earliest work on socialist productivity in the Soviet Union and the question of political economic organisation in relation to economic flexibility and rigidity. He returned to the socialist economies in the 1970s to explore how they were being transformed by the world economic crisis outside the socialist bloc (Frank 1980; 1983). This led him to explore both the internal contradictions within eastern Europe generated by the world crisis and how the former socialist economies are being transformed under the influence of the IMF, and the consequences of the destruction of COMECON (Council for Mutual Economic Assistance) on the global division of labour.

A second theme was social and anti-system movements, a project initiated with Marta Fuentes and developed further in a collective project (1990) with **Samir Amin**, Immanuel Wallerstein and Giovanni Arrighi. The argument turned on the decline of Left parties and organisations and the flowering of social movements of all sorts which carried the hopes of social transformation without the capture of state power. These movements were placed on a larger canvas of world historical mobilisation, and confronted the uncomfortable fact that much of the flowering of civil society could be markedly "uncivil" (i.e. unprogressive).

The third theme – world system development and underdevelopment – began with his move to Latin America. He first deployed the term "world system and dependence" in the mid-1960s, and these ideas shaped his understanding of Third World underdevelopment. Using case studies of Mexico, Chile and Brazil, Frank laid out a historical account of the creation of complex dependent relations linking the metropole and the periphery in the wake of the 15th-century conquest. Rejecting the idea that Latin America was or had been feudal, Frank explored the processes of incorporation into the capitalist world market and the structures of dependence (from inter-state to intra-village) so created. A number of other key theorists – especially **Fernando Cardoso** – developed these ideas through detailed case studies; the ideas were picked up and developed in other ways by the Economic Commission for Latin America (ECLA) and UNCTAD (United Nations Conference on Trade and Development), and came to have enormous influence in academic and policy circles. Dependency emerged from both the Marxist and structuralist antecedents (Love 1990), especially the trade-based theorists like **Raúl Prebisch** at ECLA, but the Marxist-inspired work of Frank (1967; 1969) and Cardoso and Faletto (1969) provided its foundations. Frank's work, in particular, unleashed a firestorm of debate from both Right and Left. His work was a direct challenge to the Latin American Communist Parties but also, of course, to the modernisation theorists. Dependency was also attacked from within Marxian theoretical circles for its "circulationist" (i.e. exchange-based) view of capitalism, its functionalism and its failure to engage with the structural Marxism of Althusser and his colleagues in France (see Brenner 1977; Laclau 1971; Bernstein and Nicholas 1983). Whatever one thinks of 1960s dependency theory (and its promotion of autarchy), Frank had identified the existence of a system of world accumulation marked by the rupture of the 14th century, and linked what he called the development of underdevelopment to its rhythms, crises and conjunctures. Here he had independently arrived at a similar point to that of **Samir Amin** (1972) and Immanuel Wallerstein (1974), who were to lay out their respective visions of the world system a little later.

By the time dependency had reached the elevated levels of policy, Frank was already troubled and was moving toward a fourth interest, namely contemporary international political economy (1978). In the early 1970s, Frank predicted a worldwide economic crisis that would both integrate the socialist bloc into its capitalist orbit and impose export intensification on the Third World through force rather than consent; he also anticipated the "oil shock" as the harbinger of a series of successive and ever-deeper recessions within a world economic crisis (see Frank 1980; Amin et al. 1982; Frank 1983). Here much of his work addressed such questions as the regionalisation of the world economy and the consequences of Thatcherism and Reaganism for the 1979–1982 world recession, and successfully predicted some of the world-historical changes in the late 20th-century political economy.

Having spent more than two decades exploring (and forecasting) contemporary political economy on the world scale, Frank returned, in his fifth theme, to world history. His earlier concern had been with 500 years of capitalist world development, but this time he sought to recast this period by breaking irrevocably with its Eurocentrism, and by pushing backwards the origins of the world system 5,000 years. In so doing, he came to abandon completely the category of capitalism itself and break irrevocably from his old adversaries Wallerstein, **Amin** and Braudel. His book *ReOrient* (1998) starts from the pathbreaking work of Janet Abu-Lughod (1989) on the 13th-century world system, and of **Jim Blaut** (1992) on the flaws of Eurocentric history (all of Weberian and Marxist history is deemed to be Eurocentric by Frank), and endeavours to reinterpret the modern age and the rise of Europe. *ReOrient* makes two broad arguments. One demolishes the pillars of Eurocentrism by showing that Asia was outpacing Europe until the mid-1700s and its decline came about because of internal crisis, not the rise of Europe. The second turns on the conjunctural circumstances that permitted the rise of Europe around 1800. His major thesis is that "there was a single global world economy with a worldwide division of labor and multilateral trade from 1500 onward" (1998: 52). But the core of the system was Asia; it dominated the intense and tight trade linkages in such a way that the whole was not a "European world system". Inevitably this book has been received in highly polarised ways (see the responses to it by Wallerstein 1999, Amin 1999 and Arrighi 1999), but the broad lineaments of the argument are consistent with some of the more important and original work emerging from the study of China (Pomerantz 2000).

Across his vast body of work, Frank's relationship to Marxism was complex and unstable. On more than one occasion he asserted that he had never claimed to be a Marxist of any stripe, but clearly he was certainly much influenced by Marxist ideas. His denial was partly a response to those Marxist critics who faulted him for some weakness in what they interpreted as his Marxist analysis. Frank's first book (1967), which launched his reputation as a radical and heterodox thinker, is dedicated to Paul Baran and Paul Sweezy, probably the two most outstanding US Marxist economists and thinkers.

Andre Gunder Frank remained as productive as ever until shortly before his death from cancer. His return to world history launched his final project, which was a tripartite analysis of first what he called the "new world [dis]order", second the "reOrienting of the Nineteenth century" (explaining why the relative powers of the Orient and the West were inverted, building upon Pomerantz's *The Great Divergence* (2000)), and finally "Bronze Age World System Cycles", in which he attempted to chart global accumulation (as he understood it) into the fourth and fifth millennia BP. Frank ended his writing career surrounded by the same sort of debate, iconoclastic style and controversy that had marked his entry into the staid world of the academy four decades earlier.

Notes

1 His bibliography covering the period 1955–95 contains 880 publications in 27 languages; by 2003 the total figure exceeded 1,000 publications. The International Institute of Social History in Amsterdam received Frank's papers in 1995; they cover 35 metres of shelf space (the inventory is available online at: http://rrojasdatabank.info/afgfrank/IISG_Pubs.html). His work has been cited in journals on average over the period 1975–present about 200 times per annum.

2 His father was a novelist and pacifist.

3 All quotations by Frank, unless otherwise indicated, are taken from his "Cold War and Me" essay and his short political memoir published on his website.

Major works

A vast amount of work in many languages is available at Frank's website: http://rrojasdatabank.info/agfrank/personal.html.

Amin, S., Arrighi, G., Frank, I.G. and Wallerstein, I. (1982) *Dynamics of Global Crisis*, New York: Monthly Review Press.

—— —— —— (1990) *Transforming the Revolution*, New York: Monthly Review Press.

Frank, A.G. (1967) *Capitalism and Underdevelopment in Latin America*, New York: Monthly Review Press.

—— (1969) *Latin America: Underdevelopment or Revolution*, New York: Monthly Review Press.

—— (1978) *World Accumulation: 1492–1789*, New York: Monthly Review Press.

—— (1980) *Crisis in the World Economy*, New York: Holmes.

—— (1983) *The European Challenge*, Nottingham: Spokesman.

—— (1998) *ReOrient: Global Economy in the Asian Age*, Berkeley, CA: University of California Press.

Further reading

Abu-Lughod, J.L. (1989) *Before European Hegemony*, New York: Oxford University Press.

Amin, S. (1972) *Accumulation on a World Scale*, New York: Monthly Review Press.
—— (1999) "ReOrientalism? History Conceived as an Eternal Cycle", *Review: A Journal of the Fernand Braudel Centre* 22(3): 36.
Arrighi, G. (1999) "ReOrientalism? The World According to Andre Gunder Frank", *Review: A Journal of the Fernand Braudel Centre* 22(3): 28.
Bernstein, H. and Nicholas, H. (1983) "Pessimism of the Intellect, Pessimism of the Will: A Response to Andre Gunder Frank", *Development and Change* 14(4): 609–624.
Blaut, J.M. (1992) *1492: The Debate on Colonialism, Eurocentrism, and History*, Trenton, NJ: Africa World Press.
Brenner, R. (1977) "The Origins of Capitalist Development", *New Left Review* 104: 25–92.
Cardoso, H. and Faletto, E. (1969) *Dependencia y desarallo en America Latina*, Mexico: Siglo XXI Editores.
Chew, S.C. and Denemark, R.A. (1996) *The Underdevelopment of Development: Essays in Honor of Andre Gunder Frank*, Thousand Oaks, CA: Sage.
Kay, C. (2005) "Andre Gunder Frank", *Development and Change* 36(6): 1177–1183.
Laclau, E. (1971) "Feudalism and Capitalism in Latin America", *New Left Review* 67: 19–38.
Love, J.L. (1990) "The Origins of Dependency Analysis", *Journal of Latin American Studies* 22(1): 143–168.
Pomerantz, C. (2000) *The Great Divergence*, Princeton, NJ: Princeton University Press.
Simon, D. (2019) *Holocaust Escapees and Postwar Development: Hidden Histories*, London: Zed Books.
Wallerstein, I. (1974) *The Modern World System*, vol. 1, New York: Academic Press.
—— (1999) "ReOrientalism? Frank Proves the European Miracle", *Review* 22(3): 18.

MICHAEL WATTS

PAOLO FREIRE (1921–1997)

In his struggle for liberation of the poor and voiceless, Paolo Freire was something of a refugee searching for a state that would accept his basic thoughts. His life's journey started in Recife, north-eastern Brazil, where he was born on 19 September 1921. In spite of his middle-class background, the prevailing economic hardships made him suffer from hunger and poverty. This personal experience made him realise the close relationship between social class and knowledge. Freire started to wonder about the apparent indifference by the poor towards their obvious oppression. The Latin American people had basically submitted to what Freire called a culture of silence, quietly accepting the superiority of the elite over the people. To be able to strike back, the people had to regain their language.

As his family's economic conditions improved, Freire was eventually able to enter the University of Recife to study law, which he initially considered as useful in his struggle for the poor. However, at the time he was also influenced by philosophy and language. His own way of thinking developed through reading Catholic ideologies, as well as **Karl Marx** and French existentialists such as Sartre. Parallel to legal studies, his interest in educational matters grew, and after completing his degree he turned to social work and education. Within the field of education he was also greatly influenced by his wife since 1944, Elza Maia Costa de Oliveira, who was a school teacher.

Working on social welfare, he came into close contact with the plight of the urban poor. It was at this early stage that the basic concepts for his dialogue method in adult education started to take shape. For his subsequent work in education at the University of Recife, he was eventually awarded a PhD degree in 1959. From his base as a Professor at that university, he was able to initiate a very successful literacy programme that was diffused to the whole of Brazil. In the midst of a radical excitement, the Freirean foundation of building his pedagogy on resistance was in demand across a wide sector of the population. In today's terms, we could say that the literacy courses offered were an instrument for empowerment. The open defiance created was not desired by the new Brazilian regime then emerging from the military coup of 1964. To Freire, the coup was caused by inconsistencies within the Left in its claim for power that was not really available to them. That scared the Right into building a force that eventually led to this military takeover (Freire 1998) and mass arrests of opponents, including Freire. This brought to a sudden end his academic-activist career in Recife. After spending a period in prison, Freire embarked on his long journey by being sent into exile in Chile via a short spell in Bolivia.

The adult education programme in which he was involved at that time was recognised internationally by UNESCO, among others. After some five years in Chile, he was invited to the USA and Harvard University as a Visiting Professor. What he now experienced provided a new input to his thinking in education, and its part in the process of social change. It was in the USA that Freire realised that exclusion was not only part of what was called the Third World or global South, but was also very much part of life in the global North. Consequently, liberation was not only a struggle confined to the people of Latin

America, Africa and Asia, but affected the USA and Europe as well. It is with these new experiences in mind that Freire (1970) finalised his most influential book, *The Pedagogy of the Oppressed*. As with his other books, this one can be seen as building directly on his own practice.

In 1970 Freire continued his journey to carry his mission further, as he ended up at the World Council of Churches in Geneva. During this period he was given an opportunity to lecture and promote his concepts regarding education in Africa and Asia. Undoubtedly Freire's intellectual work proved a major encouragement in attempts to establish a new kind of educational thinking in many of these countries. However, many regimes found the practice of this education for liberation difficult to accept. Ivan Illich (1971), his close friend and collaborator, commented on this by saying that Geneva was an exile for Freire, while his friends and followers had been imprisoned or exposed to torture. It is difficult for any government to incorporate the Freire agenda, but it is a very powerful alternative for the poor in society to decide on their own future.

In 1979 Freire was invited back to his native Brazil to take up a position at the University of São Paolo. Some years later he was even appointed Minister of Education for São Paolo state, which gave him authority to direct school reforms in a major part of the country. He passed away in São Paolo on 2 May 1997, leaving a legacy of an educational theory and an even more powerful practice. To Freire, education could never be neutral; it was always to constitute part of a political agenda.

Freire's significance is that his work has provided essential inputs to the theoretical debate on education; but the practice he has generated is even more important. The protest against the culture of silence is summarised in a critique against "banking" methodology, a conventional custom in school. According to the Banking Model of Education, teachers have all the necessary knowledge and make deposits as to a bank. The students are the depositories that receive and memorise the information given to them passively. As in a bank, information stored in this way can be withdrawn as required at any time. This pedagogy is a reflection of the oppressive society against which Freire reacted at an early stage of life. Some of the obvious contradictions in this methodology were brought out by Freire:

- The teacher teaches – the students are taught.
- The teacher knows all – the students know nothing.
- The teacher thinks – the students are thought about.

- The teacher talks – the students listen.
- The teacher disciplines – the students are disciplined.
- The teacher chooses and enforces the choice – the students comply.
- The teacher acts – the students have an illusion of acting.
- The teacher chooses the programme content – the students adapt to it.
- The teacher is the subject – the pupils are mere objects (adapted from Freire 1970).

By this methodology, students store the knowledge that in actual fact oppresses them – making them accept things as they are. Part of the teaching by means of the banking method is to conceal certain facts that could demystify our entire existence. Therefore, they will not develop a critical consciousness that would make them able to intervene in world affairs. "Banking" education does not stimulate the critical thinking needed to challenge their oppressors.

Contrary to Illich, this did not lead Freire to the conclusion that the way out is *deschooling*, but rather to a way to change education. His proposed alternative is an educational model that is *problem-posing* that would replace the banking education. A key element in this is a changed relationship between the teacher and the student, in which they are involved in a dialectical *dialogue*. In this new pedagogy, teachers and students are turned into both teachers and learners at the same time, now all being subjects exchanging and recreating knowledge. In a dialogue, a question or a concept can be raised by the teacher, which in return will be answered by another question or idea. From this a synthesis will be found that, in itself, will pose a new question or thesis. Both teachers and students will learn and teach each other, as well as critically find a new way of thinking.

From a practical point of view, the pedagogy was built on the reality of the oppressed themselves, often illustrated by using a set of pictures to spark a discussion. This was combined with selecting words that could be divided into some commonly used syllables. From this, new words were constructed, gradually going from the simple to the more complex. It must be noted here that this was easy in the case of Portuguese and Spanish, but is difficult to adapt to other languages. Finally, the words were structured in such a way that it was possible to contextualise them in terms of the social and political realities. In Freire (1973), some graphic expressions of his practical approach to education are given.

The purpose of the dialogue education was to achieve a kind of political literacy – a process of *conscientisation*. With a critical pedagogy, educators and learners were to be involved jointly in a practice to demystify reality. When oppressed persons begin looking critically at the world through a joint dialogue, it constitutes a challenge to the culture of silence. The oppressed person will no longer be an object accepting the status quo of society, but becomes a subject reacting more actively by questioning surrounding social forces. Overcoming a state of dependency is initiated by realising that dependency exits.

Freire was a contemporary of some of the early *dependistas* in Latin America, and they were inspired by a similar intellectual tradition. In both cases, a starting point could be found in **Marx**ism or neo-Marxism. We can also see how Freire used the example of Che Guevara, as a revolutionary who practised a constant dialogue with the people to achieve an authentic strategy for liberation action (Freire 1998). On the other hand, he was highly critical of most existing socialist societies, except Cuba and China, because they were unable to go beyond bourgeois education to create a critical mass of the new breed of people.

In Latin America the closed society, based on a culture of silence, took shape with the conquest by the Spanish and Portuguese. Since that time, only the centre of decision-making has changed – Spain, Portugal, England and the USA. To Freire, the economic dependence was obvious, but he also focused on the educational system that serves the status quo. This position is not created in a specialised laboratory and brought down to Latin America. It is part of the whole relationship between the dominated in the Third World and the dominant forces of the affluent North. To aid understanding, it is necessary to analyse connections between the culture of silence and the culture that has a voice (Freire 1972).

Naturally, much of the practical influence of the Freire pedagogy has been concentrated in Latin America. Apart from his native Brazil, we can note that Chile under Allende's socialist government (1970–1973) was inspired by Freire's thinking. Another important example is the Sandinista period in Nicaragua (1979–1990), which signifies one of the sincere attempts to create a revolutionary transformation, not least with an educational system appropriate to that agenda. In Africa, his ideas were first taken up by some of the former Portuguese colonies, such as Angola and Guinea-Bissau, both of which had a similarly progressive approach to their educational practice at an early stage. The practical experiences in Guinea-Bissau have been elaborated on in Freire (1978).

One of the outstanding proponents of an alternative development philosophy for Africa was **Julius Nyerere**, in his capacity as President of Tanzania. Early on, the country was praised for its achievements in adult education, which were largely a product of the Freirean concept. Even if there were some drawbacks as Tanzania was forcefully brought into mainstream neoliberal global politics, there was a return to a new progressive literacy agenda in the late 1990s – the Integrated Community Based Adult Education (ICBAE). In addition, these kinds of programmes – the Freirean approach to education – have been adopted by various grassroots organisations and liberation movements. Among others, the Oromo Liberation Front (OLF) in Ethiopia has used this pedagogy to increase awareness among its cadres. In the global North, we find that Freire has influenced the Black Panther movement and feminist groups in the USA, among others. This illustrates the usefulness of these pedagogical concepts to the industrialised world as well, and to the oppressed people found in the midst of plenty.

Indeed, he was undoubtedly one of the leading critical thinkers on education during the second half of the 20th century. Today, his ideas live on in many contexts and are studied and propagated by the Freire Institute, an organisation for transformative community-based learning, at the University of Central Lancashire in the UK (www.freire.org). This is one of a network of similar institutes in various countries in Europe, the USA, South Africa and his native Brazil. They also organise a series of International Paulo Freire Conferences.

Major works

Freire, P. (1970) *Pedagogy of the Oppressed*, New York: Seabury.
— (1972) *Cultural Action for Freedom*, Harmondsworth: Penguin Books.
— (1973) *Education for Critical Consciousness*, New York: Seabury.
— (1978) *Pedagogy in Progress: The Letters to Guinea Bissau*, New York: Seabury.
— (1985) *The Politics of Education: Culture, Power and Liberation*, London: Macmillan.
— (1996) *Letters to Cristina: Reflections on My Life and Work*, London: Routledge.
— (1998) *Pedagogy of the Heart*, New York: Continuum.
— (2005) *Teachers as Cultural Workers: Letters to Those Who Dare Teach*, Boulder, CO: Westview Press. (Expanded edition, with new commentaries.)

Further reading

Illich, I. (1971) *Celebration of Awareness: A Call for Institutional Revolution*, London: Marion Boyars.

ANDERS NÄRMAN WITH DAVID SIMON

JOHN FRIEDMANN (1926–2017)

John Friedmann was a key development thinker notable both for the quality and quantity of his contributions and for a remarkable career in which he continued to provide insights on the intellectual frontiers of development until his death after a very short illness at the age of 91. Friedmann's line of original and formative thought over half a century was distinctive in its breadth, depth and originality. His work helped define the new field of "regional development", and in addition he both originated and facilitated key development concepts such as core–periphery, urban fields, agropolitan development, world cities, hyper-urbanisation, polarised development, empowerment and civil society.

Distinctively, he was always able to bring an interdisciplinary breadth and innovative perspective to topics at the forefront of development thinking. He understood power, space, sociology and economics. He continually derived innovative ways of combining these rich intellectual concepts to elucidate new and important meanings in our understanding of development. Wander through the administrative halls of today's academia and the word "interdisciplinary" brings forth positive affirmation in every corridor. It is noteworthy that Friedmann was trained in a unique interdisciplinary tradition over half a century ago and that much of his life's work honoured that tradition.

John Friedmann was born in Vienna, Austria, in 1926. His family fled the Nazis for the USA in 1940 and, after a somewhat peripatetic existence, during which he became a US citizen in 1944 and then served with the US military in Germany after the end of World War II (Simon 2019), he earned his MA and eventually PhD from the University of Chicago in 1955. His PhD was undertaken in "a short-lived Program for Education and Research in Planning" under the direction of Harvey Perloff (Friedmann 2002: 119). Uniquely, this programme was interdisciplinary and his PhD was conferred in the fields of "planning, economics and geography". Friedmann and his mentors were among the first scholars formally to break disciplinary boundaries. Normally, planners were trained within schools of architecture. Instead, Friedmann received significant training in the social sciences and began to think of planning as an applied social science. This approach resounded throughout his life's work.

In a personal retrospective (2002), Friedmann acknowledged several important early influences. His father, Robert Friedmann, was a historian and philosopher who introduced him to the works of American urbanists Reinhold Niebuhr and Lewis Mumford, and the political theorist Hannah Arendt, and encouraged him to wander through the intellectual minefields of the philosophy of science. He also engaged with the Jewish philosopher Martin Buber. "I inherited from this a strong sense of moral purpose, a philosophical disposition, and (ultimately) a strong sense of history that questioned the possibilities of reason as an active force in history" (p. 120). Linking the large concepts of society, power and space established Friedmann as a "new thinker" in development.

Friedmann's unique degree at the University of Chicago carried a fine pedigree – the new degree programme having been started by the former Governor of Puerto Rico, Rexford Tugwell. Tugwell brought in many scholars, including planner Harvey Perloff, who was to become Friedmann's Advisor and eventual colleague at the University of California at Los Angeles (UCLA). John Friedmann also worked with urban designer Melville Branch, economic planner Julius Margolis, technology specialist Richard Meier and housing expert Martin Meyerson during his graduate years. Friedmann was also a witness to the inception of neoliberalism (a firm conservative belief in the market system) by economists Milton Friedman and Theodore W. Schultz that led the University of Chicago's Department of Economics (and the country and the globe) on a political economic course that today dominates global policy.

"Regional development" was a very new concept at the time, and Friedmann's dissertation study of the Tennessee Valley Authority was a pioneer effort to indicate how decisions could be made in "regional" ways that included multiple states, counties and local authorities. Friedmann showed that "watershed" boundaries were not necessarily appropriate to a dynamic postwar economy. The effect on cities was a key, but so too was the persistence of rural poverty.

Following the award of his PhD, Friedmann spent six years in Brazil and Korea working with US development agencies. He taught, he wrote, he absorbed, he planned (alas, his plans did not gain support from the United States Agency for International Development – USAID). In 1960, the Massachusetts Institute of Technology (MIT) – a premier institution in the planning field – recruited Friedmann to its faculty. He quickly became engaged in a joint research programme in Venezuela focused on regional development.

In 1964, Friedmann left MIT to be the Director of the Ford Foundation programme in Chile. As for most planners, the trade-off between taking action and simply thinking is a true challenge, although they can be happily and appropriately interspersed. Friedmann accomplished this. During this period he identified the concept of "hyperurbanisation", which he defined as a doubling rate in the size of urban population in less than 20 years. He accurately noted that this process would lead to a new situation of political awareness that would substantially change the status quo. The political history of Chile proves his case.

In 1965, Friedmann and John Miller (his graduate student) published an article on "The Urban Field" that extended Jean Gottmann's (1961) analysis out of a single case and showed how urban areas across the United States were coalescing in a variety of important ways in economic and spatial dimensions. They predicted that 85–90 per cent of the American populace would soon be living in an "urban field".

The "core–periphery" concept has its strongest origin, especially at the regional scale, in Friedmann's (1966) work during this period. He showed clearly that core–periphery relationships worked strongly on the regional scale, and brought space, power and society into unequal relationships. This concept eventually became a critical component for world systems theory as formulated by Immanuel Wallerstein (1974). Friedmann developed the concept of nearly immutable power relationships that transcend scales between rich and poor and translate into economic/development relationships. Indeed, his work was somewhat scale-related to regional levels. Nonetheless, these relationships soon proved to be scale-transcending, and proved to be the basis of a large corpus of local–global theory on scale relationships.

In 1969 Friedmann accepted an academic position in planning at UCLA, bringing to an end his *Wanderjahre* (years of wandering), as he subsequently referred to the first major part of his career characterised by active development engagement in successive positions. He clearly helped build UCLA into one of the best planning programmes in the USA. Graduate students were mesmerised by the onslaught of knowledge he provided, both in his classes and in the programme. Somehow he managed to have the likes of David Harvey, **Samir Amin**, Janet Abu-Lughod, Manuel Castells and other famous figures visit for a whole week. They would give a colloquium, but graduate students would also meet with them after reading their works, and they would participate in scheduled seminars. Many of the graduate students interviewed admitted they felt they were at the centre of the intellectual universe. Through this process, Friedmann not only gave many students a special education but also provided a model for how best to teach graduate students.

During this period, John Friedmann engaged the popular concept of growth centres. His work on polarised development (1972) called into question the spatial dynamics and possible patterns of growth centre implementation, and harked back to his earlier arguments on the core–periphery concept. This work was further elucidated in *Urbanisation, Planning and National Development* (1973b), which addressed questions of the role of the nationstate in urban and regional planning.

A pragmatist is someone who has an approach to the world that asks two questions: (1) Is there a problem? and (2) Can we help solve it? As a probable pragmatist, Friedmann wrote *Retracking America* (1973a). In this book he introduced the "communicative turn" into planning and emphasised the possibilities of transactive planning. One real question is whether Friedmann was a simple pragmatist or a "radical" as he claims himself to be. We will return to this.

Friedmann was always engaged with urban questions that involve space, society and economy. In *Territory and Function* (Friedmann and Weaver 1979), the concept of "agropolitan" development acknowledged the rural–urban linkage that was of vital importance in the economics of the developing world, and spoke to the importance of the agricultural hinterland that supported a city's economic existence and viability. In this concept, Friedmann refocused planning on rural development as the engine that drives urban development in small- and medium-scale urban centres. Again, he re-emphasised the region as a critical factor in development.

As Friedmann himself noted, "'World City' was an idea waiting to be born" (2002: 145). As a major development concept, "World City" received an initial formalisation by Friedmann and Wolff (1982) and was further elaborated as Friedmann's (1986) "The World City Hypothesis". Simply, the "World City" exploded the nation-state-based urban hierarchy to reflect major city roles in a new globalising economy. This is a powerful extension of urban thought, reflected, for example, in a major evaluative and prospective conference on the hypothesis and its application in 1993, at which he was guest and keynote speaker (Friedmann 1995; Knox and Taylor 1995).

"Empowerment" (1992) encapsulated Friedmann's approach to development. As the world changed with the collapse of regimes in the Soviet Union and Eastern Europe, Friedmann argued for an alternative development theory. He explicitly asserted the human development link to economic development, asking important questions that once again link political power and economic development. Disempowerment is seen as coming from poverty. This disempowerment has social, political and psychological components. This is very much in line with **Amartya Sen**'s (1999) later work on *Development as Freedom* that provided a "morally informed framework" for embedding policies that are pro-poor in the structure of mainstream development planning. Friedmann's empowerment work benefits from a "creative tension" between mainstream and poverty initiatives that outlined viable futures for the poor of the world.

Friedmann even engaged the "sustainable city" initiative with his work on Mexico City (2000). This is just another instance in which he was not following but leading the debate on important urban and regional issues.

His final phase of research focused on the power of the household in development and the significance of livelihoods research, and also examines the increasing importance of civil society. *The Prospect of Cities* (2002) continues to raise important questions, while in Friedmann (2005) he surveyed the unprecedented urban transition in China.

Friedmann not only continued to make major contributions to the field of development throughout his life, but did so with a remarkable persistence, resilience and a wise person's sense of what is needed. In his later years and following formal retirement from UCLA in 1996, he wrote more autobiographically. This vein of work commenced with "A Life in Planning" in *The Prospect of Cities* (2002), and concluded with his final publication, "Planning as a Vocation: The Journey So Far" (Friedmann 2017), which appeared shortly before his death. Further details on his lifecourse, autobiographical reflections and secondary analysis appear in Simon (2019). Friedmann's book on philosophy, *The Good Society* (1979), should be required reading by all who have some allegiance to the pragmatic tradition.

John Friedmann made the "regional" important to development. His interdisciplinary focus provided numerous signal intellectual contributions that ranged across disciplines. He was clearly a key thinker in development, leading and provoking successive shifts in thinking and in the relationship between theory and practice; but he is best recognised and remembered for his role as an ethical and moral leader in development.

Major works

Friedmann, J. (1966) *Regional Development Policy: A Case Study of Venezuela*, Cambridge, MA: MIT Press.

—— (1972) "A General Theory of Polarised Development", in N.M. Hansen (ed.), *Growth Centers in Regional Economic Development*, New York: Free Press, 82–107.

—— (1973a) *Retracking America: A Theory of Transactive Planning*, Garden City, NY: Doubleday/Anchor.

—— (1973b) *Urbanisation, Planning and National Development*, London: Sage.

—— (1979) *The Good Society: A Personal Account of its Struggle with the World of Social Planning and a Dialectical Inquiry into the Roots of Radical Practice*, Cambridge, MA: MIT Press.

—— (1986) "The World City Hypothesis", *Development and Change* 17(1): 69–84.

—— (1992) *Empowerment: The Politics of Alternative Development*, Oxford: Blackwell.

—— (1995) "Where We Stand: A Decade of World City Research", in P.L. Knox and P.J. Taylor (eds), *World Cities in a World-System*, Cambridge: Cambridge University Press.

—— (2000) *Human Settlements and Planning for Ecological Sustainability: The Case of Mexico City*, Cambridge, MA: MIT Press.

—— (2002) *The Prospect of Cities*, Minneapolis: University of Minnesota Press.

—— (2005) *China's Urban Transition*, Minneapolis: Minnesota University Press.

—— (2017) "Planning as a Vocation: The Journey So Far", in B. Haselsberger (ed.), *Encounters in Planning Thought: 16 Autobiographical Essays from Thinkers in Spatial Planning*, London and New York: Routledge.

—— and Miller, J. (1965) "The Urban Field", *Journal of the American Institute of Planners* 31(4): 312–319.

—— and Weaver, C. (1979) *Territory and Function: The Evolution of Regional Planning*, Berkeley, CA: University of California Press.

—— and Wolff, G. (1982) "World City Formation: An Agenda for Research and Action", *International Journal for Urban and Regional Research* 6(3): 309–344.

Further reading

Gottmann, J. (1961) *Megalopolis: The Urbanized Northeastern Seaboard of the United States*, New York: Twentieth Century Fund.

Knox, P.L. and Taylor, P.J. (eds) (1995) *World Cities in a World-System*, Cambridge: Cambridge University Press.

Sen, A. (1999) *Development as Freedom*, New York: Knopf.

Simon, D. (2017) "Obituary: John Friedmann, 1926–2017", *International Development Planning Review* 39(3):v–vii; Open Access at https://doi.org/10.3828/idpr.2017.14.

—— (2019) *Holocaust Escapees and Global Development: Hidden Histories*, London: Zed Books.

Wallerstein, I. (1974) *Capitalist Agriculture and the Origins of the European World Economy in the Sixteenth Century*, New York: Academic Press.

GARY GAILE WITH DAVID SIMON

CELSO FURTADO (1920–2004)

Celso Monteiro Furtado, considered by most Brazilian academics as their most renowned economist, left behind a considerable literary legacy. This included more than 30 books, primarily dealing with the concept, theory and process of development in its very broadest sense, focusing on development in Latin American countries and, especially in Brazil.

Born in 1920 in Pombal, a town in the interior of Brazil's poorest region, the northeast, which is regularly beset by severe droughts, he was fortunate to be born into a middle-class and highly cultured family, whose priority was to provide him with a good education. His schooling began in Pombal and continued in Recife, the Northeast's most important cultural centre; he finally graduated from the Faculty of Law in Rio de Janeiro, the then capital of Brazil, in 1944.

The inhuman conditions of a region that was backward, within a country that was itself backward, were engraved on his conscience. This inspired his commitment to making every effort over the course of his life to better understand the reasons for this backwardness, which would later become better known as "underdevelopment". At university, he spent more time on questions of the organisation than on law as such. The study of the organisation, both in the public and the private sectors, led him to start thinking about planning, which was his first contact with the field of economics.

Shortly after graduating, Celso was called up to join the Brazilian Expeditionary Force, a contingent of troops sent to fight Nazi fascism in Italy, until its surrender in 1945. He wrote his first book about his experiences as a soldier, entitled *From Naples to Paris: Tales of the Expeditionary Life* (Furtado 1946). When the war ended, he applied for and successfully completed a PhD in Economics at University of Paris. Entitled "The Brazilian colonial economy", his thesis signalled from the outset what the focus of all his work as a researcher and writer would eventually be, namely the process of development from a historical perspective. His readings of Marx influenced his view that Economics is a science with strong and conditioning historical roots.

In 1949, he was invited to join a team of economists at ECLA (Economic Commission for Latin America – then ECLAC, adding the Caribbean region), an organisation set up by the UN to study, analyse and offer directives to help the economies of Central and South America. From its inception, ECLAC had a leader, the great Argentinian economist **Raúl Prebisch**, who has studied (with **Hans Singer**) the necessity of the continent to fill the void of backwardness, once based on the deterioration of the terms of trade, that keeps separate the core economies from the periphery of the capitalist system. Celso Furtado was appointed by **Prebisch** as director of the institution's Economic Development Division and was able to formulate ideas that went well beyond the mere confines of economics.

After several missions and reports on Latin American countries,[1] he was invited by Nicholas Kaldor to undertaken postgraduate studies at Cambridge University's King's College, in England, where he wrote the book that brought him to the world`s attention: *The Economic Growth of Brazil* (*Formação Econômica do Brasil*), published in 1959.

Furtado's first book to meet with great international success was *Development and Underdevelopment* (1961). In this seminal book, Furtado showed clearly, through his extensive historical analysis, the major focus of his work: the idea that development and underdevelopment are interlinked historical processes, even though each has very different characteristics; the term "under" is not a stage of the same process that led developed nations to the more advanced levels of wealth and economic productivity they came to enjoy. Underdeveloped nations would not, therefore, need to make more effort to save and attract external capital to raise investment, and thus speed up the process to arrive more quickly at the same stage of the more developed. They would not have to go through the same stages of the same process, as suggested by the models of development in vogue at that time put forward by **Walt Rostow** and other modernisation theorists.

Celso Furtado studied all available mathematical theory on the subject but decided to delve more deeply into the historical reality of each case, in the search for the reasons and the factors that either helped boost or curbed the workings of its economy. He also showed that under-development was a space reserved for economies that were not yet industrialised and that had been infiltrated, ever since the 19th century, by enterprises created through the expansion of already-industrialised economies in Europe. Their main objective was to obtain the raw material and labour needed to expand their capitalist-industrial productive systems. These were enterprises with advanced characteristics that penetrated more backward economies tied to archaic socio-economic structures, thereby creating a socio-economic duality in these economies: on one side, a small "modern" sector, that quickly acquired the consumption habits of the rich nations, and on the other, a clear majority of the poor, old structures of a backward nation. The modern company, unused to the local structure, continued to make decisions and operated as if it were still in the developed economy from whence it came, only generating a new and very small workforce of a higher standard within the backward economy. It did not make any kind of significant impact on the outdated structures, which continued to block any dissemination of these new standards.

Underdevelopment was not, therefore, merely one stage of a single process that everyone needed to follow, but rather part of a specific framework within the Western capitalist-industrial system that had been expanding since the 19th century. This was a framework that created a partial "modernisation" of privileged segments of the backward society, always in the form of a duality – a hybridism within these societies and their respective economies – which tended to be perpetuated through a series of existing conditions on both sides. This included, one should add, political domination.

Celso Furtado was, above all, a great and profound student and analyst of the phenomenon of underdevelopment, a field that demanded, as he clearly perceived, a good deal of autonomous theorisation. It could not simply be described using the classical models and theories of those economists who wished to present development as a general process, one based on accumulation, on investment and on scientifically based technological progress, as seen in countries that spearheaded industrial capitalism, especially the United Kingdom:

> Development and underdevelopment are, therefore, two historical processes derived from the same initial impulse. In other words, they both have their roots in the acceleration in accumulation that took place in Europe at the end of the 18th and beginning of the 19th Centuries. In order to better understand the reasons behind the historical persistence of underdevelopment, one must view it as part of a whole that is in movement, or perhaps, as an expression of the dynamic of the global economic system engendered by industrial capitalism.
>
> (Furtado 2013: 231)

The development process in each country was, therefore, essentially a historical one and reflected the agility or stagnation specific to the socio-economic structures created throughout the history of each one. This was a process that separated, almost eternally, the nations of the historical-economic space in the West during the 20th century, into the developed and the underdeveloped (the centre and the periphery). Thus emerged the idea of so-called structuralism, a new approach to economic phenomena and one characteristic of the fertile plain of ideas produced at ECLAC, which predominated from then on in academic circles throughout Latin America.

Development, according to the ideas put forward by Furtado, became the responsibility of the state and should be based on competent planning, that is, technical-political planning but a process that was more political than technical.

Furtado studied the processes of two specific nations that had sought to escape from their condition of underdevelopment already in the 19th century, namely Germany and Japan, both of which were subjected to revolutionary and authoritarian political processes aimed at catching up economically. He also studied economic success stories in the 20th century, including South Korea and Taiwan, both similarly subjected to authoritarian political processes but with help from abroad (the US), with an eye on the progress of neighbouring countries considered as enemies (North Korea and China).

He considered, in more detail, cases where countries attempted to break the barrier of underdevelopment using a politically supported process of industrialisation. The idea here was to take advantage of the difficulties of importing goods to satisfy the consumer demands of "modernised" segments of the population, and to help steadily expand this modern contingent of the population (an exemplary case being Brazil).

His conclusion was that

> the case of a beneficial "modernisation" of the population as a whole is no more than a school theory. The reality of the facts is that the process of "modernisation" only aggravated the concentration of wealth and income that already existed, accentuating it during the phase of substitutive industrialisation.
>
> (Furtado 2013: 259)

Within the concept of development, the social dimension was, for Furtado, as vital as the economic one. Indeed, today we use the expression "development" without qualifying it as being purely economic, as was the case when it first emerged in the 1940s and 1950s. This generic expression now necessarily also includes social, cultural, political and even environmental aspects – a broader definition that owes much to the ideas of Celso Furtado.

Indeed, the science of development has definitively recouped, thanks to the efforts of Furtado and ECLAC, a historical dimension that is strong enough to become indispensable. It has also demoted the highly mathematical ideas of classical theory to the condition of mere exercises in abstract thinking, distant from the reality of the specific processes of each country or region.

Celso Furtado was called upon by the Brazilian government in 1959 to be a director at the National Bank of Economic Development (BNDE). Shortly afterwards, he was given the task of setting up and running a government agency aimed at encouraging development in the north-east of the country, its poorest region. It was a major challenge to apply his ideas to the confines of his own country, indeed to his own region, home to around a third of the Brazilian population but responsible for little more than 10 per cent of GDP. What was worse, the trend suggested that this disparity could only worsen.

Backed by the then President of Brazil, Juscelino Kubitschek, as head of SUDENE (Superintendency for the Development of the Northeast), Furtado faced strong political opposition from the longstanding special interests linked to the archaic structures that predominated in the north-east. He produced some notable successes in terms of social modernisation, managing to overcome certain obstacles and administrating the process in the correct manner, skilfully avoiding antagonising the governors of the 10 states with territories within the region. Planning, he said in his inauguration speech, is a technical and political task: more political than technical.

Furtado realised that drought was undoubtedly a serious problem, but not the prime cause of backwardness and poverty in the region. Therefore, he implemented a policy of rationalisation and modernisation of agricultural production, which was the most important activity in the north-east, and brought in a policy of tax incentives aimed at the growth and diversification of the rather backward industrial sector. A cult figure throughout the north-east today, there is general agreement among Brazilians that Celso Furtado rose very well to that challenge.

In 1962, Furtado was invited by President João Goulart to be the first Minister of Planning. He produced a proposal for the Triannual Economic and Social Development Plan, which covered the remaining years of the president's mandate. The climate of exacerbated political radicalisation that preceded the coup did not, however, allow any government initiative of rationalisation to be implemented, and the plan was attacked from all sides, from both right and left.

In 1964, the military coup that ousted President João Goulart also cancelled Furtado's political rights, and indeed those of hundreds of Brazilians who were battling for structural reforms to be implemented in

Brazil, eventually condemning him to exile abroad. Whilst in exile, he worked as a professor at the University of Paris/Sorbonne for twenty years and lectured at other prestigious universities, including Yale, Columbia and Cambridge.

He further developed his ideas on development, adding certain important dimensions linked to culture and creativity to the concept:

> The strategic goal is to ensure a development that translates into the enrichment of culture in its multiple dimensions and allows one to contribute with one's own creativity to a civilisation that is globalising. In the end, one has the desire to preserve one's own identity in the common adventure of the civilizational process.
>
> (Furtado 2013: 283)

When he returned to Brazil, which was enabled by the democratization process in 1985, Furtado was appointed ambassador to the European Economic Community. Then, from 1986 to 1988, following his works on culture, he was invited to be the Minister of Culture. In 1997, he was elected a member of the Brazilian Academy of Literature and the Brazilian Academy of Sciences.

Celso Furtado died in Rio de Janeiro, at the age of 84, in 2004. One year later, on 22 November 2005, the International Celso Furtado Center for Development Policies was inaugurated, with a principal mission to help enrich the debate regarding different strategies for development, in line with the views of Celso Furtado, with particular reference to Brazil and Latin America. In 2009, his bibliographic archives were fully catalogued and became available to the public in the Celso Furtado Library.

Note

1 One of these, *El desarrollo reciente de la economía venezolana (planteamiento de algunos problemas)*, remained unpublished until 2008 (Furtado 2008a).

Major works

Notes:

1. Rosa Freire d'Aguiar, his companion for the last 30 years of his life, has simplified the task enormously for those who want to delve into the works of Celso Furtado, putting together a collection of the author's writings that faithfully express the very heart of the ideas and the personality. This

collection comes in the form of a book *Celso Furtado: Essential*. She also organized the collection of books entitled Celso Furtado Archives and published the "definitive version" of his major works.
2. A full catalogue of Celso Furtado's works and also articles, books and theses about Furtado are available at www.centrocelsofurtado.org.br and www.bibliotecacelsofurtado.org.br.

Furtado, C. (1946) *Contos da vida expedicionária: de Nápoles a Paris*, Rio de Janeiro: Zelio Valverde.

—— (1959) *A Operação Nordeste*, Rio de Janeiro: Instituto Superior de Estudos Brasileiros.

—— (1996) *O mito do desenvolvimento econômico*, São Paulo: Paz e Terra.

—— (2000) *Teoria e política do desenvolvimento econômico*, 10th rev. edn, São Paulo: Paz e Terra.

—— (2007a) *Formação econômica do Brasil*, 34th (definitive) edn, São Paulo: Companhia das Letras.

—— (2007b) *A economia latino-americana*, definitive edn, São Paulo: Companhia das Letras.

—— (2008a) *Ensaios sobre a Venezuela: subdesenvolvimento com abundâncias de divisas*, Rio de Janeiro: Centro Internacional Celso Furtado de Políticas para o Desenvolvimento: Contraponto (Arquivos Celso Furtado).

—— (2008b) *Criatividade e dependência na civilização industrial*, definitive edn, São Paulo: Companhia das Letras.

—— (2009) *Desenvolvimento e subdesenvolvimento*, 5th edn, Rio de Janeiro: Centro Internacional Celso Furtado de Políticas para o Desenvolvimento: Contraponto (Economia política e desenvolvimento).

—— (2013) *Essencial*, São Paulo: Penguin Classics/Companhia das Letras.

—— (2014) *Obra* autobiográfica, definitive edn, São Paulo: Companhia das Letras.

Further reading

D'Aguiar, R.F. (ed.) (2013) *Celso Furtado e a dimensão cultural do desenvolvimento*, Rio de Janeiro: Centro Internacional Celso Furtado de Políticas para o Desenvolvimento (Coleção Pensamento Crítico no. 2).

Furtado, A.T. et al. (2010) *Celso Furtado: os desafios do desenvolvimento*, São Paulo: Cultura Acadêmica.

Guillén, A. and Vidal, G. (2008) *Celso Furtado*, Madri: Agencia Española de Cooperación Internacional para el Desarrollo, Ministerio de Asuntos Exteriores y de Cooperación.

Lima, M.C. and David, M.D. (eds) (2008) *A atualidade do pensamento de Celso Furtado*, Leste Vila Nova: Verbena.

Mallorquin, C. (2005) *Celso Furtado: um retrato intelectual*, trans. Célia Regina Barbosa Ramos, São Paulo: Xamã; Rio de Janeiro: Contraponto.

Saboia, J. and Carvalho, F.J.C. (eds) (2007) *Celso Furtado e o século XXI*, Rio de Janeiro: Minha Editora/Instituto de Economia da UFRJ; São Paulo: Manole.

Tavares, M. da C. (ed.) (2001) *Celso Furtado e o Brasil*, São Paulo: Fundação Perseu Abramo.

ROBERTO SATURNINO BRAGA

MOHANDAS (MAHATMA) GANDHI (1869–1948)

> If you cannot change yourself, how can you change the World?
> Gandhi

Mohandas Karamchand Gandhi, known as *Mahatma*, meaning "great-souled", was born on 2 October 1869 in western India, and went to England in 1888 to study law. He returned to India in 1891, and after practising law for some time he moved to South Africa in 1893, where he was a determined opponent of the "pass laws" and other forms of racial discrimination. There he organised *satyagraha* (non-violent action, or "passive resistance") in 1906, 1908 and 1913. With a view to serving his country of birth, he returned to India in 1914 and soon became a leading figure in the cause of Indian nationalism and development movements. In 1915 Gandhi established the Satyagraha Ashram at Ahmedabad, and in 1917 moved it to the Sabarmati River. On 30 January 1948 Gandhi was shot dead by a Hindu fanatic, while holding a prayer meeting in Delhi.

In the late 1940s, the term "development" was not in currency as it is today. Gandhi had, therefore, used the term "progress" for development, more with respect to ethics and cosmic integrity. He said: "by economic progress we mean material advancement without limit, and by real progress we mean moral progress, which again, is the same thing as progress of the permanent elements in us" (*The Collected Works of Mahatma Gandhi* (*WMG*), vol. 87: 249). He never made an academic contribution to "development"; instead he pleaded for an organised effort to change the ruling paradigm and move towards a superhuman stage. The Gandhian idea of development is based on the foundational ethics of *ahimsa* (non-violence, which he interpreted as "firmness in the truth"). Perhaps Gandhi's most important influence has been on the black civil rights movement in the USA, led by Martin Luther King.

In the cosmic-moral organisation, faith is the path of spirituality, and spirituality in its true sense is the motive force behind development. Gandhi suggested "seven Social Sins" to be avoided. Of course, these are ideals, but they are more relevant in the present era of desperation and could easily be accepted. According to Gandhi, the seven Social Sins are:

(i) *Consumption without conscience.* The means and symbols of consumption are the 3Ps, namely property, power and prestige. Without conscience or moral responsibility, consumption always turns to evil; for example, eating food is consumption, but without conscience it leads to sickness. Consumerism is one of the basic root causes of social conflict.

(ii) *Knowledge without character.* Knowledge is power, but without morality it is hypocrisy. With a lack of morality, knowledge becomes a heavy burden on humanity.

(iii) *Wealth without labour.* In the modern era, wealth is the highest symbol of projecting the level of personality and status. This promotes the tendency of the "rich becoming richer and the poor becoming poorer". Such rich people never realise their social responsibility; this leads to social crisis.

(iv) *Business without morality.* According to the philosophy of welfare economics, the basis of economic success is morality. "Economics that hurt the moral well-being of an individual or a nation are immoral and, therefore, sinful" (*WMG*, vol. 13: 317; also cf. *WMG*, vol. 75: 158).

(v) *Religious duty without sacrifice.* Compassion, service, sacrifice and altruism are the basic rules of *dharma* (righteous duty). Lacking the sense of sacrifice promotes individualism, egocentricity and selfishness. Serving the poor is service to God.

(vi) *Science without human sense.* Science has the power of both creation and destruction; its nature depends on the way in which it is used by mankind. Misconception of science as a tool and technique for the use of resources has promoted the philosophy of consumerism, which has finally resulted in environmental disorder, pollution and loss of peaceful life. Science is the means, but the end is human kindness.

(vii) *Politics without principles.* Politics without principles is the cause of the global crisis and destruction of social harmony. Gandhi emphasised the need for spirituality as the first step of politics and governance.

Whenever the ethical code of a society is lost, civilisation will end. In such a situation, a cultural tradition is unable to find a balance between the needs and expectations of society, and that is how it fails to illuminate a path of revolution. In this way, the road to progress and development comes to a dead end. Gandhi proposed a philosophy of revival and peace which he called *sarvodaya* ("well-being for all"). Enhancing personal welfare (*sva-*) to the level of communal well-being (*sarva-*) is the cultural code of *sarvodaya.* History has proved that the alternative to war and conflict is *ahimsa* ("non-violence"), i.e. peaceful agitation, creating mass awareness of cultural unity, which was derived from the traditional moral restraint of not injuring another being.

There are two ideas inherent in the philosophy of *sarvodaya*:

(i) *Democracy is a lifestyle.* Democracy is not only a way of governance, it is also a way of life. Gandhi warned that a process of dialogue and criticism must always be maintained between the public and ruling powers. The continuing trend towards loss of public awareness is a sign of the decline of democracy.

(ii) *Machines have a cultural value.* The use of *machines* (a product of science) also has a cultural value. The application of science, or its tool, the machine, depends on human intention. Science is like a machine: the effects of its use depend upon the attitude and motives of the person who has control over it. The path on which science is presently advancing will lead to a great dissolution. Cultural values are imposed upon machines by human beings (cf. *WMG*, vol. 25: 251–252).

Gandhi's view of non-violence, *ahimsa,* vegetarianism and *karma* (right action) is based on the idea of the total spiritual interconnectedness and divinity of life as a whole. All natural phenomena are therefore divine, sacred and of equal value. As human beings, we have to take the main responsibilities towards nature through a moral-ethical-religious approach. His theory of *ahimsa* was not strict like a sectarian rule; he said: "whoever believes in *ahimsa* will engage himself in occupations that involve the least possible violence."

Committed to non-violence (*ahimsa*) and self-realisation (*svachetana*), Gandhi wanted to solve India's problems from the perspective of individual conversion and with the ideology that "every man has an equal right for the necessities of life even as birds and beasts". According to him, the "right thing" is a moral order (*dharma*) operated by right action (*karma*). Gandhi's emphasis upon self-realisation and rules of conduct and virtues is essential for spiritual life and also for the maintenance of the social order. Gandhi warned politicians about the social and political evils of which they become part.

Gandhi said, "every human being has a right to live and, therefore, to find the wherewithal to feed himself and where necessary to clothe and house himself" (*WMG*, vol. 38: 197). He said, "the world has enough for everyone's need, but not enough for everyone's greed" (Pyarelal 1956: 552). In the contemporary world, Gandhi's insight is being put to the test as never before. If every little village became self-sufficient, then one would have an ideal setting for life. Therefore he did not advocate macro-economic policy. All villages should become self-sufficient. The system must work from the bottom upwards and not the top down.

As we work in different walks of life, we should continue to be inspired by his words that we must "recall the face of the poorest and the weakest man (and woman) whom you have seen and ask yourself if the step you contemplate is going to be of any use to him (and her)" (*WMG*, vol. 89: 125).

Gandhi practised what he preached. Because he was concerned about the "poorest of the poor" he adopted a life style which reflected his constituency. And because he preached "cleanliness" and the uplifting of Harijans ("untouchables"), he himself undertook the scavengers' work, and emphasised the importance of inner and outer cleanliness. To follow Gandhi is difficult, and yet millions did. He emphasised throughout that we must change ourselves before we can change others, and that our real enemies are within.

The essence of natural cures is that we learn the principles of hygiene and sanitation and abide by those laws as well as the laws relating to proper nutrition. If rural reconstruction does not include rural sanitation, our villages will remain the dung-heaps that they are today.

Gandhi's 21 years of experience in South Africa transformed his views on life and human existence, which he again experimented with in India. He started to look at the world from a poverty-trapped peasant's perspective, rather than from a middle-class bourgeois perspective. This led him to the creation of three principles of sustainable development: *sarvodaya, swadeshi* and *satyagraha.* The other very important feature of sustainable development which he propagated was the whole question of local economy, in which everybody in the area would be self-sufficient. They would be employed, and could sustain themselves and their families with dignity and work.

For Gandhi, the *swadeshi* spirit extended to all the elements composing the *desh* (community), and implied a love of not only the traditional way of life but also the natural environment and especially the people sharing it. Gandhi used the term *swaraj* to describe a society run in the *swadeshi* spirit. It meant self-rule or autonomy, and implied not only formal independence but also cultural and moral autonomy.

The removal of untouchability, an end to Hindu–Muslim enmity, economic self-sufficiency, and making non-violence effective were fundamental elements of Gandhi's vision. He offered a viable alternative for self-rule and self-government. He focused on the welfare of all – *sarvodaya* – and the method he followed was that of *satyagraha*.

Sarvodaya ("the uplifting of all") was a philosophical position that Gandhi maintained. Society must strive for the economic, social, spiritual and physical well-being of all, not just the majority. He favoured a holistic approach to well-being, and a total approach to the community. For him the well-being of every individual was an important concern.

He advocated that the locus of power must be situated in the village or neighbourhood unit. He believed that there should be an equitable distribution of resources, and that communities must become self-sustaining through reliance on local products instead of large-scale imports from outside. Gandhi was opposed to large-scale industrialisation, and favoured small local industries that promote local self-sufficiency, which he called *swadeshi*. In current terms, it means buy local, be proud of local, support local, uphold and live local. Economic equality should never mean possession of an equal amount of worldly goods by everyone. It does mean, however, that everyone should have a proper house, an adequate and balanced diet and sufficient cloth(ing). It also means that the cruel inequality that obtains today will be removed by purely non-violent means.

Finally, Gandhi's best-known theory of *satyagraha*, "truth force" or non-violent direct action, is actually a way of life, not just an absence of violence. It also entailed respect for all beings regardless of religious beliefs, caste, race or creed, and a devotion to the values of truth, love and responsibility. Mass awakening exemplified by the *satyagraha* movement, is "the perfect example of how one could confront an unjust and uncaring, though extremely superior power" (*WMG*, vol. 9: 118, and cf. Gandhi 1990: 21). The achievement of political and moral ends through *ahimsa* is what Gandhi called *satyagraha*. This notion of non-violent action is the crucial part of Gandhi's political theory. *Satyagraha*, in fact, is a theory of action. It calls for courage, strength of character and positive commitment to a righteous cause. In some circumstances, such as inhuman acts like molestation or killing, it might be better to choose violence than craven submission to injustice. Gandhi thought that "total non-violence" might be feasible only when mankind has acquired superhuman qualities.

The spectres of global warming, lack of water through deforestation and continued depletion of natural resources and diversity on Earth are some of the results of "unsustainable development" and economic growth. Should we not follow the path of Gandhi? He corrected himself over and over in his striving for moral-spiritual perfection while doing his best to reform existing institutions and social practices. He wrote in 1932, "I do not accept defeat but hope, with God's grace, to melt the stoniest heart and, therefore, continually strive to perfect myself" (*WMG*, vol. 50: 451).

We are slowly realising that reducing poverty or moving towards sustainable development is not just an economic or a technical problem or one of acquiring greater financial inputs. All these goals are important, but achieving them also needs an inner awakening, an inner transformation of man – a path already paved by Gandhi. If we ignore Gandhi, "we ignore him at our own peril", Martin Luther King once said.

Major works

Gandhi, Mohandas Karamchand (1927) *An Autobiography or the Story of My Experiments with Truth*, Ahmedabad: Navajivan.

—— (1928) *Satyagraha in South Africa*, Ahmedabad: Navajivan.

—— (1958–1994) *The Collected Works of Mahatma Gandhi*, 100 vols, New Delhi: Publication Division, Ministry of Information and Broadcasting, Govt of India.

—— (1990 [1909]) *Hind Swaraj and Other Writings*, ed. A.J. Parel, Cambridge: Cambridge University Press.

Pyarelal [Nayyar] (1958) *Mahatma Gandhi: The Last Phase* (vol. 10, part II), Ahmedabad: Navajivan.

Further reading

Adams, I. and Dyson, R.W. (eds) (2003) "Mohandas Gandhi", in *Fifty Major Political Thinkers*, London and New York: Routledge, 196–199.

Bilimoria, P. (2018) "Mahatma Gandhi", in J.A. Palmer Cooper and D.E. Cooper (eds), *Key Thinkers on the Environment*, London and New York: Routledge, 160–166.

Bondurant, J.V. (1965) *Conquest of Violence: The Gandhian Philosophy of Conflict*, Berkeley, CA: University of California Press.

Hick, J. and Hemple, L. (1989) *Gandhi's Significance for Today*, New York: St. Martin's Press.

Ivanov, A. (2017) *Gandhi*, Boston, MA: New Word City.

Naess, A. (1965) *Gandhi and the Nuclear Age*, trans. A. Hannay, Totowa, NJ: Bedminster Press.

Parekh, B. (1997) *Gandhi*, Oxford: Oxford University Press.
Parel, A.J. (2000, 2002) *Gandhi, Freedom, and Self-Rule*, Lanham, MD: Lexington Books; Indian edn, New Delhi: Vistaar.
Singh, R.P.B. (2007) "Culture and Civilisation for Human Development: Perspectives in India on the Gandhian Path", in P. Maiti (ed.) *Development Studies*, vol. 1, New Delhi: Atlantic Publishers & Distributors, 105–152.
—— (2009) "Mahatma Gandhi's Vision of Sustainable Development", in R.P.B. Singh, *Uprooting Geographic Thoughts in India: Toward Ethics, Ecology and Culture in 21st Century*, Newcastle upon Tyne: Cambridge Scholars, 192–223.
Swan, M. (1985) *Gandhi: The South African Experience*, Johannesburg: Ravan.
Tendulkar, D.G. (1951–1954) *Mahatma: The Life of M.K. Gandhi*, 8 vols, New Delhi: Ministry of Information and Broadcasting.
Terchek, R.J. (1998) *Gandhi: Struggling for Autonomy*, Lanham, MD: Rowman & Littlefield.
Weber, Thomas (2004) *Gandhi as Disciple and Mentor*, Cambridge: Cambridge University Press.

RANA P.B. SINGH

SUSAN GEORGE (1934–)

Susan George is a progressive scholar-activist, whose innovative work has examined world hunger, poverty, debt and North–South relations, as well as provide critiques of the world capitalist system and the major players within it, particularly the international financial institutions and unaccountable global corporations. A deep-seated concern throughout her writing has been global injustice and the need to campaign for change. Indeed, her unrivalled prominence as a development thinker over the past four decades reflects her ability to engage not just with academics but with politicians, activists and journalists, as well as with untold thousands of students whose lives have been changed by her writing. Her influence is not restricted to the English-speaking world: her work has been widely translated, including into four Scandinavian languages, as well as Thai, Korean, Bengali, Estonian and Japanese. While she communicates her research through books and journals, she also writes, or is interviewed frequently, for newspapers and other print and social media around the world, as well as speaking at scores of influential conferences.

Born in the USA, Susan George has lived all her adult life in France, gaining French citizenship in 1994. While her early academic career was in the USA, where she attained a BA degree in French and

Government from Smith College, Northampton, Massachusetts, in 1956, she later completed her doctorate at the Sorbonne in Paris in 1978 with a thesis on the transfer of the food system from the USA to the rest of the world (George 1981). She is currently the President of the Board of the Transnational Institute in Amsterdam, where she has been based since its foundation in 1973; this is a global network of scholar-activists who work on challenging global inequalities. She is also the honorary President of ATTAC France (Association to Tax Financial Transactions to Aid Citizens), which was founded in 1998 to promote the taxation of financial markets and transnational corporations with a view to redistributing global income.

By her own admission, Susan George is not a scholar in the conventional sense, something which has been a major strength. Free from the strictures of academia where, in her view, orthodoxy is revered, George has confronted the injustices of the global system with a passion and foresight often missing from academic writing. This talent for adopting a powerful yet unconventional approach is rooted in her first book, *How the Other Half Dies* (1976), which provides a critique of standard accounts of the prevalence of world hunger. Focusing on how hunger is part of a wider capitalist system based on unequal power relations between global North and South and between the poor and the elites within countries, she criticises a host of phenomena hitherto viewed as benign, including agribusiness, the Green Revolution and food aid. However, her most path-breaking point in the book is that famine, poverty and hunger have little to do with "natural" disasters, but rather with politics and exploitation on the part of the world's powerful (see also **Amartya Sen**). While this book identified George as someone keen to challenge contemporary opinion, the academic criticism of it also led her to study for her PhD in an attempt to counteract the "pompous, usually male, professors proclaiming that 'this woman has no credentials for saying all these terrible things she's saying'" (George 2004: 205).

With her PhD duly earned, George continued addressing issues that others were afraid or unable to identify. With her flair for presenting the facts in bold terms, and for using simple yet sophisticated arguments, she continued to expose the powerful actors benefiting from the inequalities of the global system, as well as the exploitative nature of the system itself. Ahead of her time, George began to work on debt through *A Fate Worse than Debt* (1987), highlighting the debilitating effects of IMF and World Bank lending on the global South. Caught in a vicious circle of debt and loan servicing, Southern countries had no choice but to comply with the conditions imposed on them through Structural Adjustment Programmes, with the the flow of money always favouring the powerful.

George also recognised that debt affected ordinary citizens of the North; in *The Debt Boomerang* (1992), she identified six "debt connections" in the North including the environment; the drugs industry; immigration; and conflict and war. The ecological damage of debt and capitalism, also discussed in *A Fate Worse than Debt*, was especially important. Here, George argued that transnational corporations were responsible for much environmental destruction globally, bolstered by the policies of the international financial institutions.

The role of these institutions and the evil of the capitalist system underpinned by exploitative global corporations are enduring themes of George's work. They first emerged explicitly in *Faith and Credit* with Fabrizio Sabelli (1994) on the World Bank, in particular, and subsequently in *The Lugano Report* (George 1999) and its sucessor, *How to Win the Class War: The Lugano Report II* (George 2013). *The Lugano Report* is a partly fictitious account based on the views of a "working party" commissioned to examine the threats and obstacles to the future flourishing of capitalism. The sequel revealed that the predictions came true, and that the power of the 1% (also called the "Davos Class" – George 2010) is deeply entrenched as they continue to exploit the 99% across the global South and in Europe. The insidious nature of unaccountable global corporations and their influence on global and national governance regimes also continues in her most recent book (George 2015). While some of this work is rather fanciful, it illustrates why George's work has such public appeal; she writes in innovative ways that use iconoclastic devices and political satire to make her point.

Despite the apocalyptic scenarios that sometimes afflict George's work, she is careful to offer strategies for change. In *A Fate Worse than Debt*, for instance, she outlined her "3-D solution of debt, development and democracy" otherwise known as "creative reimbursement", where debtor countries can pay back their loans in local currencies over a long period of time (1987: 243–262). However, her ideas about practical change for the future are most concisely outlined in *Another World Is Possible If …* (George 2004). While she proposes such initiatives as taxing financial markets, including the Tobin Tax (see **James Tobin**) (as in her work with ATTAC), she also provides some guidelines for those wanting to become involved in the global justice movement such as her "seven commandments" that include how to plan meetings and produce leaflets (2004: 164–166).

Ultimately however, while Susan George's work has been invaluable in highlighting the potentially catastrophic path down which capitalist development might lead the world, she remains optimistic. Not only does she have faith in people working and organising at the grassroots, she also truly believes that hope is the key, however tenuous. To end with her own words: "Despite everything, however, I don't just go about mouthing the slogan 'Another world is possible'. I actually believe it, while recognising that it's a very long shot and a fragile hope" (George 2004: 116).

Major works

George, S. (1976) *How the Other Half Dies: The Real Reasons for World Hunger*, London: Penguin.

—— (1981) *Les Stratèges de la faim*, Geneva: Grounauer with the Institut Universitaire d'Études de Développement, University of Geneva.

—— (1987) *A Fate Worse than Debt*, London: Penguin.

—— (1990) *Ill Fares the Land*, London: Penguin.

—— (1992) *The Debt Boomerang: How Third World Debt Harms Us All*, London: Pluto Press with the Transnational Institute.

—— (1999) *The Lugano Report: On Preserving Capitalism in the Twenty-First Century*, London: Pluto Press.

—— (2004) *Another World is Possible if...*, London: Verso.

—— (2010) *Whose Crisis? Whose Future?* Cambridge: Polity.

—— (2013) *How to Win the Class War: The Lugano Report II*, Paris: Transnational Institute.

—— (2015) *Shadow Sovereigns: How Global Corporations are Seizing Power*, Cambridge: Polity.

—— and Sabelli, F. (1994) *Faith and Credit: The World Bank's Secular Empire*, London: Penguin.

Further reading

George, S. (2002) "ATTAC: A Citizen's Movement for Global Justice Responds to September 11th", *Development* 45(2):97–98: www.tni.org/george.

CATHY MCILWAINE

ALEXANDER GERSCHENKRON (1904–1978)

Known as "Shura" by his family, Alexander Gerschenkron was born in 1904 in Odessa, Russia, to a religiously mixed but secular family: his mother was Russian Orthodox and his father Jewish. He grew up exposed to both traditions, although his parents were "enlightened humanists" (Dawidoff 2002: 34–35; Simon 2019). His family moved to Vienna after the Russian Revolution and he received his doctorate at the University of Vienna in 1928. He later became an academic at the Austrian Institute of Business Cycle Research (then headed by Friedrich Hayek). With his life and liberty in imminent danger, he left for the USA on the day of the German invasion of Austria in 1938. After working initially as a shipbuilder and then as an academic economist in Berkeley and subsequently in the Research and Statistics and later International Divisions of the Federal Reserve Board in Washington from 1944 to 1948 (Dawidoff 2002; Simon 2019), he became a Professor at Harvard University in 1946. For many years he directed economic research in the university's Russian Research Centre (Dawidoff 2002; Blaug 1998: 82; Simon 2019). He retired in 1974 and died in 1978. His thinking was heavily influenced by his association with, and analysis of, the economies of Russia and eastern Europe, and he stressed the need for broad geographical frameworks in the study of economic history and development. "Insularity is a limitation on comprehension," he argued in his seminal essay (Gerschenkron 1962: 6). At Harvard he was joined by other European former refugee scholars such as Bert Hoselitz and **Albert Hirschman**, whose theories of unbalanced growth have strong connections with Gerschenkron's ideas.

It is approaching three-quarters of a century (1951) since Alexander Gerschenkron wrote his seminal essay, "Economic Backwardness in Historical Perspective". The essay was published eleven years later in a slightly revised form as the first chapter of a book of his essays with the same title (Gerschenkron 1962). The essay addressed the **Marx**ian generalisation that the history of advanced industrial economies traced out the pathway to development for more backward countries. Gerschenkron (1962: 7) thought that this was a half-truth because, in many crucial respects, "the development of a backward country may, by the very virtue of its backwardness, tend to differ fundamentally from that of an advanced country".

Gerschenkron saw a tension between the actual state of economic activities in a backward country on the one hand and the great promise inherent in industrial development on the other. He asked the question: what does it take for a "latecomer" country to industrialise and become like the early industrialisers such as Britain and Germany? He took the "latecomer" economies of late 19th-century central and eastern Europe as his source, and discussed their links and involvement with an increasingly international economic system. His answer was essentially in the form of a metaphor. Latecomer countries would need "to leap the gap of knowledge and practice separating the backward economy from the advanced" (Landes 1999). Economic history revealed that relative backwardness could be overcome by novel strategies and pathways to industrialism and accelerated growth. In his analysis, history could ride roughshod over any neat recipes of prerequisites as outlined by **Walt Rostow** (1960). The rivalry between Gerschenkron's ideas and those of Rostow since the 1960s has been analysed by a number of authors (Trebilcock 1981; Gwynne 1990; Gootenberg 2001).

Gerschenkron's main concept revolved around the complex nature of, and opportunities provided by, economic backwardness and deprivation. The model was concerned with the beginnings of industrial growth, and assumed under-endowment as the starting point. Gerschenkron argued that as other countries advance and backwardness deepens, the underprivileged society will become increasingly sensitive to the contrast. Intellectuals and politicians could transfer this into national ideologies "igniting the imaginations" of people and instilling faith that the "golden age lies not behind but ahead" (Gerschenkron 1962: 24). As social tension increases, a vast effort is made to bridge the gap and a lunge for the benefits of industrial growth occurs. Dore (1990: 359) followed Gerschenkron in arguing that the sense of backwardness "provided a charter for state action to mobilise resources and to take initiatives and risks". From Dore's comparative work on Latin America and East Asia, he concluded that the urgency of the desire to catch up had been stronger among the political leaders and intellectuals of Asian countries after the Second World War due to cultural as much as economic reasons.

According to Gerschenkron, rapid industrial growth was deemed possible, as backwardness could convey a number of economic advantages as well. The more backward the country, the more sophisticated will be the industrial equipment, technology and plant it can select for its manufacturing debut. The country can import technologically advanced machines with the theoretical advantage of enjoying the most significant economies of scale available and hence lower per-unit costs. In theory, the costly stages of technological development may be skipped and new industrial systems installed.

This also provided the theoretical justification for a capital-intensive industrial pathway in contrast to a more labour-intensive one. Gerschenkron (1962: 9) emphasised the problems inherent in creating a large and efficient industrial labour force. These ideas were influential in the capital-intensive policies linked to import-substitution industrialisation (ISI) in Latin America. Bennett and Sharpe (1985: 41) pointed out that countries in Latin America entered into ISI not with Gerschenkronian producer goods but with previously imported goods. Hence they coined the phrase "late late industrialisers" for countries that had even further to catch up, and had to develop products and processes that were even more costly, risky and technologically sophisticated than the "late industrialisers" of Russia and Eastern Europe.

Gerschenkron's comparative approach can be seen 50 years later in the study of "duelling peripheries" of the global economy. How apt and useful are the development and historical comparisons between eastern Europe and Russia, east Asia and Latin America (Gwynne et al. 2003)? One example of comparative work between eastern Europe and Latin America is Love (1996), who argues for a link between Romanian development debates (in their latecomer stage, 1880–1940) and the later Brazilian push towards rapid industrialisation. Gerschenkron had observed that some industrial latecomers during both the nineteenth and twentieth centuries grew at a faster rate than their predecessors. Gerschenkron relied on the work of his colleague at Harvard, Henry Rosovsky (1966), for material on the Japanese comparison. Since his essay, the rapid growth of South Korean and Taiwanese industry in the 1960s and 1970s (Wade 1990) and of China since the mid-1980s has added more contemporary weight to this thesis that benefits can be conferred on industrial latecomers.

Another key concept focused on the role of institutions in this economic transformation. In situations of relative backwardness, the scales of institutions, firms and plants loom large. Gerschenkron saw the importance of large-scale "universal" banks dedicated to long-term investment in heavy industry. For example, he argued that these banks (in which commercial and investment banking were combined) had been the key institution in Germany's rapid industrial growth in the late 19th century. Forsyth and Verdier (2003) have recently shown that financial systems were regionally more complex in European countries than in the interpretation put forward by Gerschenkron, but they still used his framework for comparative purposes.

However, for more backward countries (and particularly his case study of Russia, where "fraudulent bankruptcy had been almost elevated to the rank of a general business practice" – Gerschenkron 1962: 20), the creation of universal banking institutions was inconceivable. Thus Gerschenkron argued that under these conditions the role of the state became fundamental for late industrialisation. He did not idealise this pivotal role of the state, and noted the inefficiencies, incompetence and corruption that had occurred in the Russian experience. However, in spite of this, the role of the state became pivotal to generate the necessary finance for and growth in large-scale heavy industry.

Indeed, as the input needed to launch industrialisation increased with backwardness, it was argued that the process required a correspondingly larger state response, particularly in terms of finance. This could be linked to the wider concept of "substitutes". If a key element for growth was lacking, such as access to finance, European societies "proved adept at developing new institutions, such as state industrial banks" (Gootenberg 2001: 57). Thus, the greater the country's initial deprivation, the more coercive and comprehensive the state's action would have to be. Wade's (1990) thesis that East Asian industrialisation can only be understood through the state "governing the market" is distinctly Gerschenkronian in this context.

Gerschenkron's theories did have an impact on the policies for promoting industrialisation in developing countries in the 1960s and 1970s, along with those of the Latin American structuralists (such as **Raúl Prebisch**) and students and former colleagues of Gerschenkron, such as **Albert Hirschman** (see also Simon 2019). Gerschenkron's perspective had the distinct advantage of recognising that the position of developing countries in relation to the world market was fundamentally different from that of the core economies (such as the UK and USA) on the eve of their industrialisation.

The subsequent industrial experience of developing countries in Asia, Africa and Latin America did reveal some practical problems with the applicability of the model. First, there was the problem of the assumed benefits to the backward country of being able to import the latest technology. The burden of development has grown over time, as the initial costs of technology and plant have increased due to rising capital intensity and the scale of modern production. Second, because these new technologies are capital-intensive, it seems irrational for them to be adopted by developing countries in which cheap labour is normally plentiful (Landes 1999).

It is interesting to ponder why the work of Alexander Gerschenkron has had such little resonance within development geography while attaining much greater significance within the realms of economic and political sociology. Going back to Keeble's (1967) early critique of models of economic development, ample room was given to a discussion of **Rostow**'s theories but none to those of Gerschenkron. Rostow was criticised for his unilinear version of development, but

Gerschenkron's more open theory of differentiated industrial transformations was basically ignored.

Such a pattern continues to this day. The fourth edition of *The Dictionary of Human Geography* (Johnston et al. 2000) can be seen as a guideline to the value of which concepts are important for contemporary geography. Gerschenkron gets but a passing mention in Michael Watts's treatment of growth theory. While Rostow has a section to himself under "stages of economic growth", there is no reference to economic backwardness, late industrialisation or "latecomer models" in the dictionary. Even in the fifth edition (Gregory et al. 2010), Gerschenkron gets only two mentions in the index, compared with five for Rostow.

One possible explanation for the lack of attention to Gerschenkron within human geography in the 1960s and 1970s was that the discipline moved strongly towards more nomothetic approaches. Meanwhile, Gerschenkron was essentially developing historically based ideas and models from the strongly idiographic vision of the economic historian. As geographers increasingly turned their back on the idiographic framework and the detailed and nuanced studies of different cultural regions, Gerschenkronian analysis may have appeared out of step.

For today's development specialists, Gerschenkron could be seen as a precursor of contingent capitalism or the idea of different pathways towards capitalism. The direction of capitalism in any one country can be seen as affected by the nature of "relevant" institutions and ideologies that will support and assist industrial and economic growth.

Since the fall of the Berlin Wall and the end of communism, academics have searched for new interpretations of development and policy initiatives in what could be called the world's peripheral economies (Gwynne et al. 2003). It might be useful to go back to Gerschenkron's nuanced view on economic backwardness and the idea of many pathways out of such a condition. He privileged historical discontinuity over linearly conceived continuities and traditions. Furthermore, he was also explicitly relational in the sense that later industrialisers should be seen as different precisely because they were affected, often through demonstration effects, by the first-comers (Gootenberg 2001: 57).

Major works

Two main collections contain Gerschenkron's major essays:

Gerschenkron, A. (1962) *Economic Backwardness in Historical Perspective: A Book of Essays*, Cambridge, MA: Harvard University Press.

—— (1968) *Continuity in History and Other Essays*, Cambridge, MA: Harvard University Press.

Further reading

Bennett, D.C. and Sharpe, K.E. (1985) *Transnational Corporations versus the State: The Political Economy of the Mexican Auto Industry*, Princeton, NJ: Princeton University Press.

Blaug, M. (1998) *Great Economists since Keynes*, Cheltenham: Edward Elgar.

Dawidoff, N. (2002) *The Flyswatter: How My Grandfather Made His Way in the World*, New York: Pantheon.

Dore, R. (1990) "Reflections on Culture and Social Change", in G. Gereffi and D.L. Wyman (eds), *Manufacturing Miracles: Paths of Industrialisation in Latin America and East Asia*, Princeton, NJ: Princeton University Press.

Forsyth, D.J. and Verdier, D. (eds) (2003) *The Origins of National Financial Systems: Alexander Gerschenkron Reconsidered*, London: Routledge.

Gootenberg, P. (2001) "*Hijos* of Dr. Gerschenkron: 'Latecomer' Conceptions in Latin American Economic History", in M.A. Centeno and F. López-Alves (eds), *The Other Mirror: Grand Theory Through the Lens of Latin America*, Princeton, NJ: Princeton University Press, 55–80. (An excellent essay on the contemporary applicability of Gerschenkron's theories in Latin America.)

Gregory, D., Johnston, R., Pratt, G., Watts, M.J. and Whatmore, S. (eds) (2009) *Dictionary of Human Geography*, 5th edn, Chichester: Wiley-Blackwell.

Gwynne, R.N. (1990) *New Horizons? Third World Industrialisation in an International Framework*, Harlow: Longman. (Analyses Gerschenkron and his policy links to Third World industrialisation.)

—— Klak, T. and Shaw, D.J.B. (2003) *Alternative Capitalisms: Geographies of Emerging Regions*, London: Edward Arnold.

Johnston, R.J., Gregory, D., Pratt, G. and Watts, M. (2000) *The Dictionary of Human Geography*, 4th edn, Oxford: Blackwell (5th edn 2010).

Keeble, D.E. (1967) "Models of Economic Development", in R.J. Chorley and P. Haggett (eds), *Socio-Economic Models in Geography*, London: Methuen, 243–302.

Landes, D. (1999) *The Wealth and Poverty of Nations*, London: Abacus.

Love, J. (1996) *Crafting the Third World: Theorizing Underdevelopment in Rumania and Brazil*, Stanford, CA: Stanford University Press.

Rosovsky, H. (1966) "Japan's Transition to Modern Economic Growth, 1868–1885", in H. Rosovsky (ed.), *Industrialization in Two Systems: Essays in Honor of Alexander Gerschenkron by a Group of His Students*, New York: John Wiley, 91–139.

Rostow, W.W. (1960) *The Stages of Economic Growth: A Non-Communist Manifesto*, Cambridge: Cambridge University Press.

Simon, D. (2019) *Holocaust Escapees and Global Development: Hidden Histories*, London: Zed Books.

Trebilcock, C. (1981) *The Industrialisation of the Continental Powers, 1789–1914*, London: Longman. (Strongly supports Gerschenkron's ideas in terms of industrialisation in eastern and southern Europe.)

Wade, R.H. (1990) *Governing the Market: Economic Theory and the Role of Government in East Asian Industrialization*, Princeton, NJ: Princeton University Press.

ROBERT GWYNNE WITH DAVID SIMON

JAYATI GHOSH (1955–)

Jayati Ghosh, born on 16 September 1955, is a leading economist in the world whose work has significantly contributed to the understanding about the way globalisation has shaped economic and sociocultural aspects of societies. She is currently a Professor at the Centre for Economic Studies and Planning, School of Social Sciences, at the Jawaharlal Nehru University (JNU) in New Delhi. Her research interests include globalisation, international trade and finance, employment patterns, macroeconomic policy, gender issues, poverty and inequality; these seemingly wide-ranging issues are connected in the way she has dealt with them, vis-à-vis the trajectory/ies that capitalism has taken in the recent past. She is currently a member of the editorial boards of *World Economic Review, Review of Social Economy, Journal of International Development, Journal of Post Keynesian Economics* and *Review of Keynesian Economics*, and on the Advisory Boards of several international institutions.

Ghosh's first major childhood influence was her father, a well-known bureaucrat and an economist; his progressive perspective on society and the range of books she grew up with exposed her to issues that she might not otherwise have been acquainted with.[1] Another notable influence in her life, both in childhood and youth, was Dr Ashok Mitra, an Indian Marxist political economist, who was the chief economic adviser to the Government of India, the Finance Minister of the state of West Bengal and a member of the Rajya Sabha, the upper house of the Parliament of India, as well as being her father's friend.[2]

> After my own father, Ashokkaka, as we called him, was probably the most important man to have shaped my approach to life and intellectual development.... Typically, he was more attracted to the rebels, the questioners, the idealists. But he held them to high standards.
>
> (Ghosh 2018)

After completing her schooling in India, Ghosh undertook her undergraduate studies in sociology at Miranda House in Delhi University. She joined the Master's programme in Economics in JNU, encouraged by Dr Mitra's advice, since he was a visiting professor there at that time. Because of her background in sociology, she remains "greatly interested in how societies and social relations are formed, evolve and transform over time".[3] Though she did not have a background in economics, which demanded some technical knowledge, including some knowledge of mathematics, she equipped herself with these skills before joining the course, since she felt strongly that social processes were deeply embedded in economic forces, and that it was not possible for her to appreciate societal changes without a comprehension of how the economy works.

Joining JNU as a master's student, according to her, was the watershed of her life, mostly because of the influence of the institution, which is one of the most progressive, politically vibrant space for higher education in India. She joined JNU during the commencement of the Emergency[4] in India, in a politically charged environment, and the department and the university had the effect of "opening [my] eyes and challenging me in ways that were completely new to me".[5] While the university and her professors in JNU collectively had a great deal of influence on her, she counts Professor Prabhat Patnaik, subsequently her colleague, as the most inspirational teacher and mentor, who probably had a lasting effect not only on her academic trajectory but also on her engagement in activism. She also greatly values the long academic partnership she has shared with her colleague, Professor C.P. Chandrasekhar. Ghosh subsequently moved to Cambridge University in the UK for her PhD, which she completed in 1984 with a thesis entitled "Non-capitalist Land Rent: Theories and the Case of North India", under the supervision of Professors G. C. Harcourt and T.J. Byres.

Before joining her alma mater, JNU, as an Assistant Professor, she held a research position at Darwin College, University of Cambridge. On returning to India she worked as a consultant to the Indian Planning Commission, before joining JNU as Assistant Professor in early 1986.

Ghosh identifies herself as a Marxian/post-Keynesian political economist, but also definitely considers herself to be a feminist and a socialist. She has become increasingly wary of the compartmentalisation of epistemological approaches, and is less concerned about what she thinks are "relatively pointless debates about the 'correct' understanding of Marx and other scholars", which often leads to "esoteric arguments about fine conceptual distinctions and purity of theoretical positions".[6]

She holds that she is comfortable using insights from different schools to frame her own understanding, as long as they contribute to a deeper understanding of economic processes. Thus, while she remains a fierce critic of the mainstream neoclassical approach, her relatively fluid methodological approach distinguishes her from many other Marxist scholars.

Ghosh's work deviates from, and is relatively unique in comparison to, the majority of the work particularly in the analysis of the financial sector, which is commonly rooted directly or tangentially in the neoclassical approach. This, as Ghosh argues in one of her interviews, is an "ahistorical" take, confined to what she considers a "much narrower and ultimately a less useful framework".[7] Her work is, in contrast, firmly anchored to development economics, and a political economy framework. She has been consistently critical of the way financial liberalisation has panned out, particularly in developing countries, on grounds of financial fragility and greater propensity to crisis, based on her work on not only India but other developing countries. She maintains that unbridled liberalisation of the financial sector leads to deflationary and dampening developmental effects, specifically on employment and environment, among other things. She states: "this is why it is increasingly recognized that liberalization can dismantle the very structures that are crucial for economic growth" (Ghosh 2005: 13). The dampening effect on employment stems from limited investments and slower growth of "real" economy, i.e. agriculture and industry, as opposed to growth stemming from the financial sector; she thus makes a clear distinction between effects of growth driven by these two sectoral categories (Ghosh 2010).

One of her central arguments is that neoliberal economics itself causes the biggest dent in its own assumptions; the state, instead of stepping aside as per the classic neoliberal prescription, in reality provides impetus to the big private players, at the expense of the masses. The increasing inequality across the world, particularly in emerging and developing countries, thus cannot be understood merely against the backdrop of increased opening up or privatisation, but through the changing role of the state that defines the way privatisation impacts the economy and people at the margins (Ghosh and Chandrasekhar 2009).

Her work on the east Asian Crisis of the late 1990s, as well as the effects of the 2008 global recession, provides a critique of the casual resumption of the growth model that caused the crisis in the first place, initiated by bailouts of large financial corporations without adequate regulations. Ghosh also establishes fairly conclusively that the popular narrative of "decoupling" of the developing and emerging economies from the advanced economies is not tenable (Ghosh 2010b). This idea, she argues, was based on the trends between 2000 and 2008, which turned out to be an aberration rather than the rule. The trends during the recession of 2008 and after it demonstrate the strong and continued affinity of the economic trends of these two sets of countries; the implication of this, she argues, is that the exhaustion of advanced capitalism is likely to have a bearing on the growth of the emerging and developing economies, which are nowhere near reaching their own potential. Additionally, her work maintains that the risk that this process holds for continued and potentially intensifying domestic and global imbalances makes it unsustainable for the other developed and developing nations to depend on the USA as the central engine of growth (Ghosh and Chandrasekhar 2009).

Ghosh's wide-ranging scholarship on employment is intrinsically connected with her work on the financial sector through the thread of the positioning of the state in embracing globalisation. The much-touted potential of the "demographic dividend" or the substantial share of youth in India's population, according to her, cannot be tapped without adequate public investments in health and education and a policy focus on employment generation. In joint work, Ghosh points out that the problem stems from the general processes of neoliberalism, rather the specificities in India, which at best partially explains the context-specific manifestations (Chandrasekhar et al. 2006). Chandrasekhar and Ghosh (2007) draw similarities between China and India, in terms of simultaneous periods of high economic growth and low employment growth. This, they argue, is not accidental and can be explained by at least three factors: trade liberalisation, increased competition and adoption of capital-intensive technologies to bring down cost of production; an agrarian crisis coupled with mechanisation of agriculture with labour displacing effects; and macro-economic policies indifferent to employment generation with an overt emphasis on macroeconomic growth and inflation control. Ghosh favours a complete reorientation of policy perspectives that visualises greater revenue mobilisation through taxation as well as even potential fiscal deficits stemming from public investments in social sectors, in order to sustain and promote an inclusive long-run growth trajectory.

Ghosh's work on gender can be situated within her research on labour and employment, in which she displays a firm conviction that a gender perspective is essential to understanding capitalism. However, according to her, the significance stems not only from a gendered division of labour and the ways in which that affects labour markets and social reproduction, but from the fact that most of the important features of contemporary globalisation rely in some way on the gender construction of societies (Ghosh 2012; 2015). This argument was also made in more specific terms in a prior work (Ghosh 2004), citing the case of export processing zones, where much of the use of women's labour in India has been in informal and unorganised workplaces, including home-based work, with associated implications for pay and working conditions, in a sector that was consistent with the copybook pathway to economic growth in the neoliberal period.

The third strand of Ghosh's work straddles issues of inequality, poverty and the social sector. This ranges over (but is not restricted to) caste-based quotas, quality of life and the potential of employment guarantee schemes, and the social impact of accumulation process. It is in issues such as this, and her work on gender, that her early background in sociology and a deep concern about the connections between social manifestations of economic processes clearly emerge. Probably her most potent contribution within this work is the understanding that private accumulation has relied upon existing social inequalities to create segmented labour markets, which in turn keep wages of certain social categories low, as well as enabling over-exploitation of nature because of the lack of political voice of dispossessed groups, leading to selective types of exclusions (Ghosh 2015).

Professor Ghosh's contributions have been recognised through several national and international prizes, including the M. Adisheshaiah Award for distinguished contributions to the social sciences in India in 2015; the International Labour Organization's Decent Work Research Prize for 2010; the NordSud Prize for Social Sciences 2010 of the Fondazione Pescarabruzzo, Italy; the Ava Maiti Award and the Satyendranath Sen Prize from the Asiatic Society, Kolkata; and the UNDP Award for excellence in analysis for the West Bengal Human Development Report 2004, of which she was the chief author. She has advised governments in India and other countries at different levels, including as Member of the National Knowledge Commission of India between 2005 and 2009. She has consulted for several international organisations including ILO, UNDP, UNCTAD, UN-DESA, UNRISD and UN Women. She is the Executive Secretary of International Development Economics Associates (www.networkideas.org), an international network of heterodox development economists. She is also one of the founders of the Economic Research Foundation in New Delhi, a non-profit trust devoted to progressive economic research.

As a scholar who recognizes herself as a progressive feminist, Ghosh believes that progressive alternatives to the current economic and social paradigm remain a distinct possibility. The spirit of the activist in this academician can be well captured in her own words:

> If I had to dream of my ideal society, I would say it is one in which someone's basic conditions and opportunities would not depend so critically the accidents of birth, determined by location, class, social group, gender: everyone's life chances would be broadly similar, but at the same time other cultural differences would be respected and even cherished.[8]

Notes

1 Based on an interview with Professor Jayati Ghosh by the author on 8 August 2018.
2 Jayati Ghosh, "Ashok Mitra", *The Wire*, 4 May 2018.
3 Interview of Jayati Ghosh by C.J. Polychronieu on 1 May 2018: https://truthout.org/articles/may-day-2018-a-rising-tide-of-worker-militancy-and-creative-uses-of-marx/.
4 "The Emergency" refers to a 21-month period from 1975 to 1977 when Prime Minister Indira Gandhi had a state of emergency declared across the country which gave her the authority to rule by decree, allowing elections to be suspended and civil liberties to be curbed. During this period, many of Mrs Gandhi's political opponents were imprisoned, the press was censored and several other human rights violations were reported. The Emergency is one of the most controversial periods of independent India's history.
5 See note 1.
6 See note 3.
7 See note 3.
8 See note 3.

Major works

Chandrasekhar, C.P. and Ghosh, J. (2001) "The Fragile Foundations of the Southeast Asian Miracle: Relocative Capital, the Proliferation of Capitalism and Industrial Growth", in K. Jomo (ed.), *Southeast Asia's Industrialization*, Basingstoke: Palgrave Macmillan, 59–85.

—— —— (2007) "Recent Employment Trends in India and China: An Unfortunate Convergence?" *The Social Scientist*: 19–46.

—— ——(2018) "The Financialization of Finance? Demonetization and the Dubious Push to Cashlessness in India", *Development and Change* 49(2): 420–436.

—— —— and Roychowdhury, A. (2006) "The 'Demographic Dividend' and Young India's Economic Future", *Economic and Political Weekly*: 5055–5064.

Ghosh, J. (1998) "Liberalization Debates", in T. J. Byres (ed.), *The Indian Economy: Major Debates Since Independence*, Delhi: Oxford University Press.

—— (2004) "Globalization, Export-Oriented Employment for Women and Social Policy: A Case Study of India", in S. Razavi, R. Pearson and C. Danloy (eds), *Globalization, Export-Oriented Employment and Social Policy*, Basingstoke: Palgrave Macmillan, 91–125.

—— (2005) "The Social and Economic Impact of Financial Liberalisation: A Primer for Developing Countries", New York: UN-DESA Working Paper.

—— (2009) "Informalization and Women's Workforce Participation: A Consideration of Recent Trends in Asia", in S. Razavi (ed.), *The Gendered Impacts of Liberalization*, London and New York: Routledge, 179–206.

—— (2010a) "Global Crisis and Beyond: Sustainable Growth Trajectories for the Developing World", *International Labour Review* 149(2): 209–225.

—— (2010b) "The Unnatural Coupling: Food and Global Finance", *Journal of Agrarian Change* 10(1): 72–86.

—— (2010c) "Global Crisis and Beyond: Sustainable Growth Trajectories for the Developing World", *International Labour Review* 149(2): 209–225.

—— (2011a) "Financial Crisis and the Impact on Women: A Historical Note", in N. Visvanathan, L. Duggan, N. Wiegersma and L. Nisonoff (eds), *The Women, Gender and Development Reader*, Chicago: University of Chicago Press, 22–27.

—— (2011b) "Dealing with 'The Poor'", *Development and Change* 42(3): 849–858.

—— (2012a) "Women, Labor, and Capital Accumulation in Asia", *Monthly Review* 63(8): 1.

—— (2012b) "Accumulation Strategies and Human Development in India", *Agrarian South: Journal of Political Economy* 1(1): 43–64.

—— (2013) "Microfinance and the Challenge of Financial Inclusion for Development", *Cambridge Journal of Economics* 37(6): 1203–1219.

—— (2014) "Financial Crises and Their Gendered Employment Impact: Emerging Trends and Past Experiences", in R. Antonopoulos (ed.), *Gender Perspectives and Gender Impacts of the Global Economic Crisis*, London and New York: Routledge, 17–48.

—— (2015) "Growth, Industrialisation and Inequality in India", *Journal of the Asia Pacific Economy* 20(1): 42–56.

—— (2018) "A Note on Estimating Income Inequality Across Countries Using PPP Exchange Rates", *Economic and Labour Relations Review* 29(1): 24–37.

Ghosh, J. (forthcoming) "Global Instability and the Development Project", *European Journal of Economics and Economic Policy*.

—— and Bharadwaj, K. (1992) "Poverty and Employment in India", in H. Bernstein (ed.), *Rural Livelihoods: Crisis and Responses*, London: Oxford University Press.

—— and Chandrasekhar, C.P. (2009) "The Costs of 'Coupling': The Global Crisis and the Indian Economy", *Cambridge Journal of Economics* 33(4): 725–739.

—— —— and Patnaik, P. (2017) *Demonetisation Decoded: A Critique of India's Currency Experiment*, New Delhi: Routledge India.

Patnaik, P. and Ghosh, J. (1991) "De-Industrialisation Without an Import Surplus: A Theoretical Note in the Context of a Stylised Colonial Economy", *Economic and Political Weekly*: 649–654.

Reinert, E.S., Ghosh, J. and Kattel, R. (eds) (2016) *Handbook of Alternative Theories of Economic Development*, Chichester: Edward Elgar.

SUCHARITA SEN

EDUARDO GUDYNAS (1960–)

Eduardo Gudynas was born in Montevideo (Uruguay) on 4 January 1960, and received a Catholic (though not liturgical, he says) education. At university he studied biology and zoology and, motivated by his passion for environmental matters, ended up in the Franciscan Centre for Human Rights and Ecology in Montevideo. There his scientific training was complemented by a perspective oriented towards human rights and poverty. His MSc thesis on Environmental Movements in Latin America was defended in 1999. From 2000 onwards he engaged in a number of consultancies in the broad Amazon region, some funded by the Friedrich Ebert Foundation and various UN bodies. He ran into the usual clashes with external funders that may have started his engagement with indigenous knowledge and the danger posed by outside experts. By now his work was focused around CLAES (Centro Latino Americano de Ecología Social – Latin American Social Ecology Centre – based in Montevideo). A key influence during his formative phase was the "popular education" tradition of **Paulo Freire**. He then drifted away from the world of consultancy to independent grassroots education work with community groups. He describes himself as a militant, not an academic. Unlike many of his generation, he is hardly at all influenced by Marxism. In Uruguay, he is not aligned with the governing Frente Amplio but considers himself part of an independent Left.

Gudynas situates himself on post-development terrain but maintains his distance from **Arturo Escobar**'s current perspective, which he sees as positing a new epoch beyond development. For Gudynas, the post-development lens should rather be seen as a post-structuralist-type device to deconstruct development discourses. It is not a self-sufficient alternative development path or strategy. That would have to be constructed through a critique of the new extractivism now dominating development prospects in Latin America, but mainly by articulating the vision of Buen Vivir (on both of these see below). The analysis Gudynas carries out on the "varieties of development" is based on the "varieties of capitalism" debate of the 1980s. Basically he argues that all the debates around different development models or theories all have as a common core a shared neglect or externalising of Nature. The "ideology of progress" which is integral to the Western notion of modernity is seen to permeate all the varieties of development thinking, to the extent that it is pointless to discuss "alternative development" models within this paradigm. Essentially, we should be looking for alternatives to development. The "varieties of capitalism" parallel could be read differently, of course, as some saw that debate as a way of exploring how not all forms of capitalism were the same. But this is an ongoing debate in and around post-development.

It is extractivism, or neo-extractivism to be precise, that is seen to dominate the current development horizon, at least in Latin America, which is the focus of Gudynas's work. He makes a clear distinction between the classic extractivism – transnational corporations exploiting natural resource enclaves – and the new or "progressive" extractivism practised by the left-of-centre governments in Latin America since 2000. This variant has seen the return of the developmental state, or what Gudynas calls the "compensatory state", which describes a state that is oriented towards economic development and is functional to capitalism, but which also needs to compensate for the negative effects of the market in terms of the social deficits it accumulates. This state seeks to regulate extraction and appropriate revenues for social redistribution. Carried out in the "national interest", neo-extractivism is given a certain degree of popular legitimacy. This project is having considerable impact in Latin America, as it drives a massive process of internal spatial transformation and power shifts. It has also been characterised by sharp political struggles which have led to the breakdown of political alliances of the ruling coalition with indigenous organizations and grassroots groups in Ecuador due to its plan to promote oil and mining extraction. This led to the resignation in 2013 of Ecuador's Minister of Energy and Mining, Alberto Acosta (a close colleague of Gudynas), from the Presidency of the Constituent Assembly. Later, the "progressive" President Correa reneged on the commitment to the ground–breaking Yasuní–ITT initiative in an area with rich diversity and home to two nomadic tribes that was to fight climate change in the Amazon rainforest by forgoing oil exploration in return for international compensation that never materialised.

It is clear that these debates are of more than academic interest in Latin America, and they set Gudynas apart from more reflexive thinkers. The alternative posed by Gudynas to the current state of affairs – his "alternative to development" as it were – is based on the ontology of Buen Vivir (living well or in harmony with nature) that has generated much interest in the Northern academy. While it has many variants and there is an ongoing discursive struggle around it, the basics of Buen Vivir are clear enough. It is a knowledge paradigm based on a harmonious spatio-temporal relationship between humans and nature. It is based, to varying degrees, on the cosmovision of the indigenous peoples of America or Abya Yala – and also on critical positions from within modernity. It has been part of a state-led "progressive" political project in the Andean countries but also, in its Sumak Kawsay variant, a project of the indigenous movements themselves. For Gudynas (and for Acosta), Buen Vivir is a "utopia to be constructed", a way of building tomorrow today. It is a post-extractivist post-development strategy that shares features with deep ecology. It is also postmodern in the sense that it breaks with the illusions of the Western modernity paradigm. It is not confined to the indigenous communities of the Americas in terms of its remit and ambition.

Gudynas is at pains to describe himself as politically independent, and he has been a fierce critic of the "progressive" governments of Bolivia and Ecuador in particular (which he describes as "extractivist") but also of the left-of-centre governments in Brazil and the Southern Cone. For Gudynas, the Left turn in Latin American politics since 2000 is, in many ways, a step backwards from the previous more open theory and practice. The "progressive" governments that dominated from around 2000 to 2015 he sees as now being in retreat on many fronts, although he does not share the common view of the "end of the Left tide", as that drive will continue for him. The previous critique of development shared across progressive circles was now being replaced under progressive governments by a quite uncritical acceptance of economic growth as the main objective. The flourishing of pluralist democracy in the struggles against dictatorships in the 1980s was now replaced by a quite authoritarian presidentialist style of government. Above all, the Left and labour movement's engagement with the "new" social movements in the 1990s was long forgotten, as these organizations returned to the sectarian or corporatist politics of the past. The "progressive" governments have tended to find the environmental, indigenous and peasant movements, in particular, a challenge for governability and not a source of strength and inspiration. There is currently a wide debate on the future of progressive politics "after" the Left tide, as the remaining governments of this ilk struggle to survive.

Appreciation

Joan Martínez Alier commented on an ill-tempered response by a research group in Ecuador to a critique by Gudynas of what he called "David Harvey's friendly colonialism" (https://entitleblog.org/2015/10/20/gudynas-and-harvey/). Gudynas was basically complaining of how notions such as "accumulation by dispossession" were being adopted uncritically in Latin America, marginalising more relevant indigenous thinking around the "new extractivism", for example. Martínez Alier admitted that "Gudynas' critique is a simplification", but goes on to show how Harvey's approach is not relevant for the Latin American theory of ecologically unequal trade. This exchange is emblematic of North–South knowledge imbalances, and of the way a thoughtless reproduction of Northern concepts in the South causes huge problems. It also highlights the importance of the work – both academic and political – by Gudynas and others on the politics of ecology in Latin America and the creative potential of the Buen Vivir vision. This is a work in progress; it does not benefit from well-funded research teams, and it is exemplary of the new Latin American critical confidence, reminiscent of dependency days in the 1960s (see the essays on **Fernando Henrique Cardoso**, **Andre Gunder Frank** and **Celso Furtado**), although in a very different time, of course.

Eduardo Gudynas represents a current of new thinking and practice in Latin America – which includes Alberto Acosta and Maristella Svampa – dedicated to renewing our understanding of development. Whereas dependency once stood the Northern construct of modernisation on its head, this new approach questions the mainstream and radical views of development alike. Its bid to "build tomorrow today" may sound utopian, but resonates with the earlier debates on pre-figurative socialism. It links with the very live tradition of **Paulo Freire**, and works to communicate and build community support for such a vision and programme for change.

Major works

For further information and updates see www.Gudynas.com.

Gudynas, E. (2013) "Debates on Development and Its Alternatives in Latin America: A Brief Heterodox Guide", in M. Lang and D. Morkani (eds),

Beyond Development: Alternative Visions from Latin America, Amsterdam: Transnational Institute.
—— (2014) *Derechos de la naturaleza ética biocéntrica y políticas ambientales*, La Paz: Plural.
—— (2015) *Extractivismos. Ecología, economia y política de un modo de entender el desarollo y la naturaleza*, Cochagamba: CEDIB.
—— (2016) "Beyond Varieties of Development: Disputes and Alternatives", *Third World Quarterly* 37(4): 721–732.
—— (2017a) *Extractivismos y corrupción. Anatomía de una íntima relación*, Lima: CooperAcción.
—— (2017b) "Post-development and Other Critiques of the Roots of Development", in H. Veltmeyer and P. Bowles (eds), *The Essential Guide to Critical Development Studies*, London: Routledge.

RONALDO MUNCK

DAG HAMMARSKJÖLD (1905–1961)

Dag Hjalmar Agne Carl Hammarskjöld was born on 29 July 1905 in the Swedish town of Jönköping, as the youngest of four sons, to Agnes and Hjalmar. His father was away, heading the delegation negotiating Norwegian independence. Appointed as governor of the Uppland region, he moved in 1907 with his family to Uppsala. He was also appointed as Prime Minister by the Swedish King during World War I. At Uppsala University, Hammarskjöld obtained a Bachelor of Arts degree in 1925 and majored in Economics in 1928, completed a Bachelor of Law degree in 1930 and finally obtained a doctoral degree in Economics at Stockholm University in 1933, with **Gunnar Myrdal** as his discussant. Both subsequently collaborated – despite differences in their approaches (Hammarskjöld was mainly a follower of Keynesian economic policy) – for decades in the Swedish government and the United Nations (UN) (Appelqvist 2008).

Despite never being a member of any political party, Hammarskjöld enjoyed a comet-like career as a high-ranking civil servant in the Swedish social democratic government. From 1936 to 1945 he was Under-Secretary in the Finance Ministry, and from 1941 to 1948 Chair of the Swedish central bank's board of governors. He was one of the architects of the welfare state, and drafted the bill for the long-term fiscal policy in 1946. From 1947 he was the main negotiator in international economic affairs, including the administration of the

Marshall Plan, as well as setting up the Committee of European Economic Co-operation (CEEC), later to become the OECD.

As from 1951 he represented Sweden in the Permanent Mission to the UN, which he headed from 1952 until his appointment as the second UN Secretary-General the following year. During his terms in office, the institution was faced with the Cold War and the decolonisation of Africa, both at the centre of Hammarskjöld's diplomacy. While he avoided mapping his approach in scholarly writings, he frequently used major speeches and lectures (preferably at internationally renowned universities), as well as his Introduction to the UN's annual reports and statements in the context of the UN, for principled deliberations, which were posthumously compiled into a total of four volumes (Cordier and Foote 1972; 1973; 1974; 1975). He was soon recognised as being more General than Secretary, and his ethics remain a reference point in the UN system among staff and successors. A posthumously published personal notebook (Hammarskjöld 1964) displayed a largely unknown spiritual side as a pillar in his guiding principles.

For Hammarskjöld, the UN was, as stated in an address at Stanford University in June 1955, "an expression of our will to find a synthesis between the nation and the world" (Cordier and Foote 1972: 521). Upon his return from a visit to more than 20 African countries during December 1959/January 1960, a journalist inquired whether the ideological trends in Africa "stem from the inner realities facing African life today or whether they reflect the often repeated clichés of foreign ideology". Hammarskjöld's clarification left no doubt:

> I do not think that the rights of man is a foreign ideology to any people … It may be that the most eloquent and the most revolutionary expressions of the rights of man are to be found in Western philosophers and Western thinking, but that certainly does not make the idea a Western idea imposed on anybody.
>
> (Cordier and Foote 1975: 533f.)

On 26 January 1960, at the second session of the Economic Commission of Africa in Tangier, he declared: "Partnership and solidarity are the foundations of the United Nations and it is in order to translate these principles into practical measures of economic cooperation that we are gathered today" (Cordier and Foote 1975: 517). In a speech on "Asia, Africa, and the West" on 4 May 1959 at Lund University in Sweden, he had reminded his audience that "nobody should forget that colonization reflected a basic approach which may have been well founded in certain limited respects, but which often mirrored false claims, particularly when it touched on spiritual development. Applied generally, it was untenable" (Cordier and Foote 1974: 381). Commenting on the Western perspectives of the early 20th century, Hammarskjöld found it striking "how much they did *not* see and did *not* hear, and how even their most positive attempts at entering into a world of different thoughts and emotions were colored by an unthinking, self-assured superiority" (Cordier and Foote 1974: 382). For Hammarskjöld, the "richest satisfaction" lay in "meeting different spiritual traditions and their representatives", provided one "approaches them on an equal footing and with a common future goal in mind". He was confident that this approach would ensure progress "in the direction of a human community which, while retaining the special character of individuals and groups, has made use of what the various branches of the family of man have attained along different paths over thousands of years" (Cordier and Foote 1974: 382).

He dismissed any claims to superiority over others based on any kind of naturalist concept of dominance, and questioned the legitimacy sought by social classes to justify their privileges: "just [as] it cannot be argued that within a community an economic upper class holds its favored position by virtue of greater ability, as a quality which is, as it were, vested in the group by nature, so it is, of course, impossible to maintain this in regard to nations in their mutual relationships" (Cordier and Foote 1974: 383). He concluded: "We thus live in a world where, no more internationally than nationally, any distinct group can claim superiority in mental gifts and potentialities of development … no nation or group of nations can base its future on a claim of supremacy" (p. 384).

In February 1956 he stressed in an address:

> the main trouble with the Economic and Social Council at present is that, in public opinion and in practice, the Council has not been given the place it should have in the hierarchy of the main organs of the United Nations. I guess that we are all agreed that economic and social problems should rank equal with political problems. In fact, sometimes I feel that they should, if anything, have priority.
>
> (Cordier and Foote 1972: 668)

The introduction to the 16th Annual Report of the UN became Hammarskjöld's last programmatic statement:

> In the Preamble to the Charter, Member nations have reaffirmed their faith "in the equal rights of men and women and of nations large and small," a principle which also has found many other expressions in the Charter. Thus, it restates the basic democratic principle of equal political rights, independently of the position of the individual or of the Member country in respect of its strength, as determined by territory, population or wealth. The words just quoted must, however, be considered as going further and imply an endorsement as well of a right to equal economic opportunities.
>
> (Cordier and Foote 1975: 544)

In his last words to his staff, on 8 September 1961, he reiterated one of his fundamental principles: "If the Secretariat is regarded as truly international, and its individual members as owing no allegiance to any national government, then the Secretariat may develop as an instrument for the preservation of peace and security of increasing significance and responsibilities" (Cordier and Foote 1975: 564). For him, as stressed in his landmark lecture "The International Civil Servant in Law and in Fact", delivered at Oxford University on 30 May 1961, the UN Secretariat had to be independent from international pressures of the Member States and the Security Council, only guided by "the principles and purposes of the Charter which are the fundamental law accepted by and binding on all states" (Cordier and Foote 1975: 487).

Hammarskjöld "developed practices of international executive rule, such as fact-finding, peacekeeping, technical assistance, and civilian administration, that were to shape the management of decolonization" (Orford 2012: 4). His defined responsibility as the world's highest international civil servant set normative standards for global development. Despite being pro-Western, his execution of office could be qualified in today's jargon as anti-hegemonic. He and the 15 others on board died in a plane crash while approaching the mining town of Ndola at the Northern Rhodesian–Congolese border in the night of 17/18 September 1961. He was to meet Moise Tshombe, the leader of the Katangese secession, to find a solution to the conflict in the Congo. Since new disturbing evidence was compiled 50 years later (Williams 2011), the UN General Assembly has adopted several resolutions tasking an eminent person to investigate anew the dubious circumstances of the crash. Hammarskjöld was laid to rest in Uppsala. To date, he has been the only recipient of the Nobel Peace Prize awarded posthumously, in 1962.

As Secretary-General of the UN, Hammarskjöld reflected upon the challenges of global governance in a world of the Westphalian nation states, the governments of which are guided by different and often colliding interests. He understood his role and that of the UN Secretariat as an advocacy, which recognises the voices of mainly those member states which do not have the same weight as the big powers. For him, the UN was in the first place *their* agency, and he constantly emphasised that a world of inequality will not create the foundation for sustainable peace and development. His legacy is a UN diplomacy rooted in the normative values of the Charter as the only legitimate guiding principles for seeking a solution to conflicts.

Major works

Cordier, A.W. and Foote, W. (eds) (1972) *Public Papers of the Secretaries-General of the United Nations*, vol. II: *Dag Hammarskjöld, 1953–1956*, New York: Columbia University Press.

—— —— (eds) (1973) *Public Papers of the Secretaries-General of the United Nations*, vol. III: *Dag Hammarskjöld, 1956–1957*, New York: Columbia University Press.

—— —— (eds) (1974) *Public Papers of the Secretaries-General of the United Nations*, vol. IV: *Dag Hammarskjöld, 1958–1960*, New York: Columbia University Press.

—— —— (eds) (1975) *Public Papers of the Secretaries-General of the United Nations*, vol. V: *Dag Hammarskjöld, 1960–1961*, New York: Columbia University Press.

Hammarskjöld, D. (1964) *Markings*, New York: Knopf. (Several subsequent editions.)

Further reading

Appelqvist, Ö. (2008) *A Hidden Duel: Gunnar Myrdal and Dag Hammarskjöld in Economics and International Politics 1935–1955*, Stockholm: Stockholm University/Department of Economic History.

Ask, S. and Mark-Jungkvist, A. (eds) (2005) *The Adventure of Peace: Dag Hammarskjöld and the Future of the UN*, Basingstoke: Palgrave Macmillan.

Fröhlich, M. (2008) *Political Ethics and the United Nations: Dag Hammarskjöld as Secretary-General*, Abingdon and New York: Routledge.

Lipsey, R. (2013) *Hammarskjöld: A Life*, Ann Arbor, MI: University of Michigan Press.

Melber, H. (2018) *Dag Hammarskjöld, the United Nations and the Decolonisation of Africa*, London: Hurst.

Orford, A. (2012) *The Past as Law or History? The Relevance of Imperialism for Modern International Law*, New York: Institute for International Law and Justice/New York University of Law.

Stahn, C. and Melber, H. (eds) (2014) *Peace Diplomacy, Global Justice and International Agency: Rethinking Human Security and Ethics in the Spirit of Dag Hammarskjöld*, Cambridge: Cambridge University Press.
Urquhart, B. (1972) *Hammarskjold*, New York: Knopf.
Williams, S. (2011) *Who Killed Hammarskjöld? The UN, the Cold War and White Supremacy in Africa*, London: Hurst. (2nd expanded edn 2016.)

HENNING MELBER

GERALD K. HELLEINER (1936–)

Gerry Helleiner's contributions have taken in academic work, support to the creation and success of a range of organisations, and policy advice. Ultimately, he is perhaps best known for his close involvement in policy advice in African countries, above all Tanzania, and for the way that this involvement has played a central role in efforts to improve the relationship between governments of low-income developing countries and international financial institutions and aid organisations.

Helleiner's intellectual formation – like that of many of his generation – was shaped by the upheavals of European and international politics from the 1930s into the post-Second World War period. He was born in Austria in 1936, to parents whose families had fallen foul of the Nazis and who were forced to emigrate. He and his mother were on the last plane out of Vienna before the war began. The family went first to England and then settled in Canada, where his father, Karl, lectured in economic history. Gerald Helleiner graduated in 1958 from the University of Toronto, studying political science and economics. He then went to Yale, where he did a "straight" economics PhD. Nonetheless, his interests were already shifting to the plight and challenges of developing countries, and in 1961 he was appointed to the new Yale Economic Growth Center, where he was based until 1965, including a year spent researching and teaching in Nigeria. In 1965 he moved to the University of Toronto, where he, his wife, Georgia, and three children have been based since, in a career interspersed with stays overseas.

His time at the Economic Growth Center at Yale produced a comprehensive economic history of Nigeria from 1900 to the mid-1960s (Helleiner 1966). The book was written just as it was becoming obvious how dramatically oil production would change the economy and just before the two coups of 1966 (cited in the preface) that initiated a cycle of political instability in Nigeria. The principal emphasis of the book was on the way in which extensive expansion of export-oriented peasant agriculture had been the key to Nigeria's economic dynamism, along with the stimulus to expansion given by government spending, especially on infrastructure (road and rail), education, and employment in the public sector.

Something of the wide sweep of the radar of Helleiner's interests is also shown by his pioneering work on intra-firm trade, organised, among other places, in *Intra-Firm Trade and the Developing Countries* (1981), pointing out that a "high proportion of the international flow in the markets for goods and services … takes place within firms", and that these transactions typically take place "in consequence of central commands rather than in response to price signals and they are recorded at prices which can be arbitrarily established and do not necessarily have anything to do with market prices". Of course an obvious source of interest in intra-firm trade as it relates to firms operating in developing countries is the scope for "transfer pricing" by transnational firms as a ruse to evade taxation. However, the book looked at a fuller range of issues than this.

Many critics of orthodox economic theory and policy are technically weak in economics. Gerry Helleiner has always – by contrast – sustained a firmly independent analytical policy stance underpinned by an extremely high level of technical capability as an economist broadly in the Keynesian tradition. It is perhaps this strength as an economist, combined with his enduring sympathy for the challenges faced by governments in low-income countries, that has enabled him to play such an important role in efforts to forge a realistic and progressive dialogue between "the West" and the developing world.

This role has partly involved serving on – and sometimes helping to create – international organisations. These have included two Canadian bodies – the International Development Research Centre and the North–South Institute; acting as a member of the advisory board of the World Institute for Development Economics Research (WIDER, part of the United Nations University and based in Helsinki, Finland); providing support to the Intergovernmental Group of 24 on International Monetary Affairs; acting as Chair of the Board of Trustees of the International Food Policy Research Institute (IFPRI) from 1990, where he was credited with resuscitating an organisation at the time in crisis; and advising the African Economic Research Consortium (AERC). He was involved in the Canadian mission to support the Democratic Movement in South Africa in its formulation of post-apartheid economic policy, and served on the advisory board of the Macro-

Economic Research Group (MERG) that produced a policy report, *Making Democracy Work: A Framework for Macroeconomic Policy in South Africa* in 1993. He has also worked on other policy reform missions, including, twice, to Uganda and also to Guyana and Nicaragua.

But above all, Helleiner is renowned for his long-term involvement in and commitment to African economic development, especially in Tanzania. From 1966 to 1968 he took leave from Toronto to become the founding director of the Economic Research Bureau at University College in Dar es Salaam. His involvement in Tanzania lasted through the years of **Julius Nyerere**'s exercise in "African socialism" and the fraught relationships with the IMF and World Bank, and into the political and economic reform years of the 1990s. Tanzania's relationship with international lenders had broken down – not for the first time – in the mid-1990s. When some kind of rapprochement began to look possible, the government and the donors, with Danish assistance, agreed to appoint an independent team of experts to analyse the relationship and to advise on a way forward. This team, headed by Professor Helleiner, produced the *Helleiner Report* in 1995. This report was not only hugely constructive in Tanzania but more broadly became a seminal contribution in the evolution of ideas that aid would be more effective if there were greater "national ownership" of the reform process and the disbursement of aid funds. After this report, the Tanzanian government and the donors agreed a series of initiatives to monitor the transparency and accountability of *both* parties to the aid relationship.

Helleiner's concern with the ideals of ownership was stamped by his earlier experiences in Tanzania, his sympathy with much of what Nyerere was trying to achieve, and distaste for the high-handed behaviour of the IMF and what he saw as its unthinking and inappropriate application of neoclassical economic theory to conditions of extreme poverty. In a talk he gave on his own reflections on the Nyerere years, at a conference celebrating Nyerere's legacy after his death, much of the humane and independent approach Helleiner has taken to economic development was clear. Noting the unpopularity of Nyerere's villagisation programme, the unfeasibility of a sweeping nationalisation programme and the over-ambition of the basic industries strategy, Helleiner nonetheless reminded the audience of Nyerere's emphasis on free education, on equity, on a leadership code for politicians and on targeting the poorest. He argued that economic failure in Tanzania ultimately had very little to do with Nyerere's socialism; rather, it was rooted in external shocks (the oil price rises of the 1970s, the cost of war against Idi Amin in Uganda and adverse weather) and in bad macro-economic policy management in response to these shocks.

This argument continued a theme Helleiner had long insisted on. His "The IMF and Africa in the 1980s" (1983) was one of the earliest critiques of IMF conditionality and lending policy in Africa. The essay was all the more effective for the calm way it built its argument. The essay, which foresaw a period of increasing conflicts between African governments and the IMF, argued from the premise that the key to African economic crisis lay in deteriorating terms of trade combined with the rising cost of borrowed capital and, therefore, the desperate shortage of foreign exchange to cover the import requirements of stability and growth. Adjustment was inevitable.

> The question is not whether there should be a difficult and painful adjustment, on both the demand and the supply side, but how it should be undertaken. With what assistance? Over what time horizon? With the burden distributed how? With what mix of policies and what sequencing? With what terms for foreign borrowing?
>
> (Helleiner 2000: 11)

His critique of the IMF focused both on the inequity of the IMF's liquidity expansion and on the analytical inadequacies behind conditionality (as well as on the gormlessness of sending missions that had little or no knowledge or experience of Africa). The first argument was that IMF finance had not grown fast enough to match the value and significance of trade for poor countries while the terms of high-conditionality loans were growing tougher. The second argument was that the IMF's analytical models were blunt, and impervious to the variety of sources of balance of payments difficulties. Helleiner argued that the "only possible professional stance in these circumstances is of considerable humility and caution in the dispensing of advice" (Helleiner 2000: 17) – a stance he maintained and emphasised in his professional memoir (Helleiner 2018). He envisaged an alarming prospect for relations between African governments and the IMF in the coming years: "badly prepared antagonists of modest ability employing data of dubious quality and entering upon a series of battles over very complex policy questions" (2000: 22). This was a pretty accurate vision of how the following decade unfolded in a number of cases, although over time there was typically less and less effective resistance within African governments to the model on offer.

Helleiner stressed a number of needs in this essay – more access to low-conditionality liquidity for the poorest African countries, more flexible conditional credit arrangements in the face of external shocks, a greater variety in the types of aid delivered by donors, institutional innovation to develop conflict resolution mechanisms to resolve tensions between the Fund and debtor governments, and a surge in training and research and data collection in Africa. Despite the HIPC (Highly Indebted Poor Country) initiative, some new IMF lending facilities, the possibility of a shift to a "post-Washington Consensus" in the World Bank, a range of data collection exercises and a shift towards prioritising "national ownership" of aid programmes and policies, many might argue that these needs remain unmet. For example, what Helleiner and many others understand by the term "ownership" appears still to differ from how the Fund sees it. In an IMF paper, ownership was defined as a "convergence of interests" between debtor countries and the IMF. Helleiner (2000: 3) argued that this effectively meant that the IFIs (international financial institutions) "now want local policy makers not simply to do what it recommends but also to believe in it".

The pragmatic and humane, moral and technically sure-footed approach that has characterised Helleiner's policy and organisational work has also stamped his academic writing, which has also been prolific. He has written or edited close to 50 volumes as well as publishing many articles in refereed journals and contributing chapters to edited books. His main academic writings have been in the fields of developing-country trade and finance and the role of transnational companies in developing countries. As with the rest of his work, this body of writing escapes glib "Left" or "Right" labelling, but is broadly within a Keynesian tradition, drawing on elements of structuralist economics, though he demurred when labelled a structuralist: "I've been allergic to dogma and religion of any kind ... Others have called me a structuralist [but] I'm not sure that means anything much" (quoted in Simon 2019: 257). He has also served on the boards of 15 academic journals.

Major works

Cornia, G.A. and Helleiner, G.K. (eds) (1994) *From Adjustment to Development in Africa: Conflict, Controversy, Convergence, Consensus?* Basingstoke: Macmillan.

Helleiner, G.K. (1966) *Peasant Agriculture, Government, and Economic Growth in Nigeria*, Homewood, IL: Richard D. Irwin.

—— (1973) "Manufactured Exports from Less Developed Countries and Multinational Firms", *Economic Journal* 83(329): 21–48.

—— (1981) *Intra-Firm Trade and the Developing Countries*, London: Macmillan.
—— (1983) "The IMF and Africa in the 1980s", Dept of Economics, Princeton, NJ: Princeton University, Dept of Economics.
—— (1989) "Transnational Corporations, Foreign Direct Investment and Economic Development", in H.B. Chenery and T.N. Srinivasan (eds), *Handbook of Development Economics*, vol. II, Amsterdam: North-Holland.
—— (1990) *The New Global Economy and the Developing Countries: Essays in International Economics and Development*, London: Edward Elgar.
—— (2000) "Towards Balance in Aid Relationships: External Conditionality, Local Ownership and Development", produced by Brian Tomlinson, as an amalgam of two earlier papers by Helleiner, for the Reality of Aid International Advisory Committee meeting, Costa Rica, September. Available at: www.devinit.org/realityofaid/jcrpapers.htm.
—— (ed.) (2001) *Non-Traditional Export Promotion in Africa: Experience and Issues*, Basingstoke: Palgrave.
—— (2018) *Toward a Better World: Memoirs of a Life in International and Development Economics*, Toronto: University of Toronto Press.

Selected mission reports

(1988) *Financing Africa's Economic Recovery*, report for UN Secretary-General.
(1989) *Guyana: The Economic Recovery Programme and Beyond*, report of a Commonwealth Advisory Group.
(1995) *The Helleiner Report: Report of the Group of Independent Advisors on Development Cooperation Issues Between Tanzania and Its Aid Donors.*

Further reading

Culpeper, R., Berry, A. and Stewart, F. (eds) (1997) *Global Development Fifty Years after Bretton Woods: Essays in Honour of Gerald K. Helleiner*, Basingstoke: Macmillan.
Simon, D. (2019) *Holocaust Escapees and Global Development: Hidden Histories*, London: Zed Books.

CHRISTOPHER CRAMER

ALBERT O. HIRSCHMAN (1915–2012)

Most of the major academic contributions of the economist Albert Hirschman fall within the political economy of development. In the community of development scholars, he was perhaps best known for his theory of unbalanced growth, whereby a pattern of quicker growth and more egalitarian development may eventually be stimulated by concentrating development efforts on key industries and locations in hitherto lagging regions, especially for the underdeveloped countries of the South. Hirschman excelled at placing economic phenomena and processes within their broader social context. His work on economic development, collective action, the political bases of economic thought and attitudes toward market society frequently addressed areas of thought unfamiliar to most economists. His writing illuminated a wide range of social behaviour that lies beyond simplistic rational-choice models, and it characteristically eschewed attempts to overgeneralise in favour of a more nuanced approach that paid special attention to the particular and the unique.

Hirschman was considered somewhat of a maverick within the field of economics, especially for his adoption of an interdisciplinary approach to development studies, unusual for contemporary economists. He was an iconoclastic trailblazer in his research on economic development, frequently stepping outside his own discipline in order to better address the complexities and particularities of local development processes in various areas of the world. He was the author of numerous influential and controversial works, often adopting novel concepts and ideas in disagreement with mainstream economic theory and major figures in the field. His work commonly exhibited an impatience with disciplinary boundaries that all too frequently separate economists from other scholars in development studies, limiting their ability to address fully the multifaceted nature of development. He was described as a playful genius, a person who loved being unconventional and did not shy away from controversy. Unsurprisingly, he was one of just ten eminent figures featured in the earliest assessment of leading "pioneers" in development (Meier and Seers 1984).

Over his long, varied and distinguished career, Hirschman worked as an economist with the US Federal Reserve Board; as an economic adviser in Bogotá, Colombia; and as a professor at Yale, Columbia and Harvard, where he was the Lucius N. Littauer Professor of Political Economy. He joined the School of Social Science at the Institute for Advanced Study in Princeton in 1975, becoming Professor Emeritus in 1985 (Simon 2019). His works have been the subject of numerous symposia in economics and related disciplines in development studies, and he has been the recipient of many awards and honours, including the 1983 Talcott Parsons Prize for Social Science from the American Academy of Sciences, the 1997–1998 Toynbee Prize and the 1998 Thomas Jefferson Medal awarded by the American Philosophical Society.

Hirschman was born on 7 April 1915 in Berlin to Jewish parents and, although baptised as an infant, to the Nazis he was Jewish. He was also a politically active opponent and therefore had to escape Germany immediately upon the Nazi ascent to power in early 1933. After attending the Sorbonne and the London School of Economics, he obtained a PhD in economic science from the University of Trieste in 1938. During and immediately following his doctoral studies, much of his life was dominated by the struggle against fascism in Europe, where he played an active role in several countries. While in Italy in the mid-1930s, he joined the underground opposition to Mussolini. In 1936 he fought with the Spanish Republican army and later with the French army until its defeat in June 1940. He remained in France for an additional six perilous months, taking part in clandestine operations based in Marseilles to assist political and intellectual refugees from Nazi-occupied Europe, mostly guiding them over the Pyrenees until his escape over the same dangerous route to the USA in January 1941 to avoid capture and arrest (Adelman 2013; Simon 2019).

Following his arrival in the USA, Hirschman's major focus in his early career concerned the structure of world trade, and especially the political implications of its asymmetries and patterns of concentration. These interests determined much of the material around which his first book (Hirschman 1945) was organised. In the immediate post-World War II era, Hirschman was particularly occupied, like many other development scholars of his generation, by economic problems (e.g. the "dollar shortage" in countries such as France and Italy) associated with efforts to smooth European reconstruction and integration. The book became immediately notable for the interdisciplinary approach it adopted, characteristic of Hirschman's work throughout his career, situated in the "grey zone" between economic and political theory. It asserted that economic power is often the handmaiden of political power between states. It also generated a good deal of controversy, another hallmark of Hirschman's career, in that it added to the growing scepticism of many development scholars over the way in which the Marshall Plan was being implemented in postwar Europe.

With the publication of his second book, *The Strategy of Economic Development* (1958a), Hirschman became widely recognised as a leading development scholar. The analysis upon which the book was based grew out of four years of practical experience in Colombia as an economic adviser to the government and several private firms. In the book, Hirschman strongly rejected the wholesale and unquestioning importation of the formal concepts and ready-made, universal strategies of conventional neoclassical economic theory, the application of which he believed inevitably sacrificed necessary attention to local developmental particularities in order to conform to the narrow strictures of modern scientific modelling. He contended that, rather than adopting conventional generalised prescriptions for economic development, the developmental needs of particular regions and countries ought to be assessed on a case-by-case basis, paying special attention to the particularities of indigenous resources and structures to achieve the desired developmental results. Efforts to impose formal universal models and strategies, regardless of the historical and structural realities of specific development contexts, were a sure-fire recipe for disastrous development policies and programmes, with many dire economic, political and socio-cultural consequences.

In the book, Hirschman argued against both laissez-faire and "rational" nationwide economic planning for developing countries. Especially in underdeveloped countries, development ought to be spurred by policies and programmes promoting "unbalanced growth", in which governments would deploy their scarce resources strategically to foster critical disequilibria important to encouraging private investment and related entrepreneurial activities in key industries and locations in lagging regions. Previously hidden and under-utilised resources would best be mobilised by targeting development efforts on key industries with strong forward/backward linkages to the local economy. Such linkages would generate "spread" effects, whereby the initial inequalities of unbalanced growth centred on industrial complexes would give way to a more egalitarian pattern of territorial deconcentration of the benefits of modern development to surrounding peripheral areas. If focused on dynamic industrial growth poles, an initial emphasis on unbalanced growth and socio-economic/spatial disequilibria would allow new governments in the South relatively quickly to overcome the common historical legacy of polarising "backwash" effects inherited from previous (neo)colonial spatial structures. These were dominated by outward-oriented primate cities delinked from much of the stagnant domestic economy and outlying peripheral regions.

Hirschman's concept of unbalanced growth underscores the particularities of resources and structures in peripheral regions upon which key industries could be constructed, thereby facilitating beneficial linkages to the local economy. His approach criticises conventional doctrinal models of economic development that neglect local particularities. In a later article, "The Rise and Decline of Development Economics" (1982a), he expanded this focus into a more general critique of the rise of "mono-economics" accompanying the ascendancy of neoliberal development strategies in the South. According to Hirschman, development economists may be divided into two basic types. On the one hand, there are those who take a mono-economic perspective, whereby orthodox neoclassical theory is equally applicable in both the global North and South. Although the South may have peculiar characteristics, such as higher levels of uncertainty, its economic agents make decisions and its markets function essentially according to the same logic as in the North. Most importantly for neoliberals, individuals or firms respond similarly to structures of incentives, whether they are (properly) established by the economic market or (improperly) set by the political market.

On the other hand, there are the followers of a "duo-economic" approach who believe that standard market economics is of limited relevance to the special problems of development in the South. Without substantial theoretical modification, they see conventional economic models and strategies as incapable of addressing problems associated with underdevelopment that arise from a series of particular Southern structural conditions. These include: massive unemployment and underemployment, despite instances of high growth and industrialisation; frequent market failures based on poorly developed circulation systems, financial networks, and other economic structures; the continuing influence of culture and tradition on common forms of behaviour that detract from utility maximisation; the persistence of extreme societal polarisation and inequalities, which are aggravated by poorly articulated social, sectoral and regional structures; and a pronounced vulnerability to externally generated crises due to high levels of foreign dependency and economic concentration within the external sector.

In the article, Hirschman was especially critical of the mono-economic approach of the conventional neoclassical paradigm. He claimed that neoliberals ought to remove their conceptual blinkers and methodological straitjackets in favour of a broader, more flexible vision of development capable of addressing the diverse realities of the South. The domination of development studies by "grand theories" has generated increasing tension between the desire to formulate universally valid principles and formal models and the need to understand the variety of actual experiences and potential alternatives of development in the South. Hirschman contends that solutions to development problems must be sought in the contextuality of development, which is a product of particular historical processes. The context of development is constantly changing in scale, over time and among societies – creating both new obstacles and new opportunities for variations. Development frameworks, especially those designed to contribute toward policy-making, need to devise ideas and methods capable of accommodating this geographical and historical diversity.

In the editor's introduction to another important book, *A Bias for Hope: Essays on Development and Latin America* (1971), Hirschman coined the term "possibilism" to describe his position against grand theorisation and overgeneralisation. In development studies, and the social sciences in general, the search for general laws often obscures the role of the particular and unique in human pursuits, which may produce much of the unpredictability that frequently confounds the scientific modelling of society. Typically, such general formulations view development progress in Southern societies linearly, based on formulations rooted in the history of the North, and according to the application of universal laws developed in and for Northern societies. The failure of development efforts, especially those of external origin, is commonly attributed to the presence of some inescapable obstacles to development unfortunately present in Southern societies. However, the development trajectories of particular societies often take paths that are a priori quite unlikely and cannot be easily anticipated; obstacles may become opportunities and vice versa, confounding rules with unexpected consequences. Hirschman's work often uses this intellectual starting point to get at the many complexities of development issues and problems, always leaving open possibilities for unanticipated phenomena and events to produce novel forms of social change.

In addition to his work in development studies, Hirschman also employed his possibilist approach to address broader problems of social theory. In *Exit, Voice and Loyalty* (1970), he offers an elegant and profound rethinking of individual choice within institutions or organisations, comparing the implications of dissatisfied clients alternatively exiting from an organisation or giving voice to their complaints. In this volume and much of his later work, including his final book, *Crossing Boundaries* (1998), Hirschman turned to social issues and problems which straddle the border between economics and politics. But he showed no inclination to follow the lead of many other economists who, via the reductionist framework of public choice theory, have essentially turned politics into economics. Instead, his work stressed that conventional models of economic behaviour based on rational choice cannot satisfactorily explain diverse forms of "public-minded" behaviour such as voicing one's convictions on public affairs, participating in demonstrations or working in support of candidates for public office.

Hirschman's disdain for reductionist, ideologically driven and invariant arguments is also clearly evident in his studies of historical views of capitalism (1977; 1982b). In later years he alluded to his penchant for "self-subversion" to stimulate intellectual inquiry (1995) – constantly and systematically revisiting his principal ideas, concepts and theories in order to add needed modifications, qualifications and complications in various ways. His studies of capitalism portray a complex and highly variable dynamic of development, one in which capitalism under certain circumstances may be a powerful civilising influence, but also may sometimes destroy a people's moral and social fabric, or may simply be too feeble to overcome the constraints of prior social forms. These competing ideological views of capitalism, in which seemingly similar processes may produce quite divergent results in different places and times, underscored Hirschman's overriding interest throughout his work to express rather than conceal the full range of human and social complexities that influence development processes, which can only be very partially addressed by using standard methods of economic science based on formal concepts and a priori models.

Major works

Hirschman, A.O. (1945) *National Power and the Structure of Foreign Trade*, Berkeley, CA: University of California Press.

—— (1958a) *The Strategy of Economic Development*, New Haven, CT: Yale University Press.

—— (1958b) *Development Projects Observed*, Washington, DC: Brookings Institution.

—— (1970) *Exit, Voice, and Loyalty: Responses to Decline in Firms, Organizations, and States*, Cambridge, MA: Harvard University Press.

—— (1971) "Introduction: Political Economics and Possibilism", in A.O. Hirschman (ed.), *A Bias for Hope: Essays on Development and Latin America*, New Haven, CT: Yale University Press.

—— (1977) *The Passions and the Interests: Political Arguments for Capitalism Before Its Triumph*, Princeton, NJ: Princeton University Press.

—— (1981) *Essays in Trespassing: Economics to Politics and Beyond*, Cambridge: Cambridge University Press.

—— (1982a) "The Rise and Decline of Development Economics", in M. Gersovitz et al. (eds), *The Theory and Experience of Economic Development: Essays in Honour of Sir W. Arthur Lewis*, London: Allen and Unwin.

—— (1982b) "Rival Interpretations of Market Society: Civilizing, Destructive, or Feeble?" *Journal of Economic Literature* 20 (December): 1463–1484.

—— (1982c) *Shifting Involvements: Private Interest and Public Action*, Princeton, NJ: Princeton University Press.

— (1984) "Against Parsimony: Three Easy Ways of Complicating Some Categories of Economic Discourse", *American Economic Review* 74(2): 89–96.
— (1986) *Rival Views of Market Society and Other Essays*, New York: Viking-Penguin International.
— (1987) "The Political Economy of Latin American Development: Seven Exercises in Retrospection", *Latin American Research Review* 22(3): 7–36.
— (1995) *A Propensity to Self-Subversion*, Cambridge, MA: Harvard University Press.
— (1998) *Crossing Boundaries: Selected Writings*, Cambridge, MA: MIT Press.

Further reading

Adelman, J. (2013) *Worldly Philosopher: The Odyssey of Albert O. Hirschman.* Princeton, NJ: Princeton University Press.
Coser, A. (1984) *Refugee Scholars in America: Their Impact and Their Experiences*, New Haven, CT: Yale University Press.
Foxley, A. et al. (eds) (1986) *Development, Democracy, and the Art of Trespassing: Essays in Honor of Albert O. Hirschman*, Notre Dame, IN: University of Notre Dame Press.
Meier, G. and Seers, D. (1984) *Pioneers in Development*, Oxford and Washington, DC: Oxford University Press for the World Bank.
Simon, D. (2019) *Holocaust Escapees and Global Development: Hidden Histories*, London: Zed Books.

JOHN BROHMAN WITH DAVID SIMON

PHILIPPE HUGON (1939–2018)

The French professor, writer and lecturer Philippe Hugon was born in Paris on 5 May 1939 and died in Versailles on 20 April 2018. He possessed all the academic credentials as well as the legitimacy of an experienced field researcher. Altogether he published 100 articles in specialised magazines and more than 30 books on development topics, Africa and international economic policy. At the University of Paris-Nanterre, where he exerted his intellectual mastery on two generations of students and researchers, he was at the heart of important scientific initiatives on the small informal economy, the analysis of agricultural value chains, the economics of education, the regionalisation of economies, the financing of development, globalisation and, last but not least, the futures of Africa.

His autobiography, published in 2013 (*Solidarity and Solitary Memories*), sets out how his political consciousness was built, from a bourgeois Christian education to a Third World commitment and then to a more academic and authoritative stance. It is a rich trajectory, faithful to the influence of some intellectual guides, in particular Albert Camus, Paul Ricoeur, Georges Balandier, Emmanuel Mounier, Father Lebret, **Albert Hirschman, François Perroux, Amartya Sen** and Stéphane Hessel.

Hugon was Professor of Economics at Paris-Nanterre University, founder of the Research Centre in the Economics of Development (CERED) at the University of Paris-Nanterre, co-founder of the Group for the Study of Globalization and Development (GEMDEV), Research Director at the Institute of Strategic International Relations (IRIS), among other roles. He was metaphorically "possessed" by Africa from his early assignment as a teacher in Cameroon (1963–1965) and then in Madagascar (1969–1974).

He took part in the great adventure of decolonisation. He affirmed his convictions as much as his approach:

> As economist, I think that it is necessary to understand development to start from a base and facts stylised for minimal quantification and consistency. But this approach must, to seize the complexity of the ground, take into account the historical trajectories, the way in which the men act within social reports, of systems of representation and cultural referents and which direction they give to this development.
>
> (Hugon 2013: 264, author's translation)

In this respect, Hugon is an economist of development, considering that the specificity of the least advanced countries requires that their history and vulnerabilities be taken into account. Thus it is essential to have a multi-disciplinary approach which avoids the logics of orthodox economic analysis, because market co-ordination alone cannot give intelligence to the facts. The perspective of *Homo economicus* cannot be the way to analyse Africa. Rather, this requires (a) taking account of the "social and cultural dimension of development"; (b) regarding poverty as the result less of an inequality of incomes than of inequalities in access to various capacities, along with the rights which determine the conditions of existence and nature of social integration; and (c) thus proposing to treat the obstacles to development as much in their institutional, cultural and economic dimensions as geopolitical. Hugon was convinced that a way exists which affirms the need for the state while seeking collaboration and a division of roles with other actors such as private-sector firms or civil-society organisations. More generally, underpinning his thought is an economic anthropological approach – see especially his famous article, "Is Homo Africanus Irrational?" (1993a) and "Anthropology and the Economy of Development or the Meeting of Robinson and Vendredi" (2001b). Specialising particularly in African economies, he published *Analysis of Underdevelopment in Black Africa: The Example of Cameroon* (1968); *Africa of Uncertainties* (1995); *The Economy of Africa* (2015, translated into English); and *Africa: Challenges and Prospects in 40 Maps to Understand the News* (2016a).

At the end of the 1980s he began publishing university textbooks and works on economic development. Particularly noteworthy are *The Economy of Development* (1989) but especially *The Economics of Development and Francophone Thought* (2009), which enriched his earlier "The Three Stages of Francophone Thought on Development" (1993b).

Hugon tackles the questions of the international economy and regionalisation of economies. According to him, it is not possible to switch between two completely opposite visions of the world – on one hand holding these views within a liberalism without limits, wanting foreign trade to be a major factor in development, and on the other, holding such views in a protectionist regime, considering that in relation to prevailing international asymmetries, it is necessary to think about an endogenous form of development. Between these two extreme positions, it is essential to see under which conditions foreign trade will or will not exert a leverage effect on economic growth and induce development. Hugon invites us to a precise analysis of situations by mobilising the tools of international political economy. His major works are *International Political Economy and Globalisation* (1997) and *The Underdeveloped Economies at the Time of Regionalisation* (2003b). Reflecting on the international political economy, he tackles the question of global public goods, because they are directly related to the context of globalisation and the "double movement" (see **Karl Polanyi**) which characterises it. On one hand is the relative decline of public space and public authorities vis-à-vis the rise of the private sphere and the powers required to produce collective goods. On the other hand is the overflow of the nation states vis-à-vis the internationalisation and a transnationalisation of many questions like epidemiological risks, climate change and water management. From this perspective, Hugon nourishes debate on two radically opposite designs of the concept of global public goods – between a neoclassical approach and an international political economy approach (see especially Hugon 2001; 2003c).

His more recent works on the geopolitics of Africa make clear the complexity of the situations, particularly in the Sahelian region (Hugon 2003a; 2016a; Courtin et al. 2018). His final work on the subject, *Africas Between Power and Vulnerability* (Hugon 2016c), is a reference synthesis which gives the reader the key to understand the possible futures of Africa and, in particular, the idea that emergence cannot be approached by reproducing the identical trajectories taken by the countries currently known as "developed". Philippe Hugon's perspective on African trajectories derives neither from Afro-pessimism nor from Afro-optimism. It is a clear perspective on the multiple exogenous and endogenous constraints confronting the continent, which explains the politico-economic cycles as fitting into a "regressive trend". International insertion strongly centred on the functioning of an *économie de rente* (rentier economy), which does not really create local added value, and is thus highly vulnerable faced with unstable world demand. Foreign actors always play a principal role in these trajectories. In these economies, population growth in Africa will remain the strongest in the world until at least 2050, with the risk of seeing the demographic dividend transforming itself into an economic dead end of transition, to the extent that the question of the employment of young people remains unsolved. The risk of seclusion in poverty traps is accompanied by a conflict trap, in particular in the West African region. Among the five scenarios for the future which Hugon formulates, most probable is that which stresses the territorial social and economic fractures, and increasing differentiation between great regions: "Africa will be, according to any increasingly contrasted and differentiated probability, with plural trajectories, coexistences of zones of prosperity and innovations and vulnerability, even of conflict" (Hugon 2016c: 255). Is it possible to design a sustainable development model for Africa? Philippe Hugon takes care not to provide a normative answer, but he provides cogent analysis that this model will not be that of the Asian emerging countries, because international conditions have changed so much since the advent of these "miracles". Because the situations are specific and complex, it is essential to create "made to measure" and not "ready to wear" solutions.

Major works

Courtin, N., Hugon, P. and Le Gouriellec, S. (2018) "Contours, entours et détours des armées en Afrique", *Afrique contemporaine* 260: 15–25.

Gabas, J.J. and Hugon, P. (2001) "Les biens publics mondiaux et la coopération internationale", *L'économie politique* 12: 19–31.

Hugon, P. (1968) *Analyse du sous-développement en Afrique noire. L'exemple du Cameroun*, Paris: Presses universitaires de France.

—— (1989) *L'économie du développement*, Paris: Dalloz.

—— (1993a) "L'Homoafricanus est-il irrationnel?" *Cahiers des sciences humaines, 1963–1992*: 57–60.

—— (1993b) "Les trois temps de la pensée francophone en économie du développement", in C. Choquet et al. (eds), *État des savoirs sur le développement. Trois décennies de sciences sociales en langue française*, Paris: Karthala, 43–74.

—— (1995) *L'Afrique des incertitudes*, Paris: Presses universitaires de France.

—— (1997) *Économie politique internationale et mondialisation*, Paris: Economica.

—— (2001) "Anthropologie et économie du développement ou la rencontre de Robinson et de Vendredi", *Journal des anthropologues* 84: 37–52.

—— (2003a) "Les conflits armés en Afrique. Apports, mythes et limites de l'analyse économique", *Revue Tiers Monde* 176: 829–855.

—— (2003b) *Les économies en développement à l'heure de la régionalisation*, Paris: Karthala.

—— (2003c) "Les biens publics mondiaux. Un renouveau pour penser l'action publique à l'échelle mondiale?" *Politiques et management public* 21(3): 55–72.

—— (2009) *L'économie du développement et la pensée francophone*, Paris: Agence universitaire de la Francophonie.

—— (2013) *Mémoires solidaires et solitaires. Trajectoires d'un économiste du développement*, Paris: Karthala.

—— (2015) *L'économie de l'Afrique*, 8th edn, Paris: La Découverte. (Translated into English as, Pretoria: Protea Book House, 2004.)

—— (2016a) *L'Afrique: défis, enjeux et perspectives en 40 fiches pour comprendre l'actualité*, Paris: Eyrolles.

—— (2016b) *Géopolitique de l'Afrique*, 4th edn, Paris: Armand Colin.

—— (2016c) *Afriques entre puissance et vulnérabilité*, Paris: Armand Colin.

JEAN-JACQUES GABAS

RICHARD JOLLY (1934–)

From his early studies of disarmament and education, through his dedication to providing water supply, sanitation and hygiene, to his recent oversight of the United Nations Intellectual History Project (UNIHP), Richard Jolly's unique combination of optimism and pragmatism has earned him widespread admiration and acclaim in the field of human development. An economist committed to a people-centred approach, he has consistently stressed meeting basic human needs in development, not only because it is morally desirable but also as a contributor to economic growth and global security. Jolly was at the centre of debates on the relationship between economic growth and social welfare, concluding – alongside others such as **Dudley Seers** and **Hollis Chenery** – that it was not necessary to choose; instead, he argued it was necessary to pursue both growth and poverty reduction *in order to produce* sustainable economic development and social stability. In a recent statement on education for all, Jolly maintained that even in periods of stagnation, huge gains in social welfare could be achieved without significant increases in expenditure, through "determined leadership, cost-consciousness and ingenuity".[1] This sentiment encapsulates much of his life's work.

Born on 30 June 1934, Richard Jolly graduated from Cambridge University in 1956 with a degree in Economics, later completing an Economics doctorate at Yale University. In 1969 he joined the Institute of Development Studies (IDS) at the University of Sussex. In 1972 he followed **Dudley Seers** as Director, a position he held for nine years. During this period (1978), he acted as special consultant on North–South issues to the Secretary General of the Organisation for Economic Co-operation and Development (OECD), and between 1978 and 1981 he had his first experiences of working in the UN system as member and rapporteur of the United Nations Committee on Development Planning.

In 1978 his co-edited volume, *Disarmament and World Development*, was published and then updated and republished (Graham et al. 1984). It explored the potential for directing resources freed up by disarmament towards poverty-focused development through targeted restructuring policies. Here, Jolly's willingness to tackle huge global issues with imagination is already evident, as is his rigorous excavation of examples that challenged conventional thinking on what was achievable within the constraints of the international system. Jolly estimated that the financial costs of providing for many of the world's basic needs within the space of a decade were less that half of one year's expenditure on armaments. Furthermore, he pointed out that a focus on social welfare could prove a useful political tool even for authoritarian and military regimes.

The themes developed in *Disarmament and World Development* recur in later works, most notably the groundbreaking *Adjustment with a Human Face* (Cornia et al. 1987), of which Jolly was co-editor. Completed during his time as Deputy Executive Director for Programmes at UNICEF (1981–1995), the context for the collection was the pattern of contradictory global forces set in motion in the early 1980s that gave rise to a severe world recession. The Bretton Woods institutions addressed these through advancing stabilisation measures and structural adjustment policies, which sought to rectify balance of payments deficits and improve macro-economic management. At the time, UNICEF was at the forefront of renewed attempts to put basic child protection and poverty alleviation measures in place. Central to the study's argument was the idea that efforts were necessary to ameliorate the effects of adjustment policies on the most vulnerable sectors of society, particularly children. Jolly's personal contribution to this work was in the field of education, following on from themes explored in his doctoral research (Jolly 1969a). He identified the goal of universal primary education as the most significant educational priority, arguing that it was possible to move towards this without significant increase in total expenditure, if existing incomes and institutions were adapted significantly.

The main thrust of *Adjustment with a Human Face* was that growth-oriented adjustment in itself was not enough to ensure protection of the vulnerable; and yet the vulnerable could be protected during adjustment if *targeted programmes* were adopted. Furthermore, the book pointed to successful examples where such programmes had been adopted and growth had not been damaged as a result: "the issue should not be adjustment or growth, but adjustment *for* growth" (Cornia et al. 1987: 5). Contrary to the received wisdom at the time, Jolly and his colleagues maintained that a "human face" in adjustment policy was, in fact, a precondition for long-term growth, insisting that a strategy "which protects the vulnerable during adjustment not only raises human welfare but is also economically efficient" (p. 290). These sentiments were consistent with Jolly's long-held view that investment in human resources exhibits high economic returns.

During his time at UNICEF, Richard Jolly was also Vice-President of the Society for International Development (SID) from 1982 to 1985, and then Chairman from 1987 to 1996. In 1995, he left UNICEF to become Special Advisor to the Administrator of the United Nations Development Programme (UNDP). Here he was principal architect of the annual Human Development Report (1996–2000). Jolly championed the "20/20 initiative", which called for developing and donor countries to devote 20 per cent of government budgets and 20 per cent of aid allocations, respectively, to services targeted at basic needs. His commitment to education remained strong. In June 1996 he delivered the concluding statement at the Mid-Decade Meeting of the International Consultative Forum on Education for All in Amman, Jordan. The purpose of the meeting was to assess the advances made since the

Jomtien Conference on Education in 1990. Acknowledging progress, he stated that education for all and greater equality were "the two essential factors for ensuring rapid economic growth and rapid advance in human development".

Jolly is considered a pioneer of the concept of Human Development. In 2001 Mark Malloch Brown, Administrator of the UNDP, said of him: "Nobody has done more to help promote the concept of human development, and provoke thoughtful discussion and debate about how best to achieve it, than Dr. Richard Jolly."[2] He was awarded the Rome Prize for Peace and Humanitarian Action in 1999, in recognition of his work on the Human Development Report and earlier contributions to shaping programmes at UNICEF. He donated the $30,000 prize money to Jubilee 2000, the international coalition campaigning for debt reduction. Two years later he was knighted (KCMG) by the United Kingdom for "long and distinguished service for international development".

Since retiring from UNDP in 2000, Richard Jolly has been Co-Director of the United Nations Intellectual History Project (UNIHP), which explores and in many cases defends in measured terms – the development legacy of the United Nations. This output constitutes part of a 12-volume history of its economic and social contribution. Some of it is normative, some informative but, in addition, the primary data are available for future scholars to puruse, investigate and make up their own minds (Eide 2011). In the first volume of this series, *Ahead of the Curve* (Emmerij, Jolly and Weiss 2001), concerns about the origin and ownership of ideas are dismissed, placing emphasis instead on their impact and spread. Across the following volumes, about half of which are co-authored or co-edited by Jolly, readers are left in no doubt that the United Nations has played an important role in this regard.

In a paper written in preparation for the Human Development Report 2003, he appealed for a more nuanced and flexible interpretation of success in terms of achieving UN goals. If setting global goals is to be valuable and successful, he contended, "it is important now to plan for *partial* success and *partial* failure, not for the extremes of either *total success or total failure*" (Jolly 2003: 18). In May 2004, during a talk at the Overseas Development Institute in London, he also defended the UN against accusations of being overpaid, resistant to reform and uncritical of developing countries. However, the best tribute Jolly could have paid to the international organisation to which he devoted much of his life is the demonstration that ideas and concepts are a driving force in human progress, a force that will be among the most important contributions of this large, complex and often much-maligned international membership organisation.

Sir Richard has continued to devote himself to his lifelong aim of achieving clean water and sanitation for the global poor. As Chair of the Water Supply and Sanitation Collaborative Council (WSSCC) from 1997 to 2003, he argued for expenditure on water supply and sanitation (most of which goes to wealthier urban areas) to be redirected towards the rural poor. "Improvements in water, sanitation and hygiene are at the core of human development and poverty alleviation," he said on World Water Day in 2001. As with his earlier work in education, Jolly has argued for greater community participation in processes to ensure success. With reference to an approach termed "VISION 21", which aimed to allow local people to plan and manage improved water and sanitation systems, he explained that "we let communities themselves drive the process and it demonstrated that people had ideas that were practical, economic and sustainable".[3] An enduring feature of Jolly's legacy will be his emphasis on the benefits of community involvement.

Jolly is likely to be remembered principally for his work as architect of the Human Development Reports in the 1990s, and for the way he has kept basic needs and poverty alleviation on the development agenda. Many of the measures he has advocated have undoubtedly had an impact. *Adjustment with a Human Face* not only steered thinking within the UN system but also influenced the World Bank's three-pronged approach to poverty reduction – addressing growth, human development and targeted social safety nets. It also strongly informed the Millennium Development Goals (MDGs), which sought to reduce poverty significantly by the year 2015, an agenda currently being taken forward in a more integrated way by the Sustainable Development Goals (SDGs), 2016–2030. The MDGs involving commitments to achieving universal primary education and reducing by half the proportion of people without sustainable access to safe drinking water by the year 2015 were particularly close to Jolly's heart. Nevertheless, many of his core passions have come under extensive critical review. A focus on basic human needs has been criticised by those wanting more emphasis on growth and by those wanting more focus on redistribution. Some welfare economists are sceptical about the efficiency and effectiveness of targeting, while the limits of community participation are increasingly well known and ventilated. However, while poverty elimination and redistribution of wealth might be nobler goals than poverty reduction and addressing basic needs, nearly half a century after the publication of

Planning Education for African Development, the MDGs were not met on target, although progress on achieving universal primary education was impressive around the world.

Even more dismal is the slow uptake of Richard Jolly's ideas on arms, conflict and well-being, a cause that he is advocating once more. Thirty-five years on, the relationship between development and global security has become more pressing and complex than ever. Dismal though this context is, Richard Jolly's optimism and pragmatism might not be misplaced. At the United Nations in New York on 30 April 2004, Ronnie Kasrils – then South Africa's Minister of Water Affairs and Forestry – said of Jolly:

> He is a man who has brought passion, commitment and immense compassion to his work in the promotion of human development and human dignity. He is a man with vision, but with the remarkable skill of turning that vision into concrete reality.

These are respectful and touching words, not least because they come from a man who was once a senior military figure in *Umkhonto we Sizwe*, the armed wing of the African National Congress. If Kasrils has exchanged the pursuit of arms for the pursuit of wells, then perhaps Jolly's pragmatic optimism is well placed and he is right about starting small and thinking big.

Notes

1 Education for All: The Vision to be Grasped" – concluding statement by Richard Jolly at the Education for All Forum, Amman, Jordan, 19 June 1996.
2 See www.undp.org/dpa/frontpagearchive/2001/january/23jan01/.
3 Message of Dr Richard Jolly on the occasion of World Water Day, 22 March 2001.

Major works

Cornia, G.A., Jolly, R. and Stewart, F. (1987) *Adjustment with a Human Face*, 2 vols, Oxford: Clarendon Press.

Eide, I. (2011) "UNIHP: Intellectually Ambitious and Politically Important", *Forum for Development Studies* 32(1): 291–294.

Emmerij, L., Jolly, R. and Weiss, T.G. (2001) *Ahead of the Curve? UN Ideas and Global Challenges*, Bloomington, IN: Indiana University Press.

Graham, M., Jolly, R. and Smith, C. (eds) (1984) *Disarmament and World Development*, 2nd edn, Oxford: Pergamon.

Jolly, R. (1983) *Third World Employment – Problems and Strategy: Selected Readings*, Harmondsworth: Penguin Education.

Streeten, P.P. and Jolly, R. (eds) (1981) *Recent Issues in World Development: A Collection of Survey Articles*, Oxford: Pergamon.

Further reading

Jolly, R. (1968) *Costs and Confusions in African Education: Some Future Implications of Recent Trends*, Brighton, UK: Institute of Development Studies, University of Sussex.

—— (1969a) *Planning Education for African Development: Economic and Manpower Perspectives*, Nairobi: East African Publishing House for the Makerere Institute of Social Research.

—— (1969b) *The Purpose of Manpower Planning and its Implications for Planning Techniques*, Brighton, UK: Institute of Development Studies, University of Sussex.

—— (1970) *Skilled Manpower as a Constraint to Development in Zambia*, Brighton, UK: Institute of Development Studies, University of Sussex.

—— (2001) *Jim Grant: UNICEF Visionary*, Florence: UNICEF Innocenti Research Centre.

—— (2003) "Global Goals: The UN Experience", www.waterweb.org/wis/wis6/papers/Global%20Goals%20v.030120.03.pdf.

UNDP Human Development Reports (1996–2000) http://hdr.undp.org/reports/view_reports.cfm?type=1.

JO BEALL

NAILA KABEER (1950–)

Naila Kabeer is one of the few economists working primarily in development studies today who is widely respected by scholars, policy-makers and activists alike. Her work has provided a strong critique of the two poles within the social sciences by bringing to the forefront the gendered subjectivities, and the various and intersecting norms, values, relations and practices to explain the specificities of women's subordination. These poles are neoclassical rational choice theory, that argues that human actions are motivated by rational calculations about self-interest and that comprised her disciplinary training, and structural narratives of "the patriarchy", that explain the subordinate status that women occupy in most societies. Therefore, she has not only been a key feminist economist but has brought her rebellious intellect to enriching critical feminist development studies.

Kabeer completed her doctorate in Economics at the London School of Economics in 1985. Her research – unusually for an economist, based on protracted fieldwork in a Bangladeshi village over 18 months – closely investigated the specificities of household economies in a peasant context in order to understand how decisions about having children were made. This allowed her to go beyond numerical data or official statistics to begin focusing on the social and cultural realities that constitute the gendered structures of constraints within which women in poorer countries are located, and that explain behaviour and motivations of rural women and men.

Her tenure at the Institute of Development Studies (IDS) at Sussex after completing her PhD was an important milestone in her research trajectory. First of all, her research became increasingly interdisciplinary. Conversations with critical anthropologists, combined with her experience of – and insights gained during – the long fieldwork period freed her from the disciplinary rigidities of her training, and allowed her to utilise diverse theoretical and methodological traditions to focus on the social relations between men and women in different contexts rather than on categories of individuals. It also began her lifelong concern with understanding reality from the perspective of the marginalised, those whose experiences and voices had been largely absent in the making of development studies.

Kabeer's contributions through her career can be summarised under two broad themes: development studies and feminist economics. Her first set of contributions consisted of integrating feminist perspectives into development studies and offering alternative visions for "doing development". *Global Perspectives on Gender Equality* (Kabeer 2008a) offers theoretical arguments that challenge established ways of thinking within development studies. Pulling together a powerful analytical framework, built on her analysis of different traditions of feminist thinking, she subjected the key concerns of the development community to critical scrutiny in order to uncover the deeply entrenched, hence barely visible biases which underpinned mainstream development theory and resulted in the consistently marginal status given to women.

Her recommendations for gender-transformative strategies which incorporated this critique sought to combine the pragmatic with the visionary. While she saw the end-goal for all policy as the construction of a more socially just world which recognised and helped realise the full potential of all human beings, she was also concerned with the challenges of getting from "here" to "there". Hence she insisted on the need to start with the practical concerns of those who had to struggle to meet their basic survival needs – but to combine this with ways which promoted their ability to question the everyday injustices of their lives and come together to act on them.

Kabeer shows how feminist researchers should engage with policy-makers at local and global levels because they command critical resources and have the capacity to make decisions which have a direct impact on key aspects of women's lives. She argues that a "*strategic* pragmatism, combining a vision of gender justice with critical assessment of what gains can be won and at what costs, given institutional constraints, may be a more effective strategy for gender advocates who engage with policy makers than the pursuit of intrinsic arguments, regardless of who is listening" (Kabeer 2008a: 256). To this end, she has been active in the synthesis and systematisation of feminist theories to make their relevance for policy and planning apparent to those who are engaged practically. An important means through which she sought to reach out to the wider development community was through the Social Relations Framework, which sought to integrate a feminist analysis of power in a way that was accessible to a wide range of policy actors (Kabeer 1999a; 1999b; 2012a). The framework was an early example of feminist institutionalist thinking in development studies. It drew attention to five aspects of social relations which were common to all institutions, namely: *rules* – how things get done officially and unofficially, explicitly or implicitly; *activities* – what is done, routine practices in the domains of production and reproduction; *resources* – both tangible and intangible, how they are used, what is produced and systems governing their allocation; *people* – how they are defined and categorised, who does what, who gets to decide; and *power* – the ability to make rules and determine priorities. The Social Relations Framework thus offered a multi-level analysis of the vexed issue of power as it played out in the lives of ordinary people and in the policy world. It subsequently became a key means through which gender analysis has been disseminated in the world of policy makers and practitioners.

Her second main contribution has been in the field of feminist economics. Her early training as an economist made her deeply suspicious of the simplifying assumptions of mainstream economics. In particular, its truncated understanding of the economy excluded large swathes of reproductive activity, done largely by women, purely because it was carried out on an unpaid basis, and its preoccupation with econometric modelling which defined the world in highly abstracted and purely measurable terms. *The Power to Choose* (Kabeer 2001) was her first attempt to offer a theoretically interdisciplinary and empirically grounded alternative. This study of how

Bangladeshi women and their families made decisions about the labour market in the very different contexts of Bangladesh and Britain seeks to construct a theoretical middle ground between structuralist approaches that were blind to the capacity of individuals to exercise agency, even in very oppressive circumstances, and rational choice theories which failed to understand how structural constraints might differentiate the capacity for choice. She also uses the voices of the women and their families to give empirical substance to decision-making processes as they were "actually" made in the face of the powerful patriarchal constraints which shaped women's options at home and in the world. Again, findings from the study challenged policies that were being made at the global level for these workers without any knowledge of their needs and priorities.

This genre of work by Kabeer has also focused on labour markets in the informal economy, where the majority of the working poor in the global South are located. Here Kabeer explores how gendered forms of power operate across the public/private divide to crowd women into a narrower range of activities than men, activities which are also more precarious and poorly paid. In addition, she explores how ideas about gender enter implicitly and explicitly into the design and implementation of social protection measures for informal workers, the extent to which they position men and women as citizens or welfare recipients, and their implications for the voice and agency of these workers.

Kabeer's ongoing concern with understanding how power relations mediate women's experiences of structure and their exercise of agency gave a corresponding importance to the idea of women's empowerment. This was first touched on in her 1994 book, *Reversed Realities*, but threaded through her subsequent work. She brought the insights of her research together in a framework for the analysis of women's empowerment which became her most widely cited work. Central to her framework is women's capacity to make choices, but the concept of choice is qualified in ways that acknowledge the significance of the broader social relations which shape the choices that women perceive to be within the realm of material and imagined possibility as well as the strategic consequences of choice for these social relations. While there is a strong emphasis on the subjective dimensions of power and empowerment, "the power within" which she discusses in *Reversed Realities*, Kabeer also emphasises collective action as a critical means by which women, often socially isolated by the patriarchal structures of their society, gain the critical consciousness and collective capabilities they need to challenge their subordination.

A great deal of her work has therefore also focused on the collective processes through which women, sometimes in co-operation with men, sometimes through their own autonomous organisations, have sought to free themselves from unacceptable constraints and to contribute to building a more just world. Here she has investigated the sources of power and organisational strategies that structurally weak groups of women workers bring to bear in pursuit of their claims, and how these differ from classic male-dominated politics of the trade union movement.

Major works

Kabeer, N. (1994) *Reversed Realities: Gender Hierarchies in Development Thought*, London: Verso.

—— (1999a) "From Feminist Insights to an Analytical Framework: An Institutional Perspective on Gender Inequality", in N. Kabeer and R. Subrahmanian (eds), *Institutions, Relations, and Outcomes: A Framework and Case Studies for Gender-Aware Planning*, New Delhi: Kali for Women, 3–48.

—— (1999b) "Resources, Agency, Achievement: Reflections on the Measurement of Women's Empowerment", *Development and Change* 30(3): 435–464.

—— (2001) *The Power to Choose: Bangladeshi Women and Labour Market Decisions in London and Dhaka*, London: Verso.

—— (2004) "Globalisation, Labour Standards and Women's Rights: Dilemmas of Collective (In)action in an Interdependent World", *Feminist Economics* 10(1): 3–35.

—— (2008a) "Passion, Pragmatism and the Politics of Advocacy: The Nordic Experience through a 'Gender and Development' Lens", in N. Kabeer and A. Stark (eds), *Global Perspectives on Gender Equality: Reversing the Gaze*, London and New York: Routledge.

—— (2008b) *Mainstreaming Gender and Social Protection in the Informal Economy*, New Delhi: Routledge.

—— (2012a) "Between Affiliation and Autonomy: Navigating Pathways of Women's Empowerment and Gender Justice in Bangladesh", *Development and Change* 42(2): 499–528.

—— (2012b) "Empowerment, Citizenship and Gender Justice: A Contribution to Locally-Grounded Theories of Change", *Ethics and Social Welfare* 6(3): 216–232.

—— (2013) "The Rise of the Female Breadwinner: Reconfigurations of Marriage, Motherhood and Masculinity in the Global Economy", in S. Rai and G. Waylen (eds), *New Frontiers in Feminist Political Economy*, London: Routledge.

—— (2014a) "Social Justice and the MDGs: The Challenge of Intersecting Inequalities", *Equal Rights Review* 13: 91–116.

—— (2014b) "The Politics and Practicalities of Universalism: Towards a Citizen-Centred Perspective on Social Protection", *European Journal of Development Research* 26: 338–354.

—— (2015) "Gender Equality, Economic Growth and Women's Agency: The 'Endless Variety' and 'Monotonous Similarity' of Patriarchal Constraints", *Feminist Economics* 22(1): 295–321.

—— (2018a) *Gender, Livelihood Capabilities and Women's Economic Empowerment: Reviewing Evidence over the Life Course*, London: Overseas Development Institute.

—— (2018b) "Women Workers and the Politics of Claims Making: The Local and the Global", *Development and Change* 49(3): 759–789.

—— Ratna Sudarshan, R. and Milward, K. (eds) (2013) *Organizing Women in the Informal Economy: Beyond the Weapons of the Weak*, London: Zed Books.

KUNTALA LAHIRI-DUTT

MICHAŁ KALECKI (1899–1970)

Along with John Maynard Keynes, the Polish economist Michał Kalecki was one of the creators of the Keynesian Revolution in Economics in the middle decades of the 20th century. Unlike Keynes, however, he devoted a considerable amount of his work to elucidating the economic problems of the developing countries. From his criticism of Keynes's ideas for the new international monetary order at Bretton Woods, and through his work on developing countries at the UN and after, Kalecki developed a set of ideas that established a "structural" approach to economic development.

Michał Kalecki was born in 1899 to a minor industrialist in the city of Łódź, within the Polish provinces of the Russian empire. A few years later, Łódź was the centre of the 1905 Revolution in the industrial centres of the empire. The business of Kalecki's father was ruined, and his family's finances deteriorated further through the First World War and the early years of Poland's independence. The young Kalecki had to give up his engineering studies, and he turned to business journalism to eke out a living. In 1929 he secured employment with the Institute for Research on Business Cycles and Prices that had just been established in Warsaw to provide market intelligence to Polish business. While there, in 1933, he wrote his "Essay on the Business Cycle Theory", which laid out the essentials of the theory that Keynes was to publish in his *General Theory* three years later (Kalecki 1933). In 1936, Kalecki was able to visit Sweden to learn more about the work of Knut Wicksell and then move to Britain on a Rockefeller Travelling Fellowship. This scheme was being used to help get vulnerable academics and scientists out of northern and central European countries beset by fascism and Nazism.

Given his Jewish origins and protests against politically motivated action against some colleagues at his Institute, he might otherwise not have survived the coming Nazi occupation and war. In the UK, he worked with Keynes, and spent the years of the Second World War researching at the Institute of Statistics in Oxford. In 1945 he joined the International Labour Office, and shortly afterwards moved to work at the United Nations in New York.

Kalecki was made aware of the political economy of development during his formative years as an economist in the 1920s. His earliest writings in economics were analyses of commodity markets and industrial development in Poland. These were rooted in the political debates in Poland, influenced by the ideas of Rudolf Hilferding, Rosa Luxemburg and Henryk Grossman, all of whom accorded an important role to developing countries in their analyses of capitalism. Rosa Luxemburg, in particular, in her book *The Accumulation of Capital*, which Kalecki knew well, examined in detail the impact of capitalism on traditional economies. In his decade of work at the United Nations, from 1945 to 1955, Kalecki examined in detail the problems of developing countries and started his consultancy work for some of their governments, notably Israel and Mexico. Following his return to Poland in 1955, he was active in seminars and research projects on developing countries, and advised the governments of India and Cuba. This work was brought to an abrupt halt in 1968, when his Polish collaborators in these projects came under political attack. His influence in Poland was marginalised, even if the institutions in which he worked continued in a more politically subservient fashion.

Kalecki's first published intervention in development economics was a review of Mihail Manoilescu's *Die Nationalen Produktivkrafte und der Assenhandel* (published in English as *The Theory of Protection and International Trade*, 1931). Manoilescu had put forward an early version of the import-substitution argument in favour of developing country protectionism. Because developing country productivity in industry is higher than in agriculture, even if that industrial productivity may be lower than in industrialised countries, improvements in efficiency can be obtained by expanding industry behind protective tariff barriers. The resulting overall improvement in productivity may result in a higher output available to an economy than might have been obtained from relying on low-productivity agricultural exports to pay for industrial imports.

Kalecki criticised one possible neoclassical objection to this theory, namely that developing countries experience capital scarcity, and that market forces would tend "naturally" to equalise the marginal productivity of capital or labour in industry or agriculture. According to Kalecki, such capital scarcity was not binding in agricultural countries, since rural unemployment or underemployment allowed investment to be undertaken unconstrained by "the supply of new saving" up to the point of full employment. At the same time, he criticised the notion that protection for new industries was a sufficient condition for economic development: "To represent free trade as the *only* obstacle to the economic progress of backward countries is to divert attention from such urgent social problems as land reform and others" (Kalecki 1938: 182).

The critical issue for developing countries was how such investment could be organised. When Kalecki returned to this question, as a government adviser to Israel in 1950, he considered two possible alternatives for countries such as Israel with minimal indigenous investment goods capacity. In this situation, countries could rely on foreign direct investment or on government-organised investments. The first possibility he considered to be of dubious benefit, because it was unlikely to be on a sufficient scale or attracted to industries that could supply the domestic market, and hence limit inflationary pressures in them. State-organised investments should therefore be undertaken with a view to maximising the exports needed to finance the import of industrial machinery, developing domestic investment capacity and minimising domestic inflation. Private-sector investment was to be regulated to ensure that it did not absorb scarce foreign exchange in ventures that were not high on the government's development priorities (Kalecki 1950).

Kalecki went on to expand on these themes in his subsequent work on economic development. In the first place, he examined the political and social formation in which economic development could take place. He had no illusions about the existence in developing countries of a "naturally" entrepreneurial capitalist class. Instead he highlighted those features of economic and social backwardness that served as obstacles to industrial development, in a way that was later taken up by the "structuralist" school of development economists. The crucial factors opening up the possibilities for such development were political leadership and land reform. Land reform was critical because upon it depended the ability of the agricultural sector to supply food to a growing industrial sector whose workers still had a large income elasticity of demand for foodstuffs. If food supplies did not increase, real wages fell, or food imports rose, alleviated perhaps by food grants from the agricultural surpluses of the industrialised countries. Without land reform, the financial accumulation from higher agricultural production or prices would be absorbed by rural debt service or rents, which would then finance luxury consumption rather than being invested in the expansion of agricultural production.

The analytical foundation of Kalecki's "structuralist" view was in his reflux theory of saving, investment and profits. In an industrialised capitalist economy, a government financial deficit expands the sales revenue of firms without increasing their costs, so that their profits and savings rise. Similarly, if firms undertake additional investment, using their own savings or bank credit, then this new expenditure ends up as additional profits for firms. In a developing economy, expenditure by the government and by the private sector on investment "leaks" out of the system into incomes of non-productive social groups (landowners, merchants and moneylenders). But the key leakage occurs through foreign trade, as a result of the increase in imports of capital equipment, raw materials and finished goods that is attendant upon an investment boom in a developing country. This, in turn, drains the scarce foreign exchange reserves of such a country.

Kalecki had little faith in the ability of foreign companies to overcome the foreign exchange constraint by means of foreign direct investment. He considered that such investment was very expensive in terms of the amount of profit that foreign companies repatriated, openly or covertly, through transfer pricing. He gave one of the earliest expositions of the effects of such pricing practices (Kalecki 1954). He considered that official loans usually came with political strings and requirements for favourable terms for companies from countries granting such loans. Instead, he favoured long-term trade agreements between countries. These, he felt, allowed governments to plan more securely with assured markets abroad.

According to Kalecki, a key political requirement for economic development was the installation in power of an "intermediate regime". This was a government, based on the "lower-middle class", the urban professional classes and small businessmen who have an interest in economic development, that is willing to push through crucial land reforms. Such a government and its supporters may not necessarily be in favour of socialism, or may be bought off by big business and foreign companies (Kalecki 1964). The "intermediate regime" is perhaps the most troublesome of Kalecki's political concepts. He had in mind Nasser's government in Egypt, playing off the two global superpowers, the Soviet Union and the United States, against each other to secure foreign aid. But that, like many of the intermediate regimes of Latin America was (initially at least) based on an alliance of military leaders and trade unions. In India, Jawaharlal Nehru's Congress Party was remarkably cautious in challenging the power of the big landowners who delivered the Congress votes in the rural areas. The thawing of the Cold War in the 1960s and the collapse of the Soviet bloc in 1990 reduced the scope for superpower rivalry in foreign aid (see Lipton and Toye 1990: 98–99).

With an intermediate regime in place, the key to economic development was planning industrialisation. Kalecki considered this to be a financial problem, even though he did not consider developing countries to be subject to a saving constraint, because the financing of investment determined whether the investment drive would be sustainable, or would dissipate into inflation and luxury consumption. Given low wages in developing countries, any substantial investment effort would result in higher demand for "necessities". If the supply of such necessities did not increase, then the resulting inflation would reduce real wages, so that the workers themselves would be largely financing the investment drive from the reduction in those real wages. At the same time, without land reform, the additional incomes of the food producers would be absorbed by higher rents and debt repayments, increasing demand for luxury goods among merchants, moneylenders and landowners. Formal finance for domestic investment could always be obtained through domestic loans. But if it became necessary to finance investment out of taxes, then Kalecki favoured taxes on profits and higher incomes, because this would have least impact on lower income groups, and might even reduce inflationary pressure by reducing luxury consumption. Capital imports (foreign credits) could ease the foreign exchange constraint, and indeed might support general government spending programmes by allowing the import of specialist health equipment, as well as investment goods (Kalecki 1963). However, foreign credits came in different forms and were frequently politically motivated. The best form of loan was one repayable in finished goods rather than in foreign currency (Kalecki and Sachs 1966).

These foreign exchange and domestic considerations made Kalecki sceptical of the benefits of installing the most capital-intensive (labour-productive) variants of particular investment projects. These had been advocated by Maurice Dobb and **Amartya Sen**, subject to a constraint of maintaining the level of real wages. In Kalecki's view, the choice of different variants in particular industries was not wide, while inflating capital requirements increased demand for wage good necessities or imports at a time when such demand was already likely to be high (Kalecki 1972: ch. 10).

Perhaps the least-known focus of Kalecki's work on developing countries is the role that those countries play in his critique of the Bretton Woods arrangement. At the time of the discussions on the proposed postwar international monetary arrangements, Kalecki was working in Oxford with **Ernst "Fritz" Schumacher** and Thomas Balogh, two other critics of the Bretton Woods proposals. All three recognised that the Keynes Plan offered the better prospects for international trade and development by reason of the more generous international reserves offered by Keynes's "bancor" proposal. But they objected to the concept of "equilibrium" in international trade that Keynes shared with the White Plan, namely a situation in which each country's trade is balanced. This they felt was particularly inappropriate to the development needs of poorer countries, which needed to sustain structural trade deficits in the process of their industrialisation. Support for those deficits should come in the form of long-term loans, provided more or less automatically on demand. Failing this, governments in developing countries would be forced into making bilateral arrangements that would inevitably restrict their development strategies and reduce the efficiency of their foreign trade (Kalecki and Schumacher 1943).

Kalecki's work on development economics won sufficient following in what were then called "Third World" countries to arouse the anxiety of the US foreign policy establishment, which recognised hostility towards American capitalism behind Kalecki's suspicion of multinational enterprises, foreign direct investment, the Washington institutions of the International Monetary Fund and the World Bank, and commercial international loans. Following his death, Kalecki's work was continued by a small but significant group of economists who had worked or studied with him. However, circumstances and the prevalent orthodoxy on economic development have shifted considerably from the issues that Kalecki examined. From the 1970s, the kind of technocratic economic planning and étatist investment that Kalecki advocated came to be widely viewed as a sink of corruption and venality. This view was reinforced the 1982 international debt crisis that reinstated the International Monetary Fund as the arbiter of government policy in developing countries. The Fund's strategy was based on trade and financial "liberalisation" and "appropriate" (i.e. pre-Keynesian) fiscal and monetary policies.

Behind this shift of circumstances and issues have been two important changes in international politics and finance. The first was the collapse of the Soviet Union and not only the system that Kalecki advocated of long-term international trade contracts used by the Communist bloc's Council for Mutual Economic Assistance, but also the bipolar international political system that fostered Kalecki's "intermediate regimes". The second major international change has been the rise of international finance, which has greatly expanded the possibilities for securing commercial foreign credits. The resulting international debt and "emerging market" crises have placed the economic and financial policies of developing countries, including successful industrialisers like the East Asian economies, under the direct supervision of the International Monetary Fund. In addition to the obvious political difficulties of pursuing sustainable development policies that effectively benefit the many living in poverty, the challenge facing any government seeking to pursue a Kaleckian development strategy is how to take advantage of the opportunities offered by international finance, while insulating the economy from the instability emanating from the international financial system.

Major works

Kalecki, M. (1950) *Report on the Main Current Economic Problems of Israel*, Jerusalem: Ministry of Finance. (Repr. in J. Osiatyński (ed.), *Collected Works of Michał Kalecki*, vol. V: *Developing Economies*, Oxford: Clarendon Press, 1993.)

—— (1955) "The Problem of Financing Economic Development", repr. in J. Osiatyński (ed.), *Collected Works of Michał Kalecki*, vol. V: *Developing Economies*, Oxford: Clarendon Press, 1993.

—— (1963) "Problems of Financing Economic Development in a Mixed Economy", repr. in J. Osiatyński (ed.), *Collected Works of Michał Kalecki*, vol. V: *Developing Economies*, Oxford: Clarendon Press, 1993.

—— (1964) "Observations on Social and Economic Aspects of 'Intermediate Regimes'", repr. in J. Osiatyński (ed.), *Collected Works of Michał Kalecki*, vol. V: *Developing Economies*, Oxford: Clarendon Press, 1993.

—— (1972) *Selected Essays on the Economic Growth of the Socialist and the Mixed Economy*, Cambridge: Cambridge University Press.

—— and Sachs, I. (1966) "Forms of Foreign Aid: An Economic Analysis", repr. in J. Osiatyński (ed.), *Collected Works of Michał Kalecki*, vol. V: *Developing Economies*, Oxford: Clarendon Press, 1993.

—— and Schumacher, E.F. (1943) "International Clearing and Long-Term Lending", *Bulletin of the Institute of Statistics Oxford Supplement* 5(5):29–33; repr. in J. Osiatyński (ed.), *Collected Works of Michał Kalecki*, vol. VII: *Studies in Applied* Economics *1940–1967: Miscellanea*, Oxford: Clarendon Press, 1997.

Further reading

Kalecki, M. (1933) "Essay on the Business Cycle Theory", repr. in J. Osiatyński (ed.), *Collected Works of Michał Kalecki*, vol. I: *Capitalism: Business Cycles and Full Employment*, Oxford: Clarendon Press, 1990.

—— (1938) "Foreign Trade and the National Forces of Production", repr. in J. Osiatyński (ed.), *Collected Works of Michał Kalecki*, vol. V: *Developing Economies*, Oxford: Clarendon Press, 1993.

Lipton, M., and Toye, J. (1990) *Does Aid Work in India? A Country Study of the Impact of Official Development Assistance*, London and New York: Routledge.

Manoilescu, M. (1931) *The Theory of Protection and International Trade*, London: P. S. King & Son.

JAN TOPOROWSKI

INGE KAUL (1944–)

Inge Annelies Kaul, who was born in Wiesbaden, Germany, on 18 August 1944, can be considered as the godmother of the term "global public good" (GPG). After a PhD at Konstanz University in 1972, she joined the UNDP in 1981 for most of her professional life. From 1989 to 1994 she was the first director of UNDP's office in charge of the *Human Development Report*. During the initial years, together with **Mahbub ul Haq**, she was the publication's team leader before becoming Director of the UNDP Office of Development Studies (1995-2005). After her retirement, she joined the Hertie School of Governance in Berlin as an adjunct professor in 2007 and is affiliated to numerous other institutions, including as a member of the advisory boards of the German Development Institute (since 2014) and the International Cooperation Council (since 2015).

The notion of public goods has long been a part of mainstream economic theory. It has also been introduced occasionally as collective good into international relations (an early example is Russett and Sullivan 1971). **Charles Kindleberger** (1986) finally translated the public good discourse more systematically into a responsibility for global governance, not least with reference to peace. He also offered a concise definition, by pointing to the inherent nature "where exclusion of consumers may be impossible, but in any event consumption of the good by one consuming unit – short of some level approaching congestion – does not exhaust its availability for others" (Kindleberger 1986: 2). Put differently: in contrast to a private good, a public good is in principle non-rival in consumption (i.e. the use of it does not affect others or their ability to consume it) and it has non-excludable benefits (i.e. they are available for everyone).

It was to a large extent Inge Kaul who introduced the notion of public goods, until then mainly limited to a core concept of economics, to the global arena with policy relevance. "If a good's benefits or costs are of nearly universal reach, spreading across all countries, or if it could potentially affect anyone anywhere, it is a GPG" (Kaul 2012: 731). While these goods are public in consumption, they differ from national public goods in terms of their multi-actor, multi-sector and multi-level nature and complexity and their policy interdependence: this results in "the compulsion for policymakers to engage in international cooperation in order to ensure the availability of a desired GPG to national constituencies" (2012: 736). She warns that neglecting or underproviding GPGs enhances the risk that, in the absence of prompt corrective action based on mutually beneficial international bargains, the latter costs are much higher. GPGs therefore should be seen as a matter of collective self-interest and as an integral part of responsible sovereignty.

Due to the growing globalisation, however, new tendencies potentially clash with the old notions of state sovereignty, and the related reluctance to delegate powers considered to be a domestic affair to new institutions and organisations. Inconsistencies in the global public domain might be the result of clashing norms and principles. Matters related to health, education, law and order, taxation and others have become more globalised challenges. This impacts on global patterns of decision-making on global issues and affects global governance. Most prominently, GPGs such as the international trade regime, the international financial architecture and environmental matters remain highly contested. They are different in nature from the classical GPGs such as the sky, or the oceans, regulated by inter-state agreements for centuries. In contrast, the new GPGs are not external to states but cross-cutting (such as property rights, investment laws, trade and finance, basic human rights, ozone layer, global climate, infectious diseases, communication and transport). Knowing no borders, these GPGs require transnational policy-making and decisions.

As a result, Kaul argues, the era of the conventional Westphalian state order has been replaced by the need for effective and fair international co-operation. But as she also observes, global governance strategies have not adjusted fully to such new realities. Policy challenges linked to GPGs risk underprovision. Since being in the public domain and possibly affecting all, they invite individual actors to free-ride by leaving others to address the challenge, while enjoying the benefits free of charge when the good is available. Challenges like climate change mitigation are particularly vulnerable to free-riding, and a potential victim of the continued preference of governments for domestic policy matters over global affairs. She calls this the "sovereignty paradox": to protect national ownership also in decision-making, states are reluctant to enter international co-operation through binding obligations and commitments, which they are afraid might imply surrendering sovereignty. At the same time, such reluctance contributes to making the problem bigger and therefore reinforces the need to react through an internationally shared policy agenda. This actually erodes policy-making sovereignty instead of enhancing it. As she suggests:

> It would be useful to clarify that the notion of mutually respectful sovereignty suggests combining the pursuit of national interests with respect for the sovereignty of other states, in line with established principles of global equity such as the principle of common but differentiated responsibility. The notion of mutually respectful sovereignty would constitute a collective approach to safeguarding national policymaking sovereignty and, as such, parallel the collective approach to the security of national borders that is already enshrined in the UN Charter. In fact, such a conceptualization of sovereignty would conform to the UN Charter and its principle of non-interference to a greater degree than the current lax attitudes towards externality management that leave considerable room for "beggar-thy-neighbour" policies and free riding.
>
> (Kaul 2015: 55f.)

In line with such an approach, she suggests the establishment of a Global Stewardship Council within the United Nations, composed of eminent personalities, "appointed in their individual capacity and mandated to support the international community in creating and maintaining a global public domain that serves all people and countries in furthering their welfare and well-being, while ensuring mutually respectful sovereignty and the carrying-capacity of the globe" (2015: 58).

A serious engagement with the GPGs beyond tokenism, as Kaul (2012: 730) warns, requires more than financial obligations and emphasises the need for a change of mindset:

> several GPGs, including climate change mitigation, pertain to another basic transformation that the international community must accomplish: a rethinking of the growth and development paradigms

> it has pursued to date, so that it can move towards a low-carbon economy.

She therefore diagnoses considerable uncertainty and the disturbance of conventional governance systems: "crises will persist and remain unresolved for quite some time – despite entailing high costs for many people, individual nations, and the world as a whole" (2012: 730).

Kaul advocates the need for fair and just international negotiation as key prerequisites for securing international agreements and ensuring effective follow-up. Davies (2017: 16–41), who calls the concept an "abstruse and surprising hit", adds to his own analysis an insightful in-depth interview with Inge Kaul. While he observes that GPGs entered the aid policy narratives in a self-serving way rather than creating obligations to be honoured outside and beyond a convenient donor-country perspective, Kaul insists that financing of GPGs as distinctive levels and different from Sustainable Development Goals calls for separate funding and implementation arrangements.

> [They] should be conceived as payment for services rendered, not aid, and should not be provided at the expense of aid. Primary responsibility for the production of global public goods, most urgently climate change mitigation, should rest with issue- and outcome-oriented "tsars" at both the global and national levels.
>
> (Davies 2017: 16)

This sounds almost like wishful thinking in times when the slogan "America first" dominates international policy matters and decisions by the administration of one of the most influential actors on the global stage. And such renewed national self-interest remains not without consequences, in return provoking responses guided by domestic policy in contradiction to global policy. Nearly a decade earlier, Long and Woolley (2009) had already bemoaned the poor definition and incoherence of the concept, which avoids analytical problems by resorting to abstraction. They conclude that despite the rhetorically effective concept of GPGs, a precise definition and conceptual disaggregation are required to advance analysis of global issues.

Notwithstanding such criticism, GPGs have introduced more than merely another catchword. The notion allows for new avenues to create further awareness of the interlinked and intertwined global realities and – following the recent radical shifts in US policies under the Trump administration – the open breach of the required global social contract.

Since the GPGs have entered the nomenclature of development-related discourses, they have inspired further contributions engaging with collective goods such as peace (Møller 2004) and human rights (Andersen and Lindsnaes 2007), to mention only two of the all-embracing, prominent reference points. They have not become obsolete only because the hegemonic policy of the government of one of the most influential states in this world could not care less.

Major works

Kaul, I. (2012) "Global Public Goods: Explaining Their Underprovision", *Journal of International Economic Law* 15(3): 729–750.

—— (2015) "What Makes Global Governance So Hard? The Sovereignty Paradox and How to Solve It", in B. Kappes and K. Seitz (eds), *Nachhaltige Entwicklung braucht Global Governance. Weltinnenpolitik für das 21. Jahrhundert*, Munich: Oekom, 37–61.

—— (ed.) (2016) *Global Public Goods*, Cheltenham: Edward Elgar.

—— Grundberg, I. and Stern, M.A. (eds) (1999) *Global Public Goods: International Cooperation in the 21st Century*, Oxford: Oxford University Press.

—— and Mendoza, R.U. (2003) "Advancing the Concept of Public Goods", in I. Kaul, P. Conceicao, K. Le Goulven and R.U. Mendoza (eds), *Providing Global Public Goods: Managing Globalization*, Oxford: Oxford University Press, 78–111.

Further reading

Andersen, E.A. and Lindsnaes, B. (eds) (2007) *Towards New Global Strategies: Public Goods and Human Rights*, Leiden: Martinus Nijhoff.

Barrett, S. (2007) *Why Cooperate? The Incentive to Supply Global Public Goods*, New York: Oxford University Press.

Davies, R. (2017) *Public Enemies: Global Public Goods in Aid Policy Narratives*, Australian National University/Development Policy Centre, Discussion Paper 57, 27 March: https://papers.ssrn.com/sol3/papers.cfm?abstract_id=2941164 [accessed 28 April 2018].

Jenks, B. (2012) "The United Nations and Global Public Goods: Historical Contributions and Future Challenges", in G. Carbonnier (ed.), *International Development Policy Aid, Emerging Economies and Global Policies*, Basingstoke: Palgrave Macmillan, 31–47.

Kindleberger, C.P. (1986) "International Public Goods Without International Government", *American Economic Review* 76(1): 1–13.

Long, D. and Woolley, F. (2009) "Public Goods: Critique of a UN Discourse", *Global Governance* 15(1): 107–122.

Møller, B. (2004) *Peace as a Global Public Good*, Aalborg, Denmark: Institut for Historie, Internationale Studier og Samfundsforhold, Aalborg Universitet.

Russett, B.M. and Sullivan, J.D. (1971) "Collective Goods and International Organization", *International Organization* 25(4): 845–865.
Sandler, T. (2004) *Global Collective Action*, Cambridge: Cambridge University Press.

HENNING MELBER

AKHTAR HAMEED KHAN (1914–1999)

As both an original thinker and influential development practitioner, Akhtar Hameed Khan was an early exponent of the modern participatory development tradition, with its emphases on grassroots perspectives, community-level self-reliance and personal critical reflection. His life and work embody the strengths as well as the limitations of such approaches. In addition to his well-known contributions to rural development through the Comilla experiment of the 1960s in East Pakistan (now Bangladesh), and to urban slum rehabilitation in the form of the Orangi Pilot Project (OPP) in Karachi from the 1980s onwards, the significance of other aspects of his ideas, life and work are only perhaps now becoming more fully apparent.

He was born in Agra on 15 July 1914 to a Pathan family, the son of a well-to-do police official, a sub-inspector, employed in the government of British India. Following undergraduate studies in history, philosophy and literature at Meerut College, he graduated in 1934 with an MA in English literature from Agra University. Influenced by his father's experience of 19th-century social reformer Sir Syed Ahmed Khan (no relation), who favoured a synthesis of Islamic thought and Western knowledge, he read widely among Hindu, Buddhist and Western texts on religion and philosophy, and was well versed in Islamic history and Indian nationalism.

Khan joined the Indian Civil Service (ICS) in 1936, and spent a period studying in Cambridge as a probationer. Khan's first posting in 1938 was to Comilla (then in Tripura State), where he served as an Assistant Magistrate. Here he gained an understanding of Bengali language and literature that gave him "new perspectives on religion, language and culture" (Khan 1983: p. xiii). He rose quickly through the ICS ranks, but grew disillusioned with life as a colonial civil servant and resigned in 1945. Distrusting worldly wealth and ambition, he felt he had learned all he could from his British rulers and worried that he would end up as "a bureaucratic bigwig, a brown *bara sahib* rich conceited and hollow" (p. xiv). He had also been shocked by the abuse of power by the British authorities during the 1943 famine, in which as many as 3 million Indians died. It was an act of personal renunciation that had also been influenced by the mystical ideas of the Persian Sufi poet Rumi and the Islamic philosopher Al Ghazali.

Khan moved to Meerut (Uttar Pradesh) and, in what he described as a two-year "Tolstoyan experiment" (Khan 1983: p. xiv), he rented a peasant house and tried to live simply, first doing day labouring and then working as a locksmith's apprentice. Settling on teaching as a new career, in 1947 he first went to Delhi to become headmaster of the Jamia Millia secondary school, then migrated to West Pakistan in 1950 to lecture in Islamic history and English literature. He left almost immediately to become Principal of Comilla College in East Pakistan, where he remained until 1958. During 1954–1955 he was seconded to become director of the US-funded national-level Village Agricultural and Development (V-AID) programme in East Pakistan, which had begun in 1953. V-AID was a government-led, top-down community development initiative that aimed to establish village councils that would modernise agriculture, improve infrastructure, health and education, and provide farmers with credit, training and marketing support. These were the early days of the Green Revolution, which aimed to increase productivity through the introduction of high-yielding seeds and new technologies (see **Norman Borlaug**). In 1958 a new rural development academy – later to be known as the Bangladesh Academy for Rural Development (BARD) – was established in Comilla as a training centre for V-AID officials, with Khan as director (1959–1964, 1965–1968) after an initial nine-month orientation at Michigan State University (MSU).

With V-AID abandoned in 1961 due to disappointing results, Khan repurposed the academy for what he hoped would be a more community-led approach. He made three changes to what had gone before: bringing a new emphasis on research-led training and education, using local people as extension agents rather than outsiders, and basing interventions on permanent organisations in the form of village co-operatives and associations. This led to the "Comilla model" of rural development, which created a local "laboratory" area around the academy (that was now to combine training with research), where officials could consult with and try to learn from the farmers, work with them to develop new solutions to identified problems and then pilot joint collaborative interventions that could eventually be evaluated, refined and scaled up in conjunction with the government. Khan designed a two-tier co-operative system, with primary local co-operatives of farmers, linked to a central *thana* (sub-district)-level co-operative association providing supervision, training, loans, technology and other services. The aims were to improve agricultural production, reduce farmer indebtedness by bypassing village moneylending and strengthen the overall position of farmers.

Central to Khan's vision was the still uncommon view that poverty was not simply the result of a lack of resources but an outcome of lack of power (see also **Amartya Sen**). He thought that people would benefit from participation in effective institutions, by claiming the right to make decisions about the issues that affected their lives, and that people could be supported in their efforts to achieve this through financial support and technical know-how. A charismatic organiser and leader, this approach was reinforced by his personal style, which emphasised grassroots rather than bureaucratic values, along with frugality and austerity, the most obvious example being his informal hand-spun *kurta pajama* workplace dress style. Family planning interventions also formed part of the Comilla experiment's agenda, and Khan worked hard to find ways to reconcile birth control with religious teaching, drawing on his own theological studies and extensive discussion with local religious leaders.

The model worked well for a while, achieved international recognition and was held up as an example for other Asian countries and beyond to follow (Karim 1985). However, by the end of the 1960s there were significant problems. The new institutions had served as contact points for external resources, but there was extensive elite capture. In the early 1970s an influential BARD study highlighted how structural factors maintained the rural poor's weak economic and social position (Wood 1976). Village-level inequalities persisted such that "sooner or later the realities of rural class structure would make themselves felt" (Blair 1978: 77). Rich farmers captured development resources at the expense of the small farming households (Van Schendel 1981). Nor did the co-operatives benefit the large number of landless people in the villages, estimated at up to 50% of households.

None of this was really news to Khan, who, with characteristic honesty, recognised many of these shortcomings. The domination by elites of efforts to introduce credit co-operatives in Bengal during the late colonial period is well documented (Iqbal 2017). While Khan noted the effectiveness of the two-tier co-operative model in providing inputs to farmers, he reflected in his *Tour of 20 Thanas* (1971: 25), "Has self-management been realised in the projects? I cannot honestly answer yes to that question."

After the 1971 war that led to East Pakistan becoming the independent country of Bangladesh, Khan was in his own words "displaced from Comilla" (1983: p. xv). He briefly went to work in Peshawar at one of the counterpart academies in West Pakistan. Unhappy in Peshawar, where he felt the force of political interference, Khan moved to Karachi University and later back to MSU, before embarking on his next major initiative, the OPP, in April 1980, based in the city's largest squatter settlement in the northwest corner of Karachi. Unlike the Comilla experiment, the OPP was conceptualised as an NGO, initially supported by the Bank of Credit and Commerce International (BCCI).

The Orangi area had first been settled in 1963 as a government township of around five square kilometres, but by the end of the 1970s there were a million people of different ethnicities packed into 32 km^2, many of whom had migrated from Bangladesh following the 1971 war. An outsider once again (he had never lived in Karachi for any length of time before), he reflected, "I set out to educate myself. For several months I wandered Orangi, looking at the settlements, talking with the people, officials, councillors, lobbyists and the chairmen of associations and clubs. I could go slowly; I was a free agent, not sent by a boss with plans and targets to achieve" (Khan 1997: 27). Discussion with residents quickly highlighted the priorities. The earliest self-help activities involved the construction of underground sewers using local labour and materials, and resulted in hundreds of kilometres of drainage pipes and auxiliary facilities. This was a diverse, predominantly poor multi-ethnic population and Khan's participatory approach successfully harnessed the high levels of energy and entrepreneurship found beneath the surface of the community in ways that the Comilla experiment had struggled to sustain. Khan's slow, careful approach made it possible for more autonomous organisations to take shape, including a research and training institute to manage the low-cost sanitation, housing and education programmes, a charitable trust organising micro-enterprise credit activities and a health and social development association overseeing health interventions.

In contrast to much of the foreign-funded short-term NGO work found in Pakistan, the OPP was acclaimed for building "the long-term relationships that are key to unlocking the main benefits of community participation" (Bano 2017: 215) by building trust over a long period of careful negotiation within the community, combining local and technical knowledge, and patient lobbying of government officials. Over time, the OPP built effective partnerships with the Karachi city authorities, mainly through its committed long-term approach, but also due to Khan's detailed knowledge of the public sector that drew on his earlier experience.

These days, of course, it is not unusual to hear of civil servants – retired or otherwise – who set up NGOs, sometimes with questionable intentions; but the OPP was a notable early example of the effective use of the NGO form as an innovative vehicle for a public-sector activist who had grown impatient with the restrictions of red tape. OPP has also informed wider practice through replication in other cities in Pakistan and beyond. Khan stepped back from his OPP leadership role in its later years and allowed second-tier managers to develop. He died in 1999, by which time the OPP had transitioned reasonably successfully into new leadership drawn from among its existing experienced staff.

Despite moving into the NGO world, Khan was always clear about the limits to their role, arguing that NGOs should not replace the functions of the state, but instead should demonstrate appropriate models, undertake social mobilisation and challenge convention. He was highly critical of the Pakistan government's increasing divestment of services to foreign-funded NGOs in the late 1980s, and refused to involve OPP in the government's donor-supported Social Action Programme (Hasan 1999). He was also a stern critic of the government's attitude to foreign aid in general, and contrasted what he saw as Pakistan's apparently insatiable appetite for aid with Japan's full repayment of US reconstruction loans within three years following the Second World War.

Khan's ideas have been influential in South Asia. For example, in 1981 the Aga Khan Foundation (AKF) initiated development work in Pakistan's Northern Areas and turned to the example of Comilla, contacting Khan to help establish what would become the Aga Khan Rural Support Programme (AKRSP). It was believed that the model was a suitable basis to work from because these remote communities had fewer inequalities and less of the rural class formation that had thwarted the Comilla model in Bangladesh. Khan did not participate directly, but approached one of his earlier colleagues and disciples, Shoaib Sultan Khan, who had been an assistant commissioner in Comilla back in the late 1950s, to take on the role. He became AKRSP's general manager in 1982. As Wood and Shakil (2006) have observed, AKRSP's community-based approach to extension "is part of a lineage deriving from early cooperative movements in the West, adapted to South Asia by Akhtar Hameed Khan in both Comilla and Peshawar, but also reflecting various colonial experiments with community development in undivided India" (p. 371).

Today Akhtar Hameed Khan is mainly remembered for the OPP, with the Comilla experiment having faded from memory. This is a shame, since the better-known Bangladesh NGO sector that largely took its place was in part inspired by his ideas. Just as he had learned from V-AID's shortcomings, emerging NGOs (also influenced by the radical ideas of **Paolo Freire** and Ivan Ilich) saw from the Comilla experience how important it was to centre their interventions on organised groups of landless rather than better-off farmers. Grameen Bank founder Muhammad Yunus has also said that his micro-credit approach owed a debt to Khan.

But Khan's significance goes beyond this. Firstly, he can be seen as an early example of what today is called a "social entrepreneur", in a sense that is often lost in today's usage. Khan's ideas were not enacted in the context of the private for-profit sector, but instead within the public and non-governmental realms. He seems to always have been in critical, reflexive mode, a cultural style that was later emphasised by alternative development practitioners. There is today an established critique of participatory development as naive about power and complicit in the perpetuation of forms of technocratic expertise (Cooke and Kothari 2001). Khan's ideas do not escape such criticisms, but provides a counterexample to the many Western participatory "experts", as a former colonial civil servant turned committed methodological populist but rooted firmly in the Global South.

Second, and related to this, his was also an early example of the power of "sector swapping", in which an individual activist builds on knowledge and experience acquired from working in one institutional sector (in his case the government) and creatively transfers this knowledge to build upon it in another (the world of NGOs). For example, his experience as a civil servant in the first part of his career has been credited with helping him to build relationships with the local authorities and officials that have contributed to the OPP's success. This phenomenon is an increasing feature of today's policy worlds, where movement between the sectors is associated sometimes with creativity and sometimes with co-option (Lewis 2008). Today's management theorists also play up the significance of such cross-boundary movements (see e.g. Lovegrove and Thomas 2013).

Finally, Khan's efforts to connect religious philosophy and practices with development ideas can be seen as ahead of its time; those with an interest in today's development challenges might be able to learn more from a person who has been characterised as "the Muslim Buddhist" (Khan 1999). For someone who reflected, writing in his *Report on Twenty Thanas*, "By temperament I am a recluse and involvement in action gives me no pleasure. To organise, to manage and administer gives me no pleasure. To observe, study and contemplate makes me happy" (1971: 1), Khan's influence and impact as a "philosopher-practitioner" (Hunzai 2006), though not as widely recognised as perhaps they should be, has been considerable.

Note on spelling

Both "Akhtar" and "Akhter" forms of the name are found in the published works, and for consistency I have used the former.

Acknowledgements

My sincere thanks for useful conversations with F.H. Abed, Masooda Bano, Stephen Biggs and Geof Wood in preparation of this article.

Major works

Khan, A.H. (1971) *Tour of Twenty Thanas: Impressions of Drainage Roads, Irrigation and Cooperative Programmes*, Comilla: Bangladesh Academy for Rural Development, Kotbari.

—— (1983) *The Works of Akhter Hameed Khan*, 3 vols, Comilla: Bangladesh Academy for Rural Development.

—— (1996) *Orangi Pilot Project: Reminiscences and Reflections*, Introduction by Arif Hasan, i–xli, Karachi: Oxford University Press.

—— (1997) "The Orangi Pilot Project: Uplifting a Periurban Settlement", ch. 2 in *Reasons for Hope: Instructive Experiences in Rural Development*, West Hartford, CT: Kumarian Press.

Further reading

Bano, M. (2017) "Buying Consent: The Achilles' Heel of Aid-Induced Community Participation", *Journal of South Asian Development* 12(3): 213–235.

Blair, H. (1978) "Rural Development, Class Structure and Bureaucracy in Bangladesh", World Development 6(1): 65–82.

Boyatzis, R. E., and Khawaja, M. (2014). "How Dr. Akhtar Hameed Khan Led a Change Process That Started a Movement", *Journal of Applied Behavior Sciences* 50(3): 284–306.

Chowdhury, A.N. (1990). *Let Grassroots Speak: People's Participation Self-Help Groups and NGOs in Bangladesh*, Dhaka, Bangladesh: South Asia Books.

Cooke, B. and Kothari, U. (2001). "The Case for Participation as Tyranny", in *Participation: The New Tyranny*, London: Zed Books, 1–15.
Hasan, A. (1999) "Akhtar Hameed Khan and the Orangi Pilot Project", *Dawn*, 13 Oct.
Hunzai, I. (2006) "The Way Forward: Fourth Dialogue", in G. Wood, A. Malik and S. Sagheer (eds), *Valleys In Transition: Twenty Years of AKRSP's Experience in Northern Pakistan*, Oxford: Oxford University Press, 454–491.
Iqbal, I. (2017) "Cooperative Credit in Colonial Bengal: An Exploration in Development and Decline, 1905–1947", *Indian Economic & Social History Review* 54(2): 221–237.
Karim, M.B. (1985) "Rural Development Projects – Comilla, Puebla, and Chilalo: A Comparative Assessment", *Studies in Comparative International Development* (Winter 1985–1986): 1–41.
Khan, T.A. (1999) "The Muslim Buddhist: Akhtar Hameed Khan (1914–1999)", *Himal* (November): 49–50.
Lewis, D. (2008) "Crossing the Boundaries Between 'Third Sector' and State: Life Work Histories from the Philippines, Bangladesh and the UK", *Third World Quarterly* 29(1): 125–141.
Lovegrove, N. and Thomas, M. (2013) "Triple-Strength Leadership", *Harvard Business Review* 91(9): 46–56.
Raper, A. F. (1970) *Rural Development in Action: The Comprehensive Experiment at Comilla, East Pakistan*, Ithaca, NY: Cornell University Press.
Uphoff, N., Esman, M.J. and Krishna, A. (1998). *Reasons for Success: Learning from Instructive Experiences in Rural Development*, West Hartford, CT: Kumarian Press.
Van Schendel, W. (1981) "After the Limelight: Longer-Term Effects of Rural Development in a Bangladesh Village", *Bulletin of Concerned Asian Scholars* 13 (4): 28–34.
Wood, G. (1976) "Class Differentiation and Power in Bandakgram: The Minifundist Case", in M.A. Huq (ed.), *Exploitation and the Rural Poor: A Working Paper on the Rural Power Structure in Bangladesh*, Comilla: Bangladesh Academy for Rural Development, ch. 3.
Wood, G. with Shakil, S. (2006) "Collective Action: The Threatened Imperative", in G. Wood, A. Malik and S. Sagheer (eds), *Valleys in Transition: Twenty Years of AKRSP's Experience in Northern Pakistan*, Oxford: Oxford University Press, 369–425.

DAVID LEWIS

CHARLES POOR KINDLEBERGER (1910–2003)

Charles Kindleberger was born in New York, the son of a lawyer. After studies at the University of Pennsylvania he completed a PhD in economics at Columbia University. From 1936 he worked with the

Federal Reserve Bank of New York, followed by the Bank for International Settlements, the Federal Reserve Board in Washington, the Joint Economic Committee of the United States and Canada, the US Office of Strategic Services and, from 1944 to 1948, the Department of State. In this last appointment he played a part in organising Marshall Aid reconstruction assistance for Europe, subsequently the prototype for large-scale aid programmes to developing countries. In 1948 he was appointed Associate Professor of Economics at the Massachusetts Institute of Technology, obtaining a full professorship three years later. He proceeded to confute the widespread stereotype of well-connected government officials honouring academic appointments by their discretion, and poured out a mass of publications on economic history, international trade and development, banking, and finance. He remained at MIT, becoming emeritus professor on his retirement in 1976 (Miall 2003). By the time of his death on 7 July 2003, he had published 29 books, excluding second and third editions.

This essay concentrates on his most original contributions to development economics. These, however, do not stand obviously out of his extensive summaries of the literatures on which he worked. Moreover, his service in government and central banking gave him a practical approach to policy and the financial mechanics of economic adjustment that excludes simple theoretical solutions. Much of the time he reads like the notorious two-handed economist, concluding too often that circumstances alter cases (e.g. Kindleberger 1965: 387; 1969: 35–36). But even this conclusion is never less than elegant or erudite, and betrays a genuine commitment to finding out about a country before assessing its condition or prospects.

Perhaps Kindleberger's most distinctive contribution to the theory of economic development processes is his argument in *Economic Development* that "economic development depends upon an open class structure and is particularly helped by the existence of a strong middle class" (Kindleberger 1965: 23). The precondition for successful industrialisation was a "commercial revolution" because "the growth of markets train(s) the entrepreneur" and "eases his tasks". Indeed,

> the commercial revolution does more. It starts the accumulation of capital. Inventories and ships are two early and easy objects of and outlets for capital accumulation. The productivity of capital becomes evident, the habit of saving spreads, and accumulations of capital capable of being converted to industrial use come gradually into existence.
>
> (1965: 163)

He concluded (p. 163):

> there is a strong argument to be made for the proposition that economic development through industrialization should be preceded by commercialisation, and the industrial by the commercial revolution. To short-cut the evolutionary process and attempt forthwith to turn a subsistence economy into an industrial economy may find a society badly equipped in capacity for transport and distribution.

For him, this was as much a political argument as it was an economic one. Early on Kindleberger expressed the view, widespread among the upper middle classes of his generation on both sides of the Atlantic, that the ruination of the Central European middle classes in the 1920s had paved the way for Hitler and fascism.

In Kindleberger's view, the "prior commercial revolution" was also necessary to ensure adequate supplies of food and wage goods, as labour shifted from agriculture to manufacturing industry, in the classic **Arthur Lewis** model of development. Kindleberger considered that this "Ricardian" model "neglects the central concern of Ricardo – how the price of food is to be held down" (Kindleberger 1965: 182). With characteristic even-handedness, he was willing to concede that the economic development of Japan after 1868 "may be put in the balance on the other side. Industrialization moved very rapidly …" He concluded: "This is a remarkable case. In my judgement, however, it provides the exception rather than the pattern" (p. 164). But it is more difficult to see how a commercial revolution would provide a country with transport infrastructure. Like many other development economists, he argued against monumental projects, including in these transport and energy projects. He felt that such projects tended to have "a lower social marginal product of capital" and were typically "under-priced" by governments (p. 164). Recent "privatisation" of public utilities might have changed attitudes towards the supply by the private sector of such infrastructure, but had yet to prove that, unaided by the state, the private sector could maintain large capital-absorbing infrastructure. Kindleberger may have had in mind the privately constructed railways and canals of Europe and North America. He could not have been unaware of the speculative and frequently fraudulent nature of these undertakings.

Kindleberger considered, however, that the cultivation of commerce is prone to two "marketing pathologies". The first is high mercantile profits with price-fixing, leading to over-capacity and under-utilisation in the retail trade, leading to petit bourgeois conservatism and economic stagnation. The other pathology is "speculative fever" (Kindleberger 1965: 164–165). His interest in the latter eventually resulted in his most popular work, *Manias, Panics, and Crashes* (1978).

Kindleberger considered that market competition was the best way of combating mercantile conservatism. Later, he put this forward as an argument in favour of liberal government policies in relation to foreign direct investment. Indeed, he viewed opposition to such investment as misguided, "nationalist" and "emotional" (Kindleberger 1969: 145). In this respect he was robustly internationalist, liberal and pro-business, while disclaiming any "doctrinaire" attachment to private enterprise as a sufficient "organiser" of economic development. He followed the liberal Keynesian consensus of his contemporaries in regarding government action as appropriate in the provision of public goods and industries with external economies and diseconomies. Even in competition with private companies, governments have the advantages of risk-spreading, better information and ease of recruitment. In the process of starting economic development in the least developed countries, Kindleberger regarded government action as indispensable. **Alexander Gerschenkron** was an important influence on Kindleberger's thinking, about not only the role of governments in economic development but also about the functions of financial institutions in that process. But, at more advanced stages of economic development, Kindleberger was sceptical of étatist planning, which he thought tended to over-centralise decision-making, resulting in excessively large projects. He felt, in particular, that governments are prone to getting diverted from "economic objectives" and towards "welfare redistribution" and the under-pricing of governments' "output" (Kindleberger 1965: 132).

His scepticism about government was explicit in Kindleberger's critique of the import-substitution strategies advocated by **Raúl Prebisch** and **Hans Singer** during the 1940s and 1950s. These strategies advocated the expansion of the domestic market in developing countries by government projects, in particular in transport, public utilities and government services, behind import barriers to ensure that the resulting economic growth was not dissipated in a surge of imports. On the face of it, such strategies would induce precisely the kind of domestic "commercial" revolution that Kindleberger believed should precede successful industrialisation. However, he argued that, notwithstanding incidental benefits, such strategies inhibit technological improvements that may be obtained by importing more advanced manufactured goods.

Above all, he considered that these strategies may divert resources away from exports and to the domestic market (Kindleberger 1963: 468). But he was typically sceptical too about arguments in favour of free trade, which he regarded as being supported by "partial equilibrium" efficiency arguments, when economic development was about improving the efficiency of the economy as a whole. This consideration led him to conclude that market prices are an unreliable guide to economic decision-making where the process of economic development alters the whole structure of the economy (Kindleberger 1965: 191–193). There is a deeper meaning to his repeated conclusion that "circumstances alter cases".

Early on, Kindleberger was a critic of the application of Keynesian policies of aggregate demand management to developing countries, which he perceived to be especially prone to inflation. This inflation is not due to too much demand in the economy at large, but is more "structural". By this he meant that the rapid expansion of employment in the sectors leading the development of the economy (in particular foreign trade, light industry and retail services) would tend to raise earnings from employment more rapidly than food production. At low levels of income, spending on food takes up a high proportion of household budgets. The rise in food prices due to increased non-agricultural employment was therefore not amenable to aggregate demand management without bringing the process of economic development to a halt (Kindleberger 1965: 227–230).

Kindleberger robustly favoured foreign direct investment. During the 1960s, he urged the establishment of an international conference to provide ways of promoting it (1969: 206–208). He believed that the absence of savings adequate to finance industrial projects is a major constraint on the economic development of poorer countries. Indeed, in his earliest writings on economic development he put forward a typology of stages of economic development, according to capital inflow from abroad. Poorer countries, he argued, use foreign capital inflows to augment their domestic saving, before eventually achieving industrial maturity and experiencing capital outflows (1963: ch. 23). This belief in a savings constraint on economic development, in large measure, reinforced his sympathy for the thrifty middle classes.

However, he was ambiguous in his attitude towards the financial liberalisation that came to characterise official US policy, and that of the World Bank and the International Monetary Fund, towards banking and finance in developing countries. During the 1950s and the early 1960s, Kindleberger took **Gerschenkron**'s side in arguing that government can more efficiently take the place of commercial banks early on in the development process. He criticised Gurley and Shaw, the leading advocates of such liberalisation in the 1950s, for underestimating the degree to which investment can be financed in advance of saving by monetary expansion (Kindleberger 1965: 234). Nevertheless, by 1970, he was arguing in favour of more lending by foreign and domestic commercial banks in developing countries, because it "depoliticises" commercial lending decisions, and because private finance "adjusts to reality" more quickly than official, or government, lenders (Kindleberger 1970).

By 1976, Kindleberger was convinced that the expansion of international bank lending in developing countries had gone too far. He argued that financial liberalisation exposes developing countries to integration with international capital markets that are prone to illiquidity, which can inflict crisis on countries without robust financial systems. He urged the establishment first of domestic banking systems before developing countries opened up their financial markets to foreign financial capital (Kindleberger 1976). It was another six years before the global 1982 debt crisis convinced international bankers and economists that financial liberalisation can have painful consequences. In 1978, Kindleberger published what is arguably his most famous book, *Manias, Panics and Crashes*. In subsequent editions, lending to developing countries, in particular from the 1970s onwards, featured as a speculative mania leading to crisis. He advocated prompt central bank lending to overcome such financial crises, and urged the creation of an international "lender of last resort" to avoid them in the future.

Kindleberger was arguably the most influential American development economist of his time. His writings on economic development are authoritative and judicious. Yet his cosmopolitan studies of banking and economic history and his ideas on economic development ventured little beyond the mainstream US academic literature of the mid-20th century. In particular, the European tradition of viewing economic development as a process of changing social and political, as well as economic, structures was foreign to an American who professed himself to be an economist in a somewhat narrower sense (Kindleberger 1969: vi). His admiration for the thrift of the middle classes echoes that of John Stuart Mill a century earlier. His advocacy of a commercial revolution preceding industrialisation was rooted in his studies in the economic history of Western Europe and his native United States of America. Alternative routes to development seemed problematic to him. Japan, he wrote, must be an "exception". The saving constraint he identified in developing countries has a distinctly pre-Keynesian air, while his notion that "mature" economies export capital is at odds with the reliance of the US and the UK on capital imports at the start of the 21st century. These limitations were redeemed by the scale of his output, his work on Marshall Aid, his early recognition that financial liberalisation leads to crisis for developing countries, and his exposure of the hubris of banking and finance at the end of the 20th century.

Major works

Kindleberger, C.P. (1953) *International Economics*, Homewood, IL: Richard D. Irwin. (2nd edn 1958, 3rd edn 1963.)

—— (1958) *Economic Development*, New York: McGraw-Hill. (2nd edn 1965.)

—— (1969) *American Business Abroad: Six Lectures on Direct Investment*, New Haven, CT: Yale University Press.

—— (1970) "Less Developed Countries and the International Capital Market", in J.W. Markham and G.F. Papanek (eds), *Industrial Organisation and Economic Development: Essays in Honor of E.S. Mason*, Boston, MA: Houghton Mifflin. (Repr. in C.P. Kindleberger, *International Money: A Collection of Essays*, London: George Allen and Unwin, 1981.)

—— (1976) "International Financial Intermediation for Developing Countries", in R.I. McKinnon (ed.), *Money and Finance in Economic Growth and Development: Essays in Honor of Edward S. Shaw*, New York: Marcel Dekker. (Repr. in C.P. Kindleberger, *International Money: A Collection of Essays*, London: George Allen and Unwin, 1981.)

—— (1978) *Manias, Panics, and Crashes: A History of Financial Crises*, Basingstoke: Macmillan. (2nd edn 1989.)

Further reading

Kindleberger, C.P. (1950) *The Dollar Shortage*, New York: The Technology Press of Massachusetts Institute of Technology and John Wiley.

—— (1964) *Economic Growth and France and Britain 1851–1950*, Cambridge, MA: Harvard University Press.

—— (1987) *International Capital Movements*, Cambridge: Cambridge University Press.

—— (1984) *A Financial History of Western Europe*, Oxford: Oxford University Press. (2nd edn 1993.)

Miall, L. (2003) Obituary: "Charles P. Kindleberger: Economic Historian and Progenitor of the Marshall Plan", *The Independent* (10 July): 16.

JAN TOPOROWSKI

SIR WILLIAM ARTHUR LEWIS (1915–1991)

William Arthur Lewis was a pre-eminent figure in the emergence of development economics from the late 1940s until his death in 1991. He made a singular contribution to its elaboration as a distinct branch of economics and as a body of ideas informing the formation of public policy. His work on problems of balanced sectoral growth, "unlimited" surplus labour, agricultural productivity and planning created an important framework and agenda which shaped much subsequent research. His stature was perhaps most succinctly summed up when Harberger called him "the patriarch" of development economics.

Arthur Lewis was born in St Lucia in 1915, the fourth of five sons of parents who were both schoolteachers. His father died when he was seven and his mother thus became the most powerful early influence in his life, imparting her strong religious convictions and her love of music to her son. Lewis left school at 14, having completed his school certificates, to work as a civil service clerk in the St Lucia Department of Agriculture. At 17, he won an island scholarship which took him to the London School of Economics (LSE). He had hoped to be an engineer, but the realities of the colonial West Indies meant that there was no employment for a black engineer. Instead he opted for a business administration degree, confessing later that "I had no idea in 1933 what economics was" (Lewis 1979). Nevertheless, he graduated in 1937 with "the highest marks ever obtained in the history of the School" (Downes 2003: 2). This opened the door to a PhD in industrial economics (which he received in 1940) and to a lectureship at the LSE, making him its first black member of staff – though not before the LSE Director had referred the appointment to the Board of Governors because "the appointment of a coloured man may possibly be open to some criticism" (Tignor 2006: 21).

Between 1938 and 1947 (by which time he had become Reader in Colonial Economics), Lewis had short attachments at the Board of Trade and the Colonial Office, allowing him research access to "reports from the colonial territories on agricultural problems, mining, currency questions, and the like". During this period he also became a member of the executive of the League of Coloured People, interacting with the likes of C.L.R. James, James Padomore and Eric Williams, and editing its journal, *The Keys*. He also published a pamphlet on economic and social problems in the West Indies for the Fabian Society, an association which would yield a number of publications. Although he professed little interest in politics, describing himself as "a social democrat who was interested in poverty eradication and economic and social development" (Downes 2003: 5), this combination of scholarship and civic activism characterised his life.

Lewis's LSE research produced two significant books which introduced themes that were to characterise his work. The first, *Economic Survey, 1919–39* (1949a), situated economic development in the context of world economic history, viewing the world economy as a single interdependent system. The second, *The Principles of Economic Planning* (1949b), written for the Fabian Society, set out his case for "planning through the market", identifying the "conditions under which the market would need to be supplemented by intelligent state action" (Bhagwati 1979: 15), and in the process developing a powerful argument against both laissez-faire and, particularly, "planning by direction". The latter he considered inefficient, inflexible, undemocratic and "procrustean", criticisms that would later be taken up by many other commentators, many of them having little in common with Lewis.

Lewis's most important intellectual achievements came after he moved to the University of Manchester in 1947 to become Stanley Jevons Professor of Political Economy. There, for the next ten years, he made a defining contribution to the elaboration of the discipline of development economics. That contribution is encapsulated in two publications. The first was a seminal paper entitled "Economic Development with Unlimited Supplies of Labour", published in *Manchester School* in May 1954. The second was an elaboration of it (and much else) in a broad treatise entitled *The Theory of Economic Growth* (1955). Wrestling with the problems of poverty and poor rates of growth in developing countries, Lewis felt increasingly that neoclassical explanations for wages, profits and savings did not work in such countries and that relative prices could not be explained within that framework. He thus took his point of departure from classical economic theory. In his autobiographical statement for the Nobel committee, he recalled:

> From my undergraduate days, I had sought a solution to the question of what determines the prices of steel and coffee. The approach through marginal utility made no sense to me. And the Heckscher-Ohlin framework could not be used, since that assumes that trading partners have the same production functions ... Another problem that troubled me was historical. Apparently during the first fifty years of the industrial revolution, real wages in Britain remained more or less constant while profits and savings soared. This could not be squared with the neo-classical framework, in which a rise in investment should raise wages and depress the rate of return on capital.
>
> One day in August 1952, walking down the road in Bangkok, it came to me suddenly that both problems have the same solution.

> Throw away the neo-classical assumption that the quantity of labour is fixed. An "unlimited supply of labour" will keep wages down, providing cheap coffee in the first case and high profits in the second case. The result is a dual (national or world) economy, where one part is a reservoir of cheap labour for the other.
>
> (Lewis 1979)

The "unlimited" supplies of labour were made possible, first, by the subsistence economy of the "traditional" agrarian sector in most developing countries and, second, by the considerable underemployment and low productivity of that sector. In a system of production where the marginal productivity of cultivators approached zero, it was possible for large ("unlimited") supplies of labour to move to the capitalist sector of the economy without either significantly raising industrial wages (since real wages were set by the consumption levels of the traditional sector rather than by supply and demand) or reducing profits – and without significantly reducing agrarian production. Thus "dualism" meant that high profits and rapid industrial growth would not be limited by the same factors as in developed economies.

What would limit the prospects for growth, Lewis concluded, was the inability of the agrarian sector in such dual economies to trade with the industrial sector so that industrialisation would need either a rapid increase in agricultural productivity or export markets. Using a Ricardo–Graham model, Lewis argued that the terms of trade (say between North and South) were determined by relative labour productivities in food. Given the low productivity of the traditional agrarian sector in developing countries, there would be what Emmanuel was later to call "unequal exchange". **Albert Hirschman** (1982: 382) describes it thus: "as long as 'unlimited supplies of labour' in the subsistence sector depress the real wage throughout the economy, any gains from productivity increases in the export sector are likely to accrue to the importing countries ...". This led Lewis to the conclusion that there was a need for "balanced growth": for rapid industrialisation and for a rapid and substantial increase in agrarian productivity so as to improve the terms of trade and to release labour from the land by reducing rural poverty. There was also a consequent need for protection to permit this process to take place.

These two models, one explaining the problems and possibilities produced by "dualism" and the other those produced by the terms of trade between economies with different production functions, set the agenda for debate, discussion and further research in development economics for a generation. As Findlay (1980: 3) observes:

> His own subsequent work, and in fact a large part of the literature of development economics, can to a large extent be seen as an extended commentary on the meaning and ramifications of this central idea. Few other instances come readily to mind of an entire field being so dominated by a single paper.

That article and the arguments it provoked have proved to be Lewis's lasting legacy. As Kirkpatrick and Barrientos (2004: 2) observe,

> Lewis's key insight into the structural dualism characterising developing countries mapped out a new and distinctive field for development economics and policy. The article has had a profound impact over the last fifty years and continues to yield valuable lessons for understanding the nature of economic transformation.

For 25 years from 1957, Lewis elaborated and reprised these ideas in various ways. A volume on development planning appeared in 1966, extending the ideas first explored in 1949 and an edited collection, *Tropical Development, 1880–1913*, published in 1970, re-examined various issues of economic history and development in a global economy. These books illustrate an important element of Lewis's work – his insistence that problems of economic development in developing countries had to be understood in the context of the world economy as a whole. This perspective was to culminate in 1978 in the publication of *Growth and Fluctuations 1870–1913* (1978a), which Findlay (1980: 73) described as "a truly remarkable work, characterised by acute theoretical insights, vast historical knowledge and painstaking labor in the collection and analysis of statistical data".

In the years after Manchester, Lewis became as much involved in administration and policy as in academic research. From 1959 to 1963 he served as the first West Indian principal of the University College of the West Indies, helping to extend its reach beyond Jamaica to Barbados and Trinidad, and negotiating its autonomy from London University to become the independent University of the West Indies. In 1967 he was appointed Chancellor of the University of Guyana, and in the mid-1970s he also lectured part-time at UWI in Barbados. Among a number of roles as adviser to the leaders of developing countries, Lewis served as UN economic adviser to President **Kwame Nkrumah** in Ghana until 1963 (advising on the production of Ghana's second national development plan, which he ultimately characterised as worthless), as deputy managing director of the UN Special Fund, and as president of the Caribbean Development Bank from 1970 to 1974. In 1963, he became James Madison Professor of Political Economy at Princeton, a tenure which lasted until his retirement in 1983. His approach to economics and economic history gave him a broad perspective and the capacity to look beyond the narrow models that dominated neoclassical economics. They also prompted him to address related issues that lay outside the disciplinary boundaries of economics. Thus, his *Politics in West Africa* (1965) explored the relationship between democracy and development in West Africa. And in 1982 he delivered the W.E.B. Du Bois Lectures, which were published as *Racial Conflict and Economic Development* (1985). Lewis was knighted in 1978 and received the Nobel Prize for Economics (jointly with Theodore Schultz) in 1979. In his paper for the Nobel Prize committee, **Jagdish Bhagwati** argued that Lewis had made a singular contribution on three levels – to political economy, to historical analysis and to the modelling of development problems.

In more sardonic vein, **Albert Hirschman** (1982: 377) observed that the Nobel Prize for economics "is often split between one person who has developed a certain thesis [here Lewis] and another who has labored mightily to prove it wrong [Schultz]". This comment highlights the fact that Lewis – and development economics in general – came under persistent attack from the start from both left and right ("the strange alliance of Neo-Marxism and Monoeconomics against Development Economics", Hirschman called it). For both sides, the relative failure of indicative planning in countries like Ghana seemingly provided confirmation of their rejection of development economics.

For many in the neoclassical orthodoxy, the Lewis models became "a privileged target", provoking criticism of his conception of wages and prices. The rise of neoliberalism and the new economic history (which in part grew out of Lewis's ideas, according to **Bhagwati**) demonstrated little sympathy for his views on dualistic markets or state intervention. For dependency and "development-of-underdevelopment" theorists, too, Lewis's contention that economic exchanges between rich and poor countries could, in some circumstances, become antagonistic missed the point. Rather, they insisted, this was the essential character of the relationship, an iron law of centre–periphery inequality in the global capitalist economy which made the hopes of balanced growth an illusion. Lewis (1984) directly rejected this dependency argument: it was appropriate, he suggested, as a description of what happened in the colonial context, in the late 19th and early 20th centuries, but it made little sense in the second half of the 20th century when independent governments sought to restructure these relationships. Nor is it likely, for all his courteous manner, that he would have had much time for the Washington Consensus and neoliberal agenda which would inevitably deepen inequality and sectoral imbalance.

Even so, there is little doubt that the tide has run against Lewis and other important figures of development economics over the last 25 years. The spiralling debt crisis, increasing social and economic inequality and widespread political instability and violence characterising the global system have undermined ideas of promoting balanced growth. What has remained is the debate about whose fault it is.

For all that, Lewis's contribution to economics and development policy remains what Kirkpatrick and Barrientos consider "lasting testimony to the author's extraordinary ability to combine theory and history into a coherent framework". The 1954 article, they suggest, made two fundamental contributions to the emergence of development economics: first, to present it as a policy science addressed to the alleviation of underdevelopment and poverty; and second, to understand development as "a multidimensional process of economic, political, social and institutional change" (Kirkpatrick and Barrientos 2004: 10–11). Thus, for all the controversies surrounding development economics, Lewis's personal and intellectual standing has endured.

Major works

Lewis, W.A. (1949a) *Economic Survey, 1919–39*, London: Allen & Unwin.

—— (1949b) *The Principles of Economic Planning*, London: Allen & Unwin.

—— (1953) *Report on Industrialisation and the Gold Coast*, Accra: Government Printer.

—— (1954) "Economic Development with Unlimited Supplies of Labour", *Manchester School* 22: 139–191.

—— (1955) *The Theory of Economic Growth*, London: Allen & Unwin.

—— (1965) *Politics in West Africa*, London: Allen & Unwin.

—— (1966) *Development Planning: The Essentials of Economic Policy*, London: Allen & Unwin.

—— (1970) *Tropical Development, 1880–1913*, London: Allen & Unwin.

—— (1978a) *Growth and Fluctuations, 1870–1913*, London: Allen & Unwin.

—— (1978b) *The Evolution of the International Economic Order*, Princeton, NJ: Princeton University Press.

—— (1985) *Racial Conflict and Economic Development*, Cambridge, MA: Harvard University Press.

Further reading
Bhagwati, J.N. (1979) "W. Arthur Lewis: An Appreciation", repr. in M. Gersovitz et al. (eds), *The Theory and Experience of Economic Development: Essays in Honor of Sir* W. *Arthur Lewis*, London: Allen & Unwin, 1982, 15–28.
Downes, A.S. (2003) "William Arthur Lewis: 1915–1991: A Biography", in paper prepared for D. Rutherford (ed.), *The Biographical Dictionary of British Economists*, Bristol: Thoemmes Press, 2004.
Findlay, R. (1980) "On W. Arthur Lewis's Contributions to Economics", repr. in M Gersovitz et al. (eds), *The Theory and Experience of Economic Development: Essays in Honor of Sir W Arthur Lewis*, London: Allen & Unwin, 1982, 1–14.
Gersovitz, M., Diaz-Alejandro, C.F., Ranis, G. and Rosenzweig, M.R. (eds) (1982) *The Theory and Experience of Economic Development: Essays in Honor of Sir Arthur Lewis*, London: Allen & Unwin.
Hirschman, A.O. (1982) "The Rise and Decline of Development Economics", in M Gersovitz et al. (eds), *The Theory and Experience of Economic Development: Essays in Honor of Sir W Arthur Lewis*, London: Allen & Unwin, 372–390.
Ingham, B. (1991) "The Manchester Years, 1947–1958: A Tribute to the Work of Arthur Lewis", Salford Papers in Economics, University of Salford.
Kanbur, R. (2017) "W. Arthur Lewis and the Roots of Ghanaian Economic Policy", in E. Aryeetey and R. Kanbur (eds.) *The Economy of Ghana Sixty Years After Independence*, Oxford: Oxford University Press.
Kirkpatrick, C. and Barrientos, A. (2004) "The Lewis Model after Fifty Years", Development Economics and Public Policy, Working Paper Series No. 9, Institute for Development Policy and Management, University of Manchester, (September 2004).
Lewis, W.A. (1979) "Autobiographical statement to the Nobel committee, Nobel Lectures, Economics 1969–80", http://nobelprize.org/economics/laureates/1979/lewis-autobio.html.
—— (1984) "Development Economics in the 1950s", in G.M. Meier and D. Seers (eds), *Pioneers in Development*, New York: Oxford University Press for The World Bank, pp. 119–37.
Tignor, R.L. (2006) *W. Arthur Lewis and the Birth of Development Economics*, Princeton: Princeton University Press.

MORRIS SZEFTEL

MICHAEL LIPTON (1937–)

Even in his eighties, Michael Lipton continues to make distinguished contributions to development research across a wide front. Most importantly, however, he is the principal contemporary theorist and advocate of the role of agriculture in development, recognised in the award, in 2012, of the prestigious Leontief Prize for Advancing the Frontiers of

Economic Thought. He has long championed the role of small-scale agricultural producers, emphasising how the agricultural sector generally, small farmers most specifically, and in consequence overall development are harmed by what he describes as "urban bias". For this he has sometimes been castigated (most memorably in Byres 1979) as a populist. But he is no dewy-eyed romantic harking back to the agrarian past. He is unusual among scholars working on development issues for his deep engagement with the hard sciences, especially plant breeding and nutrition. And he certainly does not envisage that all agricultural production is necessarily, and always, carried out most efficiently on small farms.

Lipton was born in London in 1937, into a German Jewish family that had left Hamburg when the Nazis came to power in 1933. He went to Haberdashers' Aske's School in Borehamwood, north London, and from there to Balliol College, Oxford, to read Politics, Philosophy and Economics, graduating with First Class honours, winning several university prizes for economics, and subsequently taking a prestigious Fellowship at All Souls' College in Oxford. He was taught as an undergraduate by **Paul Streeten**, and it was as a result of some notes that he wrote for Streeten on **Gunnar Myrdal**'s *Beyond the Welfare State* (1960) that Myrdal subsequently invited him to work on parts of what was to become *Asian Drama* (Myrdal 1968). Ultimately published in three large volumes, *Asian Drama* was in its time a pioneering study in which an institutional approach was applied to the analysis of the development problems in South Asia. "Mr Lipton", Myrdal writes, "wrote the first draft of Appendix 10 on climate and prepared an intensive analysis of national accounting in the South Asian countries … [and his] … most difficult and time-consuming assignment was to carry out a lengthy comparative analysis of the development plans in the countries of the region" (Myrdal 1968: xv). It was this experience that drew Michael Lipton into development economics and gave him his lifelong interest in the development of South Asia, in particular.

According to his own account, he taught himself development economics as a young lecturer at Sussex University, and he studied mathematics and econometrics for a year at MIT in 1963–4, alongside the future Nobel Prize-winners **Joseph Stiglitz** and George Akerloff. Shortly after this, in 1965, he took the very unusual step for an economist of going to live for eight months in a village in the western Indian state of Maharashtra. In the work that he had done for *Asian Drama*, he had come to realise the inadequacy of existing micro-level research for understanding problems of agricultural development, and he set about making up for this lack. His experiences in the village (called

Kavathe) gave rise to a seminal article on "The Theory of the Optimising Peasant" (1968a) and to an important critique of India's agricultural development policy that appeared in a major book, *The Crisis of Indian Planning*, published in the same year (Streeten and Lipton 1968, see Lipton 1968b). That year, 1968, was remarkable in Lipton's career, for he published (at the age of only 31) not just these two outstanding contributions to understanding peasant agriculture and its role in development, but also a book on British economic performance (Lipton 1968c) and a book of chess problems. Chess, as well as poetry (he enjoys, for example, the work of Emily Dickinson) and music, have remained passions for Michael Lipton (he published a new book on chess problems in 2002), but he has not written on British economic policy since the 1960s.

The Kavathe experience not only taught him about the particular rationality of small peasant farming but also stimulated his thinking about the significance of the urban–rural gap, and the idea of "urban bias" appeared for the first time in his paper (1968b) for *The Crisis of Indian Planning*. It appeared again in another seminal paper, on land reform, published in 1974, and it was finally elaborated, controversially, in the book *Why Poor People Stay Poor* in 1977 (see his responses to critics: Lipton 1984; Lipton and van der Gaag 1993).

Meanwhile he had sought to put his ideas about redistributive land reform to practical effect in his contributions to an influential ILO report on the development of Sri Lanka (ILO 1971). All of this work was done from the Institute of Development Studies at the University of Sussex, where he was appointed as a Professorial Fellow in 1967, and remained a leading figure for almost 30 years. His role in the establishment of academic development studies was also reflected in his contributions as a managing editor of the still very young *Journal of Development Studies* – though he staked out a more cautious view of the role of "Interdisciplinary Studies in LDCs" than some others have advocated, in an article published in the *JDS* in 1970. At the time, Lipton was directing another project that resulted from his Kavathe experience, the Village Studies Project (VSP), which had the ambitions of generating general theory from systematic comparison of village studies, and of situating economic activity in developing countries in its wider social context. Though the case for analysing institutions that are situated between the household on the one hand and the state on the other is strong, the VSP was not entirely successful, partly because of the very uneven quality of the existing village surveys on which it was based. It gave rise to several interesting outputs (e.g. Connell et al. 1976;

Connell and Lipton 1976), but it was left to others to make more sense of "villages" as institutions (see, most recently, Krishna 2002).

Underlying much of Michael Lipton's work are ideas that derive from the Kavathe experience and from his appraisal of Indian agriculture in the 1960s. At that time, thinking about peasant agriculture was influenced especially by the (later Nobel Prize-winning) Chicago economist Theodore Schultz, whose book *Transforming Traditional Agriculture* (1964) elaborated the view that peasant farmers are "efficient but poor". Schultz's argument was that peasant farmers, by maximising their profits, allocate their resources efficiently. But they are poor, and their resources are severely limited – so achieving growth of output must involve changing the conditions of production through the application of new technology and new inputs (such as chemical fertilisers).

This is what the Indian government was trying to do by the later 1960s, concentrating those resources that were devoted to agriculture on areas that were thought to have a high potential for such modern cultivation. Lipton's "theory of the optimising peasant", however, held that it would be irrational for small peasant farmers to attempt to maximise profits, for such a strategy would take no account of the considerable risks and uncertainty associated with agricultural production. In their actual agricultural practices, Lipton showed, poor farmers seek to insure themselves against failure, and they may well sacrifice potential output or profit maximisation in the process. They have, he said, a "survival algorithm", pursuing – in terms of formal game theory – an *optimising* "minimax strategy", rather than trying to maximise. The aim of public policy, therefore, should be to reduce the riskiness of agriculture. The dedication of resources to irrigation, for instance, rather than provision of chemical fertilisers to farmers operating in more favoured areas, would enable much broader-based, more egalitarian agricultural development. Given appropriate policies, the relative efficiency of small-scale, household producers in the agricultural sector – reflected in the common observation of the existence of an inverse relationship between farm size and the productivity of land, and deriving from the fact that this mode of production reduces many of the "transaction costs" involved, and so encourages intensive application of labour – can become the basis for dynamic and also well-distributed growth.

At the time, thinking about development policy was still dominated by the idea that economic growth, necessarily involving industrialisation in the longer run, was best pursued in poor countries through planned, import-substituting industrialisation. This meant that there was a strong tendency, as Lipton noted in the Indian case, for agriculture to be neglected in the allocation of public investment. Development policy was characterised by an urban, industrial bias – a bias against what could be shown to be more efficient and more equitable ways of using public resources, through investment, in the first instance, in small-scale agriculture. Urban bias was underpinned, however, by a powerful political alliance in which the urban classes, bourgeoisie and proletarians with interests alike in cheap agricultural goods and the transfer of resources out of agriculture (another aspect of urban bias) drew in also the more powerful people in the rural economy through the transfer of some subsidies and the offer of places in the political sun. This alliance worked against the possibility of redistributive land reform that, in many rural economies, had the potential to create both a more just and a more efficient rural economy. This is the case that Lipton set out in detail in the 1974 paper – revisited in 1993a – and that he tried to see implemented in Sri Lanka. Urban bias, therefore, militated against a potentially more effective strategy for development, based on the transformation of the rural economy as a whole – an "agriculture first" strategy for development.

Lipton has maintained his advocacy of small-scale producers and of redistributive land reform – laid out with great eloquence in his book *Land Reform in Developing Countries: Property Rights and Property Wrongs* (2009). This is described in a review by Albert Berry as "one of the most important books ever written about agriculture in developing countries" (2011: 637), and the overarching point it makes is that, under most conditions, the arguments advanced for redistributive land reform in the middle of the 20th century remain valid in the 21st. Drawing on a breadth and depth of scholarship, the book is argued with the utmost rigour, and Lipton is scrupulous in confronting counter-evidence.

In the later 20th century, Michael Lipton devoted much of his research to understanding the causes and conditions of poverty (see Lipton 1983a; 1983b; 1983c; 1985; 1993a; Lipton and Ravallion 1993). He brought this research and his work on agriculture together in an authoritative book, *New Seeds and Poor People* (Lipton and Longhurst 1989), on the impact of the "modern seeds" (or Higher-Yielding Varieties) that had brought the "Green Revolution". There had been a spate of influential work in the 1970s that was critical of the effects of the introduction of the new seeds, as contributing to the further impoverishment of large numbers of rural people (see especially Griffin 1974). Lipton himself had been a little sceptical in the early days (1968b: 118) – but he later showed that much of the criticism of the Green Revolution had been premature and that, most significantly because of its effects on both labour and product markets, the new technology, in South Asia certainly, had brought real benefits for poor people. The cultivation of the new seeds had absorbed more labour and helped to tighten rural labour markets, imparting a push upwards to rural wages, while increased output of cereals had tended to bring prices down. The two factors together had benefited the poor, a majority of whom depend on wages and on purchasing food rather than producing it themselves. Lipton has continued to emphasise the importance for poverty outcomes of the ways in which labour markets function, and his continued championing of small-scale agriculture is because of the ways in which this mode of agricultural production can generate productive employment while the process of the structural economic transformation takes place. Ultimately, such small-scale production may well be replaced by larger-scale commercial production, but it is mistaken to attempt to short-circuit this historical process (as he explains in the interview he gave following the award of the Leontief Prize, accessible at: www.ase.tufts.edu/gdae/about_us/leontief.html).

Interested in the relevance of Asian experience for Africa, and vice versa, Lipton took up work on rural labour in Africa, initially in Botswana in the later 1970s, and subsequently with his wife, Merle, and others, in South Africa (Lipton et al. 1996); he began to engage more thoroughly with nutrition science (reflected first in Lipton 1983c); he has continued to work on economic demography (Lipton 1983a; Lipton and Eastwood 1999; 2011) – and he promises a book on "Malthus and Africa"; he undertook studies of aid effectiveness; and he has entered into the controversy about the potentials and problems associated with the applications of biotechnology (Lipton 2007). He is firmly of the view that this technology has enormous potentials in regard to the livelihoods of poor people, and that much of the criticism – like that of the Green Revolution three decades ago – is muddle-headed. He remains an active researcher and, with his champion chessplayer's style of argument, he is as engaging and formidable in debate as ever.

Major works

Connell, M., Dasgupta, B., Laishley, R. and Lipton, M. (1976) *Migration from Rural Areas: The Evidence from Village Studies*, Oxford: Oxford University Press.

International Labour Office (1971) *Matching Employment Opportunities and Expectations: A Programme of Action for Ceylon*, vol. 2, *Working Papers*, Paper 10.

Lipton, M. (1968a) "The Theory of the Optimising Peasant", *Journal of Development Studies* 4: 327–51.

—— (1968b) "Strategy for Agriculture: Urban Bias and Rural Planning", in P. Streeten and M. Lipton (eds), *The Crisis of Indian Planning: Economic Planning in the 1960s*, London: Oxford University Press.

—— (1968c) *Assessing Economic Performance: Some Features of British Economic Development 1950–65 in Light of Economic Theory and Principles of Economic Planning*, London: Staples Press.

—— (1970) "Interdisciplinary Studies in LDCs", *Journal of Development Studies* 7: 5–18.

—— (1974) "Towards a Theory of Land Reform", in D. Lehmann (ed.), *Agrarian Reform and Agrarian Reformism*, London: Faber, 269–315.

—— (1977) *Why Poor People Stay Poor: A Study of Urban Bias in World Development*, London: Temple Smith.

—— (1983a) *Demography and Poverty*, Washington, DC: World Bank.

—— (1983b) *Labour and Poverty*, Washington, DC: World Bank.

—— (1983c) *Poverty, Undernutrition and Hunger*, Washington, DC: World Bank.

—— (1984) "Urban Bias Revisited", *Journal of Development Studies* 20(3): 139–166.

—— (1985) *Land Assets and Rural Poverty*, Washington, DC: World Bank.

—— and Longhurst, R. (1989) *New Seeds and Poor People*, London: Unwin Hyman.

—— (1993a) "Land Reform as Commenced Business: The Evidence Against Stopping", *World Development* 21(4): 641–657.

—— (1993b) "Urban Bias: Of Consequences, Classes and Causality", *Journal of Development Studies* 29(4): 229–258.

—— (2007) "Plant Breeding and Poverty: Can Transgenic Seeds Replace the 'Green Revolution' as a Source of Gains for the Poor?" *Journal of Development Studies* 43(1): 31–62.

—— (2009) *Land Reform in Developing Countries: Property Rights and Property Wrongs*, London and New York: Routledge.

—— and Eastwood, R. (1999) "The Impact of Change in Human Fertility on Poverty", *Journal of Development Studies* 36(1): 1–30.

—— —— (2011) "Demographic Transition in Sub-Saharan Africa: How Big Will the Demographic Dividend Be?" *Population Studies* 65(1): 9–35

—— Lipton, Merle, De Klerk, M. and Ellis, F. (eds) (1996) *Land, Labour and Livelihoods in Rural South Africa*, 2 vols, Durban: Indicator Press.

—— and Ravallion, M. (1993) *Poverty and Policy*, Washington, DC: World Bank.

—— and van der Gaag, J. (eds) (1993) *Including the Poor*, Washington, DC: World Bank

Further reading

Berry, A. (2011) "The Case for Redistributional Land Reform in Developing Countries", *Development and Change* 42(2): 637–648.

Byres, T.J. (1979) "Of Neo-Populist Pipe-Dreams", *Journal of Peasant Studies* 6: 210–244.

Connell, J. and Lipton, M. (1976) *Assessing Village Labour Situations in Developing Countries*, Delhi: Oxford University Press.
Griffin, K. (1974) *The Political Economy of Agrarian Change*, London: Macmillan.
Krishna, A. (2002) *Activating Social Capital*, New York: Columbia University Press.
Myrdal, G. (1960) *Beyond the Welfare State: Economic Planning in the Welfare States and its International Implications*, London: Duckworth.
—— (1968) *Asian Drama: An Inquiry into the Poverty of Nations*, 3 vols, London: Penguin.
Schultz, T. (1964) *Transforming Traditional Agriculture*, New Haven, CT: Yale University Press.

JOHN HARRISS

REVEREND THOMAS ROBERT MALTHUS (1766–1834)

Not many people widely quoted in the contemporary development literature have given their name to a familiar concept and a word, as has Malthus. "Malthusian" connotes a pessimistic and negative long-term prospect: a "Malthusian scenario" is a frequent phrase anticipating an overuse of resources driven by excess population and rapid population growth, with the likelihood of serious demographic and economic collapse and "crisis"; the "Malthusian trap" is often invoked to argue that population growth is too rapid to permit savings and investment in national economic growth and thereby to raise standards of living; "neo-Malthusian" is a term widely applied to views that identify the contemporary relevance of his ideas, first developed over 200 years ago, on the relationship between population and resources – essentially that population growth and sustainable resource use are fundamentally incompatible in the long term.

Today we would call Malthus a political economist, though he was also an Anglican clergyman. He studied mathematics at Cambridge during a period much affected by the intellectual ferment of the Enlightenment and over the period of the French Revolution, when the nature of society and the relationship between poverty and inequality were dominant in public debate. Malthus was prominent in that debate, principally through two famous essays in 1798 and 1803. The second essay, with several subsequent editions, including as a contribution to the *Encyclopaedia Britannica* in 1824 (Malthus 1830), was the more reasoned and is more widely quoted. It was first published just before the beginning of Malthus's long period as Professor of Political Economy (from 1805 till his death) at the East India Company's Staff College, Haileybury, the major training institution for rulers of British India throughout the 19th century. However, there is not much about India in the essays, and he himself never went out of Europe (Flew 1970; Mayhew 2014).

Malthus's long-lasting contribution to debates on the nature of development has been to argue that, whereas population has the potential to grow geometrically, the resource base to support that population with food and other essentials can only grow much more slowly. Since he assumed a relatively fixed relationship between population and resources, taking resources as givens so that they could not really be created, ultimately population must outstrip resources unless its growth could be controlled. Population control, Malthus argued, could be, and had largely been, achieved by preventive measures – i.e. positive checks controlling the birth rate, essentially through sexual restraint outside and within marriage (but not the use of contraception). However, where these failed (and he saw them failing in the England of his lifetime), then the balance would inevitably be set by direct constraints – i.e. negative checks on the death rate, by "natural" factors: disease, famine and (not quite "natural") warfare.

Population growth in most human societies is usually controlled by preventive measures – humans do not breed to the limits of their fecundity, and social institutions such as marriage and custom will affect family size and childbearing patterns. However, Malthus believed, given the inequality and poverty of his own times (just after the French Revolution and the Napoleonic wars, a time of rapid industrialisation in western Europe), that the poor were showing insufficient restraint and that they were poor as a result of having large families that they could not support. As a result, their mortality was raised, due to malnutrition and disease. The poor were "victims of their own passion"! These ideas represented a response to apologists for the French Revolution – commentators we would now see as developing socialist or egalitarian arguments, who claimed that poverty and degradation were due primarily to inequalities in wealth rather than the absolute lack of resources to support any population, i.e. Malthus's position (Wrigley 1986). The debate on population and poverty between Malthusian ideas and socialist, Marxian ideas persists (Woods 1989; Watts 2000).

Today we can appreciate essential failings of Malthus's arguments, to the extent that he has recently been said quite explicitly to be "wrong" (Shermer 2016; *The Economist* 2008). The most obvious weakness is on the population side in that he did not anticipate the demographic transition, that globally population growth rates would fall as populations grew richer. Global population growth peaked during the last quarter of the 20th century, but has been falling in all continents – less rapidly in Africa – since then. There is no continuing evidence of geometric population growth. Malthus has also been shown to be equally "wrong" in his argument that resources were largely given and could not really be created beyond an arithmetic ratio. That this is not the case had become evident in the early/mid-19th century, with the great increase in the availability of land, largely in the New World. This revolutionised the global food supply, with densely populated Europe being increasingly supplied with cheaper food from overseas sources: the American plains, Australia and Argentina. New resources, of technology (e.g. refrigerated ships) as well as land, were being created and were able to support the burgeoning European and global populations at rising levels of living.

Technology continued to revolutionise food production throughout the 20th century, most evidently in the very substantial increases in food production associated with Green Revolution technologies from the 1970s, to the extent that the erstwhile bread-basket "Malthusian" cases of India and Bangladesh are no longer major food-deficit countries, despite their continuing – though now falling – high rates of population growth. Malthusian attention in the food debate has now moved to Africa, where there are serious and recurring food deficits. However, are these more appropriately attributed to the rapid population growth of the last half-century (the Malthusian view) or rather to inappropriate national and global food policies and inadequate resources being allocated to agricultural development? Is the contemporary African food crisis Malthusian or a political/economic crisis?

The prominence of Malthusian ideas in Development Studies peaked during the 1960s and 1970s, a period of rapid population growth, especially in the global South, and substantial economic growth in richer countries. Population growth was seen to be a constraint on development, with many poor countries caught in a Malthusian trap. In 1990, when fertility rates were not thought to be generally declining in African countries, the World Bank's Annual Report could argue that "Population growth is the biggest single development problem for Africa". This neo-Malthusian view is very much associated with the American commentators Lester Brown (1998) and Paul Ehrlich (Ehrlich and Holdren 1971; Ehrlich and Ehrlich 1990), who, quite separately, in a series of influential books and articles from the 1970s, argued the case for the necessity for preventive checks. These views were important for the growth of the family planning movement, later "softened" by the late 1980s and 1990s into a concern for reproductive health and mother-and-child programmes, and all integrated into global thinking through United Nations declarations and objectives. These developed a widely accepted neo-Malthusian agenda in that major problems were seen to lie on the population side of the population/resource equation, and solutions were to be sought in restricting population growth as well as in expanding production and resource use. At that time, international agencies and national governments saw the need simultaneously to promote Green Revolution technologies as well as extending major support for population programmes to accelerate progress through the demographic transition from high to low population growth.

One further influential strand in the neo-Malthusian argument in the late 20th century was associated with the Club of Rome (Meadows et al. 1972). In a series of alarmist simulations of current resource depletion and increasing pollution, this largely European-based group predicted a serious overburdening of the global system by the year 2000, to the extent that economic development would be thrown sharply into reverse. Globally, population growth and increasing consumption of physical resources with polluting technologies seemed no longer sustainable. These neo-Malthusian apocalyptic scenarios have proved to be unfounded. Today, the world is much richer than 30 years ago, global agreements have been promulgated to control pollution, while new technologies have brought new resources into the economic system. Population growth rates are everywhere falling, and there is a global awareness of the need to balance population and resources at all scales. As a result of greatly increased geological knowledge and volatile oil prices, often driven upwards principally for political reasons rather than by economic shortage, there is no longer an imminent threat of the global fuel crisis forecast by the Club of Rome. New resources are being used at scale (e.g. renewable energy from wind and solar power), while others (coal, copper) are much less in demand, though they are still important fuels for China and India. As Julian Simon (1981) argued persuasively in his polemic, *Population: The Ultimate Resource*, the long-term prices for energy and mineral resources have been falling, a sign of long-term surplus and sustainability rather than depletion and degradation. His position diametrically opposes that of Malthus, in that people are themselves resources, creators and managers of natural producers of resources, rather than merely mouths to feed. For Simon, low population growth is normally preferable to high population growth, but it may also be preferable to zero population growth. Investment in people through education and training and better health, as well as better access to technology and resources, will lead to enhanced productivity for the poor as for the rich.

The perspective about population being both cause and effect – common to both Malthus and **Marx** – is also shared with the Danish agricultural economist **Ester Boserup** (Boserup 1990; Lee 1986; Gould 2015: ch. 2). She is the most prominent advocate of the view that population growth can act as a stimulus rather than an impediment to economic change. For her, population growth can be a critical stimulus for agricultural intensification, with rising outputs per unit of input of land, labour and technology. Where there is no population growth, there is no pressure to raise outputs, to introduce new technologies or to increase labour inputs.

Given this range and frequency of qualifications and counter-arguments, why does the neo-Malthusian discourse remain so prominent in Development Studies into the 21st century (Kreager 2014), as any brief moment looking up "Malthus" or "Malthusian" on a search engine would confirm? We can cite four sets of reasons:

Empirical evidence: There is ample empirical evidence that high rates of population growth, consistent over several decades, have been a major constraint on development in some circumstances. Malthus himself adduced wide-ranging empirical evidence in his essays (Wrigley 1986), and **Boserup**'s formulations on the importance of population pressure for agricultural intensification were based on her extensive experience in India. The adverse impacts of rapid population growth in difficult environments, such as the Sahel in West Africa, are all too apparent (Gould 2015: box 2.2), However, alternative development scenarios may always be suggested that seek to restore any population/resource imbalances by concentrating on the resource side of the equation rather than population. We are now aware, particularly through the work of Michael Mortimore and collaborators – first in the Kano Close-Settled Zone in Nigeria, but more generally in West Africa (Mortimore 1998) and subsequently in East Africa, most famously in Machakos, Kenya – that there can be "more people, less erosion" (Tiffen et al. 1994), that long-term environmental improvement and productivity gains are possible with the stimulus of population growth, combined with commercialisation of the economy. In the light of their evidence, any neo-Malthusian interpretation of the current African environmental, economic and political plight therefore needs serious reassessment. Considerations of the question "Is Africa over- or under-populated?" requires a multivariate perspective that cannot be answered by considering the population factor independently.

The long-term perspective: One school of thought sees the Malthusian scenario as merely having been postponed as a result of new lands and new technologies, arguing that in the long term there is an essential logic about the need to achieve a balance between the population and resources of a finite planet. There are limits to technology; there must also be limits to population growth. The threat of "standing room only", of a very large global population living at low and falling levels of consumption, is a worst-case scenario. Current UN projections are that the global population will grow from 7.6 billion in 2018 to stabilise at 11.2 billion in 2100. As rates of population growth continue to fall from their previous peak as populations grow richer, what changes in consumption pattern are needed for global and regional populations to be sustained at current (and hopefully rising) quality of life?

Environmentalist discourses: The widely accepted need to strive towards "sustainable development" does require a better balance between population and resources. As the international development agenda becomes increasingly driven by threats of global warming and of land, air and marine pollution apparently out of control, there is a distinctive neo-Malthusian strand in the global discourse that sees the growing population to be at least part of the global and local pressures: more people seem inevitably to require greater consumption of resources, with more pollution and a greater risk of global warming. However, growing pollution is more the result of growth in per capita consumption of resources, not of population growth itself. Richer people consume more and pollute more. What seems to be required is a new development ethic, with growing affluence that is based on renewable energy and more sustainable lifestyles. Since increasing poverty of the population cannot be a political or humanitarian option, lifestyles and consumption patterns of the increasingly affluent and consumerist populations of the South, as well as the North, become critical. Richer populations are likely to experience spontaneously reduced population growth, even to below replacement levels, as now prevail in many European states. In these terms, to misquote Malthus, it is the rich, not the poor, who seem now to be victims of their own passion for consumption.

Global poverty and inequality: Discussion of neo-Malthusian scenarios is most common for countries in the global South. It is typically a view from the outside – and has to be seen in the broader context of global inequality. The Malthusian threat has receded in the rich North; indeed, there may be greater demographic threat from population ageing and below-replacement fertility than from population growth, or even from excess per capita consumption. In the poor South, within and between nations in a globalising world of increasing economic inequalities, population growth appears from the outside to continue to be a problem. Yet, as in Malthus's time, a critical discourse needs to focus on production and distribution, on the ownership and use of resources, on human rights and environmental justice, more than on aggregate consumption and the crude number of consumers.

Major works

Malthus, T.R. (1798) *An Essay on the Principle of Population, as it Affects the Future Improvement of Society, with remarks on the speculations of Mr. Godwin, M. Condorcet, and other writers*, London: J. Johnson. (Open access electronic edition 1998 by Electronic Scholarly Publishing Project: www.esp.org/books/malthus/population/malthus.pdf.)

—— (1803) *An Essay on the Principle of Population, as it Affects the Future Improvement of Society; or a view of its Past and Present Effects on Human Happiness; with an Inquiry into our Prospects respecting the Future Removal or Mitigation of the Evils which it Occasions*, 2 vols, London: J. Johnson.

—— (1830) *A Summary View of the Principle of Population*, London: John Murray, from *Encyclopaedia Britannica* (1824). (Repr. in D.V. Glass, *Introduction to Malthus*, London: Watts, 1959.)

Further reading

Boserup, E. (1990) *Economic and Demographic Relationships in Development*, Baltimore, MD: Johns Hopkins University Press.

Brown, L.R. (ed.) (1998) *State of the World, 1998*, Washington, DC: Worldwatch Institute.

Economist, The (2008) "Malthus, the False Prophet: The Pessimistic Parson and Early Political Economist Remains as Wrong as Ever" (15 May).

Ehrlich, P.R. and Ehrlich, A.H. (1990) *The Population Explosion*, New York: Simon and Schuster.

—— and Holdren, J. (1971) "The Impact of Population Growth", *Science* 171: 1212–1217.

Flew, A. (1970) "Introduction", in T.R. Malthus, *"An Essay on the Principle of Population"* and "*A Summary View of the Principle of Population*", London: Penguin, 7–55.

Gould, W.T.S. (2015) *Population and Development*, 2nd edn, London and New York: Routledge.

Kreager, P. (2014) "On the History of Malthusian Thought", *Population and Development Review* 40(4): 731–742.

Lee, R.D. (1986) "Malthus and Boserup: A Dynamic Synthesis", in D. Coleman and R. Schofield (eds), *The State of Population Theory*, Oxford: Blackwell, 96–130.

Mayhew, R.J. (2014) *Malthus: The Life and Legacy of an Untimely Prophet*, Cambridge, MA: Belknap Press.

Meadows, D.H., Meadows, D.L., Randers, J. and Behrens, W.W. (1972) *The Limits to Growth*, London: Pan Books.

Mortimore, M. (1998) *Roots in the African Dust: Sustaining the Dry Lands*, Cambridge: Cambridge University Press.

Shermer, M. (2016) "Why Malthus Is Still Wrong", *Scientific American* (1 May).

Simon, J.L. (1981) *Population: The Ultimate Resource*. Princeton, NJ: Princeton University Press, and Oxford: Martin Robertson.

Tiffen, M., Mortimore M. and Gichuki, F. (1994) *More People, Less Erosion: Environmental Recovery in Kenya*, Chichester: Wiley.

Watts, M.J. (2000) "Development at the Millennium: Malthus, Marx and the Politics of Alternatives", *Geographische Zeitschrift* 88(2):67–93.

Woods, R. (1989) "Malthus, Marx and Population Crises", in R.J. Johnson and P.J. Taylor (eds), *A World in Crisis? Geographical Perspectives*, Oxford: Blackwell, 151–174.

Wrigley, E.A. (1986) "Elegance and Experience: Malthus at the Bar of History", in D. Coleman and R. Schofield (eds), *The State of Population Theory*, Oxford: Blackwell, 46–64.

W.T.S. (BILL) GOULD

MAO ZEDONG (1893–1976)

Mao (1893–1976) was co-founder of the Chinese Communist Party, first President of the People's Republic of China, prominent strategist of guerrilla warfare and people's struggle and a major communist theoretician. His ideas were important for many revolutionary struggles, particularly during the decades of the 1950s–1970s. His ideas about social change and economic development turned out to be largely disastrous for China, but some of them nevertheless had an impact on development thinking.

Mao Zedong was born in Shaoshan village, near Changsha, the capital of Hunan province. He was the son of a middle-income farmer and his father wanted him to become a farmer as well, but he showed great interest in reading and learning, something that he maintained into old age. He was forced to marry at the age of 14 but his wife died some years later. At 16, he began studying law and classical Chinese texts, first through private lessons and later in schools. He developed an interest in Chinese history and literature but he also read Western thinkers and philosophers.

Around the turn of the century, profound changes were taking place in China. Nationalists were revolting against the imperial Qing dynasty that prevented any modernisation of China's economy and rigid social structure. In 1911 the empire collapsed and China became a republic, with Sun Yat-sen as its first president. In these turbulent years of revolution, Mao served for six months in the nationalist army. However, the Republic did not succeed in bringing unity and peace to the country. The country was divided among military warlords who wielded absolute power in their respective areas.

In 1918 Mao graduated from the Changsha teachers' training college and became a history teacher. He went twice to Beijing, where he learned about Marxist socialist theories. His activities attracted the attention of revolutionaries and he was invited to Shanghai, where the Comintern, the international socialist union dominated by the USSR, initiated the Chinese Communist Party (CCP), which joined the Kuomintang nationalist party of General Chiang Kai-shek.

Mao became an organiser of peasant and industrial unions in Hunan province. There he married his second wife. In 1926 and 1927 he made several visits to the countryside, where he investigated the peasant rebellions on which he wrote in his famous *Report of an Investigation into the Peasant Movement in Hunan*, explaining that the peasantry constituted a potential revolutionary force. In the meantime, the CCP had increased in force. Chiang and the warlords considered the party a threat, and in 1927 Chiang forcefully eliminated the CCP in Shanghai. Naturally this led to a split between the Kuomintang and the CCP.

Strong disagreement existed within the CCP regarding the strategies to be pursued. Mao represented the view that the establishment of soviets in rural areas and the building up of the Red Army were indispensable in order to capture the cities and, ultimately, the country. From 1928 until 1931, he and his comrades established and consolidated the Jiangxi soviet in the border area of the southern provinces of Jiangxi and Fujian. Mao's second wife stayed behind in Changsha, where she was executed by the nationalist troops in 1930. Meanwhile in Jiangxi, Mao met his third wife, who later was exiled to Moscow, where she died. Like his second wife, she seems to have had an independent character. This corresponded with Mao's ideas about the role of women in society, which, seen against the background of Chinese traditions and Confucian thoughts, can be considered as progressive. The first law he enacted in the first Soviet Republic of China gave men and women equal rights in marriage and divorce. It would also be the first law enacted in the People's Republic 20 years later.

In Jiangxi, Mao's star was rising. In 1931, he was elected as the first Chairman of the Soviet. When Chiang's troops forced the Red Army to leave Jiangxi, Mao led the army to Yan'an in Shaanxi, north-west China, a distance of some 9,600 km. This journey became famous as the Long March of 1934–1936. It included about 90,000 people, of whom only 20,000 survived. For eleven years Yan'an became the headquarters of the CCP. Mao studied Marxist philosophy and wrote some important essays in which he applied Marxist thinking to Chinese reality. In *On Practice* (1937) he stressed the necessity to investigate empirically the socio-economic conditions in areas where the CCP wanted to be active. In *On Contradiction* (1937), he argued that the power of the human will could also determine the material base rather than seeing the latter as the key independent variable.

In Shaanxi, Mao met his fourth wife, Jiang Qing, who later became one of the leading figures of the Cultural Revolution. Mao emerged as the major Communist leader and in 1943 became chairman of the Politbureau and CCP's Central Committee. With his growing power, he became more dominant and less flexible towards people holding other ideas. A personal cult developed around his life. In the introduction to the manifesto of the CCP at the 7th Party Congress in 1945, it was stated that the CCP must take the thoughts of Mao Zedong as guidelines for all its activities, and the history of the CCP was rewritten to give him a more prominent place.

During the second Sino-Japanese War (1937–1945), both the nationalists and the CCP had to fight against the Japanese while continuing the civil war. The civil war ended with the flight of Chiang Kai-shek to Taiwan. On 1 October 1949, Mao proclaimed the People's Republic of China. He became the first President of the republic, as well as chairman of the CCP and president of the military commission. He was re-elected to these positions in 1954.

Mao and his comrades inherited a country of 9.6 million km^2 devastated by decades of revolts and war. China's population was 600 million, the majority of whom were peasants living in extreme poverty and accustomed to suppression by local landlords, regionally based warlords and Japanese occupation forces. The first tasks of the CCP were to restore order and security, and to rebuild the country. Land reform proved to be crucial. Tenants and the rural landless received land, while the landlords were executed. Nowhere in history had such a large-scale land reform ever taken place. It was only logical that, after this period, China's development policy became based on the only socialist model known, the Stalinist approach to development. Thus, the major features of the First Five Year Plan (1953–1957) were central planning, industrialisation – stressing large-scale heavy industry – and mechanisation of the countryside.

The model was not uncontested. A real break with the Stalinist model came in 1958, when Mao launched the Great Leap Forward policy in an attempt to decentralise the economy by establishing self-reliant communities. These communes each numbered 50,000 or more people and were meant to co-ordinate the local economy. This concerned the organisation of the smaller units of work brigades and teams for farm work, mass mobilisation for infrastructural works, and rural industrialisation. The aim of the policy was social as well, namely to eliminate differences in class, education, sex and age. It lasted three years and coincided with natural disasters in various parts of the country, such as droughts, typhoons and floods. The policy and natural disasters created chaos and starvation in the countryside. The real size of the disaster will never be known, but estimates of the number of its human victims range between 20 and 40 million.

This disaster revealed the existence of a serious struggle within the CCP. Mao believed in the priority of politics above economics; self-reliance; radical equality; and mass mobilisations for permanent revolutions. His opponents considered economic growth as the first priority and adhered to the classical Marxist-Leninist idea of the party as a vanguard of the proletariat. Crucially, Mao had the support of Lin Biao, army general and Minister of Defence. The latter ordered the composition of a "Little Red Book" containing quotations from the work of Mao, and subsequently he ordered every soldier to use it as a guidebook. In the Red Army, distinctions were abolished. Students followed with campaigns against "elite thinking". In 1966, this resulted in the launching of the Great People's Cultural Revolution and the organisation of students in Red Brigades to spread the thinking of Chairman Mao. They swarmed over the country, attacking intellectuals and party officials, destroying cultural heritages and forbidding "capitalist habits" such as the reading of non-revolutionary literature. The Cultural Revolution disrupted the country totally and at the end of 1967 the army had to restore order.

However, the party struggle continued. The so called "Gang of Four", with Mao's wife, Jiang Qing, as one of its members, set the tone and continued their campaigns against bourgeois habits. Mao died in September 1976. One month later the Gang of Four was arrested and with that the Cultural Revolution came to its real end. Although officially Mao's ideas have never been denounced, since his death China has followed a different development policy. In the Socialist Market Economy, households have been restored as the basic production unit, private ownership in industry is encouraged and foreign investment in Special Economic Zones is welcomed. The new policies have led to economic growth but also enlarged the difference between the new rich and the poorer parts of the population, especially in the countryside. Western China also now lags far behind the east and southeast of the country.

Three tendencies can be discerned in Mao's thinking. The first is that he succeeded in adapting Marxism–Leninism to Chinese realities. This applies in particular to his idea that peasants and not the industrial proletariat constituted the revolutionary class. Second, as a leader, Mao was merciless with his opponents within and outside the party. Third, he was a visionary, utopian thinker. Mao believed in permanent criticism and in the power of the will of people, who did not need modern technology to change the world.

Mao's ideas were important for many revolutionary struggles, particularly in Third World countries. In Vietnam, rural guerrilla warfare led to victory in 1975, but proved catastrophic in Cambodia. Today, rebel movements in the Philippines, Nepal and Peru still refer to Maoism as their source of inspiration.

Indirectly, Mao's thoughts and the Maoist development model in China, as perceived by many Western intellectuals, stimulated "alternative" models and strategies in development thinking. Concepts such as bottom-up approaches, urban bias, decentralisation, self-reliance and primary health care, that are commonplace today, find their origin in debates of the 1960s and 1970s. Maoism certainly contributed significantly to these debates.

Major works

Mao Zedong (1966) *Quotations from Chairman Mao Tse-tung*, Peking: Foreign Languages Press (also known as the Little Red Book).

—— (1971) *Selected Readings of Mao Tsetung*, Beijing: Foreign Languages Press. These selected readings include major essays such as:

—— (1926) *Analysis of the Classes of the Chinese Society.*

—— (1927) *Report of an Investigation into the Peasant Movement in Hunan.*

—— (1936) *Problems of Strategy in China's Revolutionary War.*

—— (1937) *On Practice.*

—— (1937) *On Contradiction.*

—— (1942) *Talks at the Yenan Forum in Literature and Art.*

Further reading

Jung Chang and Jon Halliday (2005) *Mao: The Unknown Story*, London: Jonathan Cape.

Short, P. (1999) *Mao: A Life*, New York: Henry Holt.

Spence, J. (1999) *Mao*, New York: Viking/Penguin.

TON VAN NAERSSEN

KARL MARX (1818–1883)

Karl Marx was a revolutionary political and economic philosopher who, together with Friedrich Engels, produced a materialist theory of the evolution of societies. The theory was powerful enough that it influenced both social theory and the geopolitical reality that theory tries to understand. Marx was born in Trier and educated in philosophy at several German universities. He served briefly as editor of the Cologne newspaper *Rheinische Zeitung* in 1842 and 1843 but was forced to resign because of his radicalism and left for Paris, and then Brussels, where he wrote *The German Ideology* in 1847 (Marx 1970, 1981), his first sustained statement of an overall philosophy. Marx and Engels also organised a network of revolutionary groups into the Communist League, for which they wrote a statement of principles called *Manifesto of the Communist Party* (1969 [1847–1848]). After being arrested and banished, Marx finally settled in London in 1849, where he was active in the First International Workingmen's Association formed in 1864. A series of works, including *A Contribution to the Critique of Political Economy* (1969b [1859]), *Wages, Price and Profit* (1969a [1865]), and the first volume of *Das Kapital* (1867, translated into English as *Capital* volume 1, 1887 (1976), formed the core of Marx's economic writings. Volumes 2 and 3 of *Capital*, edited by Friedrich Engels, were published posthumously in 1885 and 1894, and translated in 1907–1909. *Capital: A Critique of Political Economy* is regarded as Marx's mature statement on capitalist development.

Marx and Engels were Enlightenment modernists, who believed in social progress and the perfectibility of humankind. They were optimists in the sense of trust in the transformative potential of science and the material plenitude made possible by technological advance. Yet they were also highly critical of the particular social and political form taken by modernism – that is, capitalism. They saw capitalist development as a process of human emancipation but also as alienation from nature, as a process of human self-creation, but one directed by a few powerful people, as progress in material life, but progress motivated by socially and environmentally irrational drives, like competition. Marxian analysis thus became not the scholastic pursuit of truth for its own sake, and certainly not legitimisation theory for the rich and famous, but a theoretical guide to radical political practice aimed at completely changing society. Development, they thought, should meet the needs of the working class and should be socially and democratically directed. Marx and Engels came to liberate modernism, not to praise it.

In their early years, Marx and Engels were followers of the German philosopher G.W.F. Hegel. Hegel thought that a kind of transcendental reason, the Absolute Idea, or World Spirit, was the hidden source of human consciousness, and the guiding force behind historical development. Yet, whereas in religion, Spirit is all-knowing from the start, in Hegelian idealism the Spirit's uncertainties were sources of the oppositions and crises inherent in every real thing. In the Hegelian dialectic, the negative element (the anti-thesis) is the living source of dynamic process. For Hegel, negation asserts itself as opposition and criticism against everything that endeavours to persist. The antagonisms between contradictory elements (thesis and antithesis) are exacerbated to the point that they can no longer co-exist, precipitating a crisis in which the contrary elements are reabsorbed into a higher and qualitatively different unity (the synthesis). In this way, the World Spirit progresses from notion to notion, each uniting in itself a new stage of reality, spiritual and material. The dialectic is Hegel's way of showing how development follows a complex yet rationally determined path (Cornu 1957; Hegel 1967).

Marx followed Hegel's dialectical method but disagreed with Hegel's teleological and spiritual idealism. In opposition to idealism, Marx was a dialectical *materialist*. As Marx and Engels put it at the time:

> In direct contrast to German philosophy [Hegel] which descends from heaven to earth, here we ascend from earth to heaven. That is to say, we do not set out from what men say, imagine, conceive, nor from men as narrated, thought of, imagined, conceived, in order to arrive at men in the flesh. We set out from real, active men, and on the basis of their real life-process we demonstrate the development of the ideological reflexes and echoes of this life-process ... Life is not determined by consciousness but consciousness by life.
>
> (Marx and Engels 1981: 47)

Marx and Engels concurred with Hegel that the driving force in development was conflict and opposition. But Marx began his analysis of conflict by examining not transcendental consciousness, but real economic conditions. From the time of *The German Ideology*, Marx and Engels (1970) saw struggle within social relations as the driving force of history. However, Marx at first understood social relations as a "form of intercourse" between people, with "intercourse" including all kinds of relations, such as trade and commerce, as well as the class relations that he eventually came to focus on. By comparison, Marx's conception of the development of the material productive forces (human labour power and means of production) was more advanced. These different levels of conceptual understanding lent the tenor of technological determinism to Marx's first overall statement about social development. As Marx (1969b: 503–4) put it:

> In the social production of their life, men enter into definite relations that are indispensable and independent of their will, relations of production which correspond to a definite stage of development of their material productive forces. The sum total of these relations of production constitutes the economic structure of society, the real foundation, on which arises a legal and political superstructure and to which correspond definite forms of social consciousness. The mode of production in material life conditions the social, political and intellectual life process in general. It is not the consciousness of men that determines their being, but, on the contrary, their social being that determines their consciousness. At a certain stage of their development, the material forces of society come in conflict with the existing relations of production, or – what is but a legal expression for the same thing – with the property relations within which they have been at work hitherto. From forms of development of the productive forces, these relations turn into their fetters. Then begins an epoch of social revolution. With the change of the economic foundation the entire immense superstructure is more or less rapidly transformed. ... In broad outlines Asiatic, ancient, feudal, and modern bourgeois modes of production can be designated as progressive epochs in the economic formation of society.

In this statement, the development of the forces of production drives change in the entire "mode of production of material life" following a conception still influenced by Hegelian idealism (the self-development of the productive forces standing in for the unfolding rationality of the World Spirit). By the time of Marx's *Capital*, however, the forces of production were subordinated to the relations of production, and these were thought of more exactly as class relations within the labour process.

For Marx, the most essential aspect of social relations was control over the productive forces available in any society. In class societies, the production of existence was controlled by a ruling elite. This created a fundamental social cleavage, a class relation, between owners of the productive forces and the labourers who performed the work necessary for life to continue. The aspect of this most crucial for issues of economic development was the extension of the working day beyond the labour time necessary for simply reproducing the worker and his or her family. So, for Marx, societies were "exploitative" when uncompensated surplus labour, or its products, were taken from the direct producers, the exploitation process being an arena of the conflict and opposition that Marx, as a dialectical thinker, focused on. In class struggle, the dominant elite used a combination of economic, political and ideological forces against the dominated, and the dominated resisted through overt means like organisation (unions, political parties), and hidden means, like reluctant compliance (putting a spanner in the works). For Marxists, while workers produce the means of economic growth, they do not control the development process. This makes capitalism both highly productive and very unstable, a dangerous combination.

The other main social relation is within the capitalist class, among capitalists competing on markets. Capital itself came from surplus made in earlier periods of history, from merchant's capital (in which profits were made by buying cheap and selling dear), from savings and hoarding, but mostly from that fearsome source of "primitive accumulation", the raiding of non-capitalist societies for their accumulated wealth, their resources and their people:

> The discovery of gold and silver in America, the extirpation, enslavement and entombment in mines of the indigenous population of that continent, the beginnings of the conquest and plunder of India, and the conversion of Africa into a preserve for the commercial hunting of blackskins, are all things which characterize the dawn of the era of capitalist production. These idyllic proceedings are the chief moments of primitive accumulation.
>
> (Marx 1976: 915)

Capitalists were originally commercial farmers or small manufacturers who put accumulated money into productive use as capital to make profit. Under market conditions, they had to produce commodities at low prices. Competition forced capitalists to extract the maximum surplus value from workers. Competition forced capitalists to re-invest most surplus in improving the forces of production, especially tools, machines and infrastructures in an attempt at making production more efficient. Competition also forced the adoption of innovative types of organisation (corporations as compared with family firms, multinationals as compared with national corporations). Thus under capitalism, investment, productivity and development are not a matter of will, but result from compulsion forced by competition – it is a case of develop or lose in the race to survive. Each technological or organisational change forced by competition then has multiple effects running through the entire economy, society and culture – hence the theme of "progress" in modern history. For Marx, development occurred unevenly in social terms of class (the owning class becoming richer) and geographical terms of space (some countries becoming richer than others), hence the theme of uneven development in Marxism. Development was a contradictory process altogether (Marx 1976; Harvey 1982; Becker 1977; Weeks 1981; Peet and Hartwick 2015).

For Marx, social transformation essentially involved a shift from one mode of production (for example, feudalism) to another (for example, capitalism). Marx envisioned such transformations as violent episodes undertaken when all the productive possibilities of old social orders had been exhausted. Crises in material development sharpened and intensified ongoing social struggles, lending struggle the potential for broad social change. The new social system, produced through class struggle, did not materialise out of thin air, nor from utopian desire alone, but was formed out of the embryonic relations already present in the dying body of the old society. Marx saw capitalism as a necessary stage in history, developing technology and destroying old beliefs in gods, ghosts and goblins. Following from this, Marx saw capitalism and imperialism as progressive aspects of the development of non-capitalists societies, like India and China, making possible a better society in the future – a viewpoint that made Marx vulnerable to criticisms of Eurocentrism. The main point is that Marx favoured modern technology and social development because these made a better life possible for the mass of ordinary people. He just wanted the development process to be owned, controlled and directed democratically, by working people. Only then could the full potential of modern development be fully realised – as the satisfaction of needs, as education for all and as sustainality. For Marx, this would require a further transformation in mode of production, from capitalism to socialism.

Marx is important in two main ways. First, his doctrines were supposedly the political-economic basis on which the Communist countries (the Soviet Union, the East European bloc, China, etc.) were established. These countries used a development model founded on growth of the productive forces, understood narrowly as heavy industries like steel, chemicals and engineering, the idea being to increase productivity as quickly as possible. Consumer goods industries were less developed within a political culture stressing hard work for the state and sacrifice for the future. Second, Marx's writings form the intellectual basis for a range of subsequent theories such as dependency and world systems theories, as well as constituting a philosophy of development in and of themselves. This body of ideas continues to influence radical social theory up to the present time.

Major works

Marx, K. (1969a [1865]) *Wages, Price and Profit*, in K. Marx and F. Engels, *Selected Works*, vol. 2, Moscow: Progress Publishers, pp. 31–76.

—— (1969b [1859]) Preface to *A Contribution to the Critique of Political Economy*, in K. Marx and F. Engels, *Selected Works*, vol. 1, Moscow: Progress Publishers, 502–506.

—— (1976 [1887, 1907–1908]) *Capital: A Critique of Political Economy*, Harmondsworth: Penguin.

—— and Engels, F. (1969 [1847–1848]) *Manifesto of the Communist Party*, in K. Marx and F. Engels, *Selected Works*, vol. 1, Moscow: Progress Publishers, 108–137.

—— —— (1970 and 1981 [1847]) *The German Ideology*, New York: International Publishers.

Further reading

Becker, J. (1977) *Marxian Political Economy*, Cambridge: Cambridge University Press.

Cornu, A. (1957) *The Origins of Marxian Thought*, Springfield, IL: C.C. Thomas.

Harvey, D. (1982) *The Limits to Capital*, Oxford: Blackwell.

Hegel, G.W.F. (ed.) (1967) *The Phenomenology of Mind*, New York: Harper.

Peet, R. and Hartwick, E. (2015) *Theories of Development: Contentions, Arguments, Alternatives*, 3rd edn, New York: Guilford Press.

Weeks, J. (1981) *Capital and Exploitation*, London: Edward Arnold.

RICHARD PEET

MANFRED A. MAX-NEEF (1932–)

Manfred A. Max-Neef, the son of German immigrants, is a Chilean economist who has had a significant influence on grassroots movements and local development projects worldwide. Since teaching at the University of California, Berkeley, in the early 1960s, he has held several academic posts in the USA and Latin America, including Rector of the Universidad Bolivariana and subsequently, of the Universidad Austral de Chile in Valdivia (1994–2002). He holds honorary doctorates from institutions in Jordan, Colombia, Argentina and the USA. He has also worked for international organisations like the Food and Agricultural Organization (FAO) and the International Labour Organization (ILO). He is the founder and Executive Director of the Centre for Development Alternatives (CEPAUR) in Santiago, Chile, which is dedicated to transdisciplinary research and action projects in order to re-orient development by stimulating local self-reliance, satisfying fundamental human needs and promoting human-scale development. He has also been an executive council member of the Club of Rome. In 1983, Max-Neef was awarded the Right Livelihood Award in recognition of his work on development alternatives. The prize is often described as the Alternative Nobel Prize, and is awarded for outstanding vision and work on behalf of our planet and its people. In 1993, he stood as an independent candidate in the Chilean presidential election and achieved a respectable minority vote. His work and writing continue to influence and inspire grassroots organisations and local movements in search of an alternative and sustainable development.

Max-Neef's thinking originates to a large extent in his frustration with classical, conventional economics. The discipline of economics, he argues, has become obsessed with being a science, with providing technical solutions to human problems and with abstract measurements such as GNP, growth rates and balance of payments. As a result, economists have become "dangerous people" (1992: 19), whose theories and models cast a powerful spell on policy-making, but have proved dismally incapable of responding adequately to poverty and human misery. Instead, conventional economics makes the majority of poor people *invisible*, as it assigns no value to work carried out at the subsistence and domestic level. This reflects the discipline's obsession with gigantism, efficiency, measurements and abstractions. In the process, economics has, in the words of Max-Neef, lost "a good deal of its human dimension" as well as its connection to moral philosophy (1992: 20). For these reasons, he writes, "I severed my ties with the trends imposed by the economic establishment, disengaged myself from 'objective abstractions', and decided to 'step into the mud'" (1992: 21). He became a self-described "barefoot economist", someone who has a profound distrust of grandiose solutions imposed from the top down and whose main concern is instead with local action and small-scale, bottom-up development. This critique of mainstream economics is powerfully restated in the co-authored book *Economics Unmasked* (Smith and Max-Neef 2011), which also seeks to outline the foundations of new economics based on the guiding values of justice, human dignity, compassion and reverence for life.

Max-Neef's ideas and experiences as a "barefoot economist" are set out in his best-known book, *From the Outside Looking In: Experiences in "Barefoot Economics"*. Since its first publication in 1982 by the Dag Hammarskjöld Foundation in Sweden, the book has been translated into five languages and has achieved the status of a classic text on grassroots development projects. In the book, Max-Neef details his personal experiences with participatory grassroots projects in two Latin American settings, interspersing his empirical account with more theoretical reflections on the nature of development. The result is a moving and passionate book, where the voices of the poor speak with clarity and force and where conventional theories and models are shown to be lacking. His recollection of his involvement with an integrated development project among peasants in Ecuador is particularly interesting and powerful, and demonstrates both the willingness and the capacities of poor people to organise and participate, given the right circumstances and opportunities. The participants' own accounts of their development problems (presented verbatim in the book) provide a forceful antidote to the widespread perception that poor people are ignorant of their situations, and speak, as Max-Neef puts it, with a voice so vivid and so true "that no expert with all his formal knowledge could have improved upon them" (1992: 71). Sadly, the project also demonstrates the intrinsically political character of participation and empowerment, in that all the project's documents and reports from the peasants' meetings were confiscated by Ecuador's military authorities. Shortly afterwards, the project was closed, presumably because its success in organising and empowering the peasants was seen as too dangerous by the political elite.

This experience leads Max-Neef to a profoundly *political*, as opposed to technical, vision of development. To assume "goodwill on the part of governments to really improve the lot of the invisible sectors [the poor] is naïve", he argues, and in this sense, development represents the interests of the dominant class (1992: 114). In the national accounts, where growth rates and import/export ratios reign supreme, the poor are expendable, or at best, they can wait patiently until the effects of economic growth "trickle down". This scepticism of top-down, grandiose solutions leads Max-Neef to put his trust in local action: "if national systems have learned to circumvent the poor, it is the turn of the poor to learn how to circumvent the national systems" (1992: 117). His vision for a better future accordingly arises from small-scale action, where the people concerned actively participate and where self-reliance and ecological harmony are guiding principles. It is only in such environments, he argues, that "human creativity and meaningful identities can truly surface and flourish" (1992: 117).

Drawing on traditions of thought stretching back to Aristotle, Max-Neef argues that large cities lead to alienation, passivity and indifference, whereas small environments allow for a dynamic equilibrium between nature, human beings and technology, as in such settings people have the opportunity to feel directly responsible for the consequences of their actions (1992: 132). As a firm believer in the relationship between theory and praxis, Max-Neef put his ideas to the test in Tiradentes, a small city in Brazil. The project was an effort to revitalise the city, counter the trend of hyper-urbanisation and improve the livelihoods of local people. Described in *From the Outside Looking In*, the project succeeded with relatively few resources in increasing not only people's incomes but also their sense of well-being, community and control, hence confirming Max-Neef's belief that local action holds the key to development.

Max-Neef's conviction that "big problems require a great number of small solutions" (1992: 204) is given further elaboration in *Human Scale Development: Conception, Application and Further Reflections* (1991). The book's main postulate is that "development is about people, not about objects", and it proposes an "ethics of well-being" against the logic of economics which dominates modern life (1991: 16, 64). Human-scale development is based on:

> the satisfaction of fundamental human needs, on the generation of growing levels of self-reliance, and on the construction of organic articulations of people with nature and technology, of global processes with local activity, of the personal with the social, of planning with autonomy and of civil society with the state.
>
> (1991: 8)

As an aid to achieving this form of development, the book proposes a radically different conception of human needs and wants. Whereas conventional economics commonly assumes that human needs are infinite, and that human beings are driven by an insatiable desire for material possessions, Max-Neef asserts that fundamental human needs are "finite, few and classifiable". What is more, these needs are "the same in all cultures and in all historical periods" (1992: 18). What changes over time and from culture to culture is the manner in which people satisfy their needs. Apart from the fundamental need of subsistence or survival, there is no hierarchy of needs, as suggested for example by the psychologist Abraham Maslow; instead, needs are constant and complementary. This, in turn, allows for a reclassification of poverty, which can no longer be defined simply in terms of income below a certain level, as all unfulfilled needs reveal a human poverty (for example, poverty of affection or of understanding).

Max-Neef identifies nine fundamental human needs; subsistence, protection, affection, understanding, participation, leisure, creation, identity and freedom, with the possibility that transcendence (a process that he doesn't define or elaborate on) may in time become a tenth universal need (1992: 27). Needs are also defined according to the existential categories of being, having, doing and interacting, and on the basis of these Max-Neef develops a 36-cell matrix designed to help small communities enhance their awareness of their deprivations and potentials. The matrix is filled with examples of so-called satisfiers, that is, the means by which various needs are fulfilled in a specific community. Satisfiers have different characteristics, and while some are relatively straightforward (food satisfying the need for subsistence, formal education the need for understanding), others are more complicated. Some satisfiers can be destroyers of the very need they are said to fulfil – an example being excessive military expenditure, which is supposed to provide protection, but instead promotes insecurity and inhibits subsistence, participation, affection and freedom (1991: 33). Others are "synergic" satisfiers, meeting more than one particular need simultaneously. Through this redefinition of human needs and the development of the needs/satisfiers matrix, Max-Neef aims to assist communities in realising their potential for local self-reliance and develop a strategy for the actualisation of human needs. *Human Scale Development* represents a departure from the prevailing economic rationale and the equation of development with attaining the material living standards of the most industrialised countries, and suggests instead that an abundance of goods and economic resources can exist side by side with the poverty of, for example, affection and creativity. In fact, the valorisation of material possessions has, according to Max-Neef, led to alienated societies, where life "is placed at the service of artifacts, rather than artifacts at the service of life" (1991: 25). While human-scale development does not exclude conventional development goals such as economic growth, it differs from such approaches in that the aims of development are considered not only as points of arrival but as part of the process itself. Fundamental human needs must be realised throughout the entire process of development, and in this way, the realisation of needs becomes, instead of a goal, the motor of development (1991: 53).

Max-Neef has made a noticeable contribution to development thinking and practice through his pioneering critique of development economics, his early advocacy of grassroots participation and empowerment and his redefinition of human needs. His arguments for development alternatives and community participation remain relevant, perhaps even more so as civil society, participation and poverty reduction have become the contemporary buzzwords of the mainstream, top-down development models. One prominent vehicle for disseminating his ideas has been *Development Dialogue*, the journal of the Dag Hammarskjöld Foundation, with which he had a close association over many years. Overall, it is thus possible that his emphasis on participation and grassroots involvement has had a belated influence on international institutions like the World Bank, as seen, for example, in its large-scale "Voices of the Poor" project in the early 2000s. For Max-Neef, this would not necessarily signify the success of his ideas, but rather the ability of the holders of power to conceal their vested interests and pour old wine into new bottles (1992: 115; see also Wisner 1988). For him, human-scale development can arise only from local settings, not from distant development organisations and predefined development models. Accordingly, his own writings do not so much propose an alternative development model as provide flexible tools whereby communities can arrive at their own solutions to their individual problems. Unlike so-called post-development writers, however, who also focus on grassroots movements and local knowledge, Max-Neef remains within the language of development, utilising the same terms and phrases as orthodox models. Many post-development thinkers, by contrast, argue for alternatives *to* development, as opposed to development alternatives, and this entails a change of the discourse that produced the problem of development in the first place (see Esteva 1987; Escobar 1992).

Like the economist **E.F. Schumacher**'s embrace of the dictum "small is beautiful", Max-Neef's vision of human-scale development is open to the charge of romanticism. Yet, for someone who describes his political philosophy as "human eco-anarchism" (1992: 55) and who wishes to revitalise our capacity to dream (1991: 3), this is not necessarily a criticism. That said, Max-Neef denies being a smallness fanatic, recognising that small can also be ugly and evil, but maintains that smallness of scale allows for a degree of responsibility and human interaction that is impossible in larger environments (1992: 143, 160). The relationship of these local communities to their national and global surroundings remains largely outside the parameters of Max-Neef's writings; the challenge is to "think small and act small, but in as many places as possible". The path, he argues, "must go from the village to the global order", an insight that remains a powerful inspiration for local action and grassroots movements (1992: 117).

Major works

(Available in English; he has written more extensively in Spanish.)

Ekins, P. and Max-Neef, M.A. (eds) (1992) *Real-Life Economics: Understanding Wealth Creation*, London and New York: Routledge.

Max-Neef, M.A. (1992) *From the Outside Looking In: Experiences in "Barefoot Economics"*, London: Zed Books.

Max-Neef, M.A. with contributions from A. Elizalde and M. Hopenhayn (1991) *Human Scale Development: Conception, Application and Further Reflections*, New York: Apex Press.

Smith, P.B. and Max-Neef, M.A. (2011) *Economics Unmasked: From Power and Greed to Compassion and the Common Good*, Cambridge: Green Books.

Further reading

Escobar, A. (1992) "Reflections on 'Development': Grassroots Approaches and Alternative Politics in the Third World", *Futures* 24(5): 411–436.

Esteva, G. (1987) "Regenerating People's Space", *Alternatives* 12(1): 125–152.

Maslow, A.H. (1943). "A Theory of Human Motivation", *Psychological Review* 50(4): 370–396.

Wisner, B. (1988) *Power and Need in Africa*, London: Earthscan.

RITA ABRAHAMSEN

TERENCE GARY MCGEE (1936–)

Terence ("Terry") McGee is a development geographer who has made key contributions to our understanding of urbanisation and development and the structure of cities, particularly as regards the rapidly growing conurbations of Southeast Asia. He is best known for his explanation of the role in development of Southeast Asian cities characterised by poor infrastructure, rapid growth due to rural–urban migration, political instability and extensive poverty. He was a pioneer in the 1970s in emphasising the significant position in the urban economy of the "informal sector" (the extensive range of unregulated enterprises and activities in the black market or underground economy that operate without any formal or official recognition by government). Up to this point there had been hardly any acknowledgement of its overwhelming contribution to economy and society in some of the world's largest cities, the way people are employed, how and where they live and how so many of the economic and social relationships in the Asian city can be explained.

McGee was born on 5 January 1936 in Cambridge, New Zealand, a small agricultural town on the North Island situated in the heart of the Waikato farming district, where he gained early experience in milking cows and mending fences, as well as the normal New Zealand aptitude for ball skills at rugby and cricket. At the age of ten he moved with his family to Auckland, where he was at school on the city's North Shore and where, in 1954, he enrolled at Auckland Teachers' College while also attending BA classes part-time in history and geography at the University of Auckland. Graduating with a teaching diploma in 1954, he transferred to Victoria University of Wellington, New Zealand, where he majored in geography and gained his BA in 1957. In Auckland, McGee attributes his early appreciation of history and its later importance in his writing to Willis Airey (1897–1968), a leading New Zealand historian of Russia, and Keith Sinclair (1922–1993), poet and expert on New Zealand history. In Wellington his mentors were Keith Buchanan (1920–1998), brilliant lecturer and utopian socialist who wrote widely on Asia; Harvey Franklin, an economic geographer who introduced him to the work of the institutional economist Clarence Ayres (1891–1972), an American who wrote the book, *The Theory of Economic Progress*, in 1944 and who propounded a theory of "institutional lag", whereby technological changes were always several steps in front of change in sociocultural institutions; and Ray Watters, a human geographer who had adopted ethnographic and anthropological approaches in his research on the Pacific.

He continued with his MA studies in 1958–1961, where the courses taught by Buchanan, Franklin and Watters were influential and where Buchanan introduced him to the French regional geographers in Asia, notably the work in Vietnam by Pierre Gourou (1900–1999) and Charles Robequain (1897–1936). Another important influence was members of the Chicago ecological school of sociologists, including Louis Wirth (1897–1952), whose approaches towards urbanism as a way of life he used in his master's thesis on Indian migration to the City of Wellington, New Zealand, which he attained with first class honours in 1961.

His academic career began alarmingly when he took up a position as temporary assistant lecturer in the Department of Geography at Victoria University to teach a class on the geography of Europe, a region which he had not at that time visited and which he only accomplished with the help of lecture notes that Franklin had left behind when he went on leave. But 1959 was also the year he gained his first continuing appointment as Lecturer in Geography at the University of Malaya in Kuala Lumpur. He was to remain there for six years, and recognises this as a very formative period in his life. McGee went to Asia expecting to teach but considers he learned far more than he ever imparted (Presidential Address to Canadian Institute of Geographers 1991). While in Malaysia he enrolled in a PhD at Victoria University on Malay migration to Kuala Lumpur City, a subject that followed on from his master's thesis. The research involved an extensive household questionnaire survey, staying as a guest of Malay families and travelling widely throughout the country to track down migration source areas. This period also presented him with an opportunity to explore Southeast Asia extensively, and it was on one such expedition that the idea for a book on the Southeast Asian city was conceived.

He returned to New Zealand to become a Lecturer in Geography at Victoria University in 1965, and completed both the book and his PhD thesis. *The Southeast Asian City: A Social Geography of the Primate Cities of Southeast Asia* was published in 1967 and his graduation followed in 1968. McGee propounded the novel view that Asian cities were characterised by a form of "pseudo urbanisation", typified by conspicuous consumption, extraction of surplus from the countryside and a drag on development. This was a perspective that progressively diminished in importance as the developing world became more closely integrated with the global economy in the last decades of the century. The book was the first in a distinguished list of titles on Asian urbanisation. A forthcoming book will study the extended metropolitan regions of Asia and will be jointly authored with some of his former PhD students. This was also a busy period in his life with marriage, the birth of his son and building a house in Otaki (near Wellington, New Zealand) taking centre stage.

One of McGee's most influential papers on the economic structure of cities was jointly written in 1968 (with Warwick Armstrong, a colleague in Wellington) before returning to Asia, to a Senior Lectureship at the University of Hong Kong. This paper, heavily influenced by members of the Latin American dependency school of development thinkers (notably **Raúl Prebisch** and **Andre Gunder Frank**), was titled "Revolutionary Change and the Third World City: A Theory of Urban Involution" (McGee and Armstrong 1968). It was the beginning of a collaboration with Armstrong that would be renewed in Canada in the early 1980s, and a precursor to their influential *Theatres of Accumulation: Studies of Urbanization in Asia and Latin America* (Armstrong and McGee 1985). In Hong Kong, McGee was soon promoted to a personal Chair in 1970, and continued his interest in the informal sector with the first of a series of major publications on street vendors (*Hawkers in Hong Kong: A Study of Policy and Planning in the Third World City*, 1974). He also began a productive relationship with the Canadian International Development Research Centre which was to result in further titles on the same topic during the 1970s.

In 1973 McGee took up a Senior Fellowship in Human Geography at the Australian National University (ANU) in Canberra, Australia. His five years at the ANU were marked by innovative new work on the political economy of development and collaborative research in the Pacific with academic colleagues (notably David Drakakis-Smith (1942–1999), with whom he had worked in Hong Kong, and Gerard Ward). Publications of particular note at this time included "The Persistence of the Proto-proletariat" (McGee 1974) and "Conservation and Dissolution in the Third World City" (McGee 1979). It was also a time saddened by the death of his first wife and the added responsibilities of being a single parent. In 1978 he moved to Canada to take up the joint positions of Professor of Geography and Director of the newly created Institute of Asian Research at the University of British Columbia in Vancouver. In spite of a heavy administrative workload, McGee continued to contribute effectively to Asian studies through fieldwork in Malaysia investigating urban female workers in the new semi-conductor factories and their rural links expressed through sending part of their incomes back to home villages. This interest in urban–rural dynamics was to lead, late in the 1980s, to some of McGee's most important work – the identification of new extended metropolitan regions in Asia characterised by the outward expansion of urban activity into the intensely crowded agricultural hinterlands of cities like Jakarta on the Indonesian island of Java. McGee termed this novel metropolitan form *desakota*, taken from the *bahasa* Indonesian names for village and town. He argued that it was an Asian form of urbanisation distinct from that of the developed countries (apart perhaps from Holland). This was an ecological viewpoint in the tradition of cultural geographers like Carl Sauer (1889–1975), and led to the publication of *The Extended Metropolis: Settlement Transition in Asia* (Ginsberg et al. 1991).

A number of objections have been raised to the idea of extended metropolitan regions (EMRs) being distinctive or peculiarly Asian. Briefly, these critiques have argued that the book was over-generalised in its claims, and that it neglected the present evidence and inevitability of growth in market systems in Asia that will replicate patterns found in the developed countries (Tang and Chung 2000; Dick and Rimmer 1998). McGee has responded by saying that the idea was never based on Asian as opposed to Western traits, but was one which stresses the influences on the urbanisation process of the context in which it operates. Government policy, foreign investment and various local factors, like the roles of municipalities and community organisations, explain much of the character and change found in EMRs such as Jabotabek (Indonesia), Bangkok (Thailand) and the Pearl River Delta (China).

Besides contributing extensively to the academic debate on urbanisation processes in Asia, McGee's work has had a practical and applied focus. During the 1990s he forged strong links with the National Centre for Humanities and Social Sciences in Hanoi (Vietnam) as part of a long-term project looking at Socialist Transition Economies and developing research capacity in Vietnamese research institutions. Further investigations examined the fiscal crisis in Southeast Asia at the end of the 1990s for the Canadian International Development Agency, and identified a list of urban indicators for the Asian Development Bank. McGee reached compulsory retirement age in 2001, leaving his positions at the University of British Columbia to become a Professor Emeritus and spend part of each year teaching in Australia and writing on Asian urbanisation. In 2009 he was awarded the prestigious Vautrin Lud International Geography Prize for his lifetime contribution to geography.

McGee has made fundamental contributions to the study and understanding of modern urbanisation processes. His work, with its emphasis on human and institutional features of urban life and change in Southeast Asia, has followed a consistent path, enduring through the 1970s when geographers all around him were caught up in what was then termed a "quantitative revolution" based on the primacy of numerical techniques, to emerge in the new century with his ideas intact and equally relevant to the contemporary debate. A measure of the resilience of his ideas is also seen in the way his formulation of theory contributed to new concepts in allied fields other than his own. For example, views about the character of retailing in the informal sector (hawking), an activity so typical of Asian cities, were adapted by him to explain the persistence of shanty-town features in these same cities (McGee 1979). Though this was the only paper on housing he has ever written, it remains a key contribution in understanding the nature of the squatting process and the determination of urban management strategies to cope with its effects.

Major works

Armstrong, W. and McGee, T.G. (1985) *Theatres of Accumulation: Studies of Urbanization in Asia and Latin America*, London: Methuen.

Ginsberg, N., Koppell, B. and McGee, T.G. (eds) (1991) *The Extended Metropolis: Settlement Transition in Asia*, Honolulu: University of Hawaii Press.

McGee, T.G. (1967) *The Southeast Asian City: A Social Geography of the Primate Cities of Southeast Asia*, London: Bell.

—— (1971) *The Urbanization Process in the Third World: Explorations in Search of a Theory*, London: Bell.

—— (1974) *Hawkers in Hong Kong: A Study of Policy and Planning in the Third World City*, Hong Kong: Centre of Asian Studies, University of Hong Kong.

—— and Robinson, I. (eds) (1995) *The Mega-Urban Regions of Southeast Asia: Policy Challenges and Response*, Vancouver: University of British Columbia Press.

Further reading

Dick, H.W. and Rimmer, R. (1998) "Beyond the Third World City: The New Urban Geography of Southeast Asia", *Urban Studies* 35(12): 2303–2321.

McGee, T.G. (1974) *The Persistence of the Proto-proletariat: Occupational Structures and Planning for the Future of Third World Cities*, Los Angeles, CA: School of Architecture and Urban Planning, University of California. (Repr. in J. Abu-Lughod and R. Ha (eds), *Third World Urbanization* (Chicago, IL: Maaroufa, 1977), 257–270.)

—— (1979) "Conservation and Dissolution in the Third World City: The Shanty-town as an Element of Conservation", *Development and Change* 10(1): 1–22.

—— (1991) "Presidential Address. Eurocentrism in Geography: The Case of Asian Urbanisation", *Canadian Geographer* 35(4): 332–342.

—— (2002) "Reconstructing the Southeast Asian City in an Era of Volatile Globalisation", *Asian Journal of Social Science* 30(1): 8–27.

—— and Armstrong, W. (1968) "Revolutionary Change and the Third World City: A Theory of Urban Involution", *Civilisations* 18(3): 353–378.

—— Lin, G.S., Wang, M., Martin, A. and Wu, J. (eds) (2007) *China's Urban Space: Development Under Market Socialism*, Abingdon and New York: Routledge.

Tang, W.-S. and Chung, H. (2000) "Urban–Rural Transition in China: Beyond the *Desakota* Model", in S.M. Li and W.-S. Tang (eds), *China's Regions, Polity and Economy: A Study of Spatial Transformation in the Post-Reform Era*, Hong Kong: Chinese University of Hong Kong Press, 276–308.

JOHN LEA

GUNNAR MYRDAL (1898–1987)

A prolific writer and policy adviser, Gunnar K. Myrdal had a profound effect on development thinking through the 1950s and 1960s. Trained as an economist, Myrdal's writings engaged with sociology and gained a wide audience among planners, governments and academics. Moreover, the issues he raised about the institutional and social milieu for favourable development are still debated today. Born in Stockholm, Myrdal was associated with that city throughout his life, despite extensive work around the globe. Educated at Stockholm University, he gained a law degree in 1923, followed by a PhD in economics in 1927. His PhD focused on the role of expectations in price formation, although he is best known for later work on macro-economics and social relations. Following a Rockefeller fellowship in the US, and a short period at the Institute of International Studies, Geneva, he returned to Stockholm. At the University of Stockholm he was Professor of Political Economy (1933–1950) and then Professor of International Economics (1960–1967). He also assumed the Directorship of the Swedish Institute for International Economic Studies, and acted as Executive Secretary to the UN Economic Commission for Europe (1947–1957), a member of the Swedish Senate and Minister of Commerce.

Myrdal was part of an iconoclastic generation of development economists whose work extended Keynes's economics and state interventionism. Learning from wealthy Western and Eastern European countries, this group adapted Keynesian demand management and Schumpeter to view development as a structural transformation process. Rejecting neoclassical economics, these structuralists blamed countries' poverty on the lack of industrialisation (Martinussen 1997: 50ff.).

Alongside Myrdal, writers including **Ragnar Nurkse**, Thomas Balogh, **Paul Rosenstein-Rodan**, **Raúl Prebisch** and **Hans Singer** highlighted issues such as dual economies, labour surplus, unbalanced growth, the vicious circle of poverty, unequal exchange and dependency, and advocated policies for rapid industrialisation and redistribution with growth. Compared with structuralist colleagues however, Myrdal considered non-economic phenomena more systematically, themes which recurred in later studies (Martinussen 1997: 73). His major Carnegie Corporation-funded study of the situation of the USA's African-American population took seriously the legal, social and cultural factors behind the creation and perpetuation of mass poverty, and led him to question the unequal status of black Americans (*An American Dilemma: The Negro Problem and Modern Democracy*, 1944). In comparison with many structuralists at the time, Myrdal argued that capital accumulation could assist economic development, bringing him closer to "modernisation" thinkers. Myrdal's influence on development can be seen in four fields: regional development, dependency approaches, the state, and non-economic factors.

With regard to regional planning, Myrdal explained the relations between a dynamic part of a developing country – usually the capital city – and stagnation elsewhere. He attributed regional inequalities to the "spiral of cumulative causation" which results in positive spread effects in the core area, and negative backwash effects in the periphery (*Economic Theory and Under-Developed Regions*, 1957), thereby articulating a highly influential humane approach (Brookfield 1975: 99). The "process of cumulative causation" of free market economies concentrated economic activities to the detriment of other sectors, reinforced by movements of capital, goods and people. The concepts of spread and backwash effects had an immediate impact on debates in regional development and urbanisation (Reitsma and Kleinpenning 1985: 202). Although neither Myrdal nor his contemporary, **Albert Hirschman**, theorised the best initiators of positive spread effects, a commonly agreed impetus was urban industrial development as occurred in Europe. Later, however, Myrdal concluded that industrialisation could not absorb all labour, and called for agricultural development to create employment and increase productivity. Against **Hirschman**'s optimism, Myrdal believed that equal development would be unlikely without strategic state intervention; between them, they established uneven development as core to development debates, and inspired others. For example, **John Friedmann**'s core–periphery model built on Myrdal and Hirschman in relation to regional economies (Brookfield 1975: 101).

Myrdal also engaged with the international context for national wealth inequalities, particularly between European and North American countries and former colonies in the global South. He noted how economic, social and political dependency relations existed between regions at the international level (Reitsma and Kleinpenning 1985: 246). Influenced by Latin American planners such as **Raúl Prebisch** and the Economic Commission on Latin America, Myrdal regarded underdevelopment as intrinsically constituted through the pervasive effects of negative backwash. Compared with **Walt Rostow**'s stage model, Myrdal accepted the structuralist argument that post-Second World War international trade could not generate growth for developing countries, as the "trickle-up" benefits accrued to industrialised countries (Martinussen 1997: 78–79). He brought circular and cumulative causation into structuralist accounts, whereby an outflow of surplus, brain drain and the destruction of domestic crafts indicated backwash effects. The vagaries of the global economy caused stronger backwash effects in a depression than in a boom (Brookfield 1975: 100). Developing a theme subsequently explored by neo-Marxists, Myrdal drew attention to colonial legacies operating through law and order policies and tax collection in the interests of colonial powers (Martinussen 1997).

Economics in the 1950s and 1960s assumed that market failure was the norm in developing countries, meaning that the state would have to mobilise domestic and foreign savings for investment in directed industrial development. Following mainstream thinking, Myrdal viewed the state as "knowledgeable and compassionate" if not always correct (Lal 1985; Toye 1993: 100). Myrdal (1960) envisaged a state as highly organised and centralised with features such as collective bargaining, democratic participation and welfare. Policies and political determination could shape states as institutions (Martinussen 1997: 227), although a developmentally favourable state was not a foregone conclusion in poor countries. Indeed Myrdal suggested that most developing country states were "soft", exploited by powerful individuals and groups (Myrdal 1968: ch. 18; 1970). Myrdal first used the term "soft state" to describe a government incapable of instituting radical reform policies or fully implementing policies (Reitsma and Kleinpenning 1985: 332). In *Asian Drama* (combining theory and policy), Myrdal analysed the detrimental impact of inefficient political development, the weak formulation of law, corruption and the continuation of exploitative systems such as sharecropping (Martinusssen 1997). **Ester Boserup**, **Michael Lipton** and **Paul Streeten** were also involved in this large and ground-breaking study. A strong state structure, according to Myrdal, could enforce social discipline or generate a tolerance of authoritarianism. Rather than focus on political structure,

Myrdal's approach interpreted soft states as "a problematic absence of the right cultural context for development planning" (Toye 1993: 124), leading to charges of ethnocentrism (Berger and Huntington 2002). In his 1958 Storr lectures at Yale, Myrdal outlined his vision for a developing country state using national planning as a positive internal force enhanced by democratisation, economic nationalism and equalisation.

As a sociologist of development, Myrdal sought to understand the impact of non-economic factors on development and poverty. Criticising capital-centric theories, he developed a general theory of societal stagnation and transformation (Martinussen 1997: 80). Criticising previous planning models, he included attitudes to work and life, and political and cultural institutions (Myrdal 1968: appendix 3, 4; Martinussen 1997: 230). His multi-factor model ranged from output and incomes, and conditions of production, to attitudes towards work and life, institutions and policies. These were understood as interrelated, and shifting in similar directions because of development or underdevelopment (Martinussen 1997: 80), as illustrated in *Asian Drama* (1968). Here, attitudes to life and work (such as low levels of work discipline and punctuality) were seen to obstruct economic growth. Myrdal was not alone in identifying a circular causation of poverty partially attributable to social attitudes; Oscar Lewis and others were discussing the (now discredited) idea of a "culture of poverty" (Lewis 1959).

Although widely influential, Myrdal's work was not without its critics. While easily summarised, his writings failed to acknowledge work in similar fields (Brookfield 1975: 100). During the "counter-revolution" against Keynesian and state-centred development, Myrdal's work was strongly criticised for its *dirigisme* dogma (Lal 1995 [1985]: 56). Neoliberals particularly disagreed with Myrdal's assumption of an omniscient central state authority, and implicit goals of egalitarianism (Lal 1995 [1985]). Acknowledging Myrdal's concern for human capital formation, Toye criticised his claims about "productive consumption" (Toye 1993: 97–98). Myrdal's theories of the soft state emphasising cultural predilections rather than political structures brought him "embarrassingly close to Peter Bauer's linking of development prospects to cultural characteristics of ethnic groups" (Toye 1993: 124). African-American critics note his evasion of class analysis and celebration of the American way of life. Revisionist analysis of Myrdal's international humanitarianism further explores the ambivalent legacies of his "social democratic internationalism" (Morey and Martin 2017).

Myrdal's work continued to have a resonance in development thinking and policy, laying the groundwork for "fruitful analysis of interactions between the state and economic development" (Martinussen 1997: 227). In 1974, he shared the Nobel Prize for Economics with Friedrich A. von Hayek for "pioneering work in the theory of money and economic fluctuations and … penetrating analysis of the interdependence of economic, social and institutional phenomena". In collaborations with **Ester Boserup** and **Michael Lipton**, and extensive friendships with scholars and politicians, Myrdal's influence was felt through succeeding generations. His wife, Alva Myrdal (1902–1986), was a highly significant figure in her own right, working variously for UNESCO, as a Swedish MP and cabinet minister, and a disarmament negotiator (for which she won the Nobel Peace Prize). Gunnar and Alva Myrdal were prime examples of a social democrat commitment to equality and public service allied with concrete information. In his sensitivity to mass poverty's multifaceted character, Gunnar Myrdal's work contributed to an understanding of the human cost of underdevelopment, un(der)employment, poor use of resources, and income inequalities. Viewing inequality as morally unacceptable and the purpose of development as social equity, Myrdal could set out the grounds for the basic needs strategy (Martinussen 1997: 82).

Major works

Myrdal, G. (1944) *An American Dilemma: The Negro Problem and Modern Democracy*, New York: Harper. (Published as an e-book in 2017, Taylor & Francis.)

—— (1956) *An International Economy*, London: Routledge.

—— (1957) *Economic Theory and Under-Developed Regions*, London: Duckworth.

—— (1960) *Beyond the Welfare State: Economic Planning in the Welfare States and its International Implications*, London: Duckworth.

—— (1963) *Challenge to Affluence*, London: Gollancz.

—— (1968) *Asian Drama*, 3 vols, London: Penguin.

—— (1970) *The Challenge of World Poverty: A World Anti-Poverty Program in Outline*, London: Penguin.

Further reading

Berger, N. and Huntington, S.P. (eds) (2002) *Many Globalizations: Cultural Diversity in the Contemporary World*, Oxford: Oxford University Press.

Brookfield, H. (1975) *Interdependent Development*, London: Methuen.

Dickenson, J. et al. (1996) *Geography of the Third World*, London: Routledge.

Lal, D. (1985) "The Misconceptions of Development Economics", repr. in S. Corbridge (ed.), *Development Studies: A Reader*, London: Edward Arnold.

Lewis, O. (1959) *Five Families: Mexican Case Studies in the Culture of Poverty*, New York: Basic Books.

Martinussen, J. (1997) *Society, State and Market*, London: Zed Books.
Morey, M. and Martin, J. (2017) "Introduction: Dossier on Gunnar Myrdal", *Humanity: An International Journal of Human Rights* 8(1): 127–132.
Reitsma, H.A. and Kleinpenning, J.M.P. (1985) *The Third World in Perspective*, Maastricht: Van Gorcum.
Toye, J. (1993) *Dilemmas of Development*, 2nd edn, Oxford: Blackwell.

SARAH RADCLIFFE

KWAME FRANCIS NKRUMAH (1909–1972)

The *Osagyefo* (meaning "the redeemer", as he was popularly known to supporters), Dr Kwame Nkrumah, the man who led Black Africa's first independent nation, Ghana, was born in Nkroful, in Ghana's Western Region close to the Ivorian border, in mid-September 1909, the son of the senior wife of an Nzima goldsmith. Some of his detractors would later cast aspersions on his Ghanaianness because of this geographical accident of birth. According to a contemporary, he was a shy child who was highly influenced by his mother, emerging as a very determined, highly motivated and disciplined individual (Davidson 1973). Nkrumah believed that Africans must not only shed the yoke of colonial inferiority but must be prepared to take up responsibilities, as a prerequisite for retrieving power from the coloniser. One of the earliest challenges the young Nkrumah had to face was that of education, in a colony where the vast majority of children dreamed of attending school but never had the opportunity to do so. After finishing his elementary education in 1926, he became a teacher in Half Assini, a small coastal town near the Ivorian border, from where at the age of 17 he was recruited by the Reverend A.G. Fraser, the founder of Achimota College, to become a student there. This he did for the next four years. After Achimota, he taught in a number of schools before leaving for the USA in 1935.

Nkrumah's choice of the USA for further studies was influenced by the rejection of the snobbery he saw "in the British way of life" as well as by the influence of people like Nigerian nationalist leader Nnamdi Azikiwe, and the realisation that many Black pioneers were themselves luminaries of US educational institutions (Davidson 1973). During his sojourn in the USA, he attended Lincoln University, Pennsylvania and got to know many of the Black emancipators in the US, men such as W.E.B. Du Bois and Marcus Garvey, who helped in the formulation of his Pan-Africanist ideals. Despite this growing period of intellectual enlightenment, life in America was not easy for the young Nkrumah, who had to toil to pay his way through university and endured racism and imperialism, factors which impelled him to return to his native land to lead the anti-colonial struggle.

Armed with Western liberal philosophy, Nkrumah became convinced about the need for Africa to both overturn the colonial system and struggle for continental unity. These two projects would come to dominate the thinking and life of Nkrumah, and he was convinced that the latter could not be achieved without attaining the first objective. Thus his famous maxim: "Seek ye first the political kingdom and all things shall be added", which points to his ability to draw connections between the colonial subjugation and the wretched plight of the mass of the African people (Mohan 1967). Once political independence had been won, Africa and its leaders would then embark on the task of continental unity, and a clear anti-Balkanisation strategy, which he saw as the way out of colonial bondage and neocolonialism. In his political treatise *Africa Must Unite* (1963), Nkrumah argued that Ghana's independence would be meaningless without the complete liberation of the African continent.

In May 1945, Nkrumah left New York, homeward bound via London, where he met George Padmore, the West Indian journalist, and the publisher T.R. Makonnen from British Guiana, as well as attending the Fifth Pan-Africanist Congress in Manchester in the same year. During the two years he spent in London, Nkrumah made an immediate impact on London's small Black community as Vice-President of the West African Students' Union, as well as endearing himself to the city's intellectual community through lecturing and studying for the Bar. He was also instrumental in setting up a secret group, which emerged from the Manchester Congress among activists of the West African National Secretariat, called The Circle, with policies designed to foster unity and national independence in West Africa.

During his stay in London he received an invitation from his old friend from Lincoln University, Ako Adjei, and Dr J.B. Danquah to return home to assume the role of Secretary General of the newly formed party, the United Gold Coast Convention (UGCC), which was about to be launched. After much soul-searching, he left for the Gold Coast in November 1947, arriving to find a country heaving with social and constitutional changes, including a shoppers' strike and restive young men, many of whom had remained unemployed since being demobilised in 1945. Nkrumah soon fell out with the leaders of the UGCC after being accused of being a communist, which led him to form the Convention People's Party (CPP) in 1949. The guiding principle of the CPP was Positive Action, which involved mass action in order to achieve the specific goal of political liberation, and Tactical Action, a strategy involving neocolonial accommodation with the colonial power in alliance with the petty bourgeoisie in order to defeat the chiefs and the middle class. Through the vehicle of the CPP, Nkrumah won independence for Ghana in 1957 and became the leader of Black Africa's first independent sovereign state. This provided him with the opportunity to put some of his ideas into practice for his country's development. This has been referred to as Nkrumahism (Mohan 1967).

Nkrumahism was not a systematic or coherent ideology; indeed, Mohan described it as "a personalised choice of slogans and formulas" with a "hopeless attempt to superimpose an apparent order over a mass of contradictions" (Mohan 1967: 211). Nonetheless, one can unravel several strands to it: the philosophy of "Consciencism"; "socialism in one country"; Pan-Africanism; Positive and Tactical Actions; political and economic strategies for national liberation and continental unity. However, as Mohan has argued, the doctrine was riddled with contradictions.

Consciencism was a "third way" philosophical outlook designed by Nkrumah to aid Africa's social and political revolution. It is an eclectic ensemble of the triple heritage of the African continent which will enable African society to digest the Western, the Islamic and the Euro-Christian elements in Africa, and develop them in such a way that they fit into the African personality (Nkrumah 1964: 79). The African personality is defined as a cluster of humanist principles underlying traditional African society. In Nkrumah's world, the need for a new philosophical or ideological outlook for Africa was premised not just on the syncretic nature of postcolonial Africa, but also on the belief that the essence and purposes of capitalism are contrary to African society and the African *Weltanschauung*. For him, capitalism was the denial of the African personality and conscience of Africa. Not only was capitalism unjust, but it was also unworkable and alien to Africans (p. 79).

Thus in order to resituate Africa's humanist and egalitarian principles, which had been transgressed by colonialism, socialism was the answer. With regard to domestic policies, in the early days the CPP was riddled with anti-communist psychosis (Mohan 1967), designed to neutralise the radical wing of the Party and depoliticise the trade unions in order to gain colonial accommodation. Part of the CPP's project of modernity was to foster national unity (as a prelude to continental unity), and in order to achieve this Nkrumahism directed its attack on the chiefs, perceived as the symbols of the *ancien régime* and sources of "tribalism" and regionalism. However, after 1951 Nkrumah defended chieftaincy in the country (Mohan 1967), provided that they delivered the vote in a new form of postcolonial decentralised despotism. Paradoxically, the emergence of a bourgeoisie, that genre of modernity *sui generis*, was to be discouraged at all costs, as it was seen as a comprador class and as a harbinger of neocolonialism. In reality, Nkrumah was concerned about the emergence of a class that might challenge the CPP for state hegemony. Mohan has argued that the rejection of the middle class put paid to the rise of an incipient Ghanaian capitalism and transformed the large farmer-chiefs into the political opponents of the CPP's dominance (Mohan 1967). National liberation, it was argued, should be constructed with the petty bourgeoisie (the famous Veranda Boys) as the vanguard and a depoliticised worker-peasant movement (Mohan 1967). The surplus that was so necessary for social and economic reconstruction was to be squeezed from the peasantry operating within a modernised agricultural sector managed by the *apparatchiks*, via the state-controlled marketing boards.

If an indigenous bourgeoisie was to be denied, then the task of social reconstruction and industrialisation was left to the state. Nkrumah invested huge amounts in both human and physical capital: schools and universities, state farms, harbours and roads, including the Accra Tema highway and factories. The Nkrumahist state was a Bonapartist one: highly centralised, autonomous of any class or group, but a developmental one with the government playing a major role in industrial transformation. To ensure national unity and stability, the *sine qua non* for economic development, not only must *dirigisme* remain supreme, but the ruling party must also have total monopoly of power as all counter-hegemonic groups were proscribed. This alienated large sections of Ghanaian society, leading to several assassination attempts.

The economic core of Nkrumahism was proto-**Marx**ist, tinctured by Leninism on a framework of African socialism, which denied the class nature of African societies by celebrating its egalitarian past as if it were monolithic (Ottaway and Ottaway 1981). Nkrumahism had a continental dimension, which was meant to "bring home Pan-Africanism" from its diasporan origin to Mother Africa and transform it from a rather nebulous movement, concerned vaguely with black nationalism, into an expression of African nationalism. In 1963, as the main continental body, the Organisation of African Unity, was inaugurated in Addis Abba, the Ethiopian capital, Nkrumah called for Africa to unite (Nkrumah 1963). His message appeared to have been ignored by African leaders. In 1965, he published his *Neo-colonialism: The Last Stage of Imperialism*, pointing to the role that transnational corporations play in draining surplus from Africa, hence fostering the continent's underdevelopment. Following the United Nations' debacle in its peace-keeping activities in the Congo, which led to the death of that country's leader, Patrice Lumumba, Nkrumah called for an African High Command in order to prevent internal and external rogue elements from destabilising budding African governments (Zack-Williams 1997a). In order to eliminate all vestiges of colonialism in Africa, Nkrumah's Ghana became a haven for liberation fighters from several African countries.

Nkrumahism demanded that Africa should remain neutral or non-aligned: supporting neither the Western alliance nor the Warsaw Pact. Thus to counterbalance Western influences, Ghana developed close economic and political relationships with the Eastern bloc nations in addition to those already existing with the West. The country also played a major role in the Non-Aligned Movement (NAM), an assembly of radical Third World nations established at Bandung, Indonesia, in 1955. In this way, not only was Ghana able to play off one superpower against another, but it was also able to attract resources from the East (Zack-Williams 1997b) on relatively more favourable terms than could be obtained from the advanced capitalist nations. Nkrumah was removed from power on 22 February 1966 while en route to Vietnam in search of the peace that he felt might bring an end to the brutal war taking place in that country. It was then that senior military officers, in cahoots with foreign Cold Warriors, removed him from power after depressing the price of cocoa, Ghana's main export.

He returned to Africa, having gained asylum in nearby Guinea Conakry, where he was declared joint President of Guinea with Ahmed Sekou Touré. He remained in Guinea for the remainder of his life, apart from a visit to Romania for medical treatment, where he died in 1972. By the time of his death, Kwame Nkrumah was not just the *Osagyefo* of Ghana but, as *The Legon Observer* (Ghana, 18 May 1972) noted: "To the black man in all parts of the world, Nkrumah gave a new pride." Hence his remains, initially buried in his home town of Nkroful, were later reinterred with ceremonial pomp in a grand mausoleum constructed adjacent to Accra's independence arch and stadium.

Given the above statements about Nkrumah's life and the struggle for African rights and freedom, one often wonders what he would make of contemporary Africa's achievements if he were alive today. Presumably he might have been disappointed with the stalled project of African unity, African development and Africa's place in the world. He would have pointed out that he did not call for the liberation of Africa in order to empower new indigenous masters to oppress the rural and urban poor by relegating them to rural squalor and urban ghettoes. He would be shocked to see that the developmental state that he favoured has been transformed into a "Vampire State", serving the class interests of a coalition of urban and rural elites in an unholy alliance. It is true that the African leaders who followed him have managed to maintain some of his dreams of a united Africa via the successor to the OAU, the Africa Union, as some aspects of its activities – including "peer review" and peace-keeping on the continent – have been strengthened. He would have been pleased to welcome South Africa, the "rainbow nation", into the family of nations, and as marking a major blow to racism on the continent as the bastion of apartheid was dismantled.

However, there are other developments that would have appalled the *Osagyefo*: the growing poverty among the urban and rural masses as the ruling classes enriched themselves at the expense of the toiling masses that are increasingly being confined to urban ghettoes and rural squalor. He would have pointed out that the status quo, including the increasing impoverishment of the rural and urban poor, was not what he and other nationalist independence pioneers fought for. Nkrumah would have been reluctant to impose IMF- and World Bank-sponsored structural adjustment policies, as these would have been deemed to be detrimental to Africa's growth and development, as well as increasing Africa's dependence on global capitalist institutions. To save his country's institutions and independence, Nkrumah would have strengthened contacts with the socialist countries as he sought to develop Ghana's economy and independence.

Major works

Nkrumah, K. (1957) *Ghana: The Autobiography of Kwame Nkrumah*, London: Thomas Nelson.

—— (1961) *I Speak of Freedom: A Statement of African Ideology*, London: Heinemann.

—— (1963) *Africa Must Unite*, London: Panaf.

—— (1964) *Consciencism: Philosophy and Ideology for Decolonisation and Development with Particular Reference to the African Revolution*, London: Heinemann.

—— (1965) *Neo-colonialism: The Last Stage of Imperialism*, London: Thomas Nelson.

—— (1970) *Class Struggle in Africa*, London: Panaf.

—— (1990) *The Conakry Years: His Life and Letters*, London: Panaf.

Further reading

Birmingham, D. (1998) *Kwame Nkrumah: The Father of African Nationalism*, Athens, OH: Ohio University Press.

Boateng, C.A. (1995) *Nkrumah's Consciencism: An Ideology for Decolonisation and Development*, Dubuque, IA: Kendall/Hunt.

Bretton, H.L. (1966) *The Rise and Fall of Kwame Nkrumah: A Study of Personal Rule in Africa*, London: Pall Mall Press.

Davidson, B. (1973) *Black Star: A View of the Life and Times of Kwame Nkrumah*, London: Allen Lane.

Frimpong-Ansal, J.H. (1991) *The Vampire State in Africa: The Political Economy of Decline in Ghana*, London: James Currey.

Milne, J. (1990) *Kwame Nkrumah: The Conakry Years, His Life and Letters*, London: Panaf, Zed Press.

Mohan, J. (1967) "Nkrumah and Nkrumahism", in R. Miliband and J. Saville (eds), *Socialist Register 1967*, London: Merlin Press.

Ottaway, D. and Ottaway, M. (1981) *AfroCommunism*, New York: Africana.

Rathbone, R.J.A.R. (2000) *Nkrumah and the Chiefs: The Politics of Chieftaincy in Ghana, 1951–60*, Oxford: James Currey.

Sherwood, M. (1996) *Kwame Nkrumah: The Years Abroad, 1935–47*, Legon: Freedom Publication.

Zack-Williams, A.B. (1997a) "Peacekeeping and an 'African High Command'", *Review of African Political Economy* 71(3): 131–137.

—— (1997b) "Labour, Structural Adjustment and Democracy in Sierra Leone and Ghana", in R.A. Siddiqui (ed.), *Sub-Saharan Africa in the 1990s: Challenges to Democracy and Development*, Westport, CT: Praeger, 57–69.

ALFRED BABATUNDE ZACK-WILLIAMS

RAGNAR NURKSE (1907–1957)

Ragnar Nurkse belongs to the handful of economic thinkers associated with so-called high development theory, or classical development economics.[1] Other thinkers associated with these labels are **Paul Rosenstein-Rodan**, **Hans Singer**, **Arthur Lewis**, **Albert Hirschman**, and **Gunnar Myrdal**. It is not a coincidence that the post-World War II era, when Nurkse and others dominated development thinking and policy, was one of exceptionally good performance for many poor countries. During the 1980s, their blend of development economics was replaced by the so-called Washington Consensus that questioned many of the main tenets of Nurkse's (and other development pioneers') thinking – and brought dramatic reversals of growth for many developing countries. One can only agree with Paul Krugman (1994) that "[the] irony is that we can now see that high development theory made perfectly good sense after all."

Ragnar Nurkse was born on 5 October 1907 in the small village of Käru in Raplamaa county, about 100 km south of Tallinn, Estonia, when that country formed one of the westernmost provinces of the Russian empire. Tallinn, since 1285 a Hanseatic merchant and port city, had at the time multicultural character that also reflects Nurkse's background and education: his Estonian father worked at a rather typical German manor house, his mother was descended from the local Swedish minority, and he got his education from a German school in Tallinn.

In 1926 Nurkse entered the University of Tartu's Faculty of Law, which included economics at that time. However, it seems that along with his family, who emigrated to Canada in 1928, Nurkse was planning to leave Estonia early on and study abroad, as he took only one exam at Tartu. In 1929 Nurkse started to study economics at the University of Edinburgh, from where he graduated in 1932 with a first-class degree. After Edinburgh, Nurkse continued his studies in Vienna from 1932 to 1934, where he studied with Ludwig von Mises, Gottfried von Haberler and Oskar Morgenstern. In Vienna, Nurkse published his first academic publications (Nurkse 1934; 1935).

While still in Vienna, Nurkse successfully applied for a job in the Secretariat of the League of Nations in Geneva, where he worked from 1 May 1934 in the Financial Section and Economic Intelligence Service. The unit was transferred to Princeton in 1940, and Nurkse also moved to the United States. Previously, in 1938, he had enrolled as a PhD student at the University of Edinburgh. However, he never finished his PhD studies. Instead, he worked in Princeton during the war on what became *International Currency Experience: Lessons of the Inter-War Period*, published under the auspices of the League of Nations but credited mostly to Nurkse. This book went on to become one of the key publications for the Bretton Woods system. Nurkse also participated personally in the Bretton Woods meetings. With the reform of the League of Nations into the United Nations, his tenure at the organisation came to an end.

Following a brief stint as a visiting lecturer in Economics at Columbia University (from September 1945 to May 1946), Nurkse was a member of the Institute for Advanced Study in Princeton from 1946 to 1948. While he was offered a high position in the International Monetary Fund, he accepted a professorship at Columbia University instead, and moved from Princeton to New York in 1947 to become Associate Professor of International Economics. In 1949, he was promoted to full Professor. It was in this period that Nurkse's focus shifted from international monetary issues to development economics. This research found its first expression in *Problems of Capital Formation in Underdeveloped Countries*, published in 1953. In 1958–1959, he was awarded a Ford Foundation Research Scholarship, which he used mostly for working at the United Nations library in Geneva (Kukk and Kukk 2009). In April 1959, he was invited to Stockholm to deliver the Wicksell Memorial Lectures on international trade. These lectures were published posthumously (Nurkse 1959) as, after returning to Geneva from Stockholm, Nurkse collapsed and died of a heart attack during a walk on the slopes of Mont Pèlerin above Lake Geneva on 6 May. He is buried in Vevey in St Martin cemetery. He had married Harriet Berger in 1946 and they had two sons, Dennis and Peter.

It can be argued that among the classical development economists, Nurkse's contribution is the strongest in terms of economic theory; **Hirschman**'s accomplishments are, however, perhaps the farthest-reaching in terms of influencing social science and development mainstream the most. To date, it is relatively typical to juxtapose these two thinkers as representing almost exactly opposing ideas about development, namely balanced versus unbalanced growth. However, as Hirschman later acknowledged, the differences between him and Nurkse were minor at the end of the day, and in large part they shared a very similar outlook (Hirschman 1984; see also Nurkse 1961: 241ff.).

The high development theory developed by Nurkse rests on two key ideas:

1. Financing for development has to come to a large extent from the developing country itself ("Capital is made at home"; Nurkse 1961: 141).
2. The key areas to be financed need to exhibit increasing returns in order to trigger dynamics of development or, as **Myrdal** argued, virtuous circles of growth.

What makes Nurkse's contribution so important is that he is the only thinker from this group to, first, incorporate both key ideas into a coherent theory of development and, second, draw clear relationships between these notions.

According to Nurkse, the financing for development has to come mainly from within the country in question, because the financing of growth through either foreign investments or increased trade was largely a historically unique phenomenon confined to the nineteenth century and more specifically to American experience (Nurkse 1961: 134–136, 282–286). Nurkse thought that it would be nearly impossible for any developing country to repeat such a successful growth strategy based on trade and foreign financing because America was very rich in resources but at the same time populated by a workforce essentially on the same skill level as Britain (1961: 143). This unique combination made the American experience non-replicable because in any other circumstance, trade and foreign investment would engender a number of obstacles to development.

These are, first, that large parts of such financing would seek to utilise poor countries' resources and eventually lock these countries into undiversified economies with a skewed social structure (Nurkse 1961: 100, 137, 144, 248), and second, that there is a clear danger of significant amounts of foreign financing ending up funding private consumption patterns to emulate Western living standards and thus creating balance-of-payments problems (Nurkse 1953: 66–70). To sum it up, a growth strategy simply based on trade and foreign financing would leave the poor countries with negative financial flows and an undiversified production structure – just as the Washington Consensus did – and this amounted to financial instability that would eventually reverse the development gains through capital flight (Kregel 2004: 11). This, arguably, is what has made various growth efforts in developing countries so difficult to sustain in the last few development decades: many growth strategies are simply based on self- reversing logic, and this is indeed what Nurkse clearly foresaw.

According to Nurkse, any economic strategy that aims to be sustainable in the long term has to find another way of financing the development. What this means, however, is that such a development strategy has to work in a relatively confined environment in terms of capital and skills. Taking into account the financial constraints described above, it is in this context that Nurkse interprets Adam Smith's famous theorem about the size of the market being limited by the division of labour. For Nurkse, following Allyn Young's (1928) essay, this theorem indicates that the size of the market is limited by real wages that are, in turn, limited by productivity growth (Nurkse 1953: 21–25). For Nurkse, productivity growth is determined by the presence of increasing returns in an economy (1953: 8, 14; also 1961: 5–9, 27–29, 32) Thus, a viable development strategy should aim at establishing a number of increasing returns activities that would become one other's customers and generate the first virtuous circle of growth. The size of the market is thus limited by the number of increasing returns activities present in the particular market.

In sum, the driving idea behind Nurkse's balanced growth is not simply a set of reasons and/or policies for the creation of diversity in increasing returns activities, but moreover to show that, both as a theoretical foundation and as a policy strategy, balanced growth is coherent and sustainable as it shows how long-term growth with financial stability can be achieved and maintained. While the post-World War II era showed that this is indeed possible, the global financial crisis commencing in 2008 showed that we once more need development models that focus both on growth and on financial stability, something that Nurkse taught us is essential to development.

Note

1 This article draws heavily on the introductory chapter in Kattel, Kregel and Reiner (2009).

Major works

All Nurkse's works in English have been reprinted in Rainer Kattel, Jan A. Kregel and Erik S. Reinert (eds), *Ragnar Nurkse: Trade and Development*. London: Anthem, 2009.

Nurkse, R. (1934) "Ursachen und Wirkungen der Kapitalbewegungen", *Zeitschrift für Nationalökonomie* 5(1): 78–96.

—— (1935) *Internationale Kapitalbewegungen*, Vienna: Julius Springer.

—— (1952) "Growth in Underdeveloped Countries: Some International Aspects of the Problem of Economic Development", *American Economic Review* 42(2), Papers and Proceeding of the Sixty-Fourth Annual Meeting of the American Economic Association, 571–583.

—— (1953) *Problems of Capital Formation in Underdeveloped Countries*, Oxford: Oxford University Press.

—— (1958) "Trade Fluctuations and Buffer Policies of Low-Income Countries", *Kyklos* 9(2): 141–154.

—— (1959) *Patterns of Trade and Development*, Stockholm: Almqvist & Wicksell.

—— (1961) "Equilibrium and Growth in the World Economy", in G. Haberler and R.M. Stern (eds), *Harvard Economic Studies CXVIII*, Cambridge, MA: Harvard University Press.

Further reading

Drechsler, W. (2009) "Towards the Law and Economics of Development: Ragnar Nurkse (1907–1959)", *European Journal of Law and Economics* 28(1): 19–37.

Hirschmann, A. O. (1984) "A Dissenter's Confession: 'The Strategy of Economic Development' Revisited", in G.M. Meier and D. Seers (eds), *Pioneers in Development*, Oxford: Oxford University Press, 87–111.

Kattel, R., Kregel, J.A. and Reinert, E.S. (2009) *Ragnar Nurkse (1907–2007): Classical Development Economics and its Relevance for Today*, London: Anthem.

Kukk, K. and Kukk, K. (2009) "Life and Times of Ragnar Nurkse", in R. Kattel, J.A. Kregel and E.S. Reinert (eds), *Ragnar Nurkse (1907–2007): Classical Development Economics and its Relevance for Today*, London: Anthem.

Krugman, P. (1994) "The Fall and Rise of Development Economics". Available at http://web.mit.edu/krugman/www/dishpan.html.

RAINER KATTE

JULIUS KAMBARAGWE NYERERE (1922–1999)

The attempt to apply the philosophy of *Ujamaa* characterises the thinking and political work of Julius Kambaragwe Nyerere during his presidency of Tanzania. Nevertheless, his thinking was not guided solely by the ethos of African tradition in which the philosophy of *Ujamaa* was encapsulated, for he was a modernist and a practical politician who tried to synthesise the best of his people's ideas with those of the Western world in pursuance of transformation and the development of his country.

In so doing, he exhibited the highest level of moral commitment. Many critics have called his *Ujamaa* strategy an economic failure. However, it is more useful to interrogate his philosophy and practice holistically in order to distil that which was valid, at the time, from that which proved unworkable but which could be revisited and modified in order to inspire future generations. In the words of Ali Mazrui,

> When he was president of the United Republic of Tanzania, Julius K. Nyerere's vision was bigger than his victories; his perception was deeper than his performance. In global terms, he was one of the giants of the twentieth century. And like all giants he had both great insights and great blind spots. While his vision did indeed outpace his victories, and his profundity outweigh his performance, he did bestride this narrow world like an African colossus.
>
> (Mazrui 2002: 1)

Mazrui, previously a great critic of Nyerere, adds that with *Ujamaa*, which means "familyhood", Nyerere attempted to "build a bridge between indigenous African thought and modern political ideas". By making it the organising principle of Tanzania's entire economic experiment from the Arusha Declaration of 1967 to the mid-1980s, Nyerere turned it into a foundation for African socialism (2002: 5). Thus an assessment of his political and development thinking must be blended with the consideration of wider issues with which he was involved throughout his political life.

Julius Nyerere was born on 13 April 1922 in Butiama village in the Musoma District of Tanganyika's Mara Region. In 1947 he obtained his diploma in education from Makerere University in Kampala. Thereafter he taught at St Mary's Secondary School in Tabora for two years before studying for his master's degree in history, geography and economics at the University of Edinburgh. On returning to Tanganyika in 1952, he taught at Pugu Secondary School, Kisarawe, in Dar es Salaam between 1953 and 1954.

His political life began when he resigned from teaching to help found the Tanganyika African National Union (TANU) on 7 July 1954, immediately becoming its President. In 1960, he became Prime Minister of Tanganyika under "self-government", and he continued to hold the position when Tanzania became independent on 12 December 1961. Nyerere soon resigned his position and handed over the government to his deputy, Rashid Mugume Kawawa, so that he could return to consolidate the party among the ordinary membership. When he felt that work to be done, he took charge of the government once more in December 1962, after the declaration of a republic, and became Tanganyika's President until his eventual resignation in November 1985.

The inspirations of an outstanding thinker and politician like Julius Nyerere cannot be traced to a single influence or a particular phase (Green 1995: 81–82). However, several trends and directions are indicative of how his thinking about rural development, with which his *Ujamaa* ideology was most associated, evolved.

His approach was based on a strong attempt to find inspiration in traditional African society. Beyond that, he tried to mould these ideas into a new compact of economic and political theory. Some have seen the origin of his ideas in Fabian socialist ideology and Catholic social teachings (Ibhawoh and Dibua 2003: 62), while others regard his political ideas as parallel to those of Rousseau (Stöger-Eising 2000: 34–35). However, Irene and Roland Brown have argued that René Dumont's book, *False Start in Africa*, had a "particular significance" for Tanzania, especially Dumont's explanations of rural poverty as forming the root cause of lack of industrialisation (Brown and Brown 1995: 9–11).

Nyerere also read **Marx**ism and scientific socialism and later came to develop close relations with the People's Republic of China, with which he entered into several bilateral economic co-operation programmes. However, he always tried to marry new ideas gleaned from his prodigious reading to the conditions on the ground in which African traditional ideas and norms played a significant role. Joseph Warioba, one-time prime minister and confidant of Nyerere, believes that Nyerere's ideas about self-reliance were inspired very much by his conversations with ordinary people and their lived experiences. He adds that Nyerere believed very much in the idea of "man-centred" development (interview, Arusha, 28 September 2004).

We have to delve deeper to understand how he integrated traditional and intellectual ideas into a personal philosophy and political practice. For this purpose, the main texts are the Arusha Declaration and its two sequels, "Socialism and Rural Development" (1975a) and *Education for Self-Reliance* (1967b). The declaration, adopted as official doctrine by the Tanganyika African National Union (TANU) in February 1967, officially placed Tanzania on a socialist path and an associated strategy of self-reliance (Nyerere 1975b). It argued that agriculture formed the basis of Tanzania's economic development, and that only through the hard work and intelligence of the people could production levels be raised (Nyerere 1977).

Nyerere's *Ujamaa* philosophy and its ideological base are set out in *Socialism and Rural Development* (1975a). The "traditional African family" is identified as the unit through which the principle of *Ujamaa* was to be examined. In his view, *Ujamaa* constituted the "Third Way" to modern development. According to Nyerere, the traditional African family "lived and worked together because that is how they understood life". The results of their joint efforts were divided unequally among them, but this was done according to well-defined customs and traditions (Nyerere 1975a: 1).

Although Nyerere the traditionalist extolled the virtues of what he regarded as traditional African values, Nyerere the modernist simultaneously recognised their limitations. He pointed out that the result of these practices "was not the kind of life we really wish to see existing throughout Tanzania". Quite apart from the failure of individuals to live up to the ideals of the social system, he noted that two other factors prevented traditional society from "flowering". The first was gender inequality: "It is impossible to deny that the women did, and still do, more than their fair share of the work in the fields and in the homes." They were made to suffer from inequalities, which had nothing to do with their contribution to the family welfare. This, he concluded, "was inconsistent with the socialist conception of the equality of all human beings and the right of all to live in security and freedom". He added: "If we want our country to make a full and quick progress now, it is essential that our women live in terms of full equality with their fellow citizens who are men" (1975a: 3).

The other limitation of traditional society, according to Nyerere, was the prevalence of poverty. Here, Nyerere began to "graduate" from the traditional world of the African family to modern neocolonialism. He accepted that the traditional system offered an attractive degree of equality, albeit at a low level that could be raised. He blamed the prevalence of poverty on two factors, namely "ignorance" and the scale of operations of the family unit. In his view, the principles of mutual respect, sharing and joint production, and work by all were, and are, the foundation of human security, of practical human equality and of peace between members of a society. They could also form the basis of economic development "if modern knowledge and modern techniques of production are used" (1975a: 3). Here, Nyerere stood alongside all modernists. This recognition of the need to move rural production forward in modern conditions constituted the basis of *Ujamaa Vijijini* (Ujamaa Villages). The implementation programme set out in the Arusha Declaration began to confront the realities in rural areas and the way that peasants looked at the same problem. To be sure, the strategy of rural development was not entirely Nyerere's idea. It was suggested by the World Bank in their first mission report, which formed the basis of the *First Five Year Plan* adopted at Tanganyika's independence in 1961 and which came into effect in 1964 (Nabudere 1981: 98).

This plan had adopted the "transformation approach" to promote "Village Settlement Schemes", which would depend on the injection of foreign capital into rural development. The plan also had a "socialistic" ring to it, envisaging the creation of 60 "settlement villages" with each "scheme" involving 250 individual family farms. It was hoped that the schemes would bring about "a relatively abrupt transition" to modern techniques with regard to land use, land tenure, patterns of agricultural production and a change in economic attitudes. Such schemes could "also be relied upon … to relieve incipient land hunger and population pressure in certain areas" (Government of Tanganyika 1964: I.21). In the event, very little foreign capital was attracted; instead, there was capital flight, leading to the imposition of exchange controls that showed signs of reversing this trend (Loxley 1972: 104). The programme proved too costly and unproductive, while the individual traditional family unit survived, recording "impressively" increased production, diversification and productivity (Ngotyana 1972: 125).

By 1965 panic arose in official circles regarding the lack of foreign capital inflow to support the plan. From this perspective it appeared that African socialism of the *Ujamaa* type would be the answer. The Arusha Declaration's main innovations were an expansion of the settlement schemes as new villages and a programme to nationalise enterprises aimed at saving foreign exchange. However, when the *Ujamaa Vijijini* policy was implemented, the peasants and pastoralists immediately resisted what they saw as a derogation of their rights to ancestral lands. From a policy of persuasion, the government resorted to using force to implement its objectives, and this soon precipitated the collapse of the entire programme. Faced with this failure and the pressure to adopt structural adjustment programmes by the World Bank and the International Monetary Fund, Nyerere stood down as President of Tanzania in 1985, creating room for reforms that his successor, Ali Hassan Mwinyi (1985–1995), was forced to adopt in line with the demands of the Bretton Woods institutions in order to obtain new loans to "revive" the economy.

Nyerere's decision to resign was very unusual in Africa, especially within one-party states. However, his leadership had a long record of modesty. Although the son of a chief, he strove to live a simple life. He placed little significance on owning money; his salary and those of his ministers were very modest. He hated corruption and tried to banish it from his government. The party and army insisted on building him a larger house in Butiama ahead of his retirement. However, he slept in it for only two weeks before falling sick. He went to London for treatment for leukaemia, and died soon afterwards at St Thomas's Hospital, aged 77. He was buried at his new house in Butiama during a state funeral.

Nyerere dedicated the last years of his life to South–South co-operation. Serving as chair of the South Commission proved a creative experience, marking the beginning of his "international career", which lasted more than 13 years and put his vision, unique capacities and political acumen at the service of the Third World and its development. The Commission's landmark report, for which Nyerere chose the title *The Challenge to the South*, constituted the first comprehensive "southern" analysis of global development challenges. It argued forcefully that intellectual and technical preparedness was vitally important if developing countries were to confront the complexities of the modern age and the challenges posed by the North's continuing global dominance. Thus, one of the report's principal priorities was the need to establish permanent and adequate institutional support for collective global action by the South. This was not taken up directly, but in September 1995 the South Centre became an intergovernmental organisation – the first South–South institution with a comprehensive global development mandate.

Nyerere served as Chancellor of the University of East Africa (1963–1970), of the University of Dar es Salaam (1970–1985) and of Sokoine University of Agriculture (1984); he was awarded numerous honorary degrees, and received prestigious awards, including the Nehru Award for International Understanding, the Third World Prize, the Nansen Medal for Outstanding Service to Refugees and the Lenin Peace Prize.

With typical modesty, Nyerere declined to write an autobiography and did not encourage others to produce a biography. Two schools of thought have emerged in the debate about his contribution (Ibhawoh and Didua 2003: 70). The first argues that Nyerere's strategies were a complete failure, wasting economic resources in the name of slavish adherence to ideology (Nursey-Bray 1980; Von Freyhold 1979; Hyden 1980; Coulson 1985; Green 1995). The second school, while conceding that the economic performance of *Ujamaa* was modest, nevertheless points to achievements in social welfare such as the provision of health and educational facilities; a movement towards greater social equality in income distribution, the maintenance of political stability and a strong sense of Tanzanian national identity (Legum and Mmari 1995; Pratt 1999; Ishemo 2000; Mwakikagile 2002).

Nyerere's legacy therefore goes beyond developmentalism. It includes his contribution to the struggle for Pan-Africanism and the creation of an East African federation, his role as Chairperson of the Liberation Committee of the Organisation of African Unity and the Front Line States (the forerunner to the Southern African Development Community). His Pan-African contribution led to Nyerere being acknowledged as the fourth "greatest" African of all time in *New African* magazine (2004), which described him as "[a] great leader who refused to allow the trappings of power to corrupt him. He was respected by his country, Africa and the rest of the world."

Major works

Nyerere, J.K. (1966) *Freedom and Unity/Uhuru na Umoja: A Selection from Writings and Speeches 1952–1965*, London: Oxford University Press.

—— (1967a) *The Arusha Declaration*, Dar es Salaam: Government Printer.

—— (1967b) *Education for Self-Reliance*, Dar es Salaam: Government Printer.

—— (1968) *Freedom and Socialism/Uhuru na Ujamaa: A Selection from Writings and Speeches 1965–1967*, London: Oxford University Press.

—— (1973) *Freedom and Development/Uhuru na Maendeleo: A Selection from Writings and Speeches 1968–1973*, London and Dar es Salaam: Oxford University Press.

—— (1975) *Speeches in Parliament*, Dar es Salaam: Government Printer, 18 July.

—— (1975a) "Socialism and Rural Development", in L. Cliffe, P. Lawrence, W. Luttrell, S. Migot-Adholla and J.S. Saul (eds), *Rural Cooperation in Tanzania*, Dar es Salaam: Tanzania Publishing House.

—— (1975b) *The Development of Ujamaa Villages*, Presidential Circular no. 1 of 1969, Dar es Salaam: Government of Tanzania.

—— (1977) *The Arusha Declaration Ten Years After*, Dar es Salaam: Government Printer.

Further reading

Brown, I. and Brown, R. (1995) "Approach to Rural Mass Poverty", in C. Legum and G. Mmari (eds), *Mwalimu: The Influence of Nyerere*, Trenton, NJ: Africa World Press; London: James Currey; Dar es Salaam: Mkuki Na Nyota.

Coulson, A. (1985) *Tanzania: A Political Economy*, Oxford: Oxford University Press.

Dumont, R. (1966) *False Start in Africa*, London: Deutsch.

Government of Tanganyika (1964) *First Five Year Plan*, Dar es Salaam: Government Printer.

Green, R.H. (1995) "Vision of Human-Centred Development: A Study in Moral Economy", in C. Legum and G. Mmari (eds), *Mwalimu: The Influence of Nyerere*, Trenton, NJ: Africa World Press; London: James Currey; Dar es Salaam: Mkuki Na Nyota.

Hyden, G. (1980) *Beyond Ujamaa in Tanzania: Underdevelopment and an Uncaptured Peasantry*, Berkeley, CA: University of California Press.

Ibhawoh, B. and Dibua, J.I. (2003) "Deconstructing Ujamaa: The Legacy of Julius Nyerere in the Quest for Social and Economic Development in Africa", *African Journal of Political Science* 8(1): 59–83.

Ishemo, S.I. (2000) "'A Symbol That Cannot Be Substituted': The Role of Mwalimu J.K. Nyerere in the Liberation of Southern Africa, 1955–1990", *Review of African Political Economy* 27(83): 81–94. (Other tributes appeared in *ROAPE* 26(81) and 26(82), 1999.)

Legum, C. and Mmari, G. (eds) (1995) *Mwalimu: The Influence of Nyerere*, Trenton, NJ: Africa World Press; London: James Currey; Dar es Salaam: Mkuki Na Nyota.

Loxley, J. (1972) "Structural Change in the Monetary System of Tanzania", in J. Saul and L. Cliffe (eds), *Socialism in Tanzania*, vol. 2, Nairobi: East Africa Publishing House.

Mazrui, A.A. (2002) *The Titan of Tanzania: Julius K. Nyerere's Legacy*, Binghamton, NY: Institute of Global Cultural Studies, Global Publications, Binghamton University, State University of New York.

Mmari, G. (1995): "The Legacy of Nyerere", in C. Legum and G. Mmari (eds), *Mwalimu: The Influence of Nyerere*, Trenton, NJ: Africa World Press; London: James Currey; Dar es Salaam: Mkuki Na Nyota.

Mwakikagile, G. (2002) *Nyerere and Africa: End of an Era. The Biography of Julius Kambarage Nyerere (1922–1999), President of Tanzania*, USA: Protea.

Nabudere, D.W. (1981) *Imperialism in East Africa*, vol. 1: *Imperialism and Exploitation*, London: Zed Press.

New African (2004) "100 Greatest Africans of All Time", 432 (August/September): 12–32.

Ngotyana, R. (1972) "The Strategy of Rural Development", in J. Saul and L. Cliffe (eds), *Socialism in Tanzania*, vol. 2, Nairobi: East African Publishing House.

Nursey-Bray, P.F. (1980) "Tanzania: The Development Debate", *African Affairs* 79(314): 55–78.

Pratt, C. (1976) *The Critical Phase in Tanzania, 1945–67: Nyerere and the Emergence of a Socialist Strategy*, Cambridge: Cambridge University Press.

—— (1999) "Julius Nyerere: Reflections on the Legacy of Socialism", *Canadian Journal of African Studies* 33(1): 137–152.

—— (2000) "Julius Nyerere: The Ethical Foundation of his Legacy", *The Round Table* 355: 355–364.

Sahle, E. (2002) *The Legacies of Julius Nyerere: Influences on Development Discourse and Practice in Africa*. Trenton, NJ: Africa World Press.

South Commission (1990) *The Challenge to the South: The Report of the South Commission*, Oxford: Oxford University Press.

Stöger-Eising, V. (2000) "Ujamaa Revisited: Indigenous and European Influences in Nyerere's Social and Political Thought", *Africa* 70(1): 118–143.

Von Freyhold, M. (1979) *Ujamaa Villages in Tanzania: An Analysis of a Social Experiment*, New York: Monthly Review Press; London: Heinemann Educational Books.

DANI W. NABUDERE WITH DAVID SIMON

ELINOR OSTROM (1933–2012)

A political scientist, Elinor Ostrom influenced the field of development studies by theorising and communicating rational choice approaches to institutional thinking, focusing primarily on the analytical (and political) challenge of identifying the conditions under which individuals will engage in collective action to protect natural resources and the commons. In 2009, she became the first – and, so far, only – woman to win a Nobel Prize for Economics (a prize shared with Oliver Williamson). Ostrom devised and explained a new theoretical framework that conceptualised institutions as common property regimes, in which rules, sanctions and monitoring mechanisms are used to prevent individual resource users from depleting or degrading commonly owned resources. Elements of this framework have been replicated in natural resources management and climate change policy, such as the use of pastures and lakes, and within climate change policy through carbon trading (although some analysts claim that economists had more influence). She also influenced academic research by organising, with her husband, Vincent Ostrom, the University of Indiana's Workshop in Political Theory and Policy Analysis and the influential International Association for the Study of the Commons (IASC).[1]

Elinor Ostrom was born Elinor Claire Awan in Los Angeles, California in 1933, the daughter of an unemployed Hollywood set-designer father and a musician mother. She studied at University of California, Los Angeles (UCLA), where – after an earlier marriage – she met and married Vincent, one of her professors. In 1965, she took up a job at the University of Indiana, where Vincent was already working. In 1973 they founded their Workshop in Political Theory and Policy Analysis, which focused on political and economic governance through public management, including federal governance of public infrastructure such as water projects. This work later analysed community collective action to protect natural resources.

During this time, Ostrom developed new frameworks concerning political science and institutions – which she defined as rules and norms shared by rational individual actors. Like many working in American political science at the time, Ostrom adopted the analytical approach of methodological individualism, originally proposed by economic philosophers such as Ludwig von Mises and Friedrich Hayek, which holds that social phenomena can be explained by focusing on the motivations and actions of individual agents. Ostrom argued that standard positivistic political science led to universal laws that were too crude and inflexible. Instead, she sought to explain how individual people or organisations could create institutions for collective action while at the same time being motivated by rational self-interest.

Arguably, Ostrom's most important and lasting contribution was *Governing the Commons: The Evolution of Institutions for Collective Action* (1990), which explored the empirical conditions under which individuals and communities create and support institutions for collective action. In this respect, *Governing the Commons* was an important and viable alternative to Garrett Hardin's *Tragedy of the Commons*, which offered a deeply pessimistic scenario that commonly owned resources will always be degraded by individuals acting rationally. In contrast to Hardin, Ostrom argued that commonly used resources such as lakes or pastures should not be identified as open-access resources (where everyone can use the resource without restriction) but instead as common-pool resources, which indicate that the resource has a system of rules, sanctions and monitoring mechanisms for restricting and regulating access. The governance system operating on a common-pool resource is known as a common-property regime. Ostrom argued that different property rights regimes (e.g. common, private or a combination of the two) are the result of a selection process, in which "resource users" craft institutions that best meet the needs of resource users and the resource base. Whether resources fall prey to the Tragedy of the Commons is therefore not a foregone conclusion, but rather an empirical outcome dependent on the existence and effectiveness of institutions governing access, utilisation, management, exclusion, ownership and transfer of ownership.

Ostrom used various examples from locations as diverse as Switzerland, Japan, Spain, the Philippines and Turkey to show how individuals had deliberately organised new rules and property rights to exclude unauthorised free-riders from using common resources, and to regulate permitted users to increase the sustainable use of resources. She also described eight design principles for building successful institutions in different locations and at varying spatial scales: clear resource boundaries (i.e. knowing physical and ecological properties of the resource), clear rules of membership (knowing who is entitled to use the resource), congruence between rules of provision/appropriation and local conditions, arenas for "collective choice", mutual monitoring, "graduated" sanctions, mechanisms for conflict resolution (i.e. ensuring that resource users are able to monitor and sanction other resource users) and, finally, a state that is willing to recognise (or at least not challenge) local rights of organisation (Ostrom 1990: 90). The book proposed "that neither the state nor the market is uniformly successful in enabling individuals to sustain long-term, productive use of natural resource systems"; and that "communities of individuals have relied on institutions resembling neither the state nor the market to govern some resource systems with reasonable degrees of success over long periods of time" (p. 1). Starting from the position that information (about the state of the resource and intentions of resource users) is costly, Ostrom argued that rules matter because they reduce the uncertainty that stems from the unpredictable behaviour of others and resource systems:

> If everyone, or almost everyone, follows these rules, resource units will be allocated more predictably and efficiently, conflict levels will be reduced, and the resource system itself will be maintained over time.
>
> (Ostrom 1990: 43)

Ostrom's new institutionalist framework was extremely popular at a time when the environmental catastrophism of the 1970s was being replaced with more nuanced thinking about the social and human dimensions of environmental change; this was also a time when much international development policy discourses were becoming dominated by ideas of "shrinking the state" and trusting community-based initiatives. It is unsurprising that Ostrom's approach proved popular with academics, policy advisers and development agencies such as the World Bank.

However, Ostrom's framework was also a source of critical attention, which led her to modify her approach over time. One persistent theme is a methodological tension between the apparent reductionism of rational-choice theory and the interpretivism of anthropological and historical research. For many critics, the focus on theory building and methodological individualism produced a body of research that was at times too far removed from the important aspects of context, culture and meaning that have bearing on the ways in which people use and manage common-property regimes. In particular, critics argued that rational behaviour by individuals should not be defined narrowly in terms of economic interest, utility and value. Indeed, local cultural values or customary practices can sometimes be more significant in shaping how different people use common resources such as rivers, forested land or irrigation tanks. Another concern was that some of the examples in Ostrom's 1990 book might have referred to community-based conservation, but not necessarily on the principles of rational choice that Ostrom claimed. Consequently, some critics have worried that new-institutional analysis in the manner of Ostrom has often looked to provide evidence for certain types of institutions, in response to a priori assumptions about certain kinds of problems, both of which exclude various other reasons and pathways to institutions.

In response to these and other criticisms, Ostrom's later writings showed more flexibility about how institutions emerge.

In her later work, Ostrom expanded her analysis of institutional design principles to argue in favour of a multi-level form of governance, often based on polycentric institutions or on multiple governing authorities at different spatial scales. She applied this to global climate change policy, and principles of this approach can be seen in carbon-trading schemes such as Reduced Emissions from Deforestation and Forest Degradation (REDD+) or the European Union's Emissions Trading Scheme (ETS). These policy schemes resemble Ostrom's frameworks because they involve clearly defined boundaries about resources and procedures; and with a multi-tier structure that can include citizens, companies, industries, nation-states and blocs of nation-states. Economists, however, have tended to claim these policy frameworks as their own. Some economists, too, queried Ostrom's 2009 Nobel Prize on the grounds that she was a political scientist who was largely unknown to economists.

Nonetheless, Ostrom's legacy for development studies lies in her commanding influence on rational choice approaches to institutions, and the trend from the late 1980s to 1990s to integrate a form of economistic analysis with a political agenda of bypassing the state. Whether or not her work reflected neoliberal development policies, Elinor Ostrom elaborated a form of analytical political science that emphasised the mutual development of theory and empiricism, and a mistrust of what she called "value-laden" theory. Through her work on the commons, she influenced the ways in which donors, NGOs and development practitioners now understand and address environmental problems, and provided the empirical and theoretical basis for moving beyond the unhelpful dichotomy between market and state.

Note

1 The International Association for the Study of the Commons (IASC) is the leading professional association dedicated to the commons: www.iasc-commons.org.

Major works

Dietz, T., Dolsak, N., Ostrom, E., Stern, P., Stonich, P. and Weber, E. (eds) (2002) *The Drama of the Commons*, Washington, DC: National Academy Press.

Hess, C. and Ostrom, E. (2007) *Understanding Knowledge as A Commons: From Theory to Practice*, Cambridge, MA: MIT Press.

Keohane, R.O. and Ostrom, E. (1995) *Local Commons and Global Interdependence: Heterogeneity and Cooperation in Two Domains*, Thousand Oaks, CA: Sage.

Ostrom, E. (1990) *Governing the Commons: The Evolution of Institutions for Collective Action*, New York: Cambridge University Press.

—— (1991) "Rational Choice Theory and Institutional Analysis: Toward Complementarity", *American Political Science Review* 85(1): 237–243.

—— (2005) *Understanding Institutional Diversity*, Princeton, NJ: Princeton University Press.

—— (2010) "Polycentric Systems for Coping with Collective Action and Global Environmental Change", *Global Environmental Change* 20(4): 550–557.

Poteete, A.R., Janssen, M. and Ostrom, E. (2010) *Working Together: Collective Action, the Commons, and Multiple Methods in Practice*, Princeton, NJ: Princeton University Press.

Further reading

Bardhan, P. and Ray, I. (2008) *The Contested Commons: Conversations between Economists and Anthropologists*, Oxford: Blackwell.

Forsyth, T. and Johnson, C. (2014). "Elinor Ostrom's Legacy: Governing the Commons and the Rational Choice Controversy", *Development and Change* 45 (5): 1093–1110.

Izzo, P. (2009) "Economists React: Nobel Award Sends Message about Economics", *Wall Street Journal*, 12 October. http://blogs.wsj.com/economics/2009/10/12/economists-react-nobel-award-sends-message-about-economics/ (accessed January 2018).

TIM FORSYTH AND CRAIG JOHNSON

FRANÇOIS PERROUX (1903–1987)

François Perroux was born on 19 December 1903 in Lyon and died on 2 June 1987 in Stains. After being Professor of Economics in Lyon (1928–1937) and Paris, he was elected Professor of Economics at the Collège de France in 1955, a post which he held until 1974. In January 1944, he founded the Institute of Applied Economics (ISEA), under the aegis of the Free School of Political Science, predecessor of the Paris Institute of Political Studies. In 1972 ISEA became the Institut de sciences mathématiques et économiques appliquées (ISMEA) with the *Cahiers de l'ISMEA*. It was a real space for debate on political economy with many economists, including Jean Weiller, Henri Bartoli and Gérard Destanne de Bernis. Perroux steered a delicate, at times even ambiguous, political course. Although he committed to the French Resistance in early 1943 and showed an intellectual attachment to the social doctrine of the Church by contributing to the religious magazine *Esprit*, he was previously also noted for his role advising the Vichy government before 1942.

In 1957 he established the Institute of Studies and Economic and Social Development (IEDES) in Paris, serving as director from 1960 to 1969. In 1959 he was appointed to the Economic and Social Council at the advent of the Fifth Republic, and he served there until 1969. François Perroux was the founder of numerous journals: *Les cahiers de l'ISEA*, the *Revue d'économie appliquée*, the *Revue Tiers Monde* (rebranded in 2017 as *Revue internationale des études de développement*) and *Mondes en développement*. He authored more than 46 books or articles on **Karl Marx**, Josef Schumpeter, John Maynard Keynes, **Michał Kalecki** and Wasili Leontiev, as well as on international relations, development economics and other topics. Among other works, he published "The Marshall Plan or the Europe Necessary for the World" (1948) and "General Theory of Economic Progress" (*Cahiers ISEA*, 1956).

He was involved in French political life, and was associated with the work of Jean Monnet's Commissariat au Plan in the effort to defend a flexible vision of planning that was at odds with the practices of communist countries. For François Perroux, planning meant "organising". To do this, the state must gather information and statistics that enable decision-makers to act knowingly. This is the reason why Perroux invested his efforts in the French national accounts. The state must then support the action of firms by the provision of collective equipment and infrastructure. The text of the Plan consists of the announcement of this provision. It is in this logic of economic policy that he inscribed his reflections on development issues in Third World countries. The state must be an incentive provider and not a producer. The state cannot be a Keynesian state: stimulating economic revival by increasing the public deficit is illusory and harmful because it creates debts and handicaps future choices. On the other hand, a fiscal policy is relevant when it combats inflation by raising taxes and reducing excess demand. As for monetary policy, it is a means of revival through reduction of interest rates, which stimulates investment.

François Perroux came from the movement of Christian humanism. The economic cannot be reduced to commercial relations between actors that are assumed in neoclassical theory to be homogeneous sets. It is the relationships of power (constraint) as well as the relations of solidarity (the gift) that constitute these relations between economic agents. His reasoning was based on asymmetries of power, the question of economic structures as well as issues of domination. His work on the economics of development is based on the essential idea that human beings must be the object of the economy: "to develop the whole Human and all the Human – that is to say to nourish, to lodge, educating, caring for the Human, in short cover the 'costs of the Human'" (1981: 19). In his 1955 article "Three Analytical Tools for the Study of Underdevelopment", Perroux characterises underdevelopment in three points:

> Observation of the countries classified statistically as underdeveloped reveals three notable features of their economies. These are inarticulate economies. These are dominated economies. These are economies that do not cover the costs of the human status of life for all; the costs of Human, the costs of providing everyone with

> life expectancy, health, access to knowledge, compatible with the concrete conditions of the place and time, are not covered.

In the introduction to Chapter 4 of his book "For a Philosophy of the New Development" (1981: 211–214), he provides a strong indictment of colonisation and insists on the finality of development and values: "The best informed Western observers recognize that we have something to learn from these resurrected homelands in human settlements in which we willingly take the attitude of the teacher and the teacher of lessons."

When Perroux published an article entitled "Economic Spaces" in 1950, he had already disseminated it widely to American economists. This article was published in February 1950 and in English in the *Quarterly Journal of Economics*, under the title "Economic Spaces: Theory and Applications", and it was the subject of a public discussion on 2 November 1949 in the presence of Walter Isard, the Harvard economist and author of *Location and Space-Economy*. Perroux was the first French economist to introduce the notion of space into economic analysis.

In "For a Philosophy of the New Development", Perroux (1981) determined the basic contents of any development policy by emphasising two points. In the first place, it is necessary to favour dynamisms of development, that is to say, those contained in the Human: the capacity for technical progress, renovation of institutions-organisations (family, unit of production) or institutions-norms (property, distribution rules, market regulation). Second, it is necessary to organise "the medium of propagation, material and informational beneficial economic effects". On this last aspect of his thinking, Perroux was in tune with **Albert Hirschman**'s theory of unbalanced growth, which in fact opposes that of the big push (**Rosenstein-Rodan**) devising a massive development policy on all sectors simultaneously and known as the theory of balanced growth. He joined this approach of industrialising industries, developing the concept of growth poles, where induced investments become concentrated. It is the theory of the blackening of the matrix of the technical coefficients of Wasili Leontiev, giving an essential role to the interactions between firms and between branches. In the Introduction to "The Economy of the Twentieth Century" (1961: ix) he writes:

> Dominant economies are studied dynamically and with their three characteristics of dimension, bargaining power and nature of activity ... The motor industries exercise their environment-driving

> effects or stopping effects, they exert them either upstream in the direction of the inputs, or downstream in the direction of the outputs. More generally, the relatively active sets and the relatively passive sets help us to interpret the dynamics, if we do not give up our understanding of the collective competitions: interindustrial and interregional, so decisive and so badly controllable by the so-called neo-classical microeconomic analyses.

In his book "For a Philosophy of New Development" (1981: 186–207), Perroux analyses "the industrial–agricultural dialectics", showing the complexity of the relations between these two sectors in each of the national and territorial situations analysed. It is necessary, he argues (p. 189) to analyse "the asymmetries and their intersections, the role of political power, the control exercised by it over the production organisations and the organisation of their milieu of activity, the new financing required by the new development and the specific conditions of multinational enterprises".

In his approach, Perroux clearly distinguished the processes of economic growth from those of development. He considered that there can be growth without development, but no development without growth. Growth without development is achieved when the economic activities are confined around the "establishments of foreign firms or large works without radiating on the whole". He therefore emphasised the analysis of economic structures, inequalities and domination in international exchange, stressing the overly important role given to foreign firms in a development process. But according to Perroux (1981: 59–60), "Neither capitalism nor socialism, revised and corrected, promise progress to developing countries; it can only be envisaged in efforts of enlightened pragmatism to endow indigenous peoples with original institutions." Perroux (1981: 231) considers that three great designs form what he calls the "new inseparable development of the new international economic order". The first design (p. 230) is that of the "popular mobilization of wills and the formation of public opinion"; the second is that of "structural reforms in the consumption, production and (income) distribution of developed and developing countries" (p. 230); the third is that of "the progressive, general and controlled disarmament strategy" (p. 230).

François Perroux is an unclassifiable economist. He criticises neoclassical liberalism, Keynesianism and Marxism to construct his own theory inspired by the economists of the Austrian school epitomised by Josef Schumpeter. His international influence has always been built in an academic context and not as a consultant or expert to international organisations; he was recognised for his theoretical work and was awarded honorary doctorates by 15 universities (São Paulo, Porto, Liège, Frankfurt, Lisbon, Cordoba, Georgetown, Montevideo, Lima, Ica (Peru), Barranquilla, Santiago, Bogotá, Bucharest and Ottawa).

Major works

Perroux, F. (1948) *Le plan Marshall ou l'Europe nécessaire au monde*, Paris: Librairie de Médicis.

—— (1950) "Les espaces économiques", *Archives de l'ISEA* 1: 225–244.

—— (1955) "Trois outils d'analyse pour l'étude du sous-développement", *Cahiers de l'ISEA*, série F, n° 1.

—— (1956) "Théorie générale du progrès économique", *Cahiers de l'ISEA*, série I *(Le progrès économique)*, n° 1, 2, 3: 49, 89, 109.

—— (1960) *Economie et société: contrainte, échange, don*, Paris: Presses universitaires de France.

—— (1961) *L'économie du XXème siècle*, Paris: Presses universitaires de France. (3rd edn, 1969.)

—— (1962) *L'économie des jeunes nations*, Paris: Presses universitaires de France.

—— (1965) *Les techniques quantitatives de la planification*, Paris: Presses universitaires de France.

—— (1970) *Aliénation et société industrielle*, Paris: Gallimard.

—— (1973) *Pouvoir et économie*, Paris: Bordas.

—— (1975) *Unités actives et mathématiques nouvelles: révision de la théorie de l'équilibre économique général*, Paris: Dunod.

—— (1981) *Pour une philosophie du nouveau développement*, Paris: Aubier/Unesco.

—— (1982) *Dialogue des monopoles et des nations: équilibre ou dynamique des unités actives*, Grenoble: Presses universitaires de Grenoble.

—— (1990–1994) *Oeuvres complètes de François Perroux*, 11 vols, Grenoble: Presses universitaires de Grenoble.

Biographies or appreciations of François Perroux

Destanne de Bernis, G. (1978) "La dynamique de François Perroux: l'homme, la création collective, le projet humain", in *Hommage à François Perroux*, Grenoble: Presses universitaires de Grenoble, 121–163.

Hugon, P. (2003/2004) "Le concept d'acteurs du développement chez les pionniers du développement: Albert Otto Hirschman et François Perroux", *Mondes en développement* 124: 9–31. https://fr.wikipedia.org/wiki/Fran%C3%A7ois_Perroux

JEAN-JACQUES GABAS

KARL POLANYI (1886–1964)

In 1944, Karl Polanyi published *The Great Transformation: The Political and Economic Origins of Our Time*, one of the towering achievements of 20th-century scholarship. In the foreword to the 1957 edition of the book, the Canadian sociologist R.M. MacIver hailed it as "a book that makes most books in its field seem obsolete or outworn. … a fresh comprehension of the form and meaning of human affairs" (1957: ix). Block and Somers (1984: 47), for their part, argued that "Karl Polanyi's intellectual work stands among the most significant and original contributions to social science scholarship in the period since the end of World War I." Shamefully neglected by much of mainstream academe (particularly in Britain), the book nevertheless provoked considerable debate and controversy internationally. It also inspired important research in a wide number of fields: economic and social history, political and economic theory, public policy, political democracy, economic anthropology and, more recently, environmental studies and labour history. Since the turn of the 21st century, it has become a critical source for those seeking ways out of the cul-de-sac of neoliberal crisis.

Karl Polanyi was born in Budapest in 1886 into a Hungarian Jewish family of some considerable achievement. His father, a convert to Christianity, was a wealthy railroad builder. His mother, a prominent Russian intellectual, organised a salon that was an important part of the intellectual life of prewar Budapest. His brother Michael, an internationally recognised scientist and philosopher, would come to espouse free-market principles directly opposed by Karl. Polanyi's own life trajectory and intellectual concerns would reflect the turmoil of the first half of the 20th century in Europe. In his twenties, as a member of the Galileo Circle, he was prominent in the radical student politics of Budapest up to the end of the Great War. In his thirties, in the wake of the collapse of the short-lived communist government of Béla Kun, Polanyi moved to Vienna, where he worked as a journalist – becoming a senior editor of Austria's leading financial newspaper, the *Österreichischer Volkswirt* – and met his wife, Ilona Duczynska, herself a participant in the communist regime led by Béla Kun. In late 1933, as storm clouds gathered over Vienna and he was dismissed from his job, Polanyi moved to England with financial support obtained via indirect facilitation by the Academic Assistance Council, set up to assist academics and researchers dismissed and victimised by the Nazis, and which was thus unable to help a journalist directly (Simon 2012). He then found employment in workers' education. In 1940, caught in the United States while on a lecture tour and unable to return to England, he eventually found a post in economic history at Columbia University. Finally, in 1953, with McCarthyism in full sail, he left Columbia and joined Ilona in Canada. She had been refused admission to the United States because of her political history, so joined their only daughter, Kari Polanyi Levitt, who by then had settled in Toronto with her Canadian husband (Simon 2019). It was there that Karl finally published *The Great Transformation*.

These travels signposted Polanyi's intellectual journey and his accumulation of the ideas that would characterise his life's work. As a founder member of the Galileo Circle (alongside the likes of Karl Mannheim and Georg Lukacs), a group of Hungarian socialist intellectuals radicalised by the bureaucracy, corruption and privilege of the stagnant Austro-Hungarian empire and by the ravages of World War I (Block and Somers 1984: 49), Polanyi came to reject two competing philosophies. Despite a critical sympathy with Marxism – he could describe himself as "brought up in Marxian socialism" (Dale 2016: 33) – like others in that circle, he rejected the deterministic Marxism of the Social Democratic Party, and

> the Second International's belief in the inevitability of progress as a consequence of predetermined stages of historical development. Central to him and others of his generation was the idea that progress could only come through conscious human action based on moral principles.
>
> (Block and Somers 1984: 50)

Equally, he rejected the free-market theories of Ludwig von Mises and Friedrich Hayek, both of whom enjoyed a considerable prestige in Viennese economic life. In "Red Vienna", he admired its socialist municipal government (an experiment that strongly influenced his ideas about a socialist democracy) and, when that was overwhelmed in 1934 by a rising tide of fascism, developed fierce, lifelong anti-fascist views; he "became a socialist again, more strongly convinced than ever" (Dale 2016: 33). In England, Polanyi became involved in workers' education and closely associated with a group of Christian Socialists and Quakers, notably G.D.H. Cole and Richard Tawney. All these influences played an important part in shaping the argument of *The Great Transformation*.

In *The Great Transformation*, Polanyi set out to understand why Europe and the capitalist world had suffered a half-century of political and economic calamity and why, in consequence, there had been a turning away from democratic government in favour of authoritarian solutions like Stalinism and, particularly, fascism. This project he set out in the first two sentences of the book: "Nineteenth century civilization has collapsed. This book is concerned with the political and economic origins of this event, as well as the great transformation which it ushered in" (1944: 3). The cause of this, of war, depression and fascism, he located in the nature of free-market capitalism. In the process, Polanyi presented a withering critique of the ideology of the self-regulating market and its adherents (notably Malthus, Ricardo, von Mises and, to a lesser extent, Smith):

> Our thesis is that the idea of a self-adjusting market implied a stark utopia. Such an institution could not exist for any length of time without annihilating the human and natural substance of society; it would have physically destroyed man and transformed his surroundings into a wilderness. Inevitably, society took measures to protect itself, but whatever measures it took impaired the self-regulation of the market, disorganized industrial life, and thus endangered society in yet another way. It was this dilemma which forced the development of the market system into a definite groove and finally disrupted the social organization based upon it.
>
> (Polanyi 1944: 3–4)

A world based on the self-regulating market could not endure, argued Polanyi. The Industrial Revolution had brought a "miraculous improvement in the tools of production" but also "a catastrophic dislocation of the lives of the common people" (1944: 33). Polanyi argued that for the market economy to operate efficiently, it was necessary for it to function without outside interference so that prices could regulate themselves. This, in turn, completely transformed traditional relations between society and economy (p. 57):

> the control of the economic system by the market is of overwhelming consequence to the whole organization of society: it means no less than the running of society as an adjunct of the market. Instead of economy being embedded in social relations, social relations are embedded in the economic system. The vital importance of the economic factor to the existence of society

> precludes any other result. For once the economic system is organized in separate institutions, based on specific motives and conferring a special status, society must be shaped in such a manner as to allow that system to function according to its own laws.

In particular, it was necessary to turn land, labour and money into commodities traded in the market, and so to make of human beings, their environment and their means of exchange, things to be bought and sold. The inevitable consequence of this was that (p. 40):

> an avalanche of social dislocation, surpassing by far that of the enclosure period, came down upon England; that this catastrophe was the accompaniment of a vast movement of economic improvement; that an entirely new institutional mechanism was starting to act on Western society; that its dangers, which cut to the quick when they first appeared, were never really overcome; and that the history of nineteenth century civilization consisted largely in attempts to protect society against the ravages of such a mechanism. The Industrial Revolution was merely the beginning of a revolution as extreme and radical as ever inflamed the minds of sectarians, but the new creed was utterly materialistic and believed that all human problems could be resolved given an unlimited amount of material commodities.

And (p. 42):

> Machine production in a commercial society involves, in effect, no less a transformation than that of the natural and human substance of society into commodities.... [inevitably,] "the disruption caused by such devices must disjoint man's relationships and threaten his natural habitat with annihilation.

The stresses and strains imposed on society by this "stark utopia" created economic dislocation and political crises. As MacIver summarises it:

> The "satanic mills" were heedless of all human needs but one; relentlessly they began to grind society itself into its atoms. For Mr Polanyi the last byword is society. The major tragedy attendant on the Industrial Revolution was brought about not by the callousness and greed of profit-seeking capitalists – though there was

> inhumanity enough in the record – but by the social devastation of an uncontrolled *system*, the market economy.
>
> (MacIver 1957: x)

States sought to counteract these in various ways: by increasing authoritarianism at home; by deepening poverty and blaming the poor for their unemployment and low wages; by imperialist expansion abroad in order to transfer costs; by currency manipulation; and even by war. None of these palliatives worked for any length of time: attempts to restore the gold standard after 1918 intensified crisis and suffering and led to the Great Depression; that in turn led to fascism and finally to World War II. A 19th century of relative peace among the European powers was succeeded in the 20th century by a catastrophic global war, the Great Depression, and ultimately the rise of fascism and the Second World War. Each crisis, argued Polanyi, produced a double movement or counter-movement (Silver and Arrighi 2003; Block and Somers 2014: 53–54, 60–68, 201–204, and passim; Dale 2016: 4–5, 128–132, and passim) to try to alleviate the effects of the market, and each counter-movement further undermined the working of the self-regulating market. For their part, adherents of the free market took refuge in blaming the victims of the dislocation. "Nowhere has liberal philosophy failed so conspicuously as in its understanding of the problem of change" (Polanyi 1944: 33).

Polanyi (p. 43) laid the responsibility for this human catastrophe squarely on the market system and the ideology which justified it:

> previously to our time no economy has ever existed that, even in principle, was controlled by markets. … Though the institution of the market was fairly common since the later Stone Age, its role was no more than incidental to economic life.

Its origins lay in Adam Smith's notion of man's "propensity to barter, truck and exchange one thing for another" which led to the concept of Economic Man, whereof "no misreading of the past ever proved more prophetic of the future". In time, with the market economy in full swing, the assumed propensity shaped not only its economic activities but also its political, intellectual and spiritual pursuits (pp. 43–44).

> Herbert Spencer, in the second half of the nineteenth century, could, without more than a cursory acquaintance with economics, equate the principle of the division of labour with barter and

> exchange, and another fifty years later, Ludwig von Mises and Walter Lippmann could repeat this same fallacy.

Far from being a spontaneous outcome of this supposed propensity, as its apologists claimed, Polanyi argued that the self-regulating market was itself constructed and maintained by the exercise of arbitrary power.

Polanyi's work clearly distinguishes him from other economic theorists and historians. His rejection of the free-market liberalism of Malthus, Ricardo, von Mises and Hayek is clear. But his argument also stands apart from **Marx**. Despite some sympathy with Marx's location of crisis as rooted in the process of production and the exploitation of the producer, Polanyi instead located it primarily in the way in which the self-regulating market inevitably consumes the very society and environment in which it operates. He stands apart also from Keynes, despite being sympathetic to the latter's concern to find a means to manage market capitalism in a democratic society. Polanyi regarded Keynes's approach as primarily technical in nature, intended to protect market enterprise from the ravages of authoritarianism. Ultimately, Polanyi felt, this too could not succeed. Nor was there any going back to some pre-industrial utopia. As MacIver observes, Polanyi did not romanticise the agrarian societies that he studied; instead, "What our age needs is the reaffirmation, for its own conditions and for its own needs, of the essential values of human life" (1957: x). For Polanyi, therefore, there was no predetermined system to respond to either capitalist crisis or its fascist counter movement. Instead, it would need radical reform in a socialist democracy in which land, labour and money would be "taken out of the market" and in which freedom would be consistently defended (Polanyi 1944: ch. 21). It was to this end, in the hope of influencing discussions about the postwar settlement, that Polanyi rather hurriedly published *The Great Transformation* in 1944.

In the event, *The Great Transformation* did not influence the post-1945 institutions. Although Keynes and Harry Dexter White, the architects of the Bretton Woods agreement, shared some of Polanyi's concerns relating to untrammelled international markets, the influence of American business forced a somewhat watered-down version of the international system they envisaged. With the onset of the Cold War, there was little public space for Polanyi's ideas for the next four decades. Instead, "Keynesianism", bolstered by postwar reconstruction and Cold War imperatives, would ensure for the developed economies (though not for the developing world) a period of rapid economic growth, technological development and the evolution of welfare states that increased social mobility and restrained the effects of markets. By the mid-1970s, however, this system became increasingly unworkable, much as Polanyi had argued it would, as increased public spending, rising deficits and falling profits produced increasing levels of inflation and economic stagnation. Government efforts to manage these problems, for example by floating exchange rates and moving costs onto dependent third world economies, had little success. Led by Britain and America, Keynesianism was quickly abandoned as governments turned instead to the heralds of the self-regulating market, notably to the free-market economists of the Chicago School. Polanyi had believed, wrongly, that the 1930s had finally discredited his free-market opponents. Instead, Hayek, Friedman and others, working through a network of think-tanks and organisations generously funded by various business groups, had kept these ideas alive until conditions changed in their favour in the 1970s (Block and Somers 2014: 19).

The next 40 years would see the increasingly global dominance of free-market economics and the "neoliberal" ideology underpinning it. Market forces, supported by powerful states in the developed economies, were unleashed across the world – from 1991 under the imprimatur of "globalisation". In the advanced economies, this was necessarily accompanied by reduced social spending, falling real wages and a progressive weakening of trade unions and other working class organisations. For their part, "economists … vigorously promoted the doctrine that 'deregulation', labour 'flexibility', tax cuts for the wealthy, and unfettered free markets would produce unprecedented prosperity" (Block and Somers 2014: 1).

> The rise of the neoliberal counter-revolution following the economic and political turmoil of the 1970s is too well known to require elaboration. Suffice it to say that the golden age of the nineteenth-century liberal order was refurbished and presented as an inevitable trend toward a globalized world of winners and losers: a world of intensified economic competition requiring the subordination of all aspects of social and cultural life to economic success, by individuals and by nations.... Neoliberal policy prescriptions include disinflationary monetary policies, deficit reduction, the partial dismantling of the welfare state, flexible labour markets and independent central banks.
>
> (Polanyi Levitt 2013: 56)

As Kari Polanyi Levitt notes, reference to a golden past was deliberate:

> In the *World Development Report 1995*, the World Bank invited developing countries to return to the "golden age" of earlier globalization of 1850–1900. "Globalisation is unavoidable … whether a new golden age arrives for all depends mostly on the responses of individual countries".
>
> (2013: 70, fn. 3)

With a few notable exceptions (of which much was made by market enthusiasts), the experience of most developing countries resembled less a golden age, more a "stark utopia". The weaker the developing country's bargaining power, the harsher was the market medicine demanded by its creditors through the IMF and World Bank. Structural Adjustment Programmes were imposed on debtor countries requiring swingeing cuts to most social spending, large-scale currency devaluations, the shift of production from domestic to export markets, the enforced privatisation of public assets and even the surrender of any real semblance of economic sovereignty.

> Countries were judged by their success in implementing market-oriented reforms with little regard for the social consequences. The patient was considered to be improving so long as the medication was administered as prescribed. The means became ends. Economics became ideology.
>
> (Polanyi Levitt 2013: 56)

The unregulated market produced increasing inequality in all societies, rich and poor, with many experiencing unemployment levels unseen since the Great Depression. It resulted in uncontrolled financial flows that endangered banking systems everywhere but particularly those in poor countries where large flows of short-term loans increased both indebtedness and capital flight. It removed huge numbers of the young and the poor from access to education, health and housing. And, through deregulation, it intensified environmental breakdown. Furthermore, the wave of counter-movements it generated demonstrated – with only a handful of exceptions – a turning away from democratic values towards authoritarianism, ultra-nationalism, protectionism, xenophobia and intolerance. Globalisation made the phenomenon even more widespread than it had been in the 1930s, giving it a new respectability even in the core democracies. The parallels with the world Polanyi analysed are glaring; as Joseph Stiglitz noted in his foreword to the

2010 edition of *The Great Transformation*, "it often seems as if Polanyi is speaking directly to present day issues" (2010: ii).

Indeed, from the 1980s on, this stark utopia encouraged a resurgence of interest in Polanyi's work across a number of fields – for example on global markets (Lacher 1999; Kirby 2002) and contemporary working class struggles (Munck 2004; 2006) – as well as explorations of the relationship between Polanyi's ideas and aspects of Marxism – for example, concerning counter- movements (Silver and Arrighi 2003) and, most interestingly, comparisons between Polanyi's and Gramsci's approaches to democratic society (Birchfield 1999; Burawoy 2003). Debate has also been joined about how to "read" Polanyi (Block and Somers 2014; Dale 2016). The resurgence has been vigorous enough to elicit this reaction from one American economics professor in June 2008:

> History has not been kind to Polanyi's prognostications. Free market capitalism is a resilient and stable system in much of the world – particularly in English-speaking countries.... The great puzzle of Polanyi's book is thus its enduring allure, given the disconnect between his predictions and modern realities.
>
> (Dale 2010: 1)

Such hubris must surely have been severely tested just three months later, on 18 September 2008, when Lehman Brothers filed for bankruptcy and the Western financial system plunged into the deepest crisis in its history. As in the 1930s, governments, in rich and poor economies alike, imposed severe austerity on their people. And, as in the 1930s, market ideologues blamed the crisis on attempts to interfere with the self-regulation of the market and even on an excess of democracy:

> Even while emphasizing the "stark utopianism" of the self-regulating market, Polanyi's most passionate argument against free market ideology is based on the limits it places on the scope of democratic politics. The hegemony of a self-regulating economy free of government "meddling" is, in the end, an assertion that the preferences of voters in democratic politics must be ignored when they conflict with the logic of a self-correcting market economy. And, in fact, free market theorists such as Milton Friedman and Paul Weyrich have argued explicitly for a legal framework that would prevent voters from taking any actions that "undermine" the autonomous economy.... free market ideologues want a set of institutions that

> will … overrule any decisions that interfere with what is not by accident called *free* enterprise.
>
> (Block and Somers 2014: 41)

How influential this renewed attention to *The Great Transformation* will ultimately prove to be remains to be seen. The book is dense and the argument complicated, in sharp contrast to the easy polemics of Hayek's *The Road to Serfdom*, which was published in the same year. Indeed, it is instructive to compare Polanyi's fortunes with those of Hayek in the years since 1944. Backed by prominent free-market advocates like Robbins and von Mises, and by the private sector, Hayek was eventually able to establish himself at the Chicago School and, armed with a Nobel Prize, to shape, alongside Friedman, much of the economic narrative from the mid-1970s. Polanyi enjoyed no such good fortune and could bring few financial and organisational resources to the battlefield. If free-market ideology is again in disrepute, it nevertheless continues to be the mantra of business, media and government alike. Nevertheless, Polanyi's seminal contribution to economic history and theory remains both an ongoing challenge to the market ideology and the disintegrative austerity it unleashes against society and environment, and a profound rallying point for those struggling to defend democracy, development and community.

Major works

Polanyi, K. (1944) *The Great Transformation: The Political and Economic Origins of Our Time*, Boston, MA: Beacon Press.

—— (1947) "Our Obsolete Market Mentality", *Commentary* 3 (Feb.): 109–117.

—— (1966) *Dahomey and the Slave Trade*, Seattle, WA: University of Washington Press for the American Ethnological Society.

—— Arensberg, C.M. and Pearson H.W. (eds) (1957) *Trade and Market in the Early Empires*, Glencoe, IL: Free Press, especially "The Place of Economies in Society", 239–242; "The Economy as Instituted Process", 243–270.

Further reading

Birchfield, V. (1999) "Contesting the Hegemony of Market Ideology: Gramsci's 'Good Sense' and Polanyi's 'double movement'", *Review of International Political Economy* 6(1): 27–54.

Block, F. and Somers, M. R. (1984) "Beyond the Economistic Fallacy: The Holistic Social Science of Karl Polanyi", in T. Skocpol (ed.), *Vision and Method in Historical Sociology*, Cambridge: Cambridge University Press, 47–84.

—— —— (2014) *The Power of Market Fundamentalism: Karl Polanyi's Critique*, Cambridge, MA: Harvard University Press.

Burawoy, M. (2003) "For a Sociological Marxism: The Complementary Convergence of Antonio Gramsci and Karl Polanyi", *Politics and Society* 31(2): 193–261.

Dale, G. (2010) *Karl Polanyi: The Limits of the Market*, Cambridge: Polity Press.

—— (2016) *Reconstructing Karl Polanyi: Excavation and Critique*, London: Pluto Press.

Dalton, G. (1961) "Economic Theory and Primitive Society", *American Anthropologist* 63(1): 1–25.

—— (1968) "Introduction", in G. Dalton (ed.), *Primitive, Archaic and Modern Economies: Essays of Karl Polanyi*, Boston, MA: Beacon Press, ix–liv.

Hayek, F.A. (1944) *The Road to Serfdom*, London & New York: George Routledge & Sons.

Holmes, C. (2018) *Polanyi in Times of Populism: Vision and Contradiction in the History of Economic Ideas*, Abingdon and New York: Routledge.

Kirby, P. (2002) "The World Bank or Polanyi: Markets, Poverty and Social Well-Being in Latin America", *New Political Economy* 7(2): 199–219.

Lacher, H. (1999) "The Politics of the Market: Re-reading Karl Polanyi", *Global Society* 13(3): 313–326.

MacIver, R.M. (1957) "Foreword", in K. Polanyi, *The Great Transformation*, new edn, Boston, MA: Beacon Press, ix–xii.

Munck, R. (2004) "Globalization, Labor and the Polanyi Problem", *Labor History* 45(3): 251–269.

—— (2006) "Globalization and Contestation: A Polanyian Problematic", *Globalizations* 3: 175–186.

Polanyi Levitt, K. (2013) *From the Great Transformation to the Great Financialization: On Karl Polanyi and Other Essays*, London: Zed Press/ Halifax, Nova Scotia: Fernwood.

Silver, B.J. and Arrighi, G. (2003) "Polanyi's 'Double Movement': The *Belle Époques* of British and U.S. Hegemony Compared", *Politics and Society* 31(2): 325–355.

Simon, D. (2012) "To and Through the UK: Holocaust Refugee Ethnographies of Escape, Education, Internment and Careers in Development", *Contemporary Social Science* 7(1): 21–38.

—— (2019) *Holocaust Escapees and Global Development: Hidden Histories*, London: Zed Books.

Stiglitz, J. (2010) "Foreword", in K. Polanyi (ed.), *The Great Transformation*, reissued edn, Boston, MA: Beacon, vii–xvii.

MORRIS SZEFTEL

RAÚL PREBISCH (1901–1986)

Economist, teacher, central banker, civil servant and institution builder, Raúl Prebisch was a pragmatic but persistent advocate of a more egalitarian system of international trade between the more developed countries of the North (which he referred to as the "centre") and the less developed countries of the South (the "periphery"). In economic circles and beyond, he is best known for what has been labelled as the "Prebisch–Singer" thesis concerning the terms of trade between the developed and developing countries. He is certainly the most influential Latin American development economist and probably its most eminent.

Prebisch was born on 17 April 1901 in Tucumán, Argentina, and died on 29 April 1986, shortly after his 85th birthday, in Santiago, Chile. His father was of German origin and his mother came from a distinguished Argentinian family. He studied economics at the University of Buenos Aires, the main national university, graduating in 1922. From 1925 to 1948 he lectured intermittently in political economy at the School of Economics in the same University. From 1930 to 1935 he was initially Under-Secretary of Finance and Agriculture and later the adviser to the Ministers of Finance and Agriculture. More importantly, he was one of the principal founders of Argentina's Central Bank, becoming its first Director in 1935, a position he held until political changes forced him to resign in 1943. He was then invited to assist the Mexican Central Bank and went on to advise other central banks in Latin America. In 1949 he was hired as a consultant by the recently established UN Economic Commission for Latin America (ECLA),[1] located in Santiago de Chile, only to be catapulted the following year into the headship of the organisation with the title of Executive Secretary, a position he kept until 1962. From 1962 to 1964 he served as Director-General of the UN's Latin American Institute for Economic and Social Planning (ILPES), also based in Santiago, and which he was instrumental in establishing.

Before joining ECLA, Prebisch was already one of the most highly regarded economists in Latin America. But his rapid rise to the top has much to do with a document he wrote a few months after his arrival and which he presented to the institution's conference in 1949. This took place in Havana in front of a select audience, including the various country representatives of the organisation. This magisterial work, *The Economic Development of Latin America and its Principal Problems*,[2] created quite a stir among the delegates, who were most impressed not only by its ideas but also by Prebisch's forceful personality, persuasiveness and vision. Years later, the prominent development economist **Albert Hirschman** (1961: 13), was to refer to this document as "ECLA's manifesto". It urged Latin American countries to industrialise, and launched an attack on the international division of labour in which the central countries specialised in the production of manufactures while the peripheral countries specialised in the production of raw materials such as agricultural products and minerals. His proposal was highly unusual at the time; even more so was his critique of the orthodox international trade theory, which had hardly been questioned before.

Conventional international trade theory, based on David Ricardo's comparative advantage theory, argued that the current economic specialisation of the developed countries in the production of industrial commodities and of the developing countries in primary commodities worked to their mutual benefit as each enjoyed comparative advantages in these respective areas. Furthermore, this theory argued that international trade would help to reduce the income gap between rich and poor countries. Prebisch, however, observed in 1949 that incomes grew faster in the centre than in the periphery. The main reason for this widening gap was, in his view, the secular deterioration of the prices of primary goods in comparison with industrial products in the world market. To support this argument he relied on historical price data from a 1948 UN study, which did not indicate authorship but had been written by **Hans Singer**, who was then working at the UN in New York. This unexpected long-run tendency for prices of primary products to deteriorate relative to the prices of manufactured goods meant that the periphery had to export an increasing quantity of raw materials in order to continue importing the same amount of industrial commodities. Thus, Prebisch argued, the prevailing international trade system, by confining the periphery to the production of primary commodities, favours mainly the centre. He set out to explain the reasons for this trend, as it contradicted existing international trade theory.

His thesis regarding the deterioration of the periphery's terms of trade reveals Prebisch's distinctive approach to the problems of development and underdevelopment as compared with neoclassical theories. Prebisch's (1959; 1964) analysis of this process deals with both demand and supply conditions of commodity markets. On the demand side, Prebisch argues that the terms of trade deteriorated against the periphery because of the different income-elasticity of demand for imports by the centre and periphery or, as he put it, due to the "dynamic disparity of demand" between centre and periphery. This means that the centre's imports of primary products from the periphery rise at a lower rate than its national income, while the periphery's imports of industrial commodities from the centre grow at a faster rate than its income. As for the supply arguments, or the "cycle version" of the periphery's deterioration of the terms of trade, these are related to the differential effect of the world economic cycle on centre and periphery. The capitalist economic system evolves in a cyclical fashion, with the centre being the initiator of these cycles, which provoke an adaptive cyclical response by the periphery. During an economic upswing, the terms of trade generally turn in favour of primary producers but during a downswing they turn against them to an even greater degree. This results in the long-run deterioration of the periphery's terms of trade, especially as the downswings tend to last longer than the upswings.[3]

This differential impact of the world economic cycle on centre and periphery is explained by the differential behaviour of prices, profits and wages in both the centre and periphery during the phases of the cycle. During an economic upswing, wages and prices grow substantially in the centre while they hardly rise in the periphery because of the availability of surplus labour. Meanwhile during a downswing, the fall in wages and prices at the centre is limited due to the existence of trade union power and the oligopolistic structure of industry. In the periphery, by contrast, the downswing greatly reduces prices and wages, as producers are able to compress wages substantially on account of surplus labour and the nonexistence and/or weakness of trade unions.

Although Prebisch did not discover independently that the terms of trade of primary products were in secular decline, as he relied on the previous study of **Singer**, he was the first to provide an economic-cycle analysis of this phenomenon as well as setting it within his wider "centre – periphery" framework. In his seminal essay, **Singer** (1950) partly advanced arguments similar to those of Prebisch, although his analysis was done independently. Thus, the hypothesis on the deterioration in the terms of trade is known in the economic literature as the "Prebisch–Singer hypothesis". In condemning the deterioration of the commodity terms of trade, Prebisch was not arguing against international trade, nor did he ever suggest delinking as did some *dependencia* advocates like **Andre Gunder Frank** and **Samir Amin**. On the contrary, he saw international trade and foreign capital as essential elements for raising productivity and growth in the periphery. The Prebisch–Singer hypothesis created much controversy and was much criticised, particularly by neoclassical economists. But the consensus is that it has stood the test of both time and new statistical techniques remarkably well, although the reasons for it may have shifted. Thus even today, barring a few exceptional cases, the gains from trade continue to be distributed unequally between those countries exporting mainly primary products and those exporting mainly manufactures.

Prebisch was the guiding light of ECLA, which under his inspirational leadership became one of the most notable UN regional organisations. It had a major influence on development thinking and policy, especially in Latin America, mainly during its heyday of the 1950s and 1960s. The "ECLA manifesto" became the cornerstone of what was to become known as "structuralism" or the Latin American structuralist school of development. Prebisch assembled at ECLA a first-rate team of mostly young social scientists, but mainly economists, from various Latin American countries who all made, to a greater or lesser extent, a contribution to the development of structuralism and the core–periphery model. The structuralist view of development stresses the crucial significance of the differences in the economic, social and political structures between the core and peripheral countries. This core–periphery model argues that the world economy is divided into an industrial centre and an agrarian periphery. While the core or developed countries have an internally homogeneous structure with similar levels of productivity in different sectors, the periphery or underdeveloped countries are characterised by heterogeneity, as the various sectors of the economy have major differences in productivity arising from different levels of technology. This duality is also replicated within each of the economic sectors. Thus the periphery is characterised by "structural heterogeneity", or internal inequality, which is reinforced by international trade. Furthermore, unequal exchange between the centre and the periphery also reinforces the external inequality between the core and the periphery. Structuralism is thus one of the first attempts to construct a theory of development and underdevelopment from the perspective of the global South.

This structuralist analysis led Prebisch to propose a new development strategy for the periphery, in particular Latin America, so as to counteract, and possibly overcome, the negative impact of unequal exchange on the periphery. It entailed a radical change in the export-led development model based on primary products, or what he called the "outward-oriented" development path, which most peripheral countries had been following. In its place he argued for a new "inward-oriented" development strategy based on a process of import-substitution-industrialisation (ISI). This required protectionist measures such as a set of duties on manufacturing imports as well as a tax on primary exports, so as to induce a switch of resources from primary production for exports to industrial activities which at first would be directed at the domestic market but later would also be geared to exports. He also proposed allowing union activity in the primary export sector to push up wages, to defend primary commodity prices through concerted international action and to press for the reduction or elimination of protection for primary commodities in the centre (Prebisch 1959: 263). The main thrust of his argument was aimed at changing the periphery's structure of production and developing an industrial sector which would lead to higher rates of productivity growth and to a greater ability to retain the "fruits of technological progress", leading thereby to higher rates of growth and incomes which could reduce the gap between the core and periphery. While the ISI process did not fulfil all of its promises, which Prebisch was one of the first to point out, he insisted that industrialisation was a key route to development of the periphery. He was also a firm believer in the importance of the state for achieving development.

It was only natural for Prebisch to move from ECLA to the wider stage. His mission for a fairer international economic system led him to propose the creation of a new UN agency. On the strength of his report to the UN Secretary-General (Prebisch 1964), as well as his lobbying skills, the UN Conference on Trade and Development (UNCTAD) was established in 1964. He was appointed its first Secretary-General, a post he held until 1969. In his report he advocated a fairer international trading system which would tackle the problem of primary commodity prices and assist developing countries to shift to industrial exports by tariff preferences in the developed countries, among other measures (Toye and Toye 2004). Prebisch's ideas had a major influence in shaping the demands of many developing countries who were proposing a New International Economic Order (NIEO) in the 1970s.

Upon his "retirement" he remained closely associated with ECLA and ILPES in Santiago. It was only in the last years of his prolific life that Prebisch set out to develop systematically his ideas on what he termed "peripheral capitalism" in a series of articles originally published in *CEPAL Review*, the journal of ECLA, which he helped to create in 1976. He became its first Director, a position he held for several years. These articles form the basis of what was to become his last major work (Prebisch 1981). Although schooled in the neoclassical tradition, his acute mind and pragmatic sense soon led him to question it. His ideas shaped structuralism and had a major influence on dependency and world systems theory. The last years of his life coincided with the rise of neoliberalism, of which he became a major critic. As an institution builder and activist, he was able to transform some of his ideas into action, but his vision for a fairer international economic system remains to be fulfilled.

Notes

1 Now the Economic Commission for Latin America and the Caribbean (ECLAC).
2 The document was published in English a year later: see Prebisch (1950).
3 *Economic Survey of Latin America 1949*, New York: United Nations, 1951. Although the authorship is assigned to the Secretariat of the Economic Commission for Latin America (ECLA), it is known that Prebisch wrote the whole of Part One, entitled "Growth, Disequilibrium and Disparities: Interpretation of the Process of Economic Development" comprising 5 chapters, pp. 3–87.

Major works

Prebisch, R. (1950) *The Economic Development of Latin America and its Principal Problems*, Lake Success, NY: United Nations. (Trans. of *El desarrollo económico de la América Latina y algunos de sus principales problemas*, Santiago: Comisión Económica para América Latina (CEPAL), 1949.)

—— (1959) "Commercial Policy in the Underdeveloped Countries", *American Economic Review* 49(2): 251–273.

—— (1963) *Towards a Dynamic Development Policy for Latin America*, New York: United Nations.

—— (1964) *Towards a New Trade Policy for Development: Report by the Secretary General of the United Nations Conference on Trade and Development*, New York: United Nations.

—— (1976) "A Critique of Peripheral Capitalism", *CEPAL Review* 1: 9–76.

—— (1981) *Capitalismo periférico: crisis y transformación*, Mexico City: Fondo de Cultura Económica.

—— (1984) "Five Stages in My Thinking on Development", in G.M. Meier and D. Seers (eds), *Pioneers in Development*, Oxford: Oxford University Press, 175–190.

Further reading

Dosman, E.J. (2008) *The Life and Times of Raúl Prebisch, 1901–1986*, Montreal: McGill-Queen's University Press. (This is the most outstanding, detailed and captivating biographical study of Raúl Prebisch.)

Furtado, C. (1988) *La fantasía organizada*, Buenos Aires: Editorial universitaria de Buenos Aires. (This biographical essay by **Celso Furtado**, who was an ECLA insider and is one of Latin America's most original development thinkers, provides a fascinating insight into the origins and developments of Prebisch's ideas at ECLA.)

Gurrieri, A. (ed.) (1982) *La obra de Prebisch en la CEPAL*, 2 vols, Mexico City: Fondo de Cultura Económica.

Hirschman, A. (1961) "Ideologies of Economic Development in Latin America", in A. Hirschman (ed.), *Latin American Issues: Essays and Comments*, New York: Twentieth Century Fund.

Love, J.L. (1980) "Raúl Prebisch and the Origins of the Doctrine of Unequal Exchange", *Latin American Research Review* 15(3): 45–72.

Rodríguez, O. (1980) *La teoría del subdesarrollo de la CEPAL*, Mexico City: Siglo Veintiuno Editores.

Singer, H.W. (1950) "The Distribution of Gains Between Investing and Borrowing Countries", *American Economic Review* 40(2): 473–485.

Toye, J. and Toye, R. (2004) *The UN and Global Political Economy: Trade, Finance and Development*, Bloomington, IN: Indiana University Press.

Various authors (2001) "Tribute to Raúl Prebisch", *CEPAL Review* 75: 1–108.

CRISTÓBAL KAY

ANÍBAL QUIJANO (1928–2018)

Aníbal Quijano was a Peruvian sociologist who has gained respect among scholars, politicians and activists interested in decolonial[1] alternatives to development. He was a key contributor to debates around dependency theory (see also the essays on **Samir Amin**, **Fernando Henrique Cardoso**, **Andre Gunder Frank**, **Celso Furtado**, **Gunnar Myrdal** and **Raúl Prebisch**), though he was best known for introducing the concept of the "coloniality of power".

Aníbal Quijano was born in 1928 in Yanama, a district in the Peruvian Andes that forms part of Yungay Province. Not much has been written about Quijano's biography.[2] His father, Marcial Quijano, founded and directed the first primary school in Yanama, and dedicated his life to supporting local indigenous communities in their everyday struggles. The work of his father most certainly inspired Aníbal Quijano, who throughout his academic career wrote prolifically about indigenous peoples in Latin America and, in more general terms, about ethnicity and racism as key features of global capitalist development (1964; 1993; 2005).

Quijano moved to Lima in 1948 to study at the University of San Marcos. For his Bachelor's degree he enrolled into a variety of courses on Latin American history, anthropology, sociology and law. He later obtained a master's degree at the Latin American Social Sciences Institute (FLACSO) in Santiago de Chile. He subsequently returned to Lima to work as Professor at the University of San Marcos and to study for a PhD, which he completed in 1964. His PhD focused on the emergence of *cholos* (rural-to-urban indigenous migrants) in Peru, and centred on one of his lasting interests, the shifting roles of ethnic identity in Latin America (Quijano 1964).

Quijano's studies and work as young professor took place in difficult political times. In 1948, the year he arrived in Lima, a military regime led by Manuel Odría took control of Peru's national government. Opposing this regime, Quijano joined the student union movement and, because of his political activism, was prosecuted and jailed for one month. These events radicalised Quijano and, inspired by well-known Peruvian philosopher José Carlos Mariátegui, he self-taught in Marxism. This led him to adapt a neo-Marxist and, later, decolonial position which centred around the argument that social transformations should evolve organically in Latin America on the basis of local conditions, and not by applying a Eurocentric Marxist approach that focuses on class-based revolutions (Quijano 2000; 2007).

Quijano remained Professor in the University of San Marcos until 1995, though he left Lima between 1965 and 1971 to work as researcher at the United Nations Economic Commission for Latin America and the Caribbean (ECLAC) in Santiago de Chile, which in this period was an intellectual hub for dependency thinking under the dynamic team assembled by **Raúl Prebisch**. From 1986, Quijano also worked as Professor at the State University of New York at Binghamton. In 2010 he founded the Department of Latin America and the Coloniality of Power at the University Ricardo Palma in Lima. Quijano's career was further punctuated by a series of visiting appointments at universities in Latin America, Europe and the United States. He was also co-founder of the interdisciplinary "modernity/coloniality" research group which unites prominent decolonial scholars such as Enrique Dussel, **Arturo Escobar**, Ramón Grosfoguel and Walter Mignolo.

Altogether, Quijano published more than 200 articles and books. His most important works (Quijano 1974; 1975; 2000; 2007; 2016; Quijano and Wallerstein 1992), have been translated into English. To date, though, most of his *oeuvre*, including key works on identity, ethnicity and racism (Quijano 1964; 1993), modernity (Quijano 1988; 1991) and marginality and dependency (Quijano 1966; 1967; 1969; 1977; Quijano and Wefford 1973), is mainly available in Spanish. This has limited his exposure in the Anglophone world. His most important Spanish essays are published in a volume edited by Daniel Assis Clímaco (2014).

In the discipline of development studies, Quijano is perhaps best known for two contributions. The first relates to his critique of modernisation theory and its conceptualisation of countries as autonomous entities which develop over time by passing through similar stages (see also Kay 1991). According to Quijano (1966; 1975; 1977), such a unilinear perspective fails to acknowledge the centre–periphery relationships between marginal and integrated sectors within countries and, more importantly, between undeveloped and developed countries in the capitalist world-system. This observation relates to his disillusionment with post-Second World War industrialisation in Latin American cities, which mainly benefited local elites and advanced countries in the West while failing to absorb and integrate local labour reserve armies. For Quijano, then, processes of underdevelopment and development are interdependent, and constitute each other through relational processes. He therefore proposed a historical and structural approach for the study of relationships of dependence:

> The various structural orders are not only interrelated among themselves, they are also related with and through society as a whole. Hence, a given society must not be studied in isolation but in relation to all others with which it is linked and which form with it a specific unit within the world system of interdependence that is emerging. … In other words, national societies in Latin America are characterized by belonging to a specific interdependent unit and, within that unit, by the fact that they are dependent on other societies.
>
> (Quijano 1975: 133–134)

In Latin America, Quijano's writings on these topics are considered to be key contributions to the neo-Marxist school of dependency theory (Grosfoguel 2000; Kay 1991). They are less known elsewhere. Instead, in the English-speaking world, work on dependency theory is mainly associated with some of Quijano's contemporaries, particularly **Andre Gunder Frank** and **Fernando Henrique Cardoso**.

Quijano's second most notable intervention relates to the "coloniality of power" (key texts in English include Quijano 2000; 2007; Quijano and Wallerstein 1992).[3] He developed this body of work in the late 1980s and early 1990s, a period dominated by the collapse of the Soviet Union and the apparent political triumph of the West. It was also a period in which governments and new social movements in seemingly underdeveloped countries of the crumbling "Third World" started questioning politico-economic conditionality imposed by "advanced" Western countries. Finally, and perhaps most important for Quijano, in this period Latin American indigenous movements emerged which challenged prevailing ethno-racial hierarchies and associated patterns of exclusion (Quijano 2005).

Quijano's "coloniality of power" explores the historical roots of present ethno-racial hierarchies as well as their relationship with racialised patterns of global capitalist development. He claims that ethno-racial hierarchies were, for the first time in world history, established by the European colonisers after the conquest of the Americas in the early 16th century (Quijano 1993). Following the Spanish model of blood politics, the colonisers divided the peoples inhabiting the Americas into three newly established ethno-racial groups: "Indians" – referring to the different native peoples, "blacks" – referring to slaves brought to the Americas from Africa, and "whites" – referring to people who arrived in the Americas from Europe (Quijano 1993).

Another key proposition of Quijano's "coloniality of power" relates to how the European colonisers used this ethno-racial hierarchy as basis for the imposition of a systematic racist division of labour. "Indians" entered serfdom, "blacks" worked as slaves, while "whites" were recognised as independent producers of commodities, entitled to engage in wage labour and commercial exchange. At the same time, being "Indian" or "black" was also associated with backwardness, tradition, rurality and a lack of citizenship rights, while being "white" was associated with a modern urban life and citizenship status. Hence, being "Indian" or "black" meant to be set apart and excluded economically, socially, politically, spatially and culturally from "whites". Quijano argues that by imposing this ethno-centric system of domination and exploitation, the colonisers managed to gain full control not only of colonised subjects but also of global commercial traffic, with the latter leading to regional economic expansion at home (Quijano 1993). Accordingly, he claims that capitalist modernity arose in Europe as a consequence of the colonisation of the Americas and, not as argued by **Karl Marx**, because of the rise of the bourgeoisie as a revolutionary class in Europe (Quijano 2000; 2007).

Perhaps Quijano's most important claim is that social, political, economic and cultural hierarchies established during the colonial conquest outlived colonialism and continue until the present. According to Quijano (2000: 539), this is evident in the continuity of postcolonial power relations which allow "white" elites to govern and exclude "non-white" groups; it is also visible in the "attitude spread out among the white property owners of any place in the world" and in the "lower wages inferior races receive in the present capitalist centers for the same work as done by whites".

A crucial implication of the "coloniality of power", then, is that our contemporary world requires decolonisation. For Quijano (2007: 177–178), decolonisation entails a liberation from European ethnocentric power relations "organized as inequality, discrimination, exploitation and as domination" and "implies the freedom of all people to choose, individually and collectively, such relations: a freedom to produce, criticize, change, and exchange culture and society".

Aníbal Quijano was regarded as one of Latin America's foremost decolonial thinkers. His work on decolonisation influenced and shaped activist debates around possibilities for another world during the World Social Forum in Porto Alegre, Brazil, in 2001. Within the world of academia, a group of predominantly Latin American scholars associated with the "modernity/coloniality" research group have applied and refined his concept of the "coloniality of power". For example, Argentine feminist María Lugones (2007) integrated gender into debates around the concept, and highlighted that this social category, like ethnicity and race, is used for the creation of social hierarchies based on domination, exploitation and violence. From a different perspective, Walter Mignolo (2005; 2012) understands the "coloniality of power" as leading not only to the exploitation of specific communities but also to the displacement of the knowledge of these communities. He therefore argues that it is important to recover and re-articulate such knowledge in scholarly writings and through development alternatives. Building on this proposition, in recent years a group of authors, including Aníbal Quijano (2016) himself but also **Arturo Escobar** (2010; 2018) and Catherine Walsh (2008), examined existing development alternatives that emerge in some Latin American countries and which build on the knowledge of historically marginalised indigenous communities. These authors thereby outline a plurality of potential pathways towards what Quijano would consider decolonisation.

Aníbal Quijano died on 31 May 2018. His legacy is a fascinating body of work on the economic, sociocultural, and political development of Latin America, and especially his home country, Peru. In particular, though, it is his *oeuvre* on the coloniality of power, Eurocentrism and decolonisation which will continue to receive critical attention by scholars and activists, as these concepts might help to illuminate and critically analyse the struggles for inclusion, diversity and pluralism that characterise our time.

Notes

1 It is important to draw distinctions between postcolonial and decolonial scholarship (for a more detailed discussion see Bhambra 2014; Mignolo 2007). Both emerged within cultural studies but initially referred to different geohistorical contexts. They have since spread across disciplines and world regions. Postcolonial scholarship emerged within Asia and the Middle East, and treats the colonization of these regions in the 19th and 20th centuries as an analytical starting point. Prominent advocates include Edward Said, Homi Bhabha and Gayatri Spivak. In contrast, decolonial scholarship, as promoted by members of the "coloniality/modernity" research group such as Aníbal Quijano but also Enrique Dussel, **Arturo Escobar** and Walter Mignolo, treats the 16th-century colonisation of the Americas as an analytical starting point.

2 The biographical details provided here derive from Danilo Assis Clímaco's (2014) prologue to his *Aníbal Quijano. Cuestiones y horizontes.*

3 For a much longer list of key writings on the "coloniality of power" in Spanish, see Assis Clímaco (2014).

Major works

Quijano, A. (1964) *La emergencia del grupo cholo y sus implicaciones en la sociedad peruana*, PhD dissertation, University of San Marcos, Lima.

—— (1966) *Notas sobre el concepto de "marginalidad social"*, Santiago de Chile: ECLAC.

—— (1967) "Urbanización, cambio social y dependencia", in F.H. Cardoso and F. Wefford (eds), *América Latina. Ensayos de interpretacion sociologica*, Santiago de Chile: ECLAC.

—— (1969) *Redefinición de la dependencia y proceso de marginalización en América Latina*, Santiago de Chile: Universidad de Chile.

—— (1974) "Marginal Pole and Marginal Labor Force in Latin America", *Economy and Society* 1: 393–428.

—— (1975) "The Urbanization of Latin American Society", in J. Hardoy (ed.), *Urbanization in Latin America: Approaches and Issues*, New York: Anchor Books.

—— (1977) *Imperialismo y "marginalidad" en América Latina*, Lima: Mosca Azul.

—— (1988) *Modernidad, identidad y utopía en América Latina*, Lima: Ediciones Sociedad y Política.

—— (1991) "La modernidad, el capital y América Latina", *ILLA: Revista del Centro de educación y cultura* 10.

—— (1993) "Raza, etnia y nación. Cuestiones abiertas", *Estudios Latinoamericanos* 3.

—— (2000) "Coloniality of Power, Eurocentrism, and Latin America", *Nepantla: Views from South* 1(3): 533–580.

—— (2005) "El 'movimiento indígena' y las cuestiones pendientes en América Latina", *Revista Tareas* 119: 31–62.

—— (2007) "Coloniality and Modernity/Rationality", *Cultural Studies* 21(2): 168–178.

—— (2016) "'Bien Vivir': Between 'Development and the De/Coloniality of Power'", *Alternautas* 3(1): 10–23.
—— and Wallerstein, I. (1992) "Americanity as a Concept, or the Americas in the Modern World-System", *International Social Science Journal* 44(4): 549–557.
—— and Wefford, F. (1973) *América Latina. Dependencia y desarrollo*, San Jose: EDUCA.

Further reading

Assis Clímaco, D. (2014) *Aníbal Quijano. Cuestiones y horizontes*, Buenos Aires: CLACSO.
Bhambra, G.K. (2014) "Postcolonial and Decolonial Dialogues", *Postcolonial Studies* 17(2): 115–121.
Escobar, A. (2010) "Latin America at a Crossroads: Alternative Modernizations, Post-liberalism, or Post-development?", *Cultural Studies* 24(1): 1–65.
—— (2018) *Designs for the Pluriverse: Radical Interdependence, Autonomy, and the Making of Worlds*, Durham, NC: Duke University Press.
Grosfoguel, R. (2000) "Developmentalism, Modernity and Dependency Theory in Latin America", *Nepantla: Views from South* 1(2): 347–374.
Kay, C. (1991) "Reflections on the Latin American Contribution to Development Theory", *Development and Change* 22: 31–68.
Lugones, M. (2007) "Heterosexualism and the Colonial/Modern Gender System", *Hypatia* 22(1): 186–209.
Mignolo, W. (2005) *The Idea of Latin America*, Oxford: Blackwell.
—— (2007) "Delinking: The Rhetoric of Modernity, the Logic of Coloniality and the Grammar of De-coloniality", *Cultural Studies* 21(2–3): 449–515.
—— (2012) *Local Histories/Global Designs: Coloniality, Subaltern Knowledges and Border Thinking*, Princeton, NJ: Princeton University Press.
Walsh, C. (2008) "Interculturalidad, plurinacionalidad y decolonialidad. Las insurgencies politico-epistemicas de refundar el Estado", *Tabula rasa* 9: 131–152.

PHILIPP HORN

JOAN ROBINSON (1903–1983)

Born on 31 October 1903 in Camberley, Surrey, in England in a family of professional soldiers and surgeons, Joan Robinson occupies a unique position among the development thinkers. She was an economic theorist *par excellence*, but she would insist on clarity about the social setting of the theory. In 1922, she chose to read economics at Girton College, Cambridge, to understand the persistence of poverty and unemployment, but was disappointed by the prevailing Marshallian thinking (Harcourt 2006: 166). She finished the Economics Tripos in 1925 but was not awarded the degree since she was a woman. This gender discrimination continued till late in her career. She could not get a formal academic position until 1931 and a professorship until 1965. Two of her students featured in this volume, **Amartya Sen** and **Joseph Stiglitz**, won Nobel Prizes in Economics. Despite strong recommendations in 1975, she could not be the first woman to receive the honour.

In 1926, she was directly exposed to "the realities of poverty in a feudal state in Central India" (Narasimhan 1983: 217). She had gone to Gwalior with Austin Robinson, whom she had married earlier in the year. A debate raged on the economic and financial relations between these quasi-autonomous states and British India. According to Austin, "Joan got all involved with the foreign secretary and others in the presentation of the Indian case for the Indian States, but I was very busy with the tutoring of the Maharajah" (Turner 1989). A two-part report was published (*The British Crown and the Indian States*, 1929). In his autobiography, the writer of the first historical part observed: "The second part of the book dealt with finance and was written by a scholar named MRS AUSTIN ROBINSON" (Panikkar 1979: 79).[1]

The British Crown and the Indian States (1929) deliberated on the dependency relationship between the British Crown and the princely states. Underdevelopment was portrayed as a precolonial condition in an overpopulated economy. Development came from the outside as a result of the political force of colonial intervention and the economic force of the international division of labour. The colonial state-built infrastructure, especially railways, and ensured security for the movement of goods. Machine- and money-based imports undersold village industries. The transition was from the village economy to the world economy. Princely states were backward relative to British India because their relative autonomy allowed the precolonial economic formation to persist longer.

Joan Robinson left India for Cambridge in 1928. The intellectual activism and the skills of debate and advocacy acquired in India were to be a characteristic feature of her future work. Each decade saw her as an intense participant in a great debate. While finishing a major microeconomic work challenging Marshallian orthodoxy on competition (Robinson 1933), she joined the Cambridge Circus that would lead to the Keynesian Revolution in the 1930s. Unemployment in the economy rather than the behaviour of the firm seized her attention. Disguised unemployment, a concept appropriated later by development economists to reflect surplus labour (Tahir 1990a), was her creation in a Keynesian context (Robinson 1936). Surplus labour models moved from disguised unemployment to internal migration determined by the chance to be employed. Joan Robinson was an important precursor of the models of Harris and Todaro (1970) (Tahir 1999). She showed not only that the chance to be employed determined labour mobility, but that there were probable effects of migration as well (Robinson 1937a; 1937b).

As Keynes broke taboos about the role of the state in achieving full employment, **Marx** drew Joan Robinson's interest (Robinson 1941: 2). The flavour of the argument in *The British Crown* (1929) tended towards the writings of Marx (1853) on India, though Joan had not yet read Marx. In the 1940s, her writings focused on Marx and unemployment, socialism, the role of the state, planning and economic dynamics. Robinson (1953: 55) set the stage for her work on capital accumulation in the 1950s and the publication of her *magnum opus*. Her conclusion – "the outputs of saleable goods [are] not co-extensive with economic wealth, let alone with the basis of human welfare" (Robinson 1956: 386) – contained early suggestions of the concept of human development pioneered by **Mahbub ul Haq** (1995), another of her former students featured in this book.

Unlike the development economists, who saw in it a way around the difficulty of reducing the abysmally low consumption in a developing economy, she did not assign disguised unemployment the function of relaxing the capital constraint, staying away from this easy but inflationary course. To her, the nature of unemployment in these economies was not Keynesian. Nor could Keynesian policies, therefore, be its remedy. It was rather a case of "Marxian unemployment", with available labour running ahead of capital equipment in existence. Based on the lesson learned from India about overpopulation, the focus was on containing population as well as developing the productive forces. She viewed industrialisation as the key to development, with capital accumulation as its main vehicle. Capital accumulation had to be ahead of population growth, ensured by an active state. Above all, there was a recurring requirement for more "exploitation", not less, exploitation being an unmistakable expression for producing re-investible surplus. Keynes's relevance was only in regard to the decision to invest.

Employment creation *per se* was not the objective of development. Higher employment per unit of output was only the obverse of lower output per unit of labour. The argument turned on the quality and adequacy of investment, not rural-end policies to check migration. In development economics, there was place only for the rural labour surplus. Joan Robinson's work contained suggestions for modelling migration when unemployment exists at the origin as well as destination, which is closer to the present reality of a rising proportion of wage labour and the possibility of open unemployment in the rural sector as well.

A strong socialist state, she believed, would be able to solve the problem of accumulation by eliminating all consumption out of profits and saving the poor masses from unbearable austerity. In a lecture in China, she worked out an intermediate labour-absorbing technique between the extremes of labour-intensive and capital-intensive techniques (Robinson 1957). This would achieve a high rate of accumulation without causing "a Poznan-type situation" (Harcourt and Kerr 2009: 148). She, therefore, watched developments in China and Communist-ruled Kerala state in India with a keen interest. She paid eight visits to China between 1953 and 1978, and would spend a month or two in Kerala every year from 1972 until her death in 1983. She never claimed to be an expert on Chinese development (Robinson 1977: 7). Her writings on China were titled as reminiscences, letters, notes, conversations, reports, etc. The only book she wrote was on the Cultural Revolution (1970b), published in a popular paperback series. Although her articles appeared in major economic journals, she never contributed any article on Chinese development. Yet she drew considerable criticism of being influenced by propaganda and becoming a fellow traveller (Turner 1989; Gewirtz 2017: 56–58).

Joan Robinson was torn between hope and reality. As a theorist, she spent a lifetime laying bare the cruel reality of capitalism. The recent re-occupation of centre-stage by the issue of extreme inequality (Piketty 2014) resonates with the Cambridge–Cambridge debate on capital theory (Harcourt 1972), in which she outwitted stalwarts like Paul Samuelson and Robert Solow. She showed that what was profitable was not necessarily right. In the capitalist system, there was no economic justification for profits. It resulted, not from the productivity of capital, but from the distribution of power in society. Income distribution was decided in politics, not economics. She had hoped there was a socialist alternative. By her own admission, however, she had been starry-eyed about Maoist China (Tahir 1990b; Tahir et al. 2002; Harcourt and Kerr 2009: 145).

Her last major lecture pinpointed the dangers of the arms race and the consequences for development:

> Keynes was thinking narrowly in terms of the problems of the industrialised West. Nowadays we are becoming conscious of the enormous needs of the impoverished Third World. In Africa in particular the growth of numbers ahead of agricultural development is posing a huge problem which from time to time comes to the surface in outright

famine here or there in that continent. There is no lack of need for investment or of know-how for designing it. The impediments that stand in the way are political and financial, not technical or economic.

(Robinson 1982: 276)

Note

1 On 4 May 1988, a letter from Professor William Brown, Faculty of Economics and Politics, University of Cambridge, informed the author of this essay about Sir Austin Robinson's recollection that "the major economics contribution was in fact his and not Joan Robinson's." Sir Austin never granted a promised interview to the writer of this essay. However, in an interview reproduced in the manuscript of Turner (1989) reviewed by this writer, Sir Austin had observed: "Joan got all involved with the foreign secretary and others in the presentation of the Indian case for the Indian States, but I was very busy with the tutoring of the Maharajah."

Major works

Robinson, J. (1933) *The Economics of Imperfect Competition*, London: Macmillan.

—— (1936) "Disguised Unemployment", *Economic Journal* 46: 225–237.

—— (1937a) *Introduction to the Theory of Employment*, London: Macmillan.

—— (1937b) *Essays in the Theory of Employment*, London: Macmillan.

—— (1941) "Marx on Unemployment", *Economic Journal* 51: 234–248.

—— (1951) *Collected Economic Papers*, vol. 1, Oxford: Blackwell.

—— (1953) *On Re-reading Marx*, Cambridge: Students' Bookshop.

—— (1955) "Marx, Marshall and Keynes", Delhi School of Economics Occasional Paper no. 9.

—— (1956) *The Accumulation of Capital*, London: Macmillan.

—— (1957) "The Problem of Choice of Technique for a Planned Economy with Limited Industrial Resources", JVR Collection iii/5.2, Cambridge: King's Modern Archives.

—— (1960) *Collected Economic Papers*, vol. 2, Oxford: Blackwell.

—— (1962a) *Economic Philosophy*, London: C.A. Watts.

—— (1962b) *Essays in the Theory of Economic Growth*, London: Macmillan.

—— (1965) *Collected Economic Papers*, vol. 3, Oxford: Blackwell.

—— (1966) *Economics: An Awkward Corner*, London: George Allen and Unwin.

—— (1970a) *Freedom and Necessity*, London: George Allen and Unwin.

—— (1970b) *The Cultural Revolution in China*, London: Penguin Books.

—— (1971) *Economic Heresies: Some Old-Fashioned Questions in Economic Theory*, London: Macmillan.

—— (1973) *Collected Economic Papers*, vol. 4, Oxford: Blackwell.

—— (1977) *Reports from China 1953–76*, London: ACEI.

—— (1979a) *Collected Economic Papers*, vol. 5, Oxford: Blackwell.
—— (1979b) *Aspects of Development and Underdevelopment*, Oxford: Blackwell.
—— (1982) "The Arms Race", in S.M. McMurrin (ed.), *The Tanner Lectures on Human Values*, Salt Lake City, UT: University of Utah Press, 257–289.

Further reading

Gewirtz, J. (2017) *Unlikely Partners*, Cambridge, MA: Harvard University Press.
Harcourt, G. C. (1972) *Some Cambridge Controversies in the Theory of Capital*, Cambridge: Cambridge University Press.
—— (2006) *The Structure of Post-Keynesian Economics: The Core Contributions of the Pioneers*, Cambridge: Cambridge University Press.
—— and Kerr P. (2009) *Joan Robinson*, Basingstoke: Palgrave Macmillan.
Harris, J.P. and Todaro, M.P. (1970) "Migration, Employment and Economic Development: A Two Sector Analysis", *American Economic Review* 60: 14–33.
Marx, K. (1979 [1853]) "The British Rule in India". (Repr. in K. Marx and F. Engels, *Collected Works*, vol. 12, London: Lawrence and Wishart, 1979.)
Narasimhan, S. (1983) "Joan Robinson: In the Radical Vein. A Laywoman's Homage", *Cambridge Journal of Economics* 7: 213–219.
Panikkar, K.M. (1979) *An Autobiography*, Delhi: Oxford University Press.
Piketty, T. (2014) *Capital in the Twenty-First Century*, Cambridge, MA: Harvard University Press.
Tahir, P. (1990a) "Some Aspects of Development and Underdevelopment: Critical Perspectives on Joan Robinson", PhD dissertation, University of Cambridge.
—— (1990b), "Making Sense of Joan Robinson on China", paper written as Joan Robinson Memorial Lecturer, mimeo.
—— (1999) "Joan Robinson: A Neglected Precursor of Internal Migration Models", in C. Sardoni and P. Kriesler (eds), *Keynes, Post-Keynesianism and Political Economy: Essays in Honour of Geoff Harcourt*, vol. 3, London and New York: Routledge, 312–333.
—— Harcourt, G. C. and Kerr, P. (2002), "On Joan Robinson and China", in P. Kerr (ed.), *Joan Robinson: Critical Assessments of Leading Economists*, vol. 5, Abingdon: Routledge, 267–280.
The British Crown and the Indian States (1929) London: P.S. King and Son.
Turner, Marjorie S. (1989) "Joan Robinson and the Americans", manuscript.
Ul Haq, M. (1995) *Reflections on Human Development*, Oxford: Oxford University Press.

PERVEZ TAHIR

WALTER RODNEY (1942–1980)

Walter Rodney was one of the most influential amongst critical writers on history, development and underdevelopment to have emerged from the Caribbean. His intellectual and political work has had worldwide impact. In particular, his book *How Europe Underdeveloped Africa* (first published in 1972) remains influential, and nearly four decades after his untimely death, he is still widely read. Rodney was killed by an explosion in his native Guyana on 13 June 1980. He was 38. In 2016, following a two-year investigation, a Commission of Enquiry blamed the Guyanese security forces.

Rodney was born in Georgetown, Guyana, in March 1942. At this time, one key aspect of the colonial economy was booming, with the allies consuming all that could be produced of British Guyana's main export, bauxite (the ore for aluminium). Two-thirds of Allied aircraft would be made from Guyanese ores. Yet wartime shortages of consumer goods and disruption to Atlantic shipping meant that many in Guyana had to struggle to make ends meet or found themselves out of work. Just months before Rodney was born, the Americans commenced the construction of an airbase close to Georgetown. This was one of eight sites in the Western hemisphere where the British permitted the USA to establish military facilities, following a September 1940 agreement between British Prime Minister Churchill and US President Roosevelt. In exchange Britain received 50 surplus American destroyers.

In other words, Rodney was born on the cusp of a shift from Britain and Europe to the USA in terms of where the centre of Western economic and political power was seen to reside. The relative eclipse of British empire and the rise of American superpower soon became evident. Emerging too were the struggles for national liberation in Africa and Asia, and against white supremacy in the United States. As Rodney entered his teens, the birth of the Third World was under way. All this took place amidst a new phase of global polarisation involving the contest between capitalism and communism and the rise of programmes and strategies for "development". These momentous geopolitical shifts and contests constituted the backdrop to his intellectual and political formation.

Whilst Rodney grew up in Guyana and entered high school in 1953, he travelled to Jamaica in 1960 to study at the University of the West Indies (then still the University College of the West Indies). He majored in history, then moved to London in 1963 for graduate studies at the history department of the School of Oriental and African Studies (SOAS). In 1966 he earned a doctorate for a thesis on "The History of the Upper Guinea Coast, 1545–1800". While at SOAS, Rodney participated in reading and study groups on **Marx**ism. Rodney was far from an orthodox Marxist, however, and he subscribed to no formal party line. At the same time as he made his way through the school and university system, other Caribbean intellectuals were formulating radical conceptions of the region's dependence, fragmentation and marginality and reconfigurations of racism. This work certainly influenced Rodney, although his key contribution was concerned with the related problem of African underdevelopment.

Rodney was an Afro-Guyanese from a working-class family, and African and Black liberation, along with class struggle, were the other inputs into his political, intellectual and personal development. According to Lewis's (1998: xvii) survey of Rodney's intellectual and political thought, we should:

> situate his intellectual and political evolution in the transatlantic diasporic locations of Guyana, Jamaica, London and Tanzania. These geographically dispersed locations were, of course, linked by the history of the British Empire and capitalism. What emerges from Rodney's work is not only a critique of empire and capitalism in general but a dissection of the domestic political elite that assumed political authority from the colonisers in Africa and the Caribbean as well as an analysis of the process of recolonisation.

On completion of his doctorate, Rodney took up a position in the history department of the University College of Dar es Salaam, Tanzania, then chaired by Terence Ranger (who later became a distinguished historian of Southern Africa). Tanzania was then embarking on a path of "African socialism" under **Julius Nyerere**. It proved a fertile environment for Rodney and other radical scholars engaged in debates on the making of Africa's underdevelopment. However, Rodney sought to return to the Caribbean with the ambition to develop African studies back in Jamaica. He was prevented from doing so by the Jamaican government, who banned him in 1968, after he returned there briefly at a time of considerable political turmoil. As part of a combined crackdown on radical intellectuals (and the banning of texts, such as the writings of Malcolm X) and on the grassroots Rastafari movement, Rodney was declared *persona non grata*. A collection of his reflections and exchanges in Jamaica was published in London in 1969 under the title *The Groundings with my Brothers*. In these Rodney (1969: 59) asserted:

> The Government of Jamaica, which is Garvey's [Marcus Garvey, born 1887 in Jamaica, died 1940 in the UK, was a founding figure of Black Liberation/Nationalism] homeland, has seen it fit to ban me, a Guyanese, a black man, and an African. But this is not very surprising because though the composition of that Government – of its Prime Minister, the Head of State and several leading personalities – though that composition happens to be predominantly black, as the Brothers at home say, they are all white-hearted.

Fortunately, Rodney was able to continue working in Dar es Salaam, where he stayed until 1974. The pull of the Caribbean remained, however, and in 1974 Rodney returned to Guyana. Although denied a university position there (on account of his growing differences with and critiques of Burnham's authoritarian government), he continued to produce academic writings, and relied on guest lecture tours in North America and Europe and external research grants to survive financially. By this time his international reputation had grown, in large part owing to the broad readership gained by his *How Europe Underdeveloped Africa*. This was, however, little protection against the agents of Burnham's regime who assassinated Rodney on 13 June 1980, largely in response to his political activity directed against the incumbent regime.

Rodney's most influential book is a variant of what elsewhere was known as dependency theory. It is thus marked by many of the controversies (and limits) of that perspective. Beyond Latin American conceptions of dependency, with their focus on class relations, extraction and imperialism, Rodney foregrounded race. On Rodney's return to Jamaica:

> News travelled fast about the young, radical and dynamic university teacher who could spellbind his audience, and "knew all about Africa". The dependency theorists at the university had just started an intensive public campaign and Rodney became the most sought-after lecturer at black, middle-class gatherings as well as at Rasta meetings in the slums of Kingston. He was dangerous [to the established order], primarily because he described the dependence *problematique* in racial terms – a subject which had never been brought up in the Latin American debate …
>
> (Blomström and Hettne 1984: 108)

Blomström and Hettne (p. 147) go on to note how, in comparison to the bulk of dependency theory writings: "Rodney ends up far to the left on all axes." More significantly, in writing on Africa from a dependency perspective, he was a significant agent in the dissemination of dependency beyond its Latin American roots (Kay 1989). Blomström and Hettne (1984: 108–109) point out:

> Emancipation was to Rodney not so much a matter of economic rights as it was a right to cultural identity: in this context his thoughts are more reminiscent of African ideas, of *African personality* and *Négritude*, than of Latin American dependency theory. Consequently, his contacts with the Rastafari movement were well developed, which in turn, the [Jamaican] government considered to be a greater danger than communist agitators.

In particular, *How Europe Underdeveloped Africa* highlighted the roles of the slave trade and subsequent predatory colonial capitalism in producing the simultaneous flow of wealth to Europe and impoverishment of Africa. Rodney pictured African social and commercial relations being fragmented and distorted as they became increasingly orientated to Western markets. For Rodney (1972: 37) therefore:

> The question as to who and what is responsible for African underdevelopment can be answered at two levels. Firstly, the answer is that the operation of the imperialist system bears major responsibility for African economic retardation by draining African wealth and by making it impossible to develop more rapidly the resources of the continent. Secondly, one has to deal with those who manipulate the system and those who are either agents or unwitting accomplices of the said system.

The latter point relates to the agency of African (and other postcolonial) elites as intermediaries, a theme that crops up in Rodney's other writings on African and Caribbean histories, development and underdevelopment (e.g. Rodney 1967; 1970; 1981). Yet during his time in Tanzania, he celebrated some of the claims and putative achievements of the Arusha Declaration (proclaiming a strategy of socialism and self-reliance) in charting an alternative to the usual neocolonial patterns in Africa. Here Rodney was undoubtedly caught up in, and his judgements shaped by, the revolutionary optimism of the times and the sense of an alternative to the neocolonialism that he felt characterised most of Africa and the Caribbean. However, *How Europe Underdeveloped Africa* transcends that moment. By insisting that African underdevelopment was not a natural condition and that African cultures and histories were rich and not themselves the source of the continent's subsequent relative underdevelopment, his book would prove inspirational among a broad readership. In particular, for many black folk in the diaspora (hence in Europe, the Caribbean and the Americas) and others in alliance, Rodney's book was part of a new wave of expression, mobilisation and radicalisation in the 1970s. It retains this capacity.

Robert Shenton (then Professor of History at the University of Toronto) ended his review of *How Europe Underdeveloped Africa* (subsequently expanded and reprinted as an introduction to later editions of Rodney's book) with the following words inviting further questions on Africa's trajectories:

> Dr Rodney's book is important because rather than just presenting history he seeks to make it. It is doubly important because Rodney, through the use of a popular, polemical style seeks to reach those who in his eyes are truly capable of making history. My own reservation is that the use of this style sometimes cloaks the need for analysis with popular polemic and creates the impression that all the important questions regarding Africa's past and present development have been asked and answered. And this clearly is not the case.
>
> (Shenton 1975: 150)

Rodney's book is clear about the genesis of African underdevelopment. As he probably intended, it is perhaps most effective as a polemic and an invitation to think historically and comparatively about African underdevelopment. In his original preface to the text, he (1972) explains that: "the purpose has been to try and reach Africans who wish to explore further the nature of their exploitation, rather than to satisfy the 'standards' set by their oppressors and their spokesmen in the academic world" (1983 edn: 8). Rodney was not much interested in convention, theoretical abstraction or academic honours. His commitment to activism underlay his return to Guyana, and ultimately led to his early death.

Lewis (1998: 256) ends his book-length study of Rodney's thought and impact by returning to his place in a radical Caribbean tradition:

> Rodney's intellectual legacy, as a historian and as an activist, forms an important part of the unfolding Caribbean intellectual tradition.... The central feature underlying Rodney's contribution to this tradition was his positive awareness of himself as a person of

African descent in the Caribbean, the link he forged with Africa and the intellectual agenda that emerged in relation to the major challenges of decolonisation.

It is interesting to reflect on how Rodney's thought would have evolved had he not been killed. What would he have said about neoliberalism, new forms of imperialism or #BlackLivesMatter? How would his politics and writings have dealt with the collapse of the USSR, the weakening of Third Worldism and the rise of China and other Asian powers? Many things have changed since his day. Yet, more than four decades after its publication, *How Europe Underdeveloped Africa* has not lost its polemical capacity to incite reactions and to radicalise readers. There is a new book-length study of *The Enduring Relevance of Walter Rodney's How Europe Underdeveloped Africa* (Hirji 2017). A series of conferences, including one in Guyana in June 2005 marking the 25th anniversary of his death, have celebrated his emancipatory vision. Others have followed, in the University of Dar es Salaam and in leading American universities. The Walter Rodney Foundation (www.walterrodney foundation.org), a non-profit organization, was established in in Atlanta, Georgia, in 2003 and an archive of his writings, correspondance and recordings has been deposited at the Robert W. Woodruff Library of the Atlanta University Center. Atlanta is the largest black-majority city in the US South – where histories of oppression and interfaces of race, power and development connect with those of Europe, Africa and the Carribean that Rodney focused on. Such histories also connect directly with liberal forms of development that are still very much on active service (Domosh 2018).

Major works

Rodney, W. (1967) *West Africa and the Atlantic Slave-Trade*, Nairobi: East Africa Publishing House.

—— (1969) *The Groundings with my Brothers*, London: Bogle-L'Ouverture Publications.

—— (1970) *A History of the Upper Guinea Coast 1545–1800*, New York: Monthly Review Press.

—— (1972) *How Europe Underdeveloped Africa*, London: Bogle-L'Ouverture Publications. (Republished 1983.)

—— (1981) *A History of the Guyanese Working People, 1881–1905*, Kingston: Heinemann Educational Books.

Further reading

Blomström, M. and Hettne, B. (1984) *Development Theory in Transition. The Dependency Debate and Beyond: Third World Responses*, London: Zed Books.

Domosh, M. (2018) "Race, Biopolitics and Liberal Development from the Jim Crow South to Postwar Africa", *Transctions of the Institute of British Geographers* 43(2): 312–324.

Hirji, K. (2017) *The Enduring Relevance of Walter Rodney's "How Europe Underdeveloped Africa"*, Montreal: Daraja Press.

Kay, C. (1989) *Latin American Theories of Development and Underdevelopment*, London: Routledge.

Lewis, R.C. (1998) *Walter Rodney's Intellectual and Political Thought*, Kingston: The Press, University of the West Indies.

Shenton, R. (1975) Review of Walter Rodney, *How Europe Underdeveloped Africa*, *Canadian Journal of African Studies* 9(1): 146–150.

JAMES D. SIDAWAY

PAUL ROSENSTEIN-RODAN (1902–1985)

Paul Rosenstein-Rodan (1902–1985), was known to the multitude of his friends simply as "Professor Rodan" or, more affectionately, as "Rosie". He was without doubt among a handful of the "greats" who oversaw the postwar rise of development economics as a distinct field. Many lamented that he did not share the Nobel Prize for Development Economics that was awarded jointly to Arthur Lewis and Ted Schultz in 1979.

Born into a Jewish family in Kraków, Poland, in 1902, Rosie's formative intellectual years were in the Austrian School of Economics at the University of Vienna, where he obtained his PhD in 1925 under Hans Mayer. He was one of the Austrian, Polish and Hungarian luminaries (including Friedrich von Hayek, Stanisław Ulam and John von Neumann) who moved abroad in the 1930s. Rosie moved to the Department of Political Economy at University College London in 1931, thus avoiding Nazi rule in continental Europe. Bloomsbury was blooming when Rosie reached London, and he was an active member.

Lionel (later Lord) Robbins at the London School of Economics, who had taken Friedrich von Hayek and Rosie under his wing, often told the story of how he had taken Rosie to the Political Economy Club, over which Arthur Bowley presided. The Club would typically hear from a distinguished economist like Rosie, with no written paper circulated earlier. Lots would be drawn and the first six had to comment on what they had just heard.

Rosie, who had written a brilliant early article, "The Role of Time in Economic Theory" in *Economica* (1934), was called upon. He apparently said that he would illustrate the role of time by analogy. If he was asked to choose between spending time with Marlene Dietrich or Greta Garbo, his choice would depend on how much time he was given. If a weekend, he would choose the sultry Marlene Dietrich. If a lifetime, he would choose the legendary Greta Garbo. When the meeting broke up, supposedly, Arthur Bowley came up to Rosie and said: "I have heard of this statistician Dietrich but who is this economist Garbo?" Lionel Robbins, who was aghast, told Rosie later: "You got away with it this time; but if you do this again, you go right back to Vienna!"

Rosie wrote other seminal papers in London, including one on the role of balanced growth and the need for co-ordination in planning, which he argued implied the need for an across-the-board "big push" as the way to promote industrialisation in east and south-east Europe (1943), and on the rationale for foreign aid. This article is sometimes referred to as a foundation of development economics, although others date the origins of this field as having been earlier.

In 1947 his trajectory took him from London to Washington, DC, to join the World Bank, where he was the first major economist in residence, from 1947 to 1954.

From the World Bank he moved to the Center for International Studies and the Economics Department at MIT in Cambridge, Massachusetts, to join its great economists. But his approach to economics changed. He would then undertake several notable projects in the field. In the first, for Italy in the 1950s, he collaborated with Ezio Vanoni, the Italian Finance Minister, on the Venoni Plan for Italian development, with particular attention to the Mezzogiorno, the less developed southern Italy. In the mid-1960s he led a development research project in Chile, supported by Chilean President Eduardo Frei. That was followed later in the 1960s by an extended project in India. He brought to India distinguished economists such as Ian Little of Oxford and Trevon Swan of Australia. He also became one of the "nine wise men", expert advisers on the Alliance for Progress of the Organization of American States from 1961 to 1966.

After retiring from MIT, Rosie moved to the University of Texas for four years and then returned to Boston across the Charles River to a prestigious appointment at Boston University. Insights into his own perspective on his life and career can be obtained from the autobiographical essay (Rosenstein-Rodan 1984) he contributed as one of eight so-called "Pioneers in Development" honoured in a World Bank book edited by Gerald Meier and **Dudley Seers**.

He was active until the end of his career; and his wit, wisdom and fund of anecdotes drew many younger economists to his home in Boston. His wife, Margaret, who was a marvellous cook, entertained with aplomb and affection. She passed away soon after Rosie, as if (as in Japanese ritual) they had spent a beautiful life together and now was the time to meet in heaven.

Major works

Rosenstein-Rodan, P.N. (1934) "The Role of Time in Economic Theory", *Economica*, n.s. 1: 77–97.

—— (1943) "Problems of Industrialisation of Eastern and South-Eastern Europe", *Economic Journal* 53(210/211): 202–211.

—— (1961) "International Aid for Underdeveloped Countries", *Review of Economics and Statistics* 43: 107–138.

—— (1963) "Notes on the Theory of the 'Big Push' in Economic Development", in H.S.Ellis (ed.), *Proceedings of a Conference of the International Economic Association*, London: Macmillan.

—— (1984) "Paul N. Rosenstein-Rodan", in G.M. Meier and D. Seers (eds), *Pioneers in Development*, Washington, DC: World Bank, 205–221.

Further reading

Chakravarty, S. (1983), "Paul Rosenstein-Rodan: An Appreciation", *World Development* 11(1): 73–75.

JAGDISH BHAGWATI AND RICHARD ECKAUS

WALT WHITMAN ROSTOW (1916–2003)

Rostow was born in New York on 7 October 1916 to a Russian immigrant father. He graduated from high school in New Haven, Connecticut, in 1932, aged 15, and was awarded a scholarship to study economics and history at Yale. In 1936 the highly gifted Rostow won a Rhodes Scholarship and experienced his first stay in Europe at Oxford from 1936 to 1938. Following his return, he wrote his PhD thesis on "British Trade Fluctuations, 1868–1896", graduating from Yale in 1940. Having worked briefly as an economics instructor at Columbia University, Rostow was called up for military service when the USA entered the Second World War, serving from 1941 to 1945 in the Office of Strategic

Services (the forerunner of the CIA) in London, analysing aerial photographs for the planning of strategic bombing. In 1945/1946 he became the Harmsworth Professor of American History at Oxford, but he decided to return to political work in 1947. He became Assistant to **Gunnar Myrdal**, head of the UN Economic Commission for Europe, based in Geneva. Yet not even the highly prestigious UN job could hold him for long. He resigned in 1949 and became Visiting Pitt Professor of American History at Cambridge (UK). It was not until 1950, aged only 33, when he was appointed to the chair of economic history at the Massachusetts Institute of Technology (MIT), a position which he held until 1961, that he more or less settled down.

For Rostow, these were the decisive academic years, marked by the founding of the Center for International Studies (CENIS) at MIT in 1951. Its head was Max Millikan, whom he knew from his Yale days. With the Korean War raging, CENIS was to develop strategies against the spread of communism. The circle of colleagues in Cambridge and Boston, where the MIT and Harvard University are separated only by the Charles River, included such highly acclaimed people as Paul Samuelson, Robert Solow, **Paul Rosenstein-Rodan**, Everett Hagen, **Charles Kindleberger**, Benjamin Higgins, Wilfred Malenbaum, Lucian Pye, Robert Baldwin, Richard Eckaus and Daniel Lerner. They constituted an illustrious collection of the pioneers in development – economists, sociologists and political scientists, all of whom would deserve inclusion in a Who's Who of development theory. At CENIS, he collaborated in two projects on the Soviet Union and China, which he summed up in 1955 in the essay, "Marx Was a City Boy, or Why Communism May Fail". Even more important was *A Proposal* (Millikan and Rostow 1957), which presented the perception that development policy can be a political instrument in the East–West conflict. From 1956 to 1958, Rostow did the groundwork for *The Stages of Economic Growth* (1960), which is based on a series of lectures he gave in Cambridge in late 1958.

The third stage of Rostow's life began in 1960 with his appointment to John F. Kennedy's election campaign team. He took leave of absence from MIT, and after Kennedy's election was named as Deputy Special Assistant to the President for National Security Affairs in January 1961. The following autumn he became Chairman of the Policy Planning Council at the State Department. Together with Paul Schlesinger, John K. Galbraith and his brother Eugene, Rostow formed the Charles River clique, Kennedy's liberal brains trust. During this time, Rostow wrote many memoranda in which development policy was conceived as a new field of US foreign policy. After Kennedy's assassination in November 1963, Rostow continued to work for Lyndon B. Johnson. In May 1964 he became the US member of the Inter-American Committee on the Alliance for Progress with the rank of Ambassador. At the beginning of 1966, Rostow returned to the White House as National Security Adviser, and held this office until January 1969, i.e. at the height of the Vietnam War. Nixon's election as President ended Rostow's political career, which had brought him increasing criticism from the liberal public because he was seen as one of the Administration officials mainly responsible for the escalation of the war. Rostow's old faculty at MIT refused to let him return to his chair, and an attempt to win an appointment at Harvard also failed. But in February 1969, he was appointed to the specially created chair of Jr. Professor Emeritus for Political Economy at the Lyndon B. Johnson School of Public Affairs at the University of Texas, in Austin, where he ended the fourth and now purely academic stage of his life. He dropped out of the public eye and wrote many academic books. He died in Austin on 13 February 2003, aged 86.

Rostow was a convinced liberal with a missionary-like zeal which was expressed equally in his commitment to development and his anti-communism. Economic growth and the modernisation of society, according to the CENIS theory, were to prevent the spread of communism. From his studies of China and the Soviet Union, Rostow was convinced that, because Marxist theory neglected agriculture, it could not master the problem of development. Therefore he argued that force of arms should be used to assert what he believed to be sensible: the stages of economic growth that he had outlined. Thus his work revealed many parallels with that of **Marx**, against whom he fought so bitterly. These related not only to his claim to have formulated a universally historic counter-concept to the Communist Manifesto, but also to his derivation from it of his demand for political action.

However, Rostow is distinguished from Marx by one thing: whereas Marx never possessed personal political power, only supplying the concepts for his successors, Rostow did have power through direct access to two US Presidents. That applied both to the period of the "Pioneers in Development" and to the Kennedy years. Rostow succumbed to the hubris of power; he believed he could not only define the world but also change it. What influenced him? Initially, it was surely the liberal minds of his parents and the anti-communism of his Russian immigrant family. In addition, there was the academic influence of Oxford and Cambridge, Keynesianism, the influence of the German Historical School, and finally the context of MIT and Harvard with their unique group of luminaries.

His "stages" theory was so influential because of its simplicity. Like Marx, he distinguished five stages through which all countries have to pass. These are: (1) the traditional society; (2) the preconditions for take-off; (3) the take-off; (4) the drive to maturity (self-sustained growth); and (5) the age of high mass consumption. The most important are stages 2–4, because they mark the transition from traditional to modern society (Rostow 1960). Rostow claimed that, in emphasising that economic change was the result of human will, he was not formulating merely a theory of growth but one of societal development in general.

The building blocks in Rostow's theory are the modernisation-promoting use of science and technology, the sharp increase in savings and investment rates in order to achieve continual growth, the role of the innovative entrepreneur and the concept of key sectors. This is a crude synopsis of elements which can be found among prominent contemporaries: big push (**Rosenstein-Rodan**), spurt (**Gerschenkron**), linkage concept and key sectors (**Hirschman**), the role of the entrepreneur (Schumpeter), and stages theory (Fourastie).

Whether it constitutes a theory at all rather than merely a taxonomy of economic stages is one of the fundamental criticisms levelled at Rostow. It is also said that, notwithstanding his own claim, he took too little account of social and political factors. However, this objection is wrong if one looks at other works in which he emphasises the decisive role of new societal forces, whereby nationalist sentiment *vis-à-vis* more advanced countries is recognised as a motivating force for modernisation. Rostow is also criticised for his fixation on the Anglo-Saxon path, which ignores the top-down modernisation implemented in countries such as Germany, Japan, the Soviet Union and actually China. In empirical terms, much does not tally; for instance, no take-off can be verified in France or Austria-Hungary although China seems to be a recent confirmation of the take-off thesis. Even Rostow's description of the USA as a model case is only partly right because, despite its mass consumption, it is not a welfare state. The relative decline of countries such as Britain can certainly not be explained at all, and countries such as Argentina have never emerged from their alleged take-off stage. In that respect, Rostow suffers the fate of many global theorists because there are always many objections in respect of empirical details. Like other representatives of the Historical School, neoclassical authors deny that his work has the character of theory because the theoretical stringency of deductively obtained models can never be achieved in an inductive way.

The high point of criticism was reached at the Constance Conference of the International Economic Association in 1960, which Rostow edited in 1963 under the title *The Economics of Take-Off into Sustained Growth*. The main critics were Kuznets and Solow. The former found fault with the lack of an empirical basis and the latter questioned whether Rostow's work had the character of theory at all. It was not until 1978, in *The World Economy*, that he provided the empirical proof for his stages theory. He originally intended his *How It All Began* (1975) to be the introduction to that. Basically, his enormous final research programme was nothing other than an attempt to put straight his earlier critics.

The controversy over take-off marked the core of the problem. Only if the existence of take-off could be proved did it make sense to research the preconditions, and only then was the theory correct that the automatism of self-sustaining growth followed take-off. Also, only then did the normative conclusion make sense that one could orchestrate this connection and promote it externally. Less noticed but much more important were the political consequences of Rostow's stage theory in reformulating American foreign policy which, against the background of the Korean War, was fixated on the military dimension of the East–West conflict. Rostow believed that the USA should assume the leadership in a new international partnership programme for world economic growth. He aimed to broaden the understanding of Containment and make clear that the East–West conflict also had to be conducted in the countries of Asia, the Middle East, Africa and Latin America because they were in the second stage of the Rostowian model, where the preconditions for take-off are laid. If the preconditions were not in place, the economy would come to crisis. The communists then would have the chance to take over power, as had already happened in China and was in the offing in Vietnam and elsewhere. Therefore the process had to be supported from outside – by means of development assistance. Rostow calculated how much capital would be required to bring the investment rates of the countries involved up to the critical level.

The partners in the assistance programmes would be the new elites in the countries mentioned, whose nationalistic strivings are expressed in the wish for economic and social modernisation. At least in the initial phase, a strong state component would be essential; if necessary, the military could certainly become a suitable partner. This is very much how things were actually handled from the 1960s onwards, with support given to every autocratic regime provided that it was robustly anti-communist and promised development. The Alliance for Progress was to contribute to "political maturity".

Previously, Rostow had believed it necessary to convince the US foreign policy elite that development policy was definitely in their national security interest. The Republicans criticised Rostow vehemently for his programme: in fact he had to justify himself to Congress against charges that he was seeking to water down American security policy and placing too much emphasis on a planned economy. Conservative economists such as Peter Bauer and Milton Friedman even said that they suspected him of socialism because of his budgetary approach. However, Rostow and his fellow combatants were able to assert their views and convince Kennedy. To be sure, they were helped by Khrushchev's political offensive from the Sputnik shock in 1957 to the Cuban missile crisis of 1962, which signalled a substantial growth in Soviet power.

The establishment of development policy at the beginning of the 1960s, the announcement of the first development decade, the founding of USAID, the Alliance for Progress, the Peace Corps and the OECD's Development Assistance Committee, and the reorientation of the World Bank from postwar reconstruction aid to financing development – all of which meant putting the new foreign policy strategy on an operational basis – reflect Rostow's influence. But a small flaw was overlooked amid the hubris of power: despite all efforts, reality could not be shaped in the manner proposed in the Rostowian stages theory. That again links him with **Marx**. Vietnam, the chosen model, became a trauma. Because the Vietcong prevented the creation of the preconditions for take-off in South Vietnam, the war was driven ahead ever more, due not least to Rostow's advisory position under Johnson. It was not until the change to the Nixon administration that the conservative realist Henry Kissinger was able to stop the bustling activity of Rostow, the liberal missionary, by accepting the American defeat in Vietnam.

What is left of Rostow today? In the 1970s, he became a classic figure of controversy – not only because of his role in the Vietnam War, but also because he had dared to provide a counter-concept to Marx. He never mentioned the subjects of dependency, the world market, terms of trade or colonialism. He liked assistance from outside, the World Bank, direct foreign investment, multinationals and military advisers. But for Vietnam, he could have taken his place among the honoured school of the development pioneers, could have returned to the MIT without his reputation having the taint of warmonger and fanatical anti-communist. What have remained are his concepts: take-off, preconditions of growth, self-sustained growth, the age of mass consumption. From today's viewpoint, against the background of the major disaster of failed and rogue states, we must recognise that his demand for the political preconditions for take-off is more topical than ever. He was one of the decisive theorists able not only to dream up development policy and justify its necessity but also to use his position at the levers of power to ensure its practical introduction. Whether US security policy interests and combating communism were the crucial motive, or whether he knew how to package his developmental engagement cleverly in political terms, as realistic critics ascribe to him, is an open question. Either way, the countless Vietnamese and the counted American victims of the Vietnam War are also part of his appraisal. The title of his last autobiographical book, published posthumously in 2003, is *Concept and Controversy: Sixty Years of Taking Ideas to Market.*

Major works

Millikan, M.F. and Rostow, W.W. (1957) *A Proposal: Key to an Effective Foreign Policy*, New York: Harper. (Repr. Westport, CT: Greenwood Press, 1976.)

Rostow, W.W. (1960) *The Stages of Economic Growth: A Non-Communist Manifesto*, Cambridge: Cambridge University Press. (2nd edn 1971.)

—— (1971) *Politics and the Stages of Growth*, Cambridge: Cambridge University Press.

—— (1975) *How It All Began: Origins of the Modern Economy*, London: Methuen.

—— (1978) *The World Economy: History and Prospect*, London: Macmillan.

—— (1980) *Why the Poor Get Richer and the Rich Slow Down: Essays in the Marshallian Long Period*, London: Macmillan.

—— (1985) *Eisenhower, Kennedy and Foreign Aid*, Austin, TX: University of Texas Press.

—— (1990) *Theorists of Economic Growth from David Hume to the Present: With a Perspective to the Next Century*, New York: Oxford University Press.

—— (1998) *The Great Population Spike and After: Reflections on the 21st Century*, New York: Oxford University Press

—— (2003) *Concept and Controversy: Sixty Years of Taking Ideas to Market*, Austin, TX: University of Texas Press.

Further reading

Baran, P.A. and Hobsbawm, E.J. (1961) "The Stages of Economic Growth", *Kyklos* 14(2): 234–242.

Cornwell, R. (2003) Obituary: "Walt Rostow: Vietnam War Super-hawk Advising Presidents Kennedy and Johnson", *The Independent*, 17 February (www.independent.co.uk).

Economist, The (2003) Obituary: "Walt Rostow: An Adviser in the Vietnam War, Died on February 13th, Aged 86", 366(8312), 22 February: 101 (www.economist.co.uk).

Fishlow, A. (1965) "Empty Economic Stages?", *Economic Journal* 75(297): 112–125.

Foster-Carter, A. (1976) "From Rostow to Gunder Frank: Conflicting Paradigms in the Analysis of Underdevelopment", *World Development* 4: 167–180.

Hodgson, G. (2003) Obituary: "Walt Rostow: Cold War Liberal Adviser to President Kennedy Who Backed the Disastrous US Intervention in Vietnam", *The Guardian*, 17 February (www.guardian.co.uk).

Kindleberger, C.P. and Di Tella, G. (eds) (1982) *Economics in the Long View: Essays in Honour of Walt Whitman Rostow*, vol. 1: *Models and Methodology*; vol. 2: *Applications and Cases, Part I*; vol. 3: *Applications and Cases, Part II*. London: Macmillan.

Meier, G.M. (ed.) (1987) *Pioneers in Development*, 2nd series, New York: Oxford University Press.

—— and Seers, D. (eds) (1984) *Pioneers in Development*, New York: Oxford University Press.

Milne, D. (2008) *America's Rasputin: Walt Rostow and the Vietnam War*, New York: Hill and Wang.

North, D.C. (1958) "A Note on Professor Rostow's 'Take-Off' into Self-Sustained Economic Growth", *Manchester School of Economic and Social Studies* 26: 68–75.

Ohlin, G. (1961) "Reflections on the Rostow Doctrine", *Economic Development and Cultural Change* 9(4): 648–655.

Pearce, K.C. (2001) *Rostow, Kennedy, and the Rhetoric of Foreign Aid*, East Lansing, MI: Michigan State University Press.

Rosowsky, H. (1965) "The Take-off into Sustained Controversy", *Journal of Economic History* 25(2): 271–5.

Rostow, W.W. (ed.) (1963) *The Economics of Take-Off into Sustained Growth: Proceedings of a Conference Held by the International Economic Association*, New York: St. Martin's Press.

ULRICH MENZEL

ARUNA ROY (1946–)

Aruna Roy is one of a small group of civil-society activists who have made a big difference to development policy and practice in India, and whose work has much wider relevance in regard to democratic development. Roy, and others, including her close collaborator, Nikhil Dey, the economist Jean Drèze, and Harsh Mander and N.C. Saxena – like her, former senior civil servants – may be described as policy entrepreneurs. With other members of the National Advisory Council appointed under the government led by the Indian National Congress party between 2004 and 2014, and with the support both of the many local activists in the civil society campaigns in which they were engaged and of the judiciary (notably justices of the Supreme Court), Roy, Dreze, Mander and Saxena, amongst others, were instrumental in establishing what has been called India's "new rights agenda". This term has been used by Sanjay Ruparelia (2013) to describe legislation passed by the Congress-led government – legislation that established as legally enforceable rights Indians' claims to education and the means of securing decent livelihoods. The rights agenda is the outcome of a broad movement in civil society, involving many different organisations, that has sought to make the Indian state more accountable to the people, and so to deepen Indian democracy beyond the regular conduct of more or less free and fair elections. The movement was certainly not the creation of Aruna Roy alone, but hers has been a notably articulate voice in expressing a vision of democratic social justice, and in putting it into practice.

Born in the south Indian city of Madras (now Chennai) in 1946, Roy grew up mainly in India's capital city, Delhi, and it was there that she studied for her degree in English literature. Soon afterwards, in 1968, she entered the prestigious Indian Administrative Service (IAS), the elite cadre of the Indian bureaucracy. In 1970, she married Sanjit (Bunker) Roy, who has also become a well-known social reformer, and who in 1972 – influenced by Gandhian ideals, and by the idea of village self-reliance – set up what was then called the Social Work and Research Centre (SWRC) in the village of Tilonia, in the north Indian state of Rajasthan. It is now very widely known as the Barefoot College, and has become the centre of an international organisation dedicated to training poor people, especially poor women (www.barefootcollege.org). Frustrated by the ways in which the government bureaucracy functions – as an instrument for managing the status quo rather than for bringing about change – Aruna Roy resigned from the IAS in 1975. This was also a moment when she found too many of her colleagues ready to kowtow to power (as she has put it), in the context of Prime Minister Indira Gandhi's declaration of emergency rule, and suspension of democracy. Roy turned to work in Tilonia.

In her work with the SWRC, Aruna Roy found an opportunity to work with village people and to see issues from their perspective. Eventually, however, she chafed at the unwillingness of her colleagues to contemplate engagement in struggle, and when it was formally decided by the management team of SWRC to continue with the search for development alternatives, rather than to question power equations, Roy decided, in 1983, to part ways. For some time thereafter she searched for social and political movements with which she could take her ideas forward.

Finally, in 1987, Aruna Roy made a home at Devdungri, in central Rajasthan – in a mud and stone hut that belonged to a cousin of her close collaborator Shankar Singh - and it was from there that she and Singh, with Nikhil Dey and many local people, formed the *Mazdoor Kisan Shakti Sangathan* (MKSS – the name means "Workers' and Farmers' Power Organisation") in 1990. The MKSS describes itself as a people's organisation, and as part of the growing non-party political process in India – referring to the sphere of political activity that is outside regular political parties and electoral politics. It generates its own resources, receiving no funding from government or any other outside institutions. Though Roy is widely known in the English-speaking world as the "face" of the MKSS, she is not its "leader", and the organisation operates through a system of internal democracy. According to its website (www.mkssindia.org), the MKSS was "set up *by the people of the area* to strengthen participatory democratic process, so that ordinary citizens could live their lives with dignity and justice" (emphasis added). The organisation was born, the site says, "out of a struggle for community land held illegally by a feudal landlord". It then went on to fight for the minimum wages to which people are entitled (initially in government-funded drought-relief works), and for people's access to basic foodstuffs to which they are also entitled, at controlled low prices, through the official Public Distribution System (PDS). In the course of these struggles, those involved came to recognize that "transparency and accountability of systems of governance are basic to access any right (and that) access to relevant information [in official documents] is a fundamental tool for ensuring transparency and accountability of government" (website).

The MKSS, therefore, came to focus on people's rights to information from the state and its agencies. The need for legislation was taken up with senior faculty members of the National Academy of Administration, and later with the Press Council of India, which eventually took the lead in fashioning the first formal draft of the law circulated to the Prime Minister and members of parliament. These groups, from different sectors, formed the National Campaign for the People's Right to Information (NCPRI) – which was finally successful in securing the passage of the Right to Information Act by the Indian parliament, in 2005. But the MKSS also came to play an important part in struggles over the establishment of the right to work – finally realized in the passage into law of the right to employment for 100 days a year, at an officially established wage, for all rural households (in the Mahatma Gandhi National Rural Employment Guarantee Act 2005) – and of the right to food (at least partially secured in the National Food Security Act 2013). As Nikhil Dey has put it, "Accountability from a citizen's point of view is inextricably tied to basic entitlements. Who can I hold accountable if I don't have any entitlement?" So rights backed by legal sanction came to be seen as a critical means of empowering citizens to mobilise and place claims on the state – and in making those claims, information is essential. The struggle for the right to information has been foundational, therefore, in the establishment of the new rights agenda. The MKSS was not so centrally involved, however, in two other key elements of this agenda, namely the Right to Education (passed into law in 2009) and the Forest Rights Act of 2006, which is intended to secure the rights to livelihoods of people who have historically been forest dwellers.

The MKSS is not a political party, but neither is it an NGO, if this term is taken to mean an entity that is more or less bureaucratic in its organisation, and that channels externally raised funds into the delivery of services – like the Barefoot College in Tilonia. The MKSS has the character of a popular local movement – though there is no doubt that its success has also depended upon the commitment of outside activists. In an early account of it, Jenkins and Goetz (1999) referred to the significance of the network of support that MKSS had developed within the IAS, in parts of the media and among intellectuals, and they noted that the organisation's success in building this network had reflected, to a great extent, Aruna Roy's own personal and professional connections. There are tensions, no doubt, between local activists and English-speaking, cosmopolitan outsiders (on which see Cross 2014), but these are more or less frankly recognized by MKSS activists. Roy herself is adamant that middle class participation is only one part of the story of the MKSS, and that grassroot struggles have been fundamental to the birth, growth and sustenance of the movement and its campaigns (see Roy at al. 2018).

Roy's personal role in the establishment of the new rights agenda has been important. Along with Jean Drèze, she presented a draft of the National Rural Employment Guarantee Act at the first meeting of the newly instituted National Advisory Council in 2004, and then fought against attempts from within government to dilute its provisions. The establishment of the rights to information, to work and to food is extremely important in itself, but the work that Roy and the MKSS have done has a wider significance, too, in regard to the reform of government through the deepening of democracy. As Aiyar and Walton (2014: 4) have put it, "the project of building a rights based welfare state was embedded in larger vision of a radical transformation of the everyday workings of the state."

The pursuit of rights in India embodies both the aim of substantive provisioning *and* the deepening of citizenship, effectively through the mobilisation of popular discontent. From early on in its history, the MKSS started to develop the instrument of public hearings – *jan sunwai*, in Hindi – organized independently of the statutory local government bodies by local MKSS activists. These hearings are presided over by respected individuals from the locality and from outside, at which detailed accounts of government expenditure are read out to assembled villagers. Through this means people may discover – for instance – that they have been listed as beneficiaries under one or other government programme, when they have received nothing, or learn of large payments that have been made to local contractors for work that they know has not been done. In this way, local officials and those elected to local government bodies may be held directly accountable. The *jan sunwai* is described by the political philosopher Neera Chandhoke (in a personal communication), as a "training ground for citizenship". The fact that their conduct has met with fierce resistance from local authorities in Rajasthan is a mark of their impact.

The principle of such social auditing of government schemes that are intended to provide welfare for people has been incorporated, for example, into the employment guarantee programme, and into the management of the Public Distribution System. There are very many difficulties surrounding the conduct of social audits – which are found threatening by politicians and bureaucrats – but this only goes to show that the principle of citizens being empowered to hold the state to account, well beyond the ballot box, is extremely important. This Indian experience, driven to a significant extent by Aruna Roy, is of wide relevance in regard to the deepening of democratic development – for which "increasing the surface area" of the contact between citizens and the state, as Patrick Heller of Brown University put it, in a lecture given at McGill in November 2016, is essential.

Major work

Roy, A. with the MKSS Collective (2018) *The RTI Story: Power to the People*, New Delhi: Roli Books.

Further reading

Aiyar, Y. and Walton, M. (2014) *Rights, Accountability and Citizenship: Examining India's Emerging Welfare State*, Delhi: Emerging Accountability Working Paper Series, Accountability Initiative, Centre for Policy Research.

Cross, J. (2014) *Dream Zones: Anticipating Capitalism and Development in India*, London: Pluto Press.

Jenkins, R. and Goetz, A.M. (1999) "Accounts and Accountability: Theoretical Implications of the Right to Information Movement in India", *Third World Quarterly* 20(3): 603–622.

Ruparelia, S. (2013) "India's New Rights Agenda: Genesis, Promises, Risks", *Pacific Affairs* 86(3): 569–590.

JOHN HARRISS

IGNACY SACHS (1927–)

Ignacy Sachs likes to call himself an "ecosocioeconomist". He is one of the intellectual pioneers who elaborated the concept of ecodevelopment, known today as sustainable development, and among the first to call for an intensification of the political, economic and intellectual relations between Southern countries. Due to his dislocations, whether chosen or forced, the places where he lived or the encounters he made, his personal biography has thoroughly shaped his thinking on development. He is a true citizen of the world.

Sachs was born in Warsaw in 1927 into an upper-middle-class family of Polish-Jewish bankers and art collectors, but when in September 1939 the Germans invaded Poland and Soviet troops occupied its Eastern part, the family left the country. They arrived in Brazil in 1940, having crossed Europe, from Romania to France, Spain and eventually Portugal. During their stay in France, which lasted several months, Sachs was enrolled at the local school and, once in São Paulo, obtained his French baccalaureate at the Lycée Français, after which he began studying economics at the University of Rio de Janeiro. For Sachs, this was an extremely formative period: at the time, São Paulo and Rio were places of convergence for many European, and in particular French, intellectuals who had fled Europe. In 1946, having settled in Rio, he began working at the Polish consulate to help finance his studies, and was in charge of cultural and press relations. Since his intellectual left-wing engagement goes back to the years spent in Brazil, in 1954, together with his wife, Viola (a professor of Anglo-American literature), and his two sons, he chose to "return" to his native Poland, driven by the desire to contribute to the edification of socialism.

Once in Warsaw, his dynamism and linguistic skills enabled him to find employment at the Institute of International Affairs, which depended at the time on the Ministry of Foreign Affairs. He also taught at the University of Warsaw and developed an interest in academia, to the point that when, in 1957, he was offered a position at the Polish embassy in India, he set as a condition of his departure the possibility to pursue doctoral studies at the prestigious Delhi School of Economics.

The family spent three years in India, where in 1960 Sachs defended his PhD thesis, "Patterns of Public Sector in Underdeveloped Economies" (Sachs 1964).[1] His research was based on a comparative approach (between India, Japan and Brazil), which would guide his work throughout his career as a researcher. His stay in India also played a major part in the development of his critique of Eurocentrism, which pervades his work. He met several individuals who would remain very significant for him throughout the years: **Amartya Sen, Gunnar** and Alma **Myrdal**, Daniel and Alice Thorner, and **Ester Boserup**.

Back in Poland (meanwhile they had a daughter, born in India), Sachs joined the Advanced School of Statistics and Planning at the Department of International Business (aka SGPiS and later SGH), first as an assistant professor, then as a professor, under the authority of **Michał Kalecki**, one of his PhD examiners and his lifelong intellectual mentor. At the school, Sachs also co-founded the programme for education and research on the Third World, which is divided into three segments: the research centre, the course on planning and an advanced seminar.

This initiative is quite exceptional for a socialist country: the course on planning is taught in English and supported by the United Nations. The approach developed in this programme kept its distance from socialist ideological indoctrination whilst remaining strongly left-oriented and influenced by "open" Marxism. The work produced at the research centre therefore stressed the need to elaborate original strategies according to the goals and factual data of the economies concerned, and which went beyond the dichotomy between the capitalist model (limited by demand) and the socialist one (limited by supply), wherein the concepts of mixed economies and "intermediate regimes" were also given special attention.

Sachs soon became the specialist in Poland relating to "underdeveloped" countries and in 1962, he defended his professorial thesis

(*Habilitation*) on foreign trade and economic development in "underdeveloped" countries (Sachs 1963). He worked in particular on the place of Third World economies in the international division of labour, and especially on the relationship between these and the socialist countries of that time, not in order to establish a relation of dependence but, on the contrary, to foster what later came to be described as self-reliance. Furthermore, Sachs has always been an advocate of South–South co-operation, be it economic, intellectual, or political. This goes along with the necessity to promote endogenous development – that is, a development that refuses mimetism, takes the local reality as a starting point and relies on the control of technologies and resources in the interest of the local populations.

From very early on, he has combined academic work with activities as a field-based consultant, as a true heir of Kalecki, who claimed that behind every word of his theories lay a concrete problem extracted from "real life". Similarly, Sachs has always argued that theory is useful to posit the problem but that the solution can only be found in fieldwork.

For the second time, in 1968, the Sachs family was forced to leave Poland as a result of a campaign against Jews and intellectuals. He was traumatised by this departure. Sachs identified as a Polish – non-Jewish – Jew: his parents had insisted on giving him a non-religious education and on affirming their ties to Polish culture and language. In spite of this, as a child, Sachs knew he was Jewish; and later, as an adult, he was well aware of his luck, since many of his relatives were victims of the Holocaust. Like several Jewish intellectuals, atheists, distant from Jewish culture and with no affinity with the state of Israel, he keeps nevertheless "claiming [his] condition as a Jew so long as anti-Semitism will exist" (Sachs 2007c: 21; see also Simon 2019). The war, the Holocaust, and the two exiles from Poland have most certainly strengthened his rejection of racial and religious segregation. It also explains his engagement in favour of decolonisation, his quest for social justice throughout the world and his faith in the United Nations system.

As they settled in France, Sachs was welcomed by Fernand Braudel and Clemens Heller at the École Pratique des Hautes Études (6th section) (EPHE, which became the School of Advanced Studies in Social Sciences – EHESS) while his wife, Viola, found a position at the University of Vincennes.

From the 1970s onwards, Sachs started to engage with environmental issues, especially after taking part in an international conference on environment held in Tokyo in March 1970.[2] This event constituted indeed a wake-up call, in the aftermath of the tragic environmental catastrophe that affected Japan.[3] Environment thus appears as a crucial question for any economist concerned with development, and Sachs was highly involved in the preparation of two landmark United Nations conferences on the subject: the first Conference on the Human Environment held in Stockholm in June 1972, and the Conference on Environment and Development (UNCED) organised in Rio de Janeiro in June 1992. He also took part in the 2012 UN Conference on Sustainable Development ("Rio +20"). If in 1972, the question tended to be posited in terms of environment versus economic growth (Club of Rome, *The Limits to Growth*, 1972), 20 years later, by contrast, the Rio conference emphasised the necessary articulation between environment and development. In 1973, Sachs founded the International Research Centre on Environment and Development (CIRED)[4] at the EPHE, with the support of the United Nations Environmental Programme (UNEP), the French Ministry of Environment and the Maison des Sciences de l'Homme. The centre aims to work on "another" growth, based on the articulation between economic, social, and environmental issues, whilst articulating the short and the long term.

The term "ecodevelopment", coined by Maurice Strong, emerged in the context of the Stockholm Conference. It refers to a civilisation project based on the harmonisation between strategies of socio-economic growth and environmental management understood in its three dimensions: resources, space and quality of life. In other words, it is based on needs-oriented strategies and on three interdependent prerequisites: economic viability, social inclusion and environmental sustainability. For Sachs, three thinkers best illustrate this concept: Benjamin Franklin for his pragmatism and sense of observation; **Mahatma Gandhi** for his ethical stance; and René Dubos for his reflection on the relationship between the economic and the environmental dimensions. The constitutive theoretical input of ecodevelopment was often rooted in fieldwork (especially in Latin America). Planning occupies a prominent place in this project; in **Kalecki**'s definition of the concept, however, it must be understood as *thinking in terms of variant*, which is radically opposed to planning *à la* Soviet, whose absurdities Sachs personally experienced from within.

While still living in Poland, Sachs commenced work with several UN agencies, which he considers play a crucial role in the development of the Third World, and he adopted a multidisciplinary approach due to the collaborations developed there. In the process, he built an international network that turned to be out very useful when he was forced out of the country in 1968. Within the United Nations he worked, among other agencies, with FAO, ILO, UNCTAD, UNEP, UNESCO and UNU (United Nations University), as well as with its regional agencies. In addition, he played a prominent role in many organisations and think-tanks, either as a founder, expert or research director. It is impossible to do full justice to all of them or to the vast number of people he worked with, but let us mention the International Foundation for Development Alternatives (IFDA) within the Dag Hammarskjöld Foundation, in which Sachs, together with an international network led by Marc Nerfin, elaborated an ambitious report on alternative ways to satisfy basic needs (health, food, energy, housing …) that were to be articulated with the new international order (Sachs et al. 1975). This project, entitled *What Now?*, also emphasised the important role played by the organised civil society, which Sachs defines as the third power next to those of the state and the market. Ecodevelopment, Sachs argues, can only succeed on the basis of this tripartite co-operation.

As the dictatorship in Brazil gradually "loosened up" by the 1980s, the country eventually became his main research field. In 1985, he founded the Research Centre on Contemporary Brazil (CRBC) at the EHESS. This period coincided with greater attention paid to fieldwork, as Brazil represents the terrain *par excellence* in which he put to good use the concepts and methods he had developed. Beyond his family and friendly ties to this country, as a researcher working on development, his interest also lies in the fact that Brazil epitomises one of the most telling examples of misdevelopment, despite its tremendous potential and reputation as a "land of good hope": the dreadful depredation of its natural resources, in particular the large-scale deforestation of the Amazon, combined with abyssal social inequalities.

Sachs is particularly interested in the relationship between rural and urban development. In a system dominated by intensive large-scale farming inherited from the colonial latifundian model, encouraging and protecting small-scale and family farming constitutes for Sachs a step forward. Indeed, farmers have a central role to play in the energy transition in tropical regions if this is based on the articulation between biodiversity, biomass and biotechnologies. In the 1980s, he developed the Food–Energy Nexus with the support of the UNU, a programme that dealt with the risk of competition between energy and food production. He collaborated with the United Nations and with national as well as regional agencies on strategies for the protection of the Amazon forest and the implementation of biosphere reserves.

Sachs also worked on programmes of urban development, in particular with his lifelong friend, the architect and city planner Jorge Wilheim, who was in charge of these questions at the state and city of São Paulo. These programmes tackled concrete issues such as land management, means of transport, social and self-build housing, and even waste management and collection.

In parallel, Sachs intensified his collaboration with other Brazilian thinkers and researchers. In Paris, he collaborated with prominent figures, such as the geographer and diplomat Josué de Castro, the economist **Celso Furtado** and **Fernando Henrique Cardoso**, some of whom he had first met in Poland. The CRBC was also conceived as a place to welcome researchers from Brazil, and as an opportunity to foster dialogue between thinkers from different regions of the world which had similar ecosystems and configurations. In this context of international exchange, Sachs tried to set up a bilateral co-operation programme between Brazil and India – two "whale-countries", as he often describes them – which, despite undeniable cultural differences, have many things in common. This is also a good illustration of the shape that South–South co-operation can take.

Arguably, Sachs is among the first thinkers in Europe to have thought about the development of the South outside the boundaries of Eurocentrism, and to have measured the strategic stakes of an integrated and multidimensional approach at economic, social, and environmental levels. This position, combined with an action-oriented method, has led him to raise questions about the reduction of poverty, environment, and economic growth in terms of complementarity rather than opposition by looking for solutions adapted to specific contexts. Just as his early discovery of Third World thinkers and theorists has most certainly helped shaping his understanding of this necessity and reciprocally, his work remains largely influential in the South, in particular in India and Latin America.

Notes

1 Sachs and his wife were among the first – if not *the* first – Westerners to have obtained a PhD in India.
2 The conference was organised by the International Social Science Council – UNESCO.
3 Between 1950 and 1968, Minamata Bay was contaminated with mercury compounds.
4 The CIRED still exists, directed by Franck Lecocq. Jean-Charles Hourcade (a disciple of Ignacy Sachs) was the director between 1988 and 2015.

Selected works

This list gives a selection of Sachs's most significant publications, many of which are available in English. For his full bibliography, please follow the links: http://crbc.ehess.fr/docannexe/file/1359/ignacy_sachs_1953_1999_vol._1.pdf and http://crbc.ehess.fr/docannexe/file/1359/ignacy_sachs_1953_1999_vol._1.pdf [accessed 9 September 2018].

Sachs, I. (1963) *Foreign Trade and Economic Development in Underdeveloped Economies*, New Delhi: Asia Publishing House.

—— (1964a) *Drogi i manowce swiata "B"*, Warsaw: PWN.

—— (1964b) *Patterns of Public Sector in Underdeveloped Economies*, New Delhi: Asia Publishing House.

—— (1966a) *Kształt niepodległosci*, Warsaw: Wiedza Powszechna.

—— (1966b) "Du Moyen Âge à nos jours. Européo-centrisme et découverte du Tiers-Monde", *Annales. Économies, sociétés, civilisations* 3: 465–487.

—— (1972) "Approaches to a Political Economy of Environment", in *Political Economy of Environment: Problems of Method*, The Hague: Mouton, 125–137.

—— (1974) "Environment and Styles of Development", *Economic and Political Weekly* 9(21): 828–837.

—— (1976) *The Discovery of the Third World*, Cambridge, MA: MIT Press. (First published in French in 1971.)

—— (1980a) "Culture, Ecology and Development: Redefining Planning Approaches", in I. Altman, A. Rapoport and J. F. Wohlwill (eds), *Human Behavior and Environment: Advances in Theory and Research*, vol. 4: *Environment and Culture*, New York: Plenum Press, 319–343.

—— (1980b) "Energy and the Village in Daniel Thorner", in E. J. Hobsbawm, W. Kula, A. Mitra and K.N. Raj (eds), *Peasants in History: Essays in Honour of Daniel Thorner*, Oxford: Oxford University Press, 259–270.

—— (1980c) *Studies in Political Economy of Development*, Oxford: Pergamon Press. (First published in French in 1977.)

—— (1987) *Development and Planning*, trans. P. Fawcett, Cambridge: Cambridge University Press. (First published in French 1984.)

—— (1991) "Growth and Poverty: Some Lessons from Brazil", in J. Drèze and A. K. Sen (eds), *The Political Economy of Hunger*, vol. 3: *Economic Hunger*, Oxford: Clarendon Press, 93–118.

—— (1993a) *Transition Strategies Towards the 21st Century*, foreword by M. F. Strong, Delhi: Interest Publications for Research and Information System for the Non-Aligned and Other Developing Countries.

—— (1993b) "North–South Deadlock: Current State of Ecodevelopment", *African Environment* (Dakar) 31–32(8): 3, 4, 5–24.

—— (1994) "The Environmental Challenge", in J.-J. Salomon, F. Sagasti, C. Sachs-Jeantet (eds), *The Uncertain Quest: Science, Technology, and Development*, New York: United Nations University Press, 302–339.

—— (1995) "Searching for New Development Strategies: The Challenges of Social Summit", *Economic and Political Weekly* 30(27): 1682–1685, 1687–1693.

—— (1999) "Brazil and India: Two 'Whales' in the Global Ocean", in K. Sinha and P. K. Dutta (eds), *Globalisation and Developing Countries*, New Delhi: New Age International, 3–17.

—— (2000) *Understanding Development: People, Markets and the State in Mixed Economies*, New Delhi: Oxford University Press.

—— (2004) "Inclusive Development and Decent Works for All", *International Labour Review* 143(1,2): 161–184.

—— (2005) "Da civilização do petróleo a uma nova civilização verde", *Estudos avançados* (São Paulo) 19(55): 195–214.

—— (2007a) *Rumo à ecossocioeconomia. Teoria e prática do desenvolvimento*, São Paulo: Cortez.

—— (2007b) "The Biofuels Controversy", *United Nations Conference on Trade and Development* (UNCTAD/DITC/TED/2007/12).

—— (2007c) *La troisième rive. À la recherche de l'écodéveloppement*, preface by T. Paquot, Paris: Bourin.

—— (2009a) "Revisiting Development in the Twenty-First Century", *International Journal of Political Economy* 38(3): 5–21.

—— (2009b) "Brazil 2022: A Land of Good Hope?", *The Perspective of the World Review* 1(1): 69–89.

—— et al. (1970) "Economics", in *Main Trends of Research in the Social and Human Science*, pt 1: *Social Sciences*, The Hague: Mouton/UNESCO, 283–350.

—— et al. (1975) "What Now? The 1975 Dag Hammarskjöld Report", prepared for the 7th Special Session of the United Nations General Assembly, *Development Dialogue* 1, 2.

—— and Clüsener-Godt, M. (eds) (1994) *Extractivism in the Brazilian Amazon: Perspectives on Regional Development*, Paris: UNESCO/MAB Digest.

—— and Coméliau, C. (eds) (1988) *Histoire, culture et styles de développement. Brésil et Inde. Esquisse de comparaison*, Paris: L'Harmattan/Unesco/CETRAL.

—— and Kalecki, M. (1966) "Forms of Foreign Aid: An Economic Analysis", *Social Sciences Information* 5: 1.

—— and Mancebo, F. (eds) (2015), *Transitions to Sustainability*, Dordrecht: Springer.

—— and Silk, D. (1990) *Food and Energy: Strategies for Sustainable Development*, Tokyo: United Nations University Press.

—— and Vinaver, K. (1979) "Integration of Technology in Development Planning: A Normative View", in F. R. Sagasti and A. Araoz (eds), *Science and Technology for Development: Planning in the STPI Countries*, Ottawa: International Development Research Center, 117–136.

—— Wilheim, J. and Pinheiro, P.S. (eds) (2010) *Brazil: A Century of Change*, Chapel Hill, NC: University of North Carolina Press.

Further reading

Caporali, R. and Guerra Amorim, M. (2014) *Desenvolvimento, inovação e sustentabilidade. Contribuições de Ignacy Sachs*, Rio de Janeiro: Gamarond.

Freire Vieira, P., Andrés Ribeiro, M., Messias Franco, R. and Caporali Cordeiro, R. (eds) (1998) *Desenvolvimento e meio ambiente no Brasil. A contribuição de Ignacy Sachs*, Florianopolis: Editora Pallotti/APED.

Passaris S. and Vinaver, K. (eds) (1998) "Pour aborder le XXIe siècle avec le développement durable", *Économies et sociétés*, series F: *Développement, croissance et progrès*, 1, Grenoble: Presses universitaires de Grenoble.

Simon, D. (2019) *Holocaust Escapees and Global Development: Hidden Histories*, London: Zed Books.

KRYSTYNA VINAVER AND MARLÈNE MONTEIRO

E.F. (FRITZ) SCHUMACHER (1911–1977)

Fritz Schumacher's book *Small is Beautiful*, published in 1973, was undoubtedly a landmark in encouraging a redefinition of the complex relationships between people, environment, technology and development. Subtitled *A Study of Economics as if People Mattered*, the book challenged orthodoxies and raised issues that remain just as relevant today, with present concerns about such matters as climate change, pollution and sustainability.

Small iis Beautiful had a truly global impact, with over 700,000 copies sold in multiple languages. As Peter Lewis commented in the cover notes of the paperback edition (1974):

> A book of heart and hope and downright commonsense about the future ... The basic message in this tremendously thought-provoking book is that man is pulling the earth and himself out of equilibrium by applying only one test to everything he does: money, profits and therefore giant operations. We have got to ask instead, what about the cost in human terms, in happiness, health, beauty and conserving the planet?

More than 40 years after his premature death in 1977, the name of Schumacher is still universally associated with *Small is Beautiful*, and the concept of "intermediate technology", which he popularised. Yet both this book and his lesser-known, more theoretical second volume, *A Guide for the Perplexed* (1977), represented the culmination of a lifetime of deep thought and indefatigable commitment, that were stimulated by Schumacher's journeys and personal encounters, his wide reading and his impressive knowledge and understanding. By any yardstick, his lifetime achievements were considerable and varied, and since his death his ideology has been championed by numerous individuals and organisations across the world.

Ernst Friedrich (Fritz) Schumacher was born on 16 August 1911 in Bonn, Germany, the son of a charismatic and well-connected professor of economics. He followed in his father's footsteps by studying law and economics, initially at Bonn University. Then, in October 1930, he was awarded a Rhodes scholarship to study at Oxford. But even before he took up his place there, at the age of 18 and on his first visit to England, he was introduced to John Maynard Keynes, who invited the young Fritz to attend his prestigious seminars in Cambridge. England in the 1930s was not an easy place for a young German to live, and in any case he didn't like Oxford, so he requested an additional year of his scholarship to be based at Columbia University in New York. At Columbia he was much more settled, and developed a good rapport with the outspoken professor of banking, Parker Willis, who, recognising the 22-year-old's ability and potential, invited Fritz to give a course of lectures and seminars in 1933.

A year later he returned to Germany, but the deteriorating situation there and his strong anti-Nazi stance prompted Fritz and his young wife, Muschi Petersen, to leave for London in 1937. Through his many contacts, Fritz was offered a small farmworker's cottage on Lord Brand's estate at Eydon, Northamptonshire, where he lived during the wartime period, working first as a farm labourer and then, from 1942, at the Oxford Institute of Statistics. A period spent during 1940 in an internment camp at Prees Heath, Shropshire, experiencing harsh physical and emotional conditions, had a significant impact on Schumacher's personal development. Four months after entering the camp, he was elected camp leader by his fellow inmates, and it was there that he began to appreciate the relevance of **Marx**'s ideas and the importance of understanding the political and social dimensions of economic issues (see Wood 1984 and Simon 2019 for further details). As his daughter, Barbara Wood, writes in her biography of her father:

> When he was released from Prees Heath, he came out not as a man emerging from the hard conditions that go with the deprivation of liberty, but as though he was returning from a stimulating seminar, on fire with new ideas and visions. He left Prees Heath invigorated …
>
> (Wood 1984: 113)

Schumacher became a British citizen in April 1946, and in November 1949 was offered the post of Economic Adviser to the National Coal Board (NCB), which was to be his workplace until early retirement in 1971. In the meantime, and following visits to Burma (1955) and India (1961 and 1962), in 1965 he teamed up with his NCB colleague George McRobie to establish the Intermediate Technology Development Group (ITDG).

With his background in economics, Fritz Schumacher was well aware of the literature and understood the arguments sufficiently well to feel confident in challenging orthodox viewpoints. His early work focused on the question of unemployment and the interface between labour and technology. In prewar Germany he was concerned about the six million unemployed, and he talked about what he called "Fritz's World Improvement Plan", an incentive scheme for manufacturers to employ labour rather than machinery. In 1943 he wrote a pamphlet for the left-wing Fabian Society on *Export Policy and Full Employment*, and he later worked closely with Sir William Beveridge (the architect of the British welfare state), in producing the report on *Full Employment in a Free Society* (1944). These experiences, coupled with his reading of **Marx** and Lenin, engendered a deep concern in Schumacher for morality, and the nature of work that people did and how they felt about it. In the early 1940s, he said that he regarded Lenin as "more exciting and illuminating than anything I know" (Wood 1984: 137). At that time, Schumacher was intolerant of religion, believing that intelligence and Christianity were incompatible.

However, his views on religion gradually changed. Undoubtedly a key landmark in his personal development was the three months he spent in Burma in 1955, as a UN-funded Economic Adviser, on secondment from the NCB. He became fascinated with the Buddhist approach to economics, which makes a distinction between "renewable" and "non-renewable" resources. Here too there was a link with his longstanding interest in employment. In a chapter on "Buddhist Economics" in *Small is Beautiful* (1974: 47) he comments:

> The very start of Buddhist economic planning would be a planning for full employment ... While the materialist is mainly interested in goods, the Buddhist is mainly interested in liberation ... The keynote of Buddhist economics, therefore, is simplicity and non-violence ... From an economist's point of view, the marvel of the Buddhist way of life is the utter rationality of its pattern – amazingly small means leading to extraordinarily satisfactory results.

Schumacher was also fascinated with the views of Gandhi, who believed that, "the poor of the world cannot be helped by mass production, only by production by the masses" (p. 128).

The visits to Burma and India also fired his interest in technology. In relation to the latter, Schumacher believed that, "the biggest single collective decision that any country in the position of India has to take is the choice of technology" (1974: 175). Rather than unquestioningly transferring technology from rich to poor countries, he argued that what was particularly needed was *intermediate technology*, that was both superior to outdated and primitive technology and also simpler, cheaper and freer than the technology of the rich. He was sceptical about sophisticated technology, and in particular of its dubious role in alleviating poverty: "Can we develop a technology which really helps us to solve our problems – a technology with a human face?" (p. 123). At a conference of eminent economists in Cambridge in September 1964, he presented a well-argued paper on the economic motivation for intermediate technology, which caused uproar and proved to be the talking point of the entire conference.

Fritz Schumacher was also deeply concerned with the quality of human life, and the need to improve the lot of the unemployed, the poor and other marginalised elements of economies and societies. Just as technology needs to have "a human face", so, he argued, economics should demonstrate a greater awareness of the human dimension, *as if people mattered*. He suggested (1974: 217):

> Economics and the standard of living can just as well be looked after by a capitalist system, moderated by a bit of planning and redistributive taxation. But culture and, generally, the quality of life, can now only be debased by such a system.

He also questioned the role of economics in clarifying the meaning and nature of development in the context of both rich and poor countries, observing (p. 171):

> Economic development is something much wider and deeper than economics, let alone econometrics. Its roots lie outside the economic sphere, in education, organisation, discipline and, beyond that, in political independence and a national consciousness of self-reliance.

Schumacher criticised the focus of much economic writing: "We do not approach economics primarily from the point of view of people; we approach it from the point of view of the production of goods, and the people as a kind of afterthought" (McRobie 1981: 6). He frequently extolled the virtues of what he termed "non-violent economics", and in 1960 wrote a much-quoted article for the UK Sunday newspaper, *The Observer* (21 August), in which he said:

> A way of life that ever more rapidly depletes the power of earth to sustain it and piles up ever more insoluble problems for each succeeding generation can only be called violent. It is not a way of life that one would like to see exported to countries not yet committed to it.

Schumacher was also fascinated with the nature of the workplace and the work experience, and in *Small is Beautiful* he writes about "[n]ew patterns of ownership", in which large companies might be broken down into smaller worker-friendly subsidiaries, where all workers can feel a sense of ownership and receive rewards that are commensurate with their efforts, as reflected in production and profitability. He believed that "people (should) have a chance to enjoy themselves while they are working, instead of working solely for their pay packet and hoping, usually forlornly, for enjoyment solely during their leisure time" (Schumacher 1974: 16–17).

Environmental issues were a lifelong passion with Schumacher. Soon after establishing the family home at Caterham, Surrey, in 1950, he became a keen gardener and an active member of the Soil Association, experimenting with organic methods in his own back garden. On a broader scale, he argued: "In agriculture and horticulture, we can interest ourselves in the perfection of production methods which are biologically sound, build up soil fertility, and produce health, beauty and permanence. Productivity will then look after itself" (1974: 16–17). He was a keen supporter of the Soil Association, and referred to its objectives regularly in his writing and lectures. In 1970 he became the Association's President.

He was also concerned about the disproportionate use of non-renewable resources by the richer countries, referring to the case of the USA, "the 5.6% of the world population which live in the United States require something of the order of 40% of the world's primary resources to keep going" (1974: 98). Large powerful countries and organisations are also responsible for damaging the environment. Indeed, he emphasised: "Although even small communities are sometimes guilty of causing serious erosion, generally as a result of ignorance,[1] this is trifling in comparison with the devastations caused by gigantic groups motivated by greed, envy, and the lust for power" (p. 29).

Fritz Schumacher was a brilliant orator and was invited to speak on many occasions across the world. From the mid-1960s, he travelled widely and regularly, his frequent absences from the NCB being generously condoned by his boss, Lord Robens. Schumacher visited the President of Peru, then **Julius Nyerere** of Tanzania, whose Arusha Declaration on "African Socialism" greatly impressed him. Other high-level visits were made to Zambia and South Africa, and in 1977, he gave a six-week coast-to-coast lecture tour in the USA and was invited to meet President Jimmy Carter at the White House.

Schumacher's final lecture, delivered in Caux, Switzerland, on 3 September 1977, the day before he collapsed and died, was typical. Reproduced in McRobie's book *Small is Possible* (1981), the lecture was titled "Technology for a Democratic Society", with the key message being that both highly industrialised and developing countries of the world must begin to develop technologies that are more in harmony with people and the environment, and are less dependent on non-renewable resources (McRobie 1981: 1). The lecture was lively, accessible and non-technical, drawing upon fascinating personal anecdotes to illustrate specific points, and making reference to such people as Robert McNamara, the then President of the World Bank, and Adam Smith's book *The Wealth of Nations*. But so many of Schumacher's ideas, which later found their way into papers and books, were initially trailed in several lecture courses which he gave at London University. His lectures on "Crucial Problems of Modern Living" were highly popular with students and initiated much debate. In these, he called for the raising of consciousness, or an "inner development", so that the necessities for the world's survival could be worked out carefully and put into practice. This call is revisited as the very last point in *Small is Beautiful*, where Schumacher says, "Everywhere people ask: 'What can I actually *do*?' The answer is as simple as it is disconcerting: we can, each of us, work to put our own house in order" (Schumacher 1974: 249–250).

Schumacher's message has been carried forward in so many ways. The Schumacher Circle was set up in 1994 as an informal network to bring together ten organisations that were inspired by Schumacher's work. In the UK are the Centre for Alternative Technology in Machynlleth, Wales, the Jeevika Trust (formerly the India Development Group), the New Economics Foundation in London, *Resurgence and Ecologist* magazine, the Soil Association, Practical Action (formerly the Intermediate Technology Development Group, ITDG), Schumacher College in Devon, the Schumacher Institute in Bristol and the publisher *Green Books*. In the USA, the Schumacher Center for a New Economics (formerly the E.F. Schumacher Society) is based in Great Barrington, Massachusetts.

On the centenary of Fritz Schumacher's birth, 16 August 2011, many events were held to celebrate his work, notably at the House of Commons, Westminster, at Oxford University and at a Schumacher Centenary Festival held in Bristol in October 2011. Such events and the organisations which exist today, over 40 years after Schumacher's death, are themselves a testament to the power and continuing relevance of his ideas. His contribution to development studies, as part of a cohort of influential Holocaust escapees and survivors, is examined by Simon (2019). More generally, global warming, pollution, the depletion of non-renewable resources, the alleviation of poverty, sustainable development, the nature and effects of different technologies, organic farming and, not least, the quality of life all remain high on international and national agendas in the twenty-first century. This is surely indicative both of the significance of Schumacher's inspirational message and also that he was without doubt well ahead of his time.

Note

1 This assertion of causality is hotly contested, not least by political economists, who emphasise structural causes and constraints, or lack of choice in daily survival struggles.

Major works

Schumacher, E.F. (1973) *Small is Beautiful: A Study of Economics as if People Mattered*, London: Blond and Briggs. (Published in paperback by ABACUS, London. in 1974.)

—— (1977) *A Guide for the Perplexed*, London: Jonathan Cape.

—— (1979) *Good Work*, London: Harper and Row.

Further reading

ITDG (2003) *Small Is Working: Technology for Poverty Reduction*, Paris, UNESCO/ ITDG/TVE.

Kirk, G. (ed.) (1982) *Schumacher on Energy*, London: Jonathan Cape.

Kumar, S. (ed.) (1980) *The Schumacher Lectures*, London: Blond and Briggs.

McRobie, G. (1981) *Small is Possible*, London: Jonathan Cape.

Schumacher, E.F. (1943) *Export Policy and Full Employment*, London: Fabian Society.

Scott, A. (1996) "Appropriate Technology: Is It Ready for – and Relevant to – the Millennium?" *Appropriate Technology* 23(3): 1–4.

Simon, D. (2019) *Holocaust Escapees and Postwar Development: Hidden Histories*, London: Zed Books.

Smillie, I. (2000) *Mastering the Machine Revisited: Poverty, Aid and Technology*, London: ITDG.

Wood, B. (1984) *Alias Papa: A Life of Fritz Schumacher*, London: Jonathan Cape.

TONY BINNS

DUDLEY SEERS (1920–1983)

"The world is inconveniently large to cover in one lifetime," wrote Dudley Seers in the Preface to his most important work, *The Political Economy of Nationalism* (1983), completed shortly before his death. Not a gesture of self-importance, his listing of over 35 countries in which he had worked was meant to underscore several unique features of his development work: the belief in a steady engagement with real situations as a basis for reliable knowledge and policy formulation; the notion that knowledge is largely influenced by one's experiences and historical context; and an acute inclination to question established truths, including his own. The fact that those countries in which he worked happened to be small or medium-sized (e.g. Colombia, Sri Lanka, with some in the early years of radical transformation including Cuba, Chile and Portugal) made him aware of issues that were to recur throughout his life, including dependency and inequality. He was, above all, an independent thinker working within a structuralist framework that conformed neither to neoclassical nor to neo-Marxist paradigms.

Seers's interest in empirical data went back to his early days as a statistician at Oxford in the 1940s. His contributions in this area were substantial, including on topics such as social development indicators (1972; 1977); the articulation between political-economic problems, planning models and national accounting systems (see his contribution to a volume in honour of one of his collaborators, **Hans W. Singer** (Seers 1976a)); and the insistence on reliable statistics on key aspects of a country's socio-economic and cultural life that could realistically inform planning "targets for key resources and styles of consumption" (1976b: 12). Seers's interest in statistics, however, was characteristically self-reflexive; he asserted outright that numbers are artefacts, and acknowledged that underlying any accounting system there lies a view of the world. As **Richard Jolly** (1992: 494) put it, it was in the areas of national planning and statistics that Seers was "deeply radical, perhaps more constructively radical than in any other area of his work". Today we would call Seers a "constructivist", believing that statistics are not value-neutral, since they narrate power-laden stories about the world. He urged planners to develop statistics that could serve the cause of the poor effectively.

Seers is best known for two crucial interventions in development debates: the inadequacy of neoclassical economics to non-industrialised countries, and his critique of growth as a standard of development. Beginning in the early 1960s, he published a handful of wonderfully titled papers on these topics which became among the most celebrated of the time. These interests were influenced by his formative years at the Economic Commission for Latin America (ECLA, or CEPAL in Spanish, 1957–1961). These were the years when CEPAL's founder, **Raúl Prebisch**, was developing his core–periphery model of underdevelopment, resulting in the famous dependency perspective in the 1960s. Prebisch's challenges to neoclassicism and emphasis on structural factors did not go unnoticed by Seers.

"The Limitations of the Special Case" (1963) made the argument that the economic doctrines and preoccupations developed for industrial economies were largely inapplicable to developing countries because the frame of reference was fundamentally different – indeed, these doctrines applied to a "highly special" case (1967 [1963]: 4–5); what was needed was a framework that did not render invisible the defining features of less-developed economies (e.g. entrenched land tenure systems, economic and political dependency). Less noticed in Seers's work has been his pioneering epistemological analysis. At a time when appeal to paradigms and research programmes was barely thinkable or still to come, Seers's analysis constituted a lucid statement on how economic doctrines sought hegemony and replaced one another, underscoring some of the central practices through which scientific discourses are deployed, including the use of textbooks and graduate training. He found his fellow economists conveniently unaware of this fact, save for exceptions such as his mentor, Professor **Joan Robinson**. Marxist economics was vulnerable to the same critique. In "The Birth, Life, and Death of Development Economics", he added "professional convenience and career interests" (1979: 709) among the factors conditioning the hegemony of particular economic theories.

That Seers found more in common with the Latin Americans who emphasised structural factors than with those who applied allegedly universal economic models (1981) was at the root of his critique of economic growth as goal and yardstick of development. Instead, he proposed a view aptly encapsulated in an oft-quoted passage from "What Are We Trying to Measure?"

> The questions to ask about a country's development are therefore: What has been happening to poverty? What has been happening to unemployment? What has been happening to inequality? If all three of these have become less severe, then beyond doubt this has been a period of development for the country concerned.

Otherwise, rapid rates of growth notwithstanding, "it would be strange to call the result "development" "(1972: 24). With time, his proposal would influence such important transformations in development thinking as the "redistribution with growth" approaches of the 1970s and landmark policy developments in the 1980s and 1990s such as the basic-needs approach, adjustment with a human face and human development indicators (pioneered by **Hans Singer**, **Richard Jolly**, **Hollis Chenery**, **Paul Streeten**, **Mahbub ul Haq** and **Amartya Sen**). Seers also often referred to the social, cultural and political requirements that could make development meaningful. His death cut short the development of this framework, a task that remains pertinent to this date.

During the last decade of his life, Seers's work took a significant turn in two respects: the application of his development framework to Europe, emphasising nationalism, and a radicalisation of his views on the Third World. The result was an integrated framework which (a) encompassed the entire world and focused on worldwide problems, respecting specificities;

and (b) subsumed development economics into an interdisciplinary field incorporating all important dimensions of development (1977; 1979; 1983). He even suggested that dependency theory could be more pertinent to many European situations than neoclassical theories; this was the case when thinking about Europe's own peripheries (e.g. southern Europe) or underdeveloped regions within core countries, e. g. southern Italy or the Scottish highlands (Seers et al. 1979; Mándi and Seers 1980). His disappointment with internationalism – the increased interdependence of nations which he saw as equivalent to super-power domination – was accompanied by the observation that the Cold War international system had become untenable.

These were the years of the Club of Rome reports on the planet's limits to growth (e.g. Meadows et al. 1972), to which, Seers felt, both capitalism and socialism failed to provide answers. Adding to this the persistence of nationalism, in his last book he postulated a third path to development, based on a judicious form of nationalism and partially self-sufficient regional blocs. Self-sufficiency in basic food staples, technology and culture were the basis of any nationalist and regional development strategy. Changing the structures of demand (particularly patterns of consumption) was essential in this regard, which called for a curb on the influence of transnational corporations in these three areas. Seers's framework resembled positions advocated by radical Third World intellectuals, particularly **Julius Nyerere**'s concept of self-reliance and **Samir Amin**'s strategy of de-linking. Today we could restate Seers's thesis in terms of selective delinking and selective re-engagement, quite different from what came to prevail in the Thatcher–Reagan–Bush neoliberal era.

Seers's anticipatory mind was also in evidence in his discussion, often in passing, of topics rarely addressed during his time, such as race, religion and culture in development. He led five interdisciplinary country missions (particularly for the International Labour Organization's World Employment Programme), each lasting between four and six weeks and each resulting in a major conceptual and policy report. He occupied many prominent positions with academic organisations, the United Nations and non-governmental organisations such as the Society for International Development (the most influential development NGO at the time). Institutionally, he will remain best known as the first director of the Institute of Development Studies at Sussex (1967–1972). From its inception, the Institute became the most exciting place in the global North for young students eager to pursue enlightened careers in the development field. He also helped provide safe haven at the Institute for scholars fleeing military dictatorships from South America and Eastern Europe, who added to the vibrant ambience of the place.

Dudley Seers had been born on 11 April 1920, "a white, English, Christian male", as he was prone to say at the beginning of his courses to warn students about the biases in his works.

> To have been brought up in the family of an executive of General Motors, educated at a preparatory school and Rugby, then at Cambridge, and to have served as an officer of the Royal Navy [1941–1945] and in the civil service at home and abroad, is to be a lifetime captive of these institutions, even (perhaps especially) when one is reacting against the attitudes they attempt to instil.
>
> (Seers 1983: 11)

His profound awareness of the historicity of all knowledge – which today goes under the rubric of "positionality" – was exemplary for his moment. Lest this sound as though he was a postmodernist *avant la lettre*, it should be remembered that his acute self-reflexivity went side by side with a deep commitment to some of the grand narratives of modernity, certainly egalitarianism and a humanism infused with the best traditions of critical thought. He did not shy away from proposing practicable solutions to what he saw as the pressing problems of the day; his pragmatism was informed by a realistic diagnosis of power and a distrust of grandiose development objectives common in the literature of the time (witness his ironic and uncompromising critique of the Brandt Report (1980) on North–South relations, which emerged out of Seers's analysis as a paradigmatic instance of a discourse run amok).

Seers's prescient mind could be found wanting on a number of issues; he was, after all, a child of his age. Even if he was always willing to reinvent it, he believed in development perhaps because, as **Hans Singer** (1983) put it, he was "a practical visionary" who had "a knack for getting his ideas and visions translated into practice". He was not an iconoclast in the sense of, say, Ivan Illich among his more radical contemporaries; he was critical of many aspects of the Enlightenment tradition but did not call for a systematic overhaul. However, he fulfilled a tremendously important role as one of the few real *enfants terribles* – or, more precisely, critical consciences – of the development age, and he was respected by Marxist and neoclassical theorists alike, despite the fact that both were equally the target of his acerbic pen.

By the time he died on 21 March 1983, he had become an inevitable reference in the development field. He was, and remains, one of the finest minds and kindred spirits to work for the well-being of what was called – in what he characterised as "the odiously euphemistic language of the UN" – the Third World. Many of his works are as relevant today as they were when they appeared, with a few pertinent adaptations. The critical examination of orthodox economics ("deconstruction", in today's jargon) is as pressing today as at the time of "The Limitations of the Special Case", given the greater hypertrophy of econometrics and the hegemony of neoliberal approaches. Nationalisms and regional blocs capable of countering the present imperial power have, if anything, become more salient issues today; even some contemporary attempts at harnessing multiculturalism would not be inconsistent with his vision for the European case. There is much he would decry in what is happening today – the self-serving internationalism of market forces, the pernicious effect of global media on the world's cultures, and the seeming capitulation of international lending institutions to the dictates of a world run by, and for the benefit of, large transnational corporations. His development framework has by no means become part of the mainstream. This does not mean that his work has ceased to be relevant – perhaps it is now even more so; or perhaps one could say with greater pertinence that he remains, even today, a pioneer of development thought.

Acknowledgements

I would like to thank Sir Richard Jolly for his clarifying and detailed comments, and Wendy Harcourt for her constructive suggestions on an earlier draft.

Major works

De Bandt, J., Mándi, P. and Seers, D. (eds) (1980) *New Trends in European Development Studies*, New York: St. Martin's Press.

Seers, D. (1967) "The Limitations of the Special Case", in K. Martin and J. Knapp (eds), *The Teaching of Development Economics*, Chicago: Aldine, 1–27. (Originally published in *Oxford University Institute of Economics and Statistics Bulletin* 25, 1963.)

—— (1969) "The Meaning of Development", *International Development Review* 11 (4): 1–6.

—— (1972) "What Are We Trying to Measure?" *Journal of Development Studies* 8 (3): 21–36.

—— (1976a) "The Political Economy of National Accounting", in A. Cairncross and M. Puri (eds), *Employment, Income Distribution and Development Strategy: Problems of the Developing Countries. Essays in Honour of H.W. Singer*, New York: Holmes and Meier, 193–209.

—— (1976b) "A New Look at the Three World Classification", *IDS Bulletin* 7: 8–13.

—— (1977) "The New Meaning of Development", *International Development Review* 19(3): 2–7.

—— (1979) "The Birth, Life and Death of Development Economics", *Development and Change* 10: 707–719.
—— (1980) "North–South: Muddling Morality and Mutuality", *Third World Quarterly* 2(4): 681–693.
—— (ed.) (1981) *Dependency Theory: a Critical Reassessment*, London: Frances Pinter.
—— (1983) *The Political Economy of Nationalism*, Oxford: Oxford University Press.
—— and Joy, L. (eds) (1971) *Development in a Divided World*, Harmondsworth: Penguin.
—— Schaffer, B. and Kiljunen, M.-L. (eds) (1979) *Underdeveloped Europe: Studies in Core–Periphery Relations*, Brighton: Harvester Press.
—— and Vasitos, C. (eds) (1980) *Integration and Unequal Development: The Experience of the EEC*, New York: St. Martin's Press.Further reading
Anon. (1990) "Dudley Seers", in S. M. Trosky (ed.), *Contemporary Authors*, vol. 129, Detroit, MI: Gale Research, 395–396.
Brandt Commission (1980) *North–South: A Programme For Survival*, London: Pan.
Jolly, R. (1992) "Dudley Seers (1920–1983)", in P. Arestis and M. Sawyer (eds), *A Biographical Dictionary of Dissenting Economists*, Aldershot: Elgar, 491–499.
Meadows, D.H., Meadows, D.L., Randers, J. and Behrens, W. (1972) *The Limits to Growth*, London: Pan.
Singer, H. (1983) "Tribute to Dudley Seers", in *The New Internationalist: The People, The Ideas, The Action in the Fight for World Development* 125 (July), Oxford: New Internationalist Publications.
Times, The (1983) "Professor Dudley Seers: Leading Overseas Development Economist", 23 March: 12.
United Nations Intellectual History Project, www.unhistory.org.
Ward, M. (2004) *Quantifying the World: UN Contributions to Statistics*, New York: United Nations. (Analysis of the pressures and biases underlying the UN's role in development statistics.)Country reports and major missions
International Labour Office (1970) *Towards Full Employment: A Programme for Colombia*, Geneva: ILO. (Mission led by Seers.)
—— (1971) *Matching Employment Opportunities and Expectations: A Programme of Action for Ceylon*, 2 vols, Geneva: ILO. (Mission led by Seers.)
—— (1981) *First Things First: Meeting the Basic Needs of the People of Nigeria*, Geneva: ILO.
—— Bianchi, A., Jolly, R. and Nolff, M. (1964) *Cuba: The Economic and Social Revolution*, Chapel Hill, NC: University of North Carolina Press.

ARTURO ESCOBAR

AMARTYA KUMAR SEN (1933–)

Amartya Kumar Sen is a leading public intellectual of our time. Born in 1933 and brought up in Dhaka (now in Bangladesh), Sen's childhood was marked by encounters with key members of the Bengali intelligentsia and by the horrors of the Great Bengal Famine and the Partition of India. Sen continues to acknowledge the presence of these events in his life and work, large parts of which have been concerned to understand how a lack of substantial freedoms leads to a lack of "functionings", or the ability to live a different or a better life.

Sen is best known in development studies for his work on poverty and famines, and on the measurement of "human development", but the citation for his 1998 Nobel Memorial Prize for Economic Science also salutes his work on social choice theory. In addition, he is known to readers of both the New York and the London *Review of Books* for well-crafted essays on environmental sustainability, religious nationalism and violence in India, and the poetic and political careers of Rabindranath Tagore. In key respects, Sen is a descendant of the "Bengal Renaissance" which began in the early 19th century. His rapid rise to prominence, and the breadth and quality of his published work, call to mind forebears like Ranmohun Roy and Tagore (an early teacher), along with contemporaries like Partha Chatterjee and Partha Dasgupta. Sen was a Professor at Jadavpur University by the age of 23, and later taught at the Delhi and London Schools of Economics before moving in turn to Oxford, Harvard and the Mastership of his old Cambridge college, Trinity. In addition to the Nobel Prize, which he received while at Cambridge, Sen has been awarded India's highest honour, the Bharat Ratna. In 2004 he returned to teaching and research at Harvard.

Few, if any, individuals have contributed more to development studies than Amartya Sen. A version of his 1957 Prize Fellowship dissertation at Trinity College was published as his first book (Sen 1960). Thereafter, he published widely on savings and capital in developing countries, on surplus labour in peasant economies, on the relationship between farm size holdings and productivity in Indian agriculture and on the isolation paradox and collective action problems. In the 1970s he began to produce the work for which he is now better known – on the measurement and meanings of human development, on the causes and prevention of famines, on gender issues and inequality and on development as freedom. In addition, while it is a matter of pride for Sen that he has never directly advised governments, it is clear that his work has been shaped by involvements with the United Nations Industrial Development Organization (where his work with Partha Dasgupta and Stephen Marglin (1972) formed the basis of UNIDO's guidelines for project analysis), the International Labour Office (two of his key books (Sen 1975; 1981) grew out of work for the World

Employment Programme of the International Labour Organization) and the United Nations Development Programme (which incorporated aspects of Sen's work on the evaluation of development into its Human Development Index and Human Development Reports).

Notwithstanding the diversity of Sen's work, which informs an extraordinary body of research now also being carried out by former students and colleagues, there is consistency in many of his leading propositions. The first of these has to do with the proper focus of economic and moral evaluations. Sen rejects the idea that individuals act according to a narrow specification of what might count as self-interest. Most of us are not rational fools, and issues like class position, gender, family influence and a sense of fairness affect the way we behave (Sen 1977). He also rejects the idea that different policies can usefully be evaluated in terms of the classic concerns of welfare economics or utilitarianism. "[M]aximizing the sum of individual utilities [pleasure, happiness, welfare] is supremely unconcerned with the interpersonal distribution of that sum. This should make it particularly unsuitable for measuring or judging inequality" (Sen 1973: 16). This is especially the case when it is combined with an insistence on Pareto optimality, or the idea that the utility (or welfare) of anyone should not be raised if it leads to a reduction in the utility (or welfare) of someone else. Such an insistence could lead to a situation where government action in favour of the poor, or those without substantive freedoms, is disabled to the extent that better-off social actors define their utilities in opposition to those of the poor.

Sen is staunchly opposed to this conclusion. In his view we are strongly marked by the accidents of history and geography: by the fact that some of us are born to wealthy families in Beverley Hills while others are born to landless labourers in Bangladesh. He thus agrees with John Rawls (1971) that a theory of justice as fairness would wish to ensure that all humans are endowed with a minimum set of primary goods (including education and an income), subject only to a prior rule that would guarantee equal personal liberties. For Sen, however, the Rawlsian Difference Principle is good, but not good enough. He argues that a strict equation cannot be drawn between primary goods and well-being because the former cannot always be converted into the latter. "For example, a pregnant woman may have to overcome disadvantages in living comfortably and well that a man need not have, even when both have exactly the same income and other primary goods" (Sen 1995: 27).

It follows that Sen's own preference is for a model of economic evaluation which suggests that "[d]evelopment consists of the removal of various types of unfreedoms that leave people with little opportunity of exercising their reasoned agency" (Sen 2000: xii). For Sen, real freedom is defined precisely in terms of certain human and civil rights that must be guaranteed for all. To live a good life we need to be fit and healthy, educated, and exposed to different sources of information. But we must also be free to choose our own accounts of the good life and to participate in market exchange. The choices which free agents make will necessarily be influenced by the differences which constitute us as individual human beings, or which shape our personal circumstances. The conversion of personal incomes and resources into capabilities, achievements, well-being or real freedoms will be affected by personal heterogeneities, environmental diversities, variations in social climate and differences within the family (after Sen 2000: 70–71).

Notions of similarity and difference in turn feed through to the empirical and practical insights that can be claimed for Sen's account of development as freedom. These begin with the language of "functionings" and "capabilities" that Sen uses to fashion his accounts of famine or gender discrimination, among other issues. "Functionings" refer to the things that a person may value doing or being, and thus denote a freedom to achieve a certain lifestyle. "Capabilities" refer to the sets of resources (physical, mental and social) that a person might command and which give rise to various "functionings".

Sen has used this approach to understand the dynamics of famines in conditions both of "boom" (Bengal in 1943–1944) and "bust" (Wollo, Ethiopia, in 1972–1974). Sen maintains that famine deaths result from a precipitate decline in the entitlements of various persons and social groups to parts of the regional food supply. In the case of the Great Bengal Famine, then, the victims were to be found mainly among agricultural labourers, fishermen, transport workers, paddy huskers and others who faced a slackening demand for their services at a time when the demand for labour in urban Bengal was helping to push up the price of the main staple crop, rice. These groups suffered because their exchange entitlements to food were all of a sudden inconsistent with their basic needs or capability to survive. More recently, Sen has argued that "no famine has ever taken place in the history of the world in a functioning democracy" (2000: 16). This claim is at one with the "instrumental" defence that Sen offers for the virtues of freedom. If freedom consists in part, but by definition, in democratic pluralism, then that same system of governance ensures that the most basic economic freedoms are guaranteed by its major institutions, including contending political parties and a free press. As ever, the free flow of information is vital to this process, and to the process of economic evaluation that is thereby entailed.

Similar arguments are at work in Sen's accounts of gender discrimination and the evils of child labour. He has written movingly about the 100 million women who are missing from the world today as a consequence of sex-selective abortion and the comparative neglect of female health and nutrition in childhood. While the precise figure might range from 60 million to perhaps 110 million missing women, the scale of this holocaust is horrible in the extreme. Sen accounts for it in terms of the capability deprivation of women, most of whom do not enjoy the same substantive freedoms as men. This manifests itself not just in unfair patterns of food sharing and health care within a household, but also in terms of the lack of voice from which many females suffer.

Sen insists that such findings bear out the importance of an approach to social comparisons that is focused on capabilities not incomes (see Sen 2015). It also restates the importance of getting inside the "household". Sen follows many feminist scholars in insisting on the asymmetrical distribution of power and resources within the household (almost always to the disadvantage of women), and he buttresses this insistence with his continuing commitment to the freedoms of the individual human subject. This commitment is further apparent in his treatment of the child labour issue. Sen is aware that campaigning groups have drawn attention to the income gap that drives some parents to seek employment for their children. He insists, even so, that child labour is very often linked to slavery or bondage, and that it removes from a child the freedom to attend school. It is thus wrong in and of itself, as indeed are coercive population policies in a country like China. This is where the foundational nature of Sen's commitment to freedom reasserts itself.

Sen's work is widely admired for its crisp prose, common sense and for the way that it consistently foregrounds the importance of public reason in the quest to make societies less unjust (on this, see especially Sen 2009). It has also helped to change the way we think about famines (less emphasis on food availability decline), and development itself (less emphasis on aggregate economic performance, more on empowerment and basic needs). But Sen's work is not without its critics. Peter Nolan (1993) was an early and ungracious critic of Sen's account of the causation and prevention of famines, and David Keen (1994) has shown how freedom to choose often disappears amid war, famines or complex emergencies – a point that Sen sometimes misses. Other critics have charged that he has a poorly developed sense of power and politics (Corbridge 2002; Gasper 2004). Sen's liberalism inclines him to place great emphasis upon the power of ideas and reasoned debate in the promotion of public policy. While he is not unaware of the embedded nature of social and economic inequalities – far from it – he is sometimes inattentive to the political struggles that must be waged to change entrenched structures of power and to secure a greater freedom of choice for a wider range of people. This relates to his additive conception of "freedom". His most recent account of development as freedom is committed to the view that an extension of freedom in one direction is or can always be linked to an extension of freedom in all other directions. He thus has no truck with the view that relatively benign authoritarian states in East Asia might have secured high and sustained rates of economic growth and development precisely because they suspended democratic freedoms in the first decades of structural transformation. This is not the commonly accepted wisdom, however, and it can reasonably be argued that Sen underestimates the power of such regimes to solve collective action problems (Wade 1990).

Two further criticisms should be noted. First, there is a growing technical literature on Sen's accounts of functionings, capabilities and entitlements (Qizilbash 1997; Giri 2000; Devereux 2001; Alkire 2002). This is partly concerned with the ways in which he defines an agent or personhood. If his rational actors are not fools, to what extent and how are they embedded in particular collectivities, cultures or systems of belief? Second, concern has been voiced about the determinacy of some of the policy suggestions that seem to flow from Sen's work. How is greater educational and health investment to be achieved? What institutions does he have in mind, and how would he mobilise support for social investment against defenders of the status quo?

In Paul Seabright's view, "Development as freedom is curiously silent about the difficulty of devising mandates that work" (Seabright 2001: 43). Sen, however, has often pointed out that this job falls mainly to others. In any case, his work has aided the development of famine early-warning systems, just as it has encouraged others to build more rounded indices of human development. Sen has also used his Nobel Prize money to fund two trusts in India and Bangladesh which do have to get to grips with institutional and political realities. The Pratichi India Trust is mainly concerned with tackling illiteracy in West Bengal, while the Pratichi Bangladesh Trust is concerned with gender inequalities and the empowerment of women. Sen continues to lead by example and by responding vigorously, but always carefully, to the criticisms that have been levelled against his work. He remains a towering and much-loved figure not just in development studies but across the social sciences. Fittingly, in 2017, as he continued to raise his voice against poor reason and attempts to foreclose identities, Amartya Sen was awarded the Johan Syktte Prize in Political

Science. The Prize was made "for his multifaceted achievement that 'combines insights into human vulnerability with knowledge about the potential of democratic political power to redress and relieve this deprivation'".

Major works

Drèze, J. and Sen, A.K. (1989) *Hunger and Public Action*, Oxford: Oxford University Press/WIDER.

Sen, A.K. (1960) *Choice of Techniques*, Oxford: Blackwell.

—— (1973) *On Economic Inequality*, Oxford: Clarendon Press.

—— (1975) *Employment, Technology and Development*, Oxford: Oxford University Press/ILO.

—— (1981) *Poverty and Famines: An Essay on Entitlement and Deprivation*, Oxford: Clarendon Press.

—— (1982) *Choice, Welfare and Measurement*, Oxford: Blackwell.

—— (1984) *Resources, Values and Development*, Cambridge, MA: Harvard University Press.

—— (1989) "Food and Freedom", *World Development* 17: 769–781.

—— (1995) *Inequality Reexamined*, Cambridge, MA: Harvard University Press.

—— (2000) *Development as Freedom*, New York: Anchor Books.

—— (2002) *India: Development and Participation*, Oxford: Oxford University Press.

—— (2004) *Rationality and Freedom*, Cambridge, MA: Belknap Press.

—— (2009) *The Idea of Justice*, London: Allen Lane.

—— (2015) *The Country of First Boys*, Delhi: Oxford University Press.

Further reading

Alkire, S. (2002) *Valuing Freedoms: Sen's Capability Approach and Poverty Reduction*, Oxford: Oxford University Press.

Atkinson, A. (1999) "The Contributions of Amartya Sen to Welfare Economics", *Scandinavian Journal of Economics* 101: 173–190.

Corbridge, S. (2002) "Development as Freedom: The Spaces of Amartya Sen", *Progress in Development Studies* 2: 183–217.

Dasgupta, P., Marglin, S. and Sen, A.K. (1972) *Guidelines for Project Evaluation (UNIDO)*, New York: United Nations.

Devereux, S. (2001) "Sen's Entitlement Approach: Critiques and Countercritiques", *Oxford Development Studies* 29: 244–263.

Drèze, J. and Sen, A.K. (1995) *India: Economic Development and Social Opportunity*, Oxford: Clarendon Press.

—— —— and Hussain, A. (eds) (1995) *The Political Economy of Hunger*, Oxford: Oxford University Press/WIDER.

Gasper, D. (2004) *The Ethics of Development*, Edinburgh: Edinburgh University Press.

Giri, A.K. (2000) "Rethinking Human Well-Being: A Dialogue with Amartya Sen", *Journal of International Development* 12: 1003–1018.

Keen, D. (1994) *The Benefits of Famine: A Political Economy of Famine and Relief in Southwestern Sudan, 1983–1989*, Princeton, NJ: Princeton University Press.

Nolan, P. (1993) "The Causation and Prevention of Famines: A Critique of A.K. Sen", *Journal of Peasant Studies* 21: 1–28.

Nussbaum, M. and Sen, A.K. (eds) (1993) *The Quality of Life*, Oxford: Clarendon Press.

Qizilbash, M. (1997) "A Weakness of the Capability Approach with Reference to Gender Justice", *Journal of International Development* 9: 251–262.

Rawls, J. (1971) *A Theory of Justice*, Cambridge, MA: Harvard University Press.

Seabright, P. (2001) "The Road Upward: Review of Development as Freedom", *New York Review of Books* 48(5): 41–43.

Sen, A.K. (1970) "The Impossibility of a Paretian Liberal", *Journal of Political Economy* 78 (1): 152–157.

—— (1977) "Rational Fools: A Critique of the Behavioural Foundations of Economic Theory", *Philosophy and Public Affairs* 6: 317–344.

—— (1990) "More than One Hundred Million Women Are Missing", *New York Review of Books*, 20 December: 61–66.

—— (1993) "The Causation and Prevention of Famines: A Reply", *Journal of Peasant Studies* 21: 29–40.

—— (1996) "Secularism and its Discontents", in K. Basu and S. Subrahmanyam (eds), *Unravelling the Nation: Sectarian Conflict and India's Secular Identity*, New Delhi: Penguin, 11–43.

—— and Williams, B. (eds) (1982) *Utilitarianism and Beyond*, Cambridge: Cambridge University Press.

Vanhuysee, P. (2000) "On Sen's Liberal Paradox and its Reception within Political Theory and Welfare Economics", *Politics* 20: 25–31.

Wade, R. (1990) *Governing the Market: Economic Theory and the Role of Government in East Asian Industrialization*, Princeton, NJ: Princeton University Press.

STUART CORBRIDGE

HANS WOLFGANG SINGER (1910–2006)

Hans (later Sir Hans) Singer was one of the best-known and most respected development economists. His work was recognised in awards, honorary doctorates and a knighthood in 1994 "for services to economic issues". Along with **Albert Hirschman**, he was included as one of the ten "pioneers in development" in a World Bank book published in 1984 (Meier and Seers 1984). Six *Festschriften* were published in his honour by the time of his death in 2006 (Cairncross and Puri 1976; Clay and Shaw 1987; Chen and Sapsford 1997; Sapsford and

Chen 1998; Sapsford and Chen 1999; Hatti and Tandon 2004). They show the depth and breadth of Singer's influence in development economics, and the esteem and affection in which he was held. Singer produced 450 publications after his first paper appeared in 1935. A complete list up to then appears in Shaw (2002; 2004), but he continued publishing until his death. He addressed more issues than any other development economist. The downside was that he dispersed his efforts so widely that he never produced the one definitive work that he carried in him.

Singer was born on 29 November 1910 in Elberfeld, now part of Wuppertal, in the German Rhineland, into a strongly assimilated, largely secular, middle-class Jewish family. Two personalities, in particular, helped to shape his outlook on life: his father, a hard-working doctor with a large practice, who often treated the poor without charge, and the local rabbi, a highly respected member of society who gave him instructions in ethics, moral values and civic responsibility.

Entering Bonn University in 1929 with the intention of studying medicine, Singer switched to economics after attending a lecture by Joseph Schumpeter and came under his spell, and that of his economic masterpiece, *The Theory of Economic Development*. Singer was forced to leave Germany immediately the Nazis came to power in 1933. Fortunately, on the recommendation of Schumpeter, he received a two-year scholarship (1934–1936) at Cambridge University to complete his PhD work. At Cambridge, he came under another spell, that of John Maynard Keynes, precisely at the time that Keynes was producing his economic masterpiece, *The General Theory of Employment, Interest and Money*. Singer's PhD dissertation was entitled "Materials for the Study of Urban Ground Rent", which covered the period 1845 to 1913. This showed his aptitude for analysing long-term data series. He was among the first group of students to receive doctorates in economics from Cambridge University.

Singer's first employment after university was on a major two-year (1936–1938) study of long-term unemployment in the depressed areas of the United Kingdom, organised by the Pilgrim Trust under the chairmanship of Archbishop William Temple, with Sir William (later Lord) Beveridge as its main adviser (Pilgrim Trust 1938). These two personalities and the study made a strong impression on him. There followed lectureships at Manchester (1938–1944) and Glasgow (1946–1947) universities. The former was interrupted by a frustrating period of internment as an "enemy alien" during 1940–1941 since his naturalisation papers had not yet come through. He was eventually released through the intervention of Keynes and other influential interlocutors able to demonstrate his usefulness to the war effort (Simon 2019). During brief employment at the Ministry of Town and Country Planning in London (1945–1946), Singer was involved in the calculation of compensation for owners as part of the Labour Party's intended programme for the nationalisation of development rights in urban land, drawing on his PhD work at Cambridge.

In 1947, there occurred an event that was to dramatically change the course of Singer's career. David Owen, who had worked with Singer on the unemployment study, was appointed the first head of the Department of Economic Affairs at the newly created United Nations. Owen sought Singer's services to strengthen his new department. Singer accepted a two-year secondment, which was to last for 22 years (1947–1969). He was assigned to work in a small section that dealt with the problems of developing countries. Simultaneously, he was appointed a professor in the graduate faculty of the New School for Social Research in New York. He welcomed the opportunity of maintaining his links with the academic world. It also gave him the opportunity to think about the theoretical framework of economic development within which his work at the UN could be placed. And he was able to publish papers in his own name (a practice not allowed at the UN), which were widely acknowledged and quoted.

Analysis of the long-run terms of trade between industrialised and developing countries is perhaps Singer's best-known contribution to development economics (Singer 1949; 1950; UN 1949). He showed that the net barter terms of trade between primary products and manufactures were subjected to a long-term downward trend. This contradicted the widely held belief at the time that the long-term trend favoured primary products. Singer argued that economic history had been unkind to developing countries. Most of the secondary and cumulative effects of investment had been removed from the developing country to the investing, industrial country. Developing countries had also been diverted into types of activities that offered less scope for technical progress, withholding a central factor of "dynamic radiation" that had revolutionised society in the industrial world.

Singer emphasised that his work on the terms of trade was meant to be more a policy guide than a long-term projection. He advised developing countries to diversify out of primary exports through developing domestic markets and industrialisation. Subsequently, he put more emphasis on relations between types of countries than between types of commodities, and on the distribution of "technological power" in more contemporary terms in answer to some of the criticisms of his original work (**Rostow** 1990). Singer also emphasised the objective of "distributive" justice, pointing to inbuilt, long-term inequalities between industrialised and developing countries. He constantly argued that unless there are fundamental changes in the world economic order, divergence rather than convergence will continue between them, threatening global economic and social advancement, stability and peace. Singer found strong allies, especially **Raúl Prebisch** at the UN Economic Commission for Latin America, who used Singer's original work (Prebisch 1950; Toye and Toye 2003). The Prebisch–Singer thesis, as it came to be known, has created a growth industry in the development economics literature. Increasingly sophisticated statistical and econometric analyses have vindicated the thesis, making it one of the very few hypotheses in economics that have stood the test of time (Sapsford and Chen 1998: 27–34).

What are some of the special attributes that set Singer apart from other pioneering development economists? Singer came under the direct influence of both Schumpeter and Keynes, the two great beacons in his intellectual development. The combination of these forces had a decisive influence on Singer's ideas: Schumpeter with his views on the importance of technology and innovation and the role of the entrepreneur; and Keynes' thesis that economics is not a universal truth applicable to all countries and conditions. They provided Singer with an enabling framework for much of his work.

Research on the terms of trade was the fulcrum for many of the other issues Singer took up in order to achieve "distributive justice" for the developing countries, including industrialisation, science and technology, human investment, planning, trade and aid, technical assistance and preinvestment activities, regional co-operation and a new international economic order with multilateral global governance.

Long service in the United Nations during its formative years was also to benefit Singer's work. He carried out a wide range of assignments that enabled him to see at first hand conditions in the developing world, discuss with leaders and planners their development problems and aspirations and set his own theoretical and conceptual framework against the background of concrete reality. Out of this milieu emerged Singer's "perspectives" on development including: the mechanics of development, the role of the public sector in economic growth, a balanced view of balanced growth, the concept of pre-investment financing, the notion of human investment, the interaction of population growth and education with economic growth, his concept of "redistribution from growth" to tackle poverty and his view that development is "growth *plus* change", culturally, socially and economically.

Singer was also one of the founders of the structural analysis of development. He maintained that there can be no "blueprint" for development (Schiavo-Campo and Singer 1970). However, a number of themes permeated his work. He considered that the starting point should be people, not money and wealth, which gave a whole new perspective on the development process. Sustained and equitable development depended not on the creation of wealth but on the capacity of people to create wealth. This underpinned Singer's insistence on the importance of the human factor in economic development, and on health, education and training, the well-being of children, and food security, science and appropriate technology, employment, income distribution and the conquest of poverty, and planning and sound institutions. All of these needed to be viewed in an international context in which trade and aid are conducted with distributive justice and efficiency so that all countries, developing and developed, might flourish and converge, not diverge. Many of these perspectives have since become conventional wisdoms of progressive development work.

Not being content only to formulate the theoretical underpinnings of the problems of developing countries, Singer became involved in many pioneering ventures to overcome them. A number of organisations that he helped to set up or strengthen in order to tackle Third World problems are still operating today. These include the United Nations Development Programme, the UN Children's Fund, the UN World Food Programme, the UN Research Institute for Social Development, the UN Industrial Development Organization, and the African Development Bank. He was also indirectly involved in the creation of the International Development Association (IDA), the soft-loan arm of the World Bank. The IDA was established in part as a foil to prevent attempts to set up a Special UN Fund for Economic Development, which Singer and others tried to establish in the United Nations (Meier and Seers 1984: 296–303).

On retiring from the UN in 1969, Singer returned to the Institute of Development Studies (IDS) as a professorial fellow, and professor (later emeritus), at the University of Sussex. During this period his output was prolific and he was still engaged in writing a paper at the time of his death on 26 February 2006. He made a significant contribution to the activities of IDS, which is now recognised as one of the world's leading institutes of development studies. He is remembered with respect and affection as a source of unlimited help and inspiration by the cohorts of students. And he was in constant demand by governments, multilateral and bilateral aid agencies and non-governmental organisations. Despite his advancing years, Singer's commitment to the case for a more just and equitable world economic order remained undiminished. His life and work serve as an inspiration for the next generation of development economists.

Major works

The following is a representative selection of Singer's many publications:

Ansari, J.A., Ballance, R.H. and Singer, H.W. (1982) *The International Economy and Industrial Development: Trade and Investment in the Third World*, Brighton: Harvester Press.

—— and Singer, H.W. (1977) *Rich and Poor Countries: Consequences of International Economic Disorder*, London: Unwin Hyman.

Jolly, R. and Singer, H.W. (1972) *Employment, Incomes and Equality: A Strategy for Increasing Productive Employment in Kenya*, Geneva: International Labour Office.

Raffer, K. and Singer, H.W. (1996) *The Foreign Aid Business: Economic Assistance and Development Co-operation*, Cheltenham: Edward Elgar.

—— —— (2001) *The Economic North–South Divide: Six Decades of Unequal Development*, Cheltenham: Edward Elgar.

Roy, S. and Singer, H.W. (1993) *Economic Progress and Prospects in the Third World: Lessons of Development Experience since 1945*, Aldershot: Edward Elgar.

Schiavo-Campo, S. and Singer, H.W. (1970) *Perspectives in Economic Development*, Boston, MA: Houghton Mifflin.

Singer, H.W. (1964) *International Development, Growth and Change*, New York: McGraw-Hill.

—— (1975) *The Strategy of International Development: Essays in the Economics of Backwardness by H.W. Singer*, ed. A. Cairncross and M. Puri, Basingstoke: Macmillan.

—— (1998) *Growth, Development and Trade: Selected Essays of Hans Singer*, Cheltenham: Edward Elgar.

—— (2001) *International Development Co-operation: Selected Essays by H.W. Singer on Aid and the United Nations System*, ed. D.J. Shaw, Basingstoke: Palgrave Macmillan.

—— Wood, J. and Jennings, T. (1987) *Food Aid: The Challenge and the Opportunity*, Oxford: Clarendon Press.

Further reading

Cairncross, A. and Puri, M. (eds) (1976) *Employment, Income Distribution and Development Strategy: Problems of Developing Countries. Essays in Honour of H.W. Singer*, Basingstoke: Macmillan.

Chen, J. and Sapsford, D. (eds) (1997) "Economic Development and Policy: Professor Sir Hans Singer's Contribution to Development Economics", *World Development* 25(11): 1853–1956.

Clay, E. and Shaw, D.J. (eds) (1987) *Poverty, Development and Food: Essays in Honour of H.W. Singer on his 75th Birthday*, Basingstoke: Macmillan.

Hatti, N. and Tandon, R. (eds) (2004) *Trade and Technology in a Globalizing World: Essays in Honour of H.W. Singer*, New Delhi: BPRC (India) Ltd.

Meier, G.M. and Seers, D. (eds) (1984) *Pioneers in Development*, New York: Oxford University Press for the World Bank, 273–311.

Pilgrim Trust, The (1938) *Men Without Work: A Report Made to the Pilgrim Trust*, Cambridge: Cambridge University Press. (Repr. New York: Greenwood Press, 1968.)

Prebisch, R. (1950) *The Economic Development of Latin America and its Principal Problems*, Santiago, Chile: United Nations Commission for Latin America.

Rostow, W.W. (1990) *Theorists of Economic Growth from David Hume to the Present with a Perspective on the Next Century*, New York: Oxford University Press.

Sapsford, D. and Chen, J. (eds) (1998) *Development Economics and Policy: The Conference Volume to Celebrate the 85th Birthday of Professor Sir Hans Singer*, Basingstoke: Macmillan.

—— —— (1999) "The Prebisch–Singer Thesis: A Thesis for the New Millennium", *Journal of International Development* 11(6): 863–916.

Shaw, D.J. (2002, 2004) *Sir Hans Singer: The Life and Work of a Development Economist*, Basingstoke: Palgrave Macmillan (hardback); New Delhi: BRPC (India) Ltd (paperback).

Simon, D. (2019) *Holocaust Escapees and Global Development: Hidden Histories*, London: Zed Books.

Singer, H.W. (1949) "Economic Progress in Under-Developed Countries", *Social Research* 16(1): 236–266.

—— (1950) "The Distribution of Gains between Investing and Borrowing Countries", *American Economic Review* 40(2): 473–485.

Toye, J. and Toye, R. (2003) "The Origin and Interpretation of the Prebisch–Singer Thesis", *History of Political Economy* 35(3): 437–467.

UN (1949) *Relative Prices of Exports and Imports for Under-Developed Countries: A Study of Post-War Terms of Trade between Under-Developed and Industrialized Countries*, 1949/II.B.3, New York: United Nations.

JOHN SHAW WITH DAVID SIMON

FRANCES STEWART (1940–)

The work of Frances Stewart, a pre-eminent figure in development economics and interdisciplinary development studies, has exercised a decisive influence in the field of international development since the 1970s, encompassing poverty, inequality, human development and conflict. As a pioneer of the human development approach that revolutionised development thinking and underpinned the UN Millennium and Sustainable Development Goals, Stewart served as the President of the Human Development and Capability Association (2008–2010) and chaired the United Nations Committee for Development Policy (2010–2012). She received UNDP's **Mahbub ul Haq** Award for outstanding contributions to human development in 2009. She is known for advocating structural adjustment with a human face and for taking issue with development policy inspired by neoclassical economic thinking and economic models based on the individual. The hallmark of her approach is an emphasis on collective norms and the social character of human development as well as the inherently political and contested nature of development policy. Her research is grounded in the empirical realities of developing countries, and she has always been involved in practical policy application. She has been described as a "norm entrepreneur", for "the promotion of a set of moral values as the basis for international development policy" (Toye 2011: 38–39).

Stewart's mother, Clarissa Goldschmidt, was a history graduate from Oxford and a Labour Party supporter, whose deep political commitment inspired Stewart's own principled interventions in international development debates. Her father was the renowned economist Nicholas Kaldor. Born on 4 August 1940 and educated at Oxford, gaining a First in Philosophy, Politics and Economics (PPE) from Somerville College, Stewart started her career in the UK Treasury. Her interest in development emerged in the late 1960s when she went to Kenya. Her husband, Michael, was the Economic Adviser to the Finance Minister, and Stewart became a Lecturer in Economics at the University of East Africa, Nairobi. This experience prompted her to join debates in development economics by embarking on doctoral research at Oxford, under the supervision of **Paul Streeten**. In her research on technology adoption, Stewart (1977) questioned the neoclassical assumption of labour and capital being substitutable factors of production. She held that developing countries needed to avoid mass unemployment and had a more restricted range of options for capital–labour combination. Drawing upon **E.F. Schumacher**'s seminal work, Stewart advanced the notion of "intermediate technology", wherein developing countries would fashion indigenous labour-intensive technology of their own, while selectively using foreign technology, as suited to the nature of the product and scale of production.

In the early 1970s, Stewart became a Fellow of Somerville College and was appointed to an academic post at Oxford's International Development Centre, Queen Elizabeth House (QEH), where she was to spend the rest of her academic career until retirement as Professor of Development Economics in 2010. As the Director of QEH (1993–2003), she transformed it into the leading academic department of International Development in the UK. In her early days at Oxford,

Stewart intervened in the debate on social cost–benefit analysis (SCBA) for choice of development projects by governments. She argued (1975) that SCBA was based on the questionable assumption that governments would objectively allocate weights and value judgements to different project options. Stewart contended that government decisions were the result of political pressure and lobbying by different socio-political classes. Toye (2011: 24) notes that this view "puts a premium on the economist acting as an advocate, consciously operating as part of the political fray, rather than trying to stand above it. This key insight informed the rest of Frances's career."

In the late 1970s, it was increasingly apparent that growth-oriented economic policies had failed to benefit the poor. This inspired new thinking on poverty. Along with other economists, including **Paul Streeten**, Stewart (1985) argued that the poor had basic needs that could not be met only through income-generation measures, but required the provision of public goods and services by the government. The Basic Needs Approach (BNA) influenced donor policy by shifting funding towards the social sector and poverty alleviation, and stimulated targeted intervention for universal primary healthcare and education. BNA was adopted by the ILO, the World Bank, UN bodies and bilateral aid agencies.

BNA would come to constitute a strand of the human development approach in the 1990s, but meanwhile in the 1980s, the debt crisis and the rise of neoliberal ideology led to the promotion of structural adjustment programmes (SAPs) by international financial institutions as an aid conditionality. The impact was to deepen inequality and exacerbate poverty. Stewart intervened at this critical juncture, when she was a special adviser on adjustment at UNICEF. In a two-volume study entitled *Adjustment with a Human Face*, Cornia, Jolly and Stewart (1987) revealed the deleterious effects of SAPs, particularly on children and other vulnerable groups, and advocated policies to minimise SAPs' social costs. This radically recast the debate on SAP, and the 1990 *World Development Report* identified poverty as a main development objective. Later on, Ranis, Stewart and Ramirez (2000) examined data on economic growth and social indicators and concluded that growth and human development were mutually interdependent. They emphasised that human development must be taken into consideration from the outset in any economic reform programme.

The human development (HD) approach became increasingly influential from the 1990s, in which Stewart played a critical role, most notably in collaborating with **Mahbub ul Haq** on the first Human Development Report (UNDP 1990). The HD approach resolutely rearticulated the rationale for international development as a project of change from a "human" perspective, instead of a focus on the economy: "The basic objective of development is to create an enabling environment for people to live long, healthy and creative lives" (p. 9). **Amartya Sen** had defined development as freedom to enhance individual capability and human flourishing. Stewart's approach was less philosophical and more pragmatic, aimed to shape development interventions, which prompted her to work with Ul Haq on the Human Development Index (HDI) that specified indicators to measure human development, namely health, education and income. In subsequent writing, however, Stewart has come to view the HDI as "reductionist" and extended the analysis to eleven dimensions of human development, including political freedoms, human rights and self-respect, that better capture all-round human development than the HDI (Stewart et al. 2018). Stewart now emphasises the importance of social institutions, highlighting the role of social cohesion, trust, shared norms of reciprocity and co-operation, collective well-being and collective action, moving away from an individualistic perspective, such as that of Sen. Stewart concurs with Sen that democracy is a conducive condition for human development, but does not find much merit in Sen's idea of a "democratic consensus" being integral to human development, and instead draws attention to a variety of political contexts that can advance HD. In her recent work, Stewart has placed particular emphasis on environmental sustainability as a key factor in HD, as well as the well-being of future generations, advocating the institutional representation of their interests in development policy deliberations.

Stewart's distinctive approach to HD, with a non-individualistic orientation, had developed in tandem with her research on groups in development from the 1990s, with an initial statement in an article in *Oxford Development Studies* (1996) and subsequently further elaborated in a book (2002), jointly edited with Judith Heyer and Rosemary Thorp. Collaborating with scholars from a number of cognate disciplines – politics, sociology, anthropology and history – they compiled a wide variety of case studies, ranging from producer and credit associations to community organisations and grassroots movements, to show the various ways in which groups advance the cause of the poor. With a similar multidisciplinary collaborative approach, Stewart also explored ethnic conflict, by setting up the Centre for Inequality, Human Security and Ethnicity (CRISE), funded by a large grant from the UK government's

Department for International Development. Resonating with Ted Gurr's notion of "relative deprivation", Stewart formulated the concept of "horizontal inequality" (2008) and explained that political conflict and violence occurred due to different ethnic groups being unequally placed in relation to each other in terms of economic resources and assets, political power and social goods. Stewart's interest in violent conflict had also led her, in collaboration with Valpy Fitzgerald, to examine the effects of ongoing civil war and conflict that were, until then, not studied at all. This shed major new light on the economic and social costs of war (Stewart 1993; Stewart, Fitzgerald and Associates 2001), contributing to a shift in aid policy from humanitarian relief to minimising such costs.

Stewart was awarded an honorary doctorate by the University of Sussex (2010) and named one of 50 outstanding technological leaders in 2003, for her contribution to "economic development strategies for promoting anti-poverty campaigns and to help quell armed conflicts in developing countries". She received the Leontief Prize for Advancing the Frontiers of Economic Thought in 2013, from the Global Development and Environment Institute, Tufts University. It was aptly noted at the award ceremony that her research and advocacy have always been for those most in need of protection from conflict, internationally imposed policies and local inequalities.

Major works

Cornia, G.A., Jolly, R. and Stewart, F. (1987) *Adjustment with a Human Face*, 2 vols, Oxford: Oxford University Press.

Heyer, J.F., Stewart, F. and Thorp, R. (eds) (2002) *Group Behaviour and Development: Is the Market Destroying Cooperation*? Oxford: Oxford University Press.

Ranis, G.F., Stewart, F. and Ramirez, A. (2000) "Economic Growth and Human Development", *World Development* 28(2): 197–219.

—— —— and Samman, E. (2006) "Human Development: Beyond the Human Development Index", *Journal of Human Development* 7(3): 323–358.

Stewart, F. (1975). "A Note on Social Cost–Benefit Analysis and Class Conflict in LDCs", *World Development* 3(1): 31–39.

—— (1977) *Technology and Underdevelopment*, London: Macmillan.

—— (1985) *Planning to Meet Basic Needs*, London: Macmillan.

—— (1993) "War and Underdevelopment: Can Economic Analysis Help Reduce the Costs?", *Journal of International Development* 5(4): 357–380.

—— (1996) "Groups for Good or Ill", *Oxford Development Studies* 24(1): 9–25.

—— (ed.) (2008) *Horizontal Inequalities and Conflict: Understanding Group Violence in Multiethnic Societies*, Basingstoke: Palgrave Macmillan.

—— Fitzgerald, V. and Associates (2001) *War and Underdevelopment*, 2 vols, Oxford: Oxford University Press.
—— Ranis, G. and Samman, E. (2018) *Advancing Human Development: Theory and Practice*, Oxford: Oxford University Press.

Further reading

Toye, J. (2011) "Social Wellbeing and Conflict: Themes from the Work of Frances Stewart", in V. FitzGerald, J. Heyer and R. Thorp (eds), *Overcoming the Persistence of Inequality and Poverty*, Basingstoke: Palgrave Macmillan.
UNDP (United Nations Development Programme) (1990) *The Human Development Report*, New York: Oxford University Press.

NANDINI GOOPTU AND AMOGH DHAR SHARMA

JOSEPH STIGLITZ (1943–)

Born in Gary, Indiana, on 9 February 1943, Joseph Stiglitz is reputed to have been dubbed by a tongue-in-cheek Paul Samuelson as the best economist ever to have originated from the declining steel town. This is, however, the hometown of the first Nobel Laureate for the discipline. Stiglitz's early career was astonishing. As a graduate student he edited Samuelson's Collected Papers. At 27, he became full professor at Yale, subsequently taking appointments at Stanford, Oxford, Princeton, Columbia and elsewhere. He received the J.B. Clark award for outstanding economist under the age of 40. His later career arguably projected him to become the leading economist as public intellectual. He joined Clinton's Council of Economic Advisors, serving as Chair and a member of Cabinet. In 1997, he became the World Bank's Chief Economist, promoting the Comprehensive Development Framework with President James Wolfensohn. In the putative turn away from the discredited neoliberal Washington Consensus, he launched the post-Washington Consensus. This helped to restore legitimacy to the Bretton Woods Institutions as more state-, people- and development-friendly. In 2001, he was awarded the Nobel Prize in Economics in part for his role in founding the New Development Economics, an element in his more general information-theoretic approach. In the volume of 2003 honouring his 60th birthday, his list of publications runs to almost 50 pages, including multiple editions of textbooks in economics, micro-economics, macro-economics and public-sector economics. He was lead author in the UN's response to the Global Financial Crisis. He is the go-to mainstream economist for progressive causes against an intolerant neoliberal elite. He has played a leading role in the Institute for New Economic Thinking, the initiative to promote alternative economics, funded by George Soros (https://www.ineteconomics.org/).

Despite these astonishing achievements, Stiglitz would not otherwise have been exceptional within development studies and economics except for an event that catapulted him to prominence. He was forced to resign from the World Bank at the instigation of Larry Summers, one of his predecessors and, at the time, US Treasury Secretary of State. He insisted that Stiglitz must go if Wolfensohn were to remain in office. The reason is that, from the outset of his appointment at the World Bank, Stiglitz had used his position to criticise the IMF. This became unacceptable once it threatened policy rather than merely offering rhetoric, and Stiglitz had become particularly outspoken over his opposition to privatisation in the former Soviet Union and the austere macro-economic policies and financial liberalisation imposed upon South Korea following the Asian crisis.

Close examination of Stiglitz's economics reveals paradoxes. First, he is not particularly radical in policy and theory, nor are his conclusions particularly original. Close attention to market imperfections leads him to seek balance between market and state, safety nets, proper sequencing and regulation of financial liberalisation in particular and other policies in general as opposed to the extremes of neoliberal shock therapy. But from the 1980s onwards, literature critical of the Washington Consensus, especially that concerned with the developmental state and adjustment with a human face, drew similar, if not more radical, conclusions. His emphasis on market and institutional imperfections as especially characteristic of developing countries represents a partial, reduced and micro-based rediscovery of the old or classic development economics of what might be termed the pre-Washington consensus of the McNamara period. Thus, Stiglitz's novelty lies less in the conclusions that he draws than in the information-theoretic approach that he uses to obtain them, and the positions from which he has been able to promote them with considerable and rare intellectual integrity, given the personal consequences.

Second, over his career, he seems to have become more radical, especially after his break with the World Bank. He has set up an Initiative for Policy Dialogue to promote alternatives to what he now terms neoclassical economics, having redefined it as the idea that markets work perfectly in order to distance himself from it. His critique of the Bretton Woods Institutions, especially the IMF, has intensified. His book *Globalization and Its Discontents*, published in 2002, which is mainly concerned with a critique of the IMF and his own experiences in promoting debate over his disagreements with it, became a best-seller. He has been hailed as the champion of the poor for his stances. But there is no increasing radicalisation in Stiglitz's economics apart from the loss of a naïve innocence about the power of his own ideas. This is someone who has remained consistently committed to Keynesianism at the macro-level and the correction of market imperfections by capable government at the micro-level. It is the intellectual environment that has changed, rather than his position within it.

Third, then, within mainstream (development) economics, Stiglitz has been at the forefront of the turn against neoliberalism since the 1990s. **Marx** argued of Ricardo that, for his analytics, it was sufficient to read the first two chapters of his *Principles*. The first deals with the labour theory of value as average labour time of production in industry, and the second, inconsistent with the first by defining value at the margin of cultivation, constructs a theory of rent. All the rest is application with no further theoretical advance. While this is in part a put-down, Marx also viewed Ricardo as the bourgeois political economist *par excellence*. This is because he applied his principles and his own version of the labour theory of value ruthlessly to whatever economic phenomena he cared to address.

What has this to do with Stiglitz? There are some striking parallels and contrasts to be drawn between him and (Marx's view of) Ricardo. First, Stiglitz can also be understood as relying upon two principles, to the first of which he is obsessively committed. With the second, as will be seen, it is inconsistent although it is mainly implicit, even unconscious. Stiglitz understands the capitalist economy in terms of what he himself dubs a new paradigm based on imperfect information. It deploys the idea that exchanges take place between agents who are imperfectly and asymmetrically informed. For this reason, markets are imperfect, and three potentially inefficient sorts of outcome result. Markets may clear (supply equals demand) but be inefficient (some trades beneficial to both parties do not occur). This is so if potential buyers (sellers) of lower- (higher-) quality goods do not wish to trade at a price reflecting a higher (lower) average quality.

Second, markets may not clear but prices not then adjust in favour of the short side (price increases if excess demand and decreases for excess supply). An example is efficiency-wages where, despite surplus labour, wages do not fall because employers anticipate a more than compensating loss in productivity, loyalty and work intensity. Third, markets can be absent, as in health insurance for the aged or sick. Any level of premium would attract an average level of health risk that is too high as the healthier choose to risk self-provision rather than pay premiums that reflect the sicklier.

For Stiglitz, these are general principles that apply to all markets, their significance depending upon the nature of the imperfections, the goods, the agents and so forth specific to each market. His voluminous writings reflect one application after another. But a difference from Ricardo is that Stiglitz's horizons extend beyond capitalism to deal with development and transitional economies. Further, Stiglitz also addresses the non-economic or non-market responses to market imperfections. Thus, institutions are perceived to be a (not necessarily efficient) response to market imperfections. His theory of share-cropping summarily dismisses both customary behaviour and exploitation as explanations in favour of monitoring and incentive problems in case tenants know more than landlords about production possibilities and work effort. Share-cropping is also used to argue that informational imperfections in one market, land, can have knock-on effects on others, such as availability of credit, choice of technique and capital intensity.

But other social sciences and political economy are notable for their absence from Stiglitz's work. His point of departure is the neoclassical economics of perfectly working markets, which he carefully picks to pieces from his own perspective. His own paradigm, however, displays considerable continuities with the neoclassical approach. Like Ricardo's, his method is deductive, even more explicitly drawing conclusions from axioms in the form of mathematical models. These are embedded in the optimising behaviour of individuals from which micro-foundations underpin macro-outcomes in the form of equilibrium. As with Ricardo, the logic of his deductive system is used to confront chosen topics with scant regard to methodology (deductivism wedded to naïve empiricism and falsifiability) or history of economic thought (with occasional token misinterpretations of Adam Smith as anticipating general equilibrium theory and Hayek the information-theoretic approach, for example).

Stiglitz's second analytical principle is to abandon his first and appeal to social, structural or other factors than optimising individuals. The need to do so arises in three ways. First, the deductive method always depends upon the exogenous as parameters. How are these to be explained other than to push out the boundaries further only to recreate them anew? Second, the real world has a nasty habit of revealing exploitation and power, and inventiveness and culture, not reducible to a calculus of imperfectly informed optimising agents engaged in individual acts of exchange with one another. As **Marx** said of

Ricardo's theory of rent, his assumption of free exchange and flow of capital onto the land would have been incomprehensible to his contemporaries anywhere else in the world. By the same token, can sharecropping be understood as free of exploitation and cultural factors, and development or lack of it the consequence of imperfections, and independent of power and associated conflicts?

Third, Stiglitz's own experience outside the academic world of neoclassical economics leads him to recognise the failure of his logic to prevail, most notably in the acknowledged rise to power of a financial system with its own compelling interests or the irrational dogma of his opponents. Significantly, these analytically residual categories leap to prominence in his more reflective contributions. *Globalization and Its Discontents* has no theory, scarcely a concept, of globalisation (other than reduction in transport and communication costs) and, by contrast, the term is effectively absent from all of his earlier work. This and the account of his period as adviser to Clinton are entirely free of the information-theoretic approach. But, if he is ultimately drawn to recognise social powers and vested interests and ideologies, these should surely serve as analytical starting point rather than conclusion of informationally-constrained optimisation.

In short, in the intellectual assault upon neoliberalism, Stiglitz has played the leading role from within mainstream economics. He has not promoted alternatives other than his own. His command of these, methodology, history of economic thought and the other social sciences is limited. For some, this puts him in the vanguard; for others he holds back a more radical and wide-ranging critique of neoliberalism. Indeed, a neat way to see him is as the self-appointed leader of the World Bank in exile. In his retrospective of the information-theoretic revolution, Stiglitz (2017: 22) suggests:

> For a long time, the Bank predicated its advice on an economic model which ignored the role of imperfect information. Fortunately, for the past two decades, the Bank has been at the forefront in raising questions about that model and enhancing our understanding of the implications of alternative frameworks—like those discussed here—for development policy.

This closes a contribution that scarcely mentions development at all, outrageously claiming to leverage huge analytical insights on the narrowest of principles (imperfect information for him as opposed to increasing returns for Krugman and non-rational behaviour for Akerlof, nudge and more). It is surely a moot point whether the post-Washington Consensus has revolutionary potential, let alone the capacity to live up to whatever that might constitute within the confines of the Bank.

Major works

For Joseph Stiglitz's cv, including list of publications, see https://sipa.columbia.edu/sites/default/files/migrated/documents/faculty/cv/Stiglitz_CV%201.27.17_0.pdf.

Chang, H.-J. (ed.) (2001) *Joseph Stiglitz and the World Bank: The Rebel Within*, London: Anthem Press. (A collection of his essays promoting the post-Washington Consensus.)

Stiglitz, J. (1986) "The New Development Economics", *World Development* 14(2): 257–265. (A concise and early account of his approach to economics and development.)

—— (1994) *Whither Socialism?* Cambridge, MA: MIT Press. (A fuller account with application to transitional economies.)

—— (2001) "Information and the Change in the Paradigm in Economics", available at http://nobelprize.org/economics/laureates/2001/stiglitz-lecture.pdf. (His own overview of his work, provided in his Nobel Prize Lecture.)

—— (2002) *Globalization and Its Discontents*, New York: W.W. Norton. (His critique of the IMF, updated in a 2nd edn, 2007.)

—— (2003) *The Roaring Nineties: Seeds of Destruction*, New York: W.W. Norton. (A narrative of his time on the Council of Economic Advisers.)

—— (2010) *The Stiglitz Report: Reforming the International Monetary and Financial Systems in the Wake of the Global Crisis*. New York: The New Press, for the UN.

—— (2017) "The Revolution of Information Economics: The Past and the Future", NBER Working Paper no. 23780, September, www.nber.org/papers/w23780.

——, A. Sen and J.-P. Fitoussi (2010) *Mismeasuring Our Lives: Why GDP Doesn't Add Up*, New York: The New Press, for President Sarkozy of France.

Further reading

For an appreciative assessment on his award, see "Markets with Asymmetric Information", available on the Nobel website, www.nobel.se/economics/laureates/2001/press.html, "Advanced Information".

Bayliss, K. et al. (eds) (2011) *The Political Economy of Development: The World Bank, Neoliberalism and Development Research*, London: Pluto Press. (On the limitations and flaws of World Bank research in a post-Washington Consensus world.)

Fine, B. (2002) "Economics Imperialism and the New Development Economics as Kuhnian Paradigm Shift", *World Development* 30(12): 2057–2070. (Critical assessments of the information-theoretic approach as paradigm.)

—— et al. (eds) (2001) *Development Policy in the Twenty-First Century: Beyond the Post-Washington Consensus*, London: Routledge. (A critique of the post-Washington Consensus.)
—— and Jomo, K. (eds) (2005) *The New Development Economics: A Critical Introduction*, Delhi: Zed Books.
Fridell, G. (2011) "Joseph Stiglitz: The Citizen-Bureaucrat and the Limits of Legitimate Dissent", *New Political Science* 33(2): 169–188. (Stiglitz as public intellectual.)
Wade, R. (2001) "Showdown at the World Bank", *New Left Review* 7: 124–137. (On his "resignation" from the World Bank.)
—— (2012) "Return of Industrial Policy?" *International Review of Applied Economics* 26(2): 223–239. (On the limited purchase of the milder, erstwhile successor Justin Lin at the Bank, let alone Stiglitz's post-Washington Consensus.)

BEN FINE

PAUL PATRICK STREETEN (1917–2019)

It would be a mistake to confine Paul Streeten to a small box labelled "development economics" and to seek to assess him within the limits of a subdiscipline when, like Keynes, **Marx** or Adam Smith himself, he was not only a total economist but a person of many disciplines who focused all his intelligence on the great economic issues of his day. He is one of that small band – which includes his two mentors and colleagues, **Gunnar Myrdal** and Thomas Balogh – who were not only widely regarded in Europe and North America as highly sophisticated economists, albeit dissenters, but who also thought long and hard about poorer parts of the world.

It has been observed that Streeten's style "did not develop, it appeared like Minerva from the head of Zeus, fully grown and armed" (Stretton 1986: 1). He burst on the scene in 1949 with articles on the theories of profit and pricing. These "dense, laconic, lucid papers" (p. 2) were followed during the next four years by seven more (one written in German, one translated into French) on the inadequacy of the price mechanism; exchange rates and national income; the inappropriateness of simple elasticity concepts in the theory of international trade (with Balogh); reserve capacity and the kinked demand curve; the modern theory of welfare economics; the effect of taxation on risk taking; plus a translation from German (with a thoughtful appendix) of Myrdal's (1953) *The Political Element in the Development of Economic Theory*. And all the while he was tutoring across the full range of economics at Balliol

College, Oxford, where he had arrived as a wounded soldier-student near the end of World War II.

For the next 50 years a steady stream of articles, books and policy reports was to emerge from the head of Zeus, all beautifully written, clear, erudite, witty, modest and fully armed. For Paul Streeten had an unrivalled mastery of economics. His next essay was on "Keynes and the Classics" (1954a). Then he edited Myrdal's essays and gave expression to his concern with methodology and with values in the social sciences: "Values are not something to be discarded, nor even something to be made explicit in order to be separated from empirical matter, but are ever-present and permeate empirical analysis through and through" (Streeten 1958: xiii). And he saw the importance of Heisenberg's Uncertainty Principle, concerning the mutual impact of the observer and the observed, with all its implications (1958: xv). It is an essay, published in the *Quarterly Journal of Economics* (1954b), which he regarded as his most genuinely innovative contribution but which made no impact on the profession. It remains required reading. While he continued to generate new insights as his experience widened, Streeten was to remain remarkably consistent down the years in his analytical style and focus.

"My interest in social justice", he wrote in his mid-seventies, "dates back to my childhood" (Streeten 1994: preface). It was a childhood far removed from the quiet Balliol of his early writing and teaching. For Paul Hornig, as he was born in the last days of the Austro-Hungarian empire, grew up in the hectic society of Vienna between the wars. His father died before he was two but his mother and his aunt moved in intellectual circles, where the young Paul met and was influenced by a wide range of thinkers, including Max Adler, Karl Popper, several psychologists and the leading Utopian socialist, Otto Bauer. As a teenager he was active in the socialist youth movement, the *Roten Falken* (Red Falcons), battling fiercely with the radical right. But all this ended abruptly in March 1938, when the Nazis marched into Vienna:

> I was on several lists and the combination of being a Jew and politically active on the left would have been enough. But though enquiries were made at our old flat, I was not given away. I believe there was an SS officer who was a friend of mine (vintage Streeten!) and who had deleted my name from the list of those to be arrested. I witnessed the hysterical city on the day of the Anschluss. I walked through the streets, and saw the armoured Nazi cars cheered by the crowds. The Viennese, reputed for their *Gemütlichkeit*, revealed

> faces distorted by hate as they shouted hysterically "Heil Hitler", "Ein Volk, ein Reich, ein Führer". It was a deplorable and frightening sight.
>
> (Streeten 1985: 47)

If Austria in the 1930s had been turbulent, the next five years were to take the young man on a dizzying roller-coaster. "Attempts to leave were entirely haphazard", but with the help of English friends he managed to cross the Channel, where he was looked after by a group of Christians including a family in Cambridge and two sisters in Sussex, Marjorie and Dorothy Streeten. Arriving just before May Week, "The transition from the turbulence, hysteria, fear and ghastliness of Vienna to the peace, tranquillity, and sunshine of the Master's lodgings in a Cambridge College … was an extraordinary experience" (Streeten 1985: 48). But more was to follow. He went to Aberdeen University, where he studied Political Economy for nearly two years until he was interned for being an enemy alien, even though he had volunteered for the Royal Air Force. For the next year he was shunted from pillar to Canada, where he and his companions, some of whom later became very distinguished, were held behind layers of barbed wire, towers, armed sentries and searchlights. Like the prisoners on Robben Island 25 years later, they started a university at which Paul overcame his deficiency in mathematics at the feet of Hermann Bondi, cosmologist and subsequently co-creator of the steady-state theory of the universe. He learned enough to be able to view with a beady eye the subsequent work of the mathematical model-builders in economics. Forty years on he was to remark, with his special combination of self-deprecation and rapier wit that,

> Of course, without a thorough training in mathematics, one feels nowadays like a handloom weaver in the days after the invention of the power loom. But the thought is made bearable by the fact that the power loom seems to be weaving the Emperor's clothes.
>
> (Streeten 1985: 56)

After five months behind the Canadian barbed wire, this motley crew was returned to England for another few months of cold, hungry, overcrowded internment. But eventually, as a long-distance runner, he managed to attract the attention of the authorities who were looking for men able to handle the arduous hush-hush work of the Inter-Allied Commando. Here, during training, they were told to pick English names in case they were taken prisoner behind enemy lines and shot as traitors. And thus, "in a few seconds" the young man changed from Hornig to Streeten and assumed the cover story of his adopted family. The maiden sisters were very pleased. Subsequently he was trained for the invasion of Sicily where, in the summer of 1943, he was badly wounded. It was a watershed day. "[N]o more punting, rock climbing, skiing, running" (1985: 54). Indeed, he was not expected to survive but, after hospitalisation in Cairo, he arrived in Balliol late the following year to read Politics, Philosophy and Economics. The roller-coaster ride was over (see Simon 2019 for further details and analysis).

Twenty years later he was still in Oxford, teaching and writing hard. During the 1960s two things happened to determine his future direction. The critique of "balanced growth" by **Rosenstein-Rodan**, **Nurkse** and others became a focus of debate in the profession and Streeten found that his own thinking about Europe (Streeten 1961) was relevant. Second, **Gunnar Myrdal** asked him to collaborate in the research and writing of *Asian Drama*. "And underlying these two tributaries, there was my … interest in the world community and my objections to the nation state and nationalism as a form of heresy" (1961). And so the die was cast.

In 1968 *Asian Drama* was published and Streeten co-edited *The Crisis of Indian Planning* (Streeten and Lipton 1968). Three years later he co-edited *Commonwealth Policy in a Global Context* (Streeten and Corbet 1971). This was followed by a small gem, *Aid to Africa* (Streeten 1972a). A generation later it is still worth reading. This was accompanied by *The Frontiers of Development Studies* "on the problems of development in which rich and poor countries co-exist and where the presence and policies of the rich crucially affect the development efforts and prospects of the poor" (Streeten 1972b). In Oxford in 1973 (at the invitation of the later-notorious publisher Robert Maxwell), he became founding editor of the journal *World Development*, which was to become influential as a monthly forum for debate. "Our goal", he wrote in what might almost be his testament, "is to learn from one another, regardless of nation or culture, income, academic discipline, profession or ideology. We hope to set a modest example of enduring global co-operation through maintaining an international dialogue and dismantling barriers to communication."[1] For the next 25 years a steady stream of articles, reviews and "essays in biography" sprang from the editorial pen. It was not until 2003, in his mid-eighties, that he stepped down as chairman of the editorial board.

In June 1976 the International Labour Organization held a seminal world employment conference in Geneva, where a strategy to "satisfy the most basic needs of all people in the shortest possible time" was endorsed (ILO 1977). Shortly after this, Streeten was asked by **Mahbub ul Haq**, then director of policy planning at the World Bank, to lead the Bank's strategic thinking about "Basic Needs". He did a brilliant job and, in his dual role of missionary and interpreter, converted everybody within the Bank so effectively that the strategy of Basic Needs is widely perceived to be that institution's major contribution to thinking about development (Streeten et al. 1981). But even then his questioning mind could see the unanswered questions, the ambiguities. In a paper delivered in apartheid South Africa in 1984, he observed: "It is not at all clear whether the basic needs approach mobilizes the power of the poor to improve radically their situation or whether it reinforces the oppressive existing order." But he stated pointedly that, "the freedom to define one's needs is itself a basic need" (Streeten 1984: 3).

During the 1980s, Paul Streeten (then based at Boston University) was at his most effective as scholar-statesman. After his success at the World Bank, he was asked by the ILO to be joint leader of a mission to Tanzania to help devise a basic needs-oriented development strategy for that country (ILO 1982). Coming at a moment of great crisis when President **Julius Nyerere**'s idealistic economic and social policies, with their consistent goal of slaying the postcolonial dragons of *Umaskini, Ujinga na Maradhi*,[2] seemed to have come unstuck, the report was a masterly document. And Streeten's fingerprints were all over it. While expressing deep empathy for the vision of one of Africa's great leaders, the report examined, unflinchingly, what had gone wrong and what could be done to put it right. From what happened subsequently in Tanzania, it would seem that the report was heard. A few years later, the IMF and the Indian Council for Research on International Economic Relations convened a seminar in Mumbai (Streeten 1988) to tackle the hot topic of structural adjustment in Asia. Once again Streeten was to play a crucial role as enabler and mediator. Observing that his criterion for a good seminar, which it had been, was to meet again one old friend and to make one new one, he went on to edit the book and write the summary of the discussion with its characteristic conclusion:

> Perhaps the most important general lesson that emerged was that there are no general lessons, and that each case has to be treated separately and on its merits. This pragmatic approach made it possible to discuss policies without getting bogged down in ideological or dogmatic positions.
>
> (Streeten 1984: 17)

Streeten was now in his seventies, an age when most people are beginning to slow down, but his greatest book was still to come. Five long lectures (plus appendices), delivered in Italy, form the core of *Thinking about Development* (Streeten 1995). It is a potent distillation, with even now new insights, of a lifetime focused on the great global questions of the past 60 years (see also Simon 2019). It is full of what **Michael Lipton** calls "Streetenesque subtlety" (Lipton 1977: 214). As Peter Timmer of Harvard has said, he was truly "one of our wisest scholars".[3] While the lectures in Italy were focused on "the eradication of poverty in the world", a small book (Streeten 1994) published in Denmark at much the same time focused more particularly on unemployment. But he was not yet finished. In 2001, in his mid-eighties, Streeten published *Globalisation: Threat or Opportunity?*, where he was able to clarify the two contradictory aspects and to consider the implications of a phenomenon with which he had been wrestling all his life.

But by 2008 he was too old, even for Paul Streeten, to be writing about the catastrophic banking crash and world depression which followed. One may not, of course, assume what he would have said were he 10 or 20 years younger, but it is safe to recognize that he would not have been amongst those economists who, bolstered by their clever mathematical models, declared at the turn of the century that the "central problem of depression-prevention had been solved". Streeten knew his mathematics but also the insights of history which have much to teach about the foolishness, blindness and irrationality of human beings; and about the dangers of prediction. And, as preparation of this second edition commenced, he celebrated his centenary in July 2017 in reasonably good health and continued to take a keen interest in current affairs despite the inevitable limitations of old age. Thereafter his health gradually declined and his wife's death was another setback. He eventually passed away peacefully on 6 January 2019, half-way to his 102nd birthday.

Paul Streeten was truly a world economist, not limited by the blinkers of the hidden assumptions of those confined to any one nation-state or region; someone who was able to see problems both from the inside, at the grassroots, and also from an overall perspective. Drawing on his mastery of economics, on his wide reading in many other subjects and on his own vast experience, he has increased the understanding and enhanced the capacity of scholars and policy-makers around the globe.

Notes

1 *World Development*: Notes for contributors.
2 "Poverty, Ignorance and Disease in kiSwahili", personal observation, Dar es Salaam, July 1963.
3 Cover comment on Streeten (1987).

Major works

Streeten, P. (1949) "The Theory of Profit", *Manchester School* 17(3). (For a detailed bibliography of his work (up to 1991) see Streeten 1995: 357–80.)

—— (1954a) "Keynes and the Classics", in K. Kurihara (ed.), *Post-Keynesian Economics*, New Brunswick, NJ: Rutgers University Press.

—— (1954b) "Programmes and Prognoses, Unbalanced Growth and the Ideal Plan", *Quarterly Journal of Economics* 68(3: 355–376. (Repr. as Introduction to Gunnar Myrdal, *Value in Social Theory*, London: Routledge and Kegan Paul, 1958.)

—— (1955) "Reformed Capitalism in Britain", in R. L. Heilbroner (ed.), *The Great Economists: Their Lives and their Conceptions of the World*, London: Eyre and Spottiswoode.

—— (1958) "Introduction", in Gunnar Myrdal (ed.), *Value in Social Theory*, London: Routledge and Kegan Paul, p. xiii.

—— (1961) *Economic Integration: Aspects and Problems*, ch. 5, Leiden: A.W. Sythoff.

—— (1964) *Economic Integration: Aspects and Problems*, ch. 5, Leiden: A.W. Sythoff, 2nd enlarged edn.

—— (ed.) (1970) *Unfashionable Economics: Essays in Honour of Lord Balogh*, London: Weidenfeld and Nicolson.

—— (1972a) *Aid to Africa: A Policy Outline for the 1970s*, New York: Praeger.

—— (1972b) *The Frontiers of Development Studies*, London: Macmillan.

—— (1984) "Basic Needs: Some Unsettled Questions", Carnegie Conference Paper no. 8, University of Cape Town.

—— (1985) *An Autobiographical Fragment*, Oxford: Balliol College Annual Record.

—— (1987) *What Price Food? Agricultural Price Policies in Developing Countries*, Ithaca, NY: Cornell University Press.

—— (ed.) (1988) *Beyond Adjustment: The Asian Experience*, Washington, DC: IMF.

—— (1994) *Strategies for Human Development: Global Poverty and Unemployment*, Preface, Copenhagen: Handelshojskolens Forlag.

—— (1995) *Thinking About Development*, Cambridge: Cambridge University Press.

—— (2001) *Globalisation: Threat or Opportunity?* Copenhagen: Handelshojskolens Forlag.

—— and Corbet, H. (eds) (1971) *Commonwealth Policy in a Global Context*, London: Frank Cass.

—— and Lipton, M. (eds) (1968) *The Crisis of Indian Planning: Economic Policy in the 1960s*, London: Oxford University Press.

—— et al. (1981) *First Things First: Meeting Basic Human Needs in the Developing Countries*, New York: Oxford University Press for the World Bank.Further reading

Balogh, T. (1963) *Unequal Partners*, 2 vols, Oxford: Basil Blackwell.

International Labour Office (1977) *Employment, Growth and Basic Needs: A One World Problem*, New York: Praeger.

—— (1982) *Basic Needs in Danger: A Basic Needs Oriented Development Strategy for Tanzania*, Addis Ababa: ILO.

Lipton, M. (1977) *Why Poor People Stay Poor: A Study of Urban Bias in World Development*, London: Temple Smith.

Myrdal, G. (1953) *The Political Element in the Development of Economic Theory*, London: Routledge and Kegan Paul.

—— (1958) *Value in Social Theory: A Selection of Essays on Methodology*, ed. P. Streeten, London: Routledge & Kegan Paul.

—— (1968) *Asian Drama: An Inquiry into the Poverty of Nations*, New York: Random House.

Simon, D. (2019) *Holocaust Escapees and Global Development: Hidden Histories*, London: Zed Books.

Stretton, H. (1986) "Paul Streeten: An Appreciation", in S. Lall and F. Stewart, *Theory and Reality in Development*, London: Macmillan.

FRANCIS WILSON

JAMES TOBIN (1918–2002)

James Tobin is one of the few economists working primarily in the mainstream of that discipline who has gained wide respect among development scholars, politicians and activists. Paradoxically, perhaps, he is best known for the proposal in 1972 to tax speculative electronic financial transactions and that, thanks to the recent international campaign, now bears his name. By contrast, his simultaneous contribution to measuring human welfare in broader terms than by per capita GNP alone is now less well remembered.

James Tobin, Nobel Laureate and widely regarded as the USA's foremost Keynesian economist of the 20th century, was born in Champaign, Illinois, on 5 March 1918, where he grew up with his younger brother, attending neighbourhood schools and then the University High School in Urbana, Champaign's twin city. There, according to the autobiography that he delivered on the occasion of his Nobel Prize award in 1981, he obtained a "marvellous" education from the university's student teachers and their trainers and was able to play in the university basketball team, "fulfilling ambitions that had seemed beyond reach in my childhood" (Tobin 1981: 1). His father, Louis, a journalist and publicity director for University of Illinois athletics, had a profound influence on his development: "My father also happened to be an intellectual, as learned, literate, informed, and curious as anyone I have known. Unobtrusively and casually, he was my wise and gentle teacher" (Tobin 1981: 1).

Tobin also attributes to his father the decision to apply to Harvard rather than the local University of Urbana-Champaign (where he had expected to land up studying law). Success in the entrance exams saw him head for Cambridge, Massachusetts, in late 1935, leaving the Midwest for the first time. There he spent six years, four of them as an undergraduate, in a period which he described as Harvard's "golden age" in Economics before and during World War II. Among his most influential teachers he counts Joseph Schumpeter, Alvin Hansen, Seymour Harris, Edward Mason, Gottfried Haberler and Wassily Leontief, while young faculty and fellow graduates included Paul Samuelson, Lloyd Metzler, Paul Sweezy, John Kenneth Galbraith, Richard Musgrave and Richard Gilbert, all of whom later became prominent economists.

Following US naval service after Pearl Harbor, Tobin returned to complete his PhD at Harvard in 1946–1947 on one of his abiding interests, the theory and statistics of the consumption function, by then persuaded that he would follow an academic career. Shortly after returning, he met, and in September 1946 married, Elizabeth (Betty) Fay Ringo, a former student of Samuelson. A postdoctoral fellowship enabled Tobin to remain at Harvard – with a stint at the Department of Applied Economics in Cambridge (UK) – until 1950. Thereafter, he moved to an associate professorship in economics at Yale University in New Haven, Connecticut, gaining a full professorship in 1955 and assuming the named Sterling Professorship of Economics two years later. There he directed the Cowles Foundation for Research in Economics from 1955 to 1961, chaired the Economics Department from 1974 to 1978 and remained a distinguished figure until retirement in 1988 and subsequently as emeritus professor until his death on 11 March 2002, just six days after his 84th birthday.

Tobin's career was punctuated by a series of visiting appointments at other universities. However, the absence from Yale of which he was most proud was to serve on President John F. Kennedy's Council of Economic Advisors in Washington, DC from 1961 to 1962 alongside fellow luminaries like Robert Solow and Kenneth Arrow. Although he soon returned to academia, he continued as a consultant to the Council for several more years, and the experience influenced much of his subsequent work (Tobin 1966; 1974; 1987; Tobin and Wallis 1968; Tobin and Weidenbaum 1988; Perry and Tobin 2000). Tobin's enduring interest in financial markets and investment decision-making is reflected in another string of books (Hester and Tobin 1967a; 1967b; 1967c; Tobin 1970; 1982; Tobin and Golub 1998) and journal articles. His proposal for a tax on speculative foreign exchange transactions originated from this strand of work (Tobin 1974).

Altogether, Tobin published some 400 journal articles, the most important of which were subsequently republished in a four-volume set of *Essays in Economics* (1972; 1975; 1982; 1996a), reflecting the principal thematic phases of his output. A fifth set, embracing his later work on global markets and finance, was published posthumously (Tobin 2003). Unlike most academic economists, Tobin was as comfortable – and forthright – penning topical potboilers for newspapers as in preparing scholarly treatises.

His reputation in the development world rests principally on three contributions, none of which was directed specifically at developing countries at the time. Indeed, he was very much a liberal American international economist in the Keynesian tradition. The first pertinent contribution reflects this directly, namely his sustained critique of the monetarist dogma popularised by Milton Friedman in Chicago and later adopted in part by President Reagan. Among those impressed by Friedman's monetarism were a group of young Chilean economics students who subsequently undertook its "purest" implementation under Augusto Pinochet's military dictatorship after 1973. Combined with the political repression, this caused much hardship to the poor but heralded what was widely known internationally as Chile's "economic miracle", one of the inspirations for the international financial institutions in formulating their neoliberal policies to tackle the 1980s debt crisis.

Tobin's other two most notable development-oriented interventions date from the early 1970s. "Is Economic Growth Obsolete?" (Nordhaus and Tobin 1972) initiated what became a sustained critique of using GNP (per capita) as the principal measure of human welfare when, in fact, it is an indicator of marketed production. This idea, and the several deductions and additions required to approximate a more useful indicator, the Measure of Economic Welfare (MEW), quickly entered economic textbooks and informed a generation of economics undergraduates. To those, myself included, learning their craft amid the daily realities of the global South, where most people engaged at least partially in non-marketed production and trade, this made eminent sense and represented an all-too-rare attempt to relate conventional economic theory to familiar conditions. This line of critique spawned experiments with compound indicators to avoid the limitations of reliance on a single variable. Ultimately the Human Development Index (HDI) was introduced by the United Nations Development Programme (UNDP) in 1990, and has become the most widely used index for international comparisons of development as well-being.

Tobin's most widely known contribution to development thinking is the "Tobin Tax" proposal for what he called "sand in the wheels" of international financial markets to reduce volatility and the profits from short-term speculative transactions relative to longer-term productive flows. Many years later, Tobin (1996b: x) observed that his early effort on the subject (1974; 1978) "did not make much of a ripple. In fact, one might say that it sank like a rock. The community of professional economists simply ignored it." That fate he attributed to the inherent objection of most economists and bankers to any market interference, accompanied by claims that it would drive financial markets to offshore tax havens while not reducing exchange rate fluctuations and speculative attacks. However,

> Most disappointing and surprising, critics seemed to miss what I regarded as the essential property of the transactions tax – the beauty part – that this simple, one-parameter tax would automatically penalize short-horizon round trips, while negligibly affecting the incentives for commodity trade and long-term capital investments. A 0.2% tax on a round trip to another currency costs 48% if transacted every business day, 10% if every week, 2.4% if every month. But it is a trivial charge on commodity trade or long-term foreign investments.
>
> (Tobin 1996b: xi)

Tobin viewed the tax as the principal objective, with use of the funds thereby raised for multilateral purposes "as a by-product". However, the resurgence of interest that commenced at the World Social Summit in Copenhagen in 1994, and which is attributable to the extreme currency speculation of the mid-1990s, regards this revenue as no less important. Even a tax rate of only 0.1 per cent (half of Tobin's original proposal), could raise between US$50 and 300 billion annually, thereby matching existing levels of official development assistance (War on Want n.d.). By 1996, the IMF's chief economist was also quite positive; the campaign has gathered momentum since, led in part by the development NGOs, War on Want and Oxfam, while several countries, notably Canada, France and Belgium, support the plan and/or have introduced enabling legislation. It is important to note, though, that supporters' motives vary: some, like Tobin, focus principally on the need for reduced financial market volatility and the efficiency of the proposed tax; many development lobbyists see stability and the funds accruing as being equally important; while antiglobalisation protesters hope that it will retard financial globalisation. Whether tactically or genuinely, Tobin (1996b; Patomäki 2001: 124–125) expressed uneasiness at the political agendas of some of the new protagonists of the tax. Today, concerns over some practical implementation problems and the loss of governmental sovereignty thereby implied and a lack of political will remain the major obstacles to its implementation. However, Tobin (1996b; 1996c) rejected most of these claims, pointing out, for instance, that the tax would restore to governments some economic sovereignty already lost to international financial markets. Moreover, Patomäki (2001) argues that universal agreement is no longer a prerequisite. His important contribution also takes a broader view of financial markets as co-responsible for the widening of global disparities, and of the current global financial institutional arrangements as inadequate and inefficient. Hence Patomäki advocates the tax on the additional grounds of global-scale distributive justice, the shared ideal of democracy, and human emancipation. A regional Tobin taks was reportedly adopted by the Bank of the South in 2007 after an initiative by the then Argentinian and Venezuelan presidents. Pleas and proposals for various forms of tax on speculative or other financial transactions continue to be made by experts, political figures and pressure groups in various contexts, including most recently the European Commission, which undertook considerable research on the subject from 2010 to 2013, and Hillary Clinton, the Democratic candidate in the 2016 US presidential election.

James Tobin's robust commentaries on contemporary economics also embraced the international arena, as illustrated by the conclusion to a 1993 paper on the challenges facing nation states in our increasingly interdependent world:

> Our economies are increasingly interdependent. Our opportunities are increasingly worldwide. Our maladies are contagious and intertwined. The leaders of the G7 countries have yet to show appreciation of the seriousness of the problems they jointly face, let alone enough imagination and initiative to seek joint remedies. In comparison with their predecessors who confronted the tasks of world economic recovery after World War II and responded with the Marshall Plan, the Bretton Woods institutions and the GATT, our present leaders are pygmies.
>
> (Tobin 1996d: 189)

Tobin's reputation spawned invitations to deliver a bewildering array of named lectures worldwide. Between 1967 and 1996, he also received no fewer than 21 honorary doctorates (mainly LLDs and Doctorates of Humane Letters), from some of the USA's leading universities and colleges (including Syracuse, Illinois, Dartmouth, Swarthmore, New York, Colgate, Harvard and Wisconsin-Madison) and three in Europe. He became a member of the National Academy of Sciences in 1972, and in 1981 was awarded the Nobel Prize in Economic Science in recognition of his analysis of financial markets and their relations to expenditure decisions, employment, production and prices.

These honours, his prolific published output and their influence will ensure that his reputation endures. However, outside the USA at least, he will probably best be remembered for a modest proposal in 1972, long deprecated but ultimately resurrected in the form of a global campaign for the Tobin Tax about which he claimed to feel uneasy. As for the man himself, fellow economist Paul Krugman (2002) paid him this tribute:

> He was a great economist and a remarkably good man; his passing seems to me to symbolize the passing of an era, one in which economic debate was both nicer and a lot more honest than it is today ... and I mourn not just his passing, but the passing of an era when economists of such fundamental decency could flourish, and even influence policy.

Major works

Hester, D.D. and Tobin, J. (eds) (1967a) *Studies of Portfolio Behavior*, New York: Wiley.

—— —— (eds) (1967b) *Risk Aversion and Portfolio Choice*, New York: Wiley.

—— —— (1967c) *Financial Markets and Economic Activity*, New York: Wiley.

Nordhaus, W. and Tobin, J. (1972) "Is Economic Growth Obsolete?" *National Bureau of Economic Research, Fifth Anniversary Colloquium V*, New York: NBER. (An earlier version appeared as *Cowles Foundation Discussion Papers 319* in 1971.)

Perry, G.L. and Tobin, J. (2000) *Economic Events, Ideas and Policies: The 1960s and After*, Washington, DC: Brookings Institution Press.

Tobin, J. (1966) *National Economic Policy*, New Haven, CT: Yale University Press.

—— (1970) *Asset Accumulation and Economic Activity: Reflections on Contemporary Macroeconomic Theory*, New Haven, CT: Yale University Press. (Repr. University of Chicago Press, 1982.)

—— (1972) *Essays in Economics*, vol. 1: *Macroeconomics*, Chicago, IL: Markham. Repr. MIT Press, 1987.)

—— (1974) *The New Economics a Decade Older*, Princeton, NJ: Princeton University Press.

—— (1975) *Essays in Economics*, vol. 2: *Consumption and Econometrics*, Amsterdam: North-Holland. (Repr. MIT Press, 1987.)

—— (1978) "A Proposal for International Monetary Reform", *Eastern Economic Journal* 4: 153–159.

—— (1982) *Essays in Economics*, vol. 3: *Theory and Policy*, Cambridge, MA: MIT Press.

—— (1987) *Policies for Prosperity: Essays in a Keynesian Mode*, Brighton: Wheatsheaf Books.

—— (1996a) *Essays in Economics*, vol. 4: *National and International*, Cambridge, MA: MIT Press.

—— (1996b) "Prologue", in M. ul Haq, I. Kaul and I. Grunberg (eds), *The Tobin Tax: Coping with Financial Volatility*, New York: Oxford University Press.

—— (1996c) "A Currency Transactions Tax: Why and How", *Open Economies Review* 7: 493–499.

—— (1996d) *Full Employment and Growth: Further Keynesian Essays on Policy*, Cheltenham: Edward Elgar.

—— (2003) *World Finance and Economic Stability: Selected Essays of James Tobin*, Aldershot: Edward Elgar.

—— and Golub, S.S. (1998) *Money, Credit and Capital*, Boston, MA: Irwin/McGraw-Hill.

—— and Wallis, W.A. (1968) *Welfare Programs: An Economic Appraisal*, Washington, DC: American Enterprise Institute.

—— and Weidenbaum, M. (eds) (1988) *Two Revolutions in Economic Policy: the First Economic Reports of Presidents Kennedy and Reagan*, Cambridge, MA: MIT Press.

Further reading

Krugman, P. (2002) "Missing James Tobin", *New York Times*, 3 December. (Also at www.pkarchive.org/column/031202.html, accessed 9 February 2005.)

Patomäki, H. (2001) *Democratising Globalisation: The Leverage of the Tobin Tax*, London: Zed Books.

Schmidt, R. (2007) *The Currency Transaction Tax: Rate and Revenue Estimates*, Ottawa: North–South Institute: www.nsi-ins.ca, accessed 8 June 2018.

Tobin, J. (1981) Autobiography: www.nobelprize.org/economics/laureates/1981/tobin-autobio.html, accessed 9 February 2005.

War on Want (n.d.) *The Tobin Tax: Win–Win for the World's Poor*, London: War on Want: www.tobintax.org.uk.

DAVID SIMON

MAHBUB UL HAQ (1934–1998)

> [The] UNDP and the United Nations system as a whole owe Mahbub a big debt of gratitude. Perhaps more than any other individual, he changed forever the way we think about development.
>
> (Dijibril Diallo 1998: 2)

When he passed away on 16 July 1998, the Pakistani economist Mahbub ul Haq ("Loved by God" as the name means in Arabic), had achieved international acclaim as a scholar-administrator who had pioneered new visions of human development and initiated the UNDP's *Human Development Report* series. Ul Haq served as Chief Economist in the Pakistan Planning Commission (1957–1970) and as Director of the World Bank's Policy Planning Department (1970–1982). Mahbub's influence on global development policy debates was also tangible; he was the Chairman of the North–South Roundtable (1979–1984), an eminent adviser to the Brandt Commission (1980–1982) and a Governor of the IMF (1985) and also the World Bank (1988). Mahbub served on the governing boards of numerous international institutions and think-tanks, including the Earth Council, the World Commission on Culture and Development and the Institute of International Economics (Speth 1998). UN Secretary General Kofi Annan described Ul Haq's's untimely death from pneumonia as "a loss to the world" (cited in HDC 1998: 1).

Having studied Economics at Government College in Lahore, Ul Haq moved to Cambridge University, where he studied for a master's degree and began a longstanding friendship with **Amartya Sen**. In the memorial lecture for Ul Haq on 15 October 1998, Sen recalled the discussions they had after arriving at Cambridge about conventional economics and its potential contribution in India and Pakistan. Sen explained that they both needed to learn Economics to be heard "but not use it much … who really wants to know what determines the price of toothpaste?" (Sen cited in Rosenfield 1998: 1). At 21, Ul Haq left Cambridge and continued his education studying for a PhD at Yale, from where he returned to Pakistan to work for the Federal Government as Chief Economist of the National Planning Commission until 1970. Working on the formulation and implementation of Pakistan's five-year development plans, Mahbub later declared, represented "happy days. My sights were set, my horizon was clear, and there was no hesitancy in my views about economic development" (Ul Haq 1976: 3).

While at Harvard, Ul Haq set to work in 1960 on *The Strategy of Economic Planning: A Case Study of Pakistan* (Ul Haq 1963). The book attempted to express themes of poverty and economic development in Pakistan and was concerned with the potential barriers to progress: the deeply unequal pattern of landholding, the pervasive illiteracy and the "warped development" that favoured a privileged minority (Ul Haq 1973: 1). In April 1968, Mahbub made a speech in Karachi about the 22 "industrial family groups" that had come to warp, skew and dominate national economic and political affairs. Not surprisingly, the speech created major shock-waves in Pakistan and beyond. Although **Julius Nyerere** wrote to Ul Haq with a letter of appreciation and Indira Gandhi used sections of Mahbub's speeches in her own policy presentations, Dr Haq later recalled that "[t]he academic community in the Western world reacted in shocked disbelief: one of its own products had suddenly gone berserk" (Ul Haq 1976: 8). Keen to avoid doctrinaire economics and planning philosophies, Amartya Sen has noted that this book was "informed by a general recognition that while a poor economy may take a very long time to become a rich country through GNP growth, the conditions of human living can be changed much more rapidly through intelligent policy making" (Sen cited in Rosenfield 1998: 1). Calling for properly targeted social intervention, the book began a longstanding commitment to interrogate GNP as a measure of progress, focusing on non-economic methods of securing positive change and improving the quality and targeting of development policy. In reference to this work, Ul Haq later reflected: "Though I have written much else since then my detractors have seldom allowed me to forget my original writings, perhaps believing that the evolution of ideas is an unforgivable sin" (cited in HDC 1998: 2).

Ul Haq's writings and speeches about inequality and economic growth brought him to the attention of World Bank President Robert McNamara.

During his tenure at the World Bank (1970–1982), Ul Haq was credited with making a major contribution to the Bank's philosophy of development economics. Upon his death in 1998, the President of the World Bank, James Wolfensohn, wrote in a letter to Mahbub's wife, Khadija, that "probably more than anyone else, [Ul Haq] provided the intellectual impetus for the Bank's commitment to poverty reduction in the early 1970s" (cited in HDC 1998: 1). As Director of the Bank's Policy Planning Department, Ul Haq set out to steer more attention towards poverty alleviation programmes, nutrition, water supply, education, social welfare and increased allocations for small farm production. According to one tribute, through this work Mahbub was able to "help sensitise a cold commercial lending institution to the concerns of the poor in the Third World" (Jabbar 1998: 2). Drawing upon his involvement with the Bank, Ul Haq wrote *The Poverty Curtain: Choices for the Third World* (1976), which highlighted the neglect of human resources in development planning and was a seminal study that provided an important precursor for the Bank's later development of the basic needs and human development approaches of the 1980s (Ul Haq and Burki 1980; HDC 1998). In the book, the "seven sins" committed by the "priesthood of development planners" are highlighted. These included playing "numbers games", constructing excessive economic controls, being constantly preoccupied with "investment illusions", the addiction to "development fashions", the divorcing of planning and implementation and the growing "mesmerisation" of planners with high growth rates in GNP.

During the early 1970s, Mahbub shifted his attention towards the intellectual self-reliance of the "Third World" and became more concerned at the simplicity of some intellectual dialogues about development at the international level (Ul Haq 1976). This feeling was sharpened by his experience of the UN Conference on the Human Environment in Stockholm in 1972 and a growing awareness of North–South differences of environmental perspective: "what were we really doing sitting through endless seminars and conferences where our own voice was neither solicited nor heard?" (Ul Haq 1976: 84). Along with **Samir Amin** and others who had shared these concerns and had also been present at Stockholm, Ul Haq founded the Third World Forum, an action group of around 100 leading intellectuals from the South, which first met in Santiago (Chile) and formalised its own constitution in Karachi in 1975.

In the early 1980s, Mahbub left the World Bank to return to Pakistan to work as Minister of Finance, Planning and Commerce (1982–1988). Under the regime of General Zia-Ul-Haq (no relation), Mahbub is credited with a major acceleration in social spending and with instigating significant tax reforms, new initiatives for poverty reduction, economic deregulation and an increased emphasis on human development (HDC 1998). Nonetheless, in reflecting some years later on his time in cabinet, Ul Haq spoke of his lack of independence and the fact that he was "not able to do very much" (cited in Brazier 1994: 4). As he put it in an interview in 1994: "I did accomplish some things but I was part of a very élitist system dominated by landlords in the Assembly, by élitist groups in the government, and by the Army, which would not let any trade-off take place between military and social expenditure" (cited in Brazier 1994: 4). It could be argued that Mahbub's experience of these restrictions and "trade-offs" between the economic and social realms of development were to have a big impact on his ideas in the years ahead and particularly when his attention turned to the measurement and conceptualisation of "human development".

In 1989 Ul Haq and his wife and intellectual partner, Khadija, moved to New York, where he became Special Adviser to the UNDP Administrator (1989–1995) (see also **Inge Kaul**). In doing so, he was reunited with **Amartya Sen** and also came together with **Frances Stewart**, **Richard Jolly** and Meghnad Desai to prepare the annual *Human Development Reports* (HDRs). According to Sen, Mahbub explained, "We need a measure of the same level of vulgarity as the GNP – just one number – but a measure that is not as blind to social aspects of human lives as the GNP is" (cited in Rosenfield 1998: 2). The Human Development Index (HDI) is now a central and recognised tool for the UNDP, and although Sen admits to initially finding the HDI a bit "coarse", it helped bring about a "major change in the understanding and statistical accounting of the process of development" (Sen, cited in HDC 1998: 2). Critics of the HDI would come back to this issue of the coarse and oversimplified nature of the Index, and have continued to highlight the costliness and numerous errors of HDRs as well as the poor quality of the underlying data. The initiation of the HDRs has been seen by some as a way for Ul Haq (who was their chief architect) to atone for his failure with human development issues in Pakistan (Brazier 1998: 4). According to Amartya Sen, however, the special achievement of the HDRs is that they bring "an inescapably pluralist conception of progress to the exercise of development evaluation" (Sen 2000: 18).

Following his retirement from the UNDP, Mahbub initiated a new series of HDRs in South Asia published by the Human Development Centre (HDC) he established in Islamabad. In his final years, Ul Haq began to deepen and further his interest in human security issues and in the regional crisis of governance in Asia. Through the HDC in Islamabad, Mahbub was able to champion "an earnest debate on restoring a better balance between arms and the people in South Asia" (Ponzio 1998: 10). Taking its lead from the UNDP's 1995 HDR on *Gender and Human Development*, the South Asia Human Development Report of 2000 (HDC 2000) opens with Ul Haq's often-cited injunction: "Human Development, if not engendered, is fatally endangered" (HDC 2000: 1). At one of the last meetings of the HDC attended by Mahbub, he quoted Bernard Shaw in telling colleagues: "You see things that are, and ask why? I dream of things that never were and ask why not?". Over several decades, Ul regularly raised the question of cuts in global military spending and consistently connected this to the possibility of improved social provision, human development and welfare. This was somewhat ironic given that he once served under a military dictatorship in Pakistan in the 1950s and 1960s. The goal of his diplomatic efforts in the 1990s was a desire for a more aggressive stance on disarmament, suggesting that the UN system needed new rules of intervention for the superpowers in order to avoid multilateral imperialism: "Otherwise we're back to the whole philosophy of colonialism which was 'the natives can't handle it, let's go in and teach them'" (cited in Brazier 1994: 2).

In the foreword to Ul Haq's *Reflections on Human Development* (1995), **Paul Streeten** recalls how he met Mahbub while working for the British Overseas Development Administration (ODA), arguing that dealing with "hardened British civil servants" gave Ul Haq a resilience, a healthy scepticism and also an awareness that development did not always involve the mobilisation of additional resources but could also be achieved through a re-allocation of existing resources. Interestingly, the book notes the unholy alliance of "shock therapists" and Bretton Woods institutions which impedes the provision and redistribution of basic needs. Nonetheless, Mahbub agreed with many of the Bank's core tenets of economic liberalism, placing an emphasis on privatisation and a belief in market competition. This comes through clearly in a speech he gave about Africa's future shortly before he died (Ul Haq 1998: 2). Having been dealt a "cruel hand" by history, African countries, he argued, should develop new technical skills and follow the Indian example, becoming "a major supplier of computer software and electronic goods to Europe" (p. 3).

Amartya Sen has recalled Mahbub's general "impatience with theory" (Sen 2000: 22) but also notes the intense pragmatism that he brought to the UNDP's formulation of "human development" and his "open minded approach, his scepticism and his perpetual willingness to listen to new suggestions" (Sen 2000: 23). Ul Haq regularly advocated an integrated UN system with a strong human development message, and in some of his final writings, speeches and interviews he talked of how the World Bank had "seriously misled policy makers" in focusing attention "on the symptoms, not the causes", noting that "to ignore the poor upstream and to count them endlessly downstream is merely an intellectual luxury" (Ul Haq 1997: 1). This was one indulgent sin committed by the "priesthood of development planners" that Mahbub considered unforgivable.

Major works

Ul Haq, M. (1963) *The Strategy of Economic Planning: A Case Study of Pakistan*, Karachi: Oxford University Press.

—— (1976) *The Poverty Curtain: Choices for the Third World*, New York: Columbia University Press.

—— (1995) *Reflections on Human Development*, New York: Oxford University Press.

—— (1999) "Human Rights, Security and Governance", in *Worlds Apart: Human Security and Global Governance*, London: IB Tauris.

—— and Burki, S.J. (1980) *Meeting Basic Needs: An Overview*, Washington, DC: World Bank.

Further reading

Brazier, C. (1994) "The New Deal: Interview with Mahbub ul Haq", *New Internationalist* 262(12):1–4: www.newint.org/issue262/, accessed 4 July 2004.

Desal, N. (1998) "Heartbeat: Tribute to Development Economist Mahbub Ul Haq – Obituary", *UN Chronicle*, 22 September: 1: www.findarticles.com/p/articles/mi_m1309/is_3_35/ai_54259329.

Diallo, D. (1998) "From the Editor in Chief", *Choices: The Human Development Magazine* 7(4): 2.

Human Development Centre (HDC) (1998) "A Tribute to Dr Haq": www.un.org.pk/hdc/Tribute/, accessed 7 July 2004.

—— (2000) *Human Development in South Asia 2000: The Gender Question*, Oxford: Oxford University Press.

Jabbar, J. (1998) "Two South Asian Stalwarts: Tribute [Nikhil Chakravarty and Mahbub ul Haq]", *Himal* 11(8): www.himalmag.com/Aug98/stalwart.htm/, accessed 4 July 2004.

Minna, M. (1999) "Tribute to Mahbub ul Haq", speech given at the Mahbub ul Haq commemorative conference "People and Poverty: Human Development into the Next Millennium", Ottawa, Canada, 13 October: www.acdicida.gc.ca/cida_ind.nsf/, accessed 4 July 2004.

Ponzio, R. (1998) "Disarmament and Human Development: The Legacy of Dr Mahbub ul Haq", *Bonn International Centre for Conversion: Bulletin* 9: 1–2.

Rosenfield, S.S. (1998) "The Not So Dismal Economist", *Washington Post*, 23 October: www.wright.edu/~tdung/mahbub.html, accessed 8 July 2004.

Sen, A. (2000) "A Decade of Human Development", *Journal of International Development* 1(1): 17–23.

Speth, J.G. (1998) "UNDP Mourns Loss of Mahbub ul Haq, Economist and Development Visionary: Administrator James Gustave Speth Praises Haq's Unwavering Commitment to Social Justice", UNDP press release: www.undp.org/dpa/pressrelease/releases/P980717e.html/, accessed 4 July 2004.

Ul Haq, M. (1973) "System is to Blame for the 22 Wealthy Families", *Times* (London), 22 March: www.un.org.pk/speeches, accessed 4 July 2004.

—— (1997) "Poverty Is Cancer, Not Flu", introductory remarks at the Special Event on Poverty Eradication arranged by UNDP, 20 May 1997: www.un.org.pk/hdc/speeches/accessed 3 July 2004.

—— (1998) "Does Africa Have a Future?", Opinion, *Earth Times News Service*: www.meltingpot.fortunecity.com/lebanon/254/ulhaq.htm, accessed 9 July 2004.

United Nations Development Programme (UNDP) (1990) *Human Development Report 1990: Concept and Measurement of Human Development*, New York: UNDP.

MARCUS POWER

SWAMI VIVEKANANDA (1863–1902)

> All differences in this world are of degree, and not of kind, because oneness is the secret of everything. Be sincere and do your duty. Everything will come all right. Truth must triumph.
>
> Swami Vivekananda (*CwV* vol. 4: p. xxiv)

Swami Vivekananda, family name Narendranath Datta, an Indian Hindu monk and chief disciple of the 19th-century Indian mystic Ramakrishna, was born into an aristocratic Bengali family on 12 January 1863 in Kolkata. Under the tutelage of Sri Ramakrishna, Vivekananda gained the spiritual knowledge and wisdom for which he later became famous. After Ramakrishna's death, Vivekananda founded the Ramakrishna Mission and an ashram in 1897, which later helped him become the major face in the revival of Hinduism worldwide. He is credited with raising interfaith awareness, bringing Hinduism to the status of a major world religion during the late 19th century, and also with reviving

Hinduism in the service of nationalism in colonial India. He studied Western philosophies and developed a comparative analytic frame of Hinduism that inspired many thinkers in conceiving ideologies of human development. He was popularly known as *Swami-ji* ("a yogi and spiritual teacher").

After the death of his guru Ramakrishna in 1886, as a wandering monk Vivekananda visited several countries in Europe and America. While travelling to the United States, representing India at the 1893 Parliament of the World's Religions, he delivered a speech that envisioned the base of interfaith dialogue and harmony. He returned to India in January 1897, and delivered a series of lectures in different parts of the country, which created a great nationwide stir. Through these inspiring and profoundly significant lectures, Vivekananda attempted to rouse the religious consciousness of the people and create in them pride in their cultural heritage; to bring about unification of Hinduism by pointing out the common bases of its sects; to focus the attention of educated people on the plight of the downtrodden masses, and to expound his plan to uplift them by the application of the principles of Practical Vedanta.

He said, "I want to teach you to live the truth, to reveal the light within your own soul" (cf. *CwV* vol. 4: 419). He believed that a country's future depends on its people, and his teachings focused on human development (Vivekananda 1996: 1–2). He wanted "to set in motion a machinery which will bring noblest ideas to the doorstep of even the poorest and the meanest" (*CwV* vol. 4: 422). Vivekananda died at the young age of 40 on 4 July 1902 in his own ashram, and was cremated there. His teachings and life philosophy have influenced several notable personalities in India and other parts of the world, like Subhas Chandra Bose, Aurobindo Ghose, **Mahatma Gandhi**, Rabindranath Tagore, Jawaharlal Nehru, Bal Gangadhar Tilak, Jamsetji Tata, Jagadish Chandra Bose, Nikola Tesla, Sarah Bernhardt, Emma Calvé, Annie Besant, Romain Rolland, the current Indian social activist of the anti-corruption movement and peace movement Anna Hazare (b. 1937) and the present Prime Minister of India, Narendra Modi (b. 1950).

The central monistic philosophy of Hindu tradition, Vedanta, recognises that "fundamentally all life is one, that in essence everything is reality, and that this oneness finds its natural expression in a reverence for all things". "Environment" and "sustainability" are increasing concerns in both traditional "Hindu" communities and their pioneering fringe of New Religious Movements, that Vivekananda expounded though his programmes at the Ramakrishna

Mission and the more scientific interpretation of the Vedanta and Yoga philosophies. He had already cast the image and frame of the Global Family and union within the environmental conscience in his four lectures in the World Parliament of Religion at Chicago in 1893 (see *CwV* vol. 1: 1–25); in fact, his speech presented the Hindu tradition as promoting the unity of all people and as providing the heart of all religion (Haigh 2010: 3502).

The ethic of "sustainable development", to which almost everyone subscribes today, requires this generation to use the world's environmental resources in ways that do not jeopardise the ability of future generations to meet their own needs. To be successful, this principle requires another dimension – reverence and revelation (ecospirituality). This involves a way of life and action determined by the deeper principle of realising the intrinsic value one's actions serve.

Obviously, "since the roots of [ecological] trouble are so largely religious, the remedy must also be essentially religious, whether that or not. We must re-think and re-feel our nature and destiny" (White 1967: 7). From this viewpoint, Vivekananda's teaching should be re-assessed. In the words of Swami Vivekananda, it is already envisioned that:

> Really speaking, therefore, law does not exist in nature. Practically it is an error to say that gravitation exists in the earth, or that there is any law existing objectively anywhere in nature. Law is the method, the manner in which our mind grasps a series of phenomena; it is all in the mind. Certain phenomena, happening one after another or together, and followed by the conviction of the regularity of their recurrence – thus enabling our minds to grasp the method of the whole series – constitute what we call law.
>
> (*CwV* vol. 1, *Karma Yoga*: vii)

Ethical values are the moral force in the sustainable existence, progress, maintenance and continuance of the human being. To *Swami-ji*, nothing could be divorced from *dharma* (natural order of moral duties/religion) because "religion" was "the essence of existence in each man which is spirituality" (Singh 2017: 158) and in this broad sense, it had nothing to do with dogmas and doctrines, with rituals and ridiculous argumentation over which is the correct direction or position for worshipping God. He exhorted his countrymen to show by their lives that religion does not mean words or names or sects, but spiritual realisation. Also it should be remembered that as early as the 1890s, Vivekananda was the first to talk about inter-faith harmony, peace and universal spiritual brotherhood, far before it became essential for the survival of humankind in the 20th and 21st centuries.

For *Swami-ji* the Hindu concept of *dharma*, right action as the inherent order, is extended to service of the God in every human. He proclaims: "Man has wanted to look beyond, wanted to expand himself; and all that we call progress, evolution, has been always measured by that one search, the search for human destiny, the search for God" (*CwV* vol. 2, pt 4: 302). He has also made the bridge between realisation (*anubhava*) and revelation (*anubhuti*), and challenged: "You, as body, mind, or soul, are a dream, but what you really are, is Being, Consciousness, Bliss. You are the God of this universe. You are creating the whole universe and drawing it in" (*CwV* vol. 1: 112).

Both *Swami-ji* and **Mahatma Gandhi** (cf. Singh 2018) considered religion (faith-based rules) to be the prime essence, the very lifeblood of the nation as well of all life – the aim of which was to realise God as Truth. In these inspiring words, *Swami-ji* said, "The best way to serve and seek God is to serve the needy, to feed the hungry, to console the stricken, to help the fallen and the friendless, to attend upon and serve those who are ill and require service" (Iyer 1983: 87).

Historically, there have been deep and complex tensions between Hindu reactionaries and extreme Islamists that continue even today in different contexts. These stem largely from the clash between the Muslim monotheistic and Hindu polytheistic outlooks. Nevertheless, Hinduism reveals a strong acceptance of other faiths. *Swami-ji* illustrated this in his four lectures (Vivekananda 1893) in the World Parliament of Religions at Chicago (USA), which encouraged acceptance and understanding of Hinduism and its teachings. Since the mid-20th century, with the driving forces of globalisation, the ethics and basic Hindu philosophy of human–nature interrelationship once envisioned by *Swami-ji* has been in the process of wider acceptance (Singh 2012: 180; Singh and Aktor 2015: 1926).

Swami-ji adds:

> the process of Self-realization involves self development, which may be achieved – by controlling nature: external and internal. Do this either by work, or worship, or psychic control, or philosophy – by one, or more, or all these – and be free. This is the

> whole of religion. Doctrines, or dogmas, or rituals, or books, or temples, or forms, are but secondary details.
>
> (Haigh 2010: 3502)

Concerning environmental conscience, *Swami-ji* says:

> Where would these *Karmas* go when I become a god? The answer is that desires can only manifest themselves in proper environments. Only those desires will come out for which the environment is fitted; the rest will remain stored up. In this life we have many godly desires, many human desires, many animal desires. If I take a god body, only the good desires will come up, because for them the environments are suitable. And if I take an animal body, only the animal desires will come up, and the good desires will wait. What does this show? That by means of environment we can check these desires. Only that *Karma* which is suited to and fitted for the environments will come out. This shows that the power of environment is the great check to control even *Karma* itself.
>
> (*CwV* vol. 1: 8)

Modernisation and messages contained in ancient wisdom should be allowed to co-exist to help realise humanity's link with nature as put forward by *Swami-ji*. This ideology should be taken as an environmental ethic having sacrality and deep attachment at its core. One may also hope for a revival of the sense of belongingness in the light of "eco-ethics", preserving the age-old intangible spirit of sustainability and recognising our ecological identity in the cosmic web of interconnectedness of all beings – a message of ecospirituality.

Change is the law of nature. Humanity's progress is transformation. Evolution is gradual change. Revolution is accelerated change. Revelation is eternal change. We can very well see the following words of Vivekananda coming true: "one vision I see clear as life before me; that the ancient Mother has awakened once more, sitting on Her throne – rejuvenated, more glorious than ever. Proclaim Her to all the world with the voice of peace and benediction" (*CwV* vol. 4: 353). In the purview of the *Katha Upanishad* (1.3.14): *Uttiṣthata Jāgrata Prāpya Barannibodhata* ("Arise! Awake! Approach the great and learn"), *Swami-ji* proclaims: "Arise, awake, and stop not till the goal is reached"; this is a call for "self-realisation" and "eternal awakening" that ultimately make the path of ecospiritual sustainability. He adds:

> Learn that truth from us, understands it, and realise it and then spread that idea broadcast, in cities, in towns, and in villages. Go and preach to all, "Arise, awake, sleep no more; within each of you there is the power to remove all wants and all miseries. Believe this, and that power will be manifested." Teach this to all, and, with that, spread among the masses in plain language the central truths of science, philosophy, history, and geography.
>
> (*CwV* vol. 6: 1)

If all crises arise from the human mind, then solutions will naturally come from the mind, as Lynn White (1967) affirmed, and as *Swami-ji* explained long ago. It is quite promising and natural that modern Hindus should derive their knowledge of Hinduism from Swami Vivekananda, directly or indirectly (Bharati 1970: 288). In this way all the humanities would realise the unitary principle of Nature, which is also a call of Gaia: "Heal the mother Earth, heal yourself!"

Major works

CwV: Vivekananda, Swami (1989) *The Complete Works of Swami Vivekananda*, 9 vols, Kolkata: Advaita Ashrama. (Available at: www.vivekananda.net/BooksByS wami/Complete Works/, accessed 15 December 2017.)

Naraimhanand, Swami (ed.) (2012) *Vivekananda: Reader*, Kolkata: Advaita Ashrama.

Pradhan, A, Banerjee, A., Singh, H., Lahiri B., Gaur, R.K., Bhattacharya, M., Singh, A.K. and Singh, V.K. (eds) (2017) *Swami Vivekananda's Contribution to India*, Varanasi: Ramakrishna Mission Ashram, with Pilgrims Book House.

Vivekananda, Swami (1893) *Four Lectures in the World Parliament of Religions*: "Response to Welcome" (11 Sept.); "Why We Disagree" (15 Sept.); paper on Hinduism (19 Sept.); "Religion Not the Crying Need of India" (20 Sept.), Chicago: World Parliament of Religions. (As in *CwV* vol. 1: 1–25. Available at www.vivekananda.net/BooksBySwami/Complete Works/, accessed 15 December 2017.)

—— (1996) *My India: The India Eternal*, ed. Swami Lokeswarananda, Kolkata: Ramakrishna Mission Institute of Culture.

Other works

Bharati, A. (1970) "The Hindu Renaissance and its Apologetic Pattern", *Journal of Asian Studies* 29: 267–288.

Haigh, M. 2010. "Education for a Sustainable Future: Strategies of the New Hindu Religious Movements", *Sustainability* (e-journal, ISSN 2071-1050) 2: 3500–3519.

Iyer, S. (1983) *Reminiscences of Swami Vivekananda*, Kolkata: Advaita Ashrama.

Singh, R.P.B. (2012) "Hinduism and Globalization: Pilgrimage Landscapes, Spirituality and Global Message", in C. Bulfoni (ed.), *Religious Traditions and Social Transformations of Contemporary Asia/Tradizioni religiose e trasformazioni dell"Asia contemporanea*, Rome: Bulzoni, 159–191.

—— (2017) "Ecospirituality and Sustainability: Vision of Swami Vivekananda", in A. Pradhan et al. (eds), *Swami Vivekananda's Contribution to India*, Varanasi: Ramakrishna Mission Ashram, with Pilgrims Book House, 157–168.

—— and Aktor, M. (2015) "Hinduism and Globalization", in S.D. Brunn (ed.), *Changing World Religion Map: Sacred Places, Identities, Practices and Politics*, Dordrecht: Springer Science, 1917–1932.

White, L. T. Jr. (1967) "The Historical Roots of Our Ecologic Crisis", *Science* 155: 1203–1212.

RANA P.B. SINGH

WANG HUI (1959–)

Wang Hui was born in 1959 in Yangzhou, Jiangsu Province, China, into a family of a mid-level party cadre and a teacher. Growing up during the Cultural Revolution in "socialist China", he read many "red" classics, including **Marx**, Lenin, and **Mao**, as well as 19th- and 20th-century European and Russian writers, helping form his emotional attachments to the working class and peasants.

Entering Yangzhou University in 1978 following the Cultural Revolution, Wang's generation was the first in China exposed to Western ideas challenging the standard curriculum of a socialist education. What ensued, and has continued to the present, is an intense societal struggle over the direction and path for China's development. Central to this ongoing debate are the relative roles and historically evolving meanings of the state, market and civil society – all prominent features of Wang's intellectual work.

Wang continued his graduate studies at Yangzhou University, receiving his MA at Nanjing University and his PhD in Chinese Literature from the Chinese Academy of Social Sciences (CASS) in 1988. Subsequently, he worked at the Institute of Literature in CASS before joining the Chinese Literature department at Tsinghua University in 2002. He is currently Professor of Chinese Language, Literature and History at Tsinghua University, and Dean of the Tsinghua Institute for Advanced Studies in Humanities and Social Sciences (TIAS). From 1996 to 2007, he edited *Du Shu*, an intellectual journal of social theory, in collaboration with Huang Ping; under their leadership it became the most influential academic publication in China.

Wang began his intellectual project on the Chinese philosopher Lu Xun (1881–1936), the most famous intellectual and social critic of early 20th-century China, exploring the enduring tensions within China's intellectual tradition between individual and nation, and progressive and cyclical history. He drew upon Lu Xun as a thinker who countered the increasing professionalisation of Chinese intellectuals. Wang refocused attention on how Lu Xun revealed key connections between the production of knowledge and inequality in contemporary China, re-awakening his legacy amongst critical intellectual thinkers.

Following China's 1978 reforms, the Chinese state and society underwent great transformations, culminating in the 1989 social movement. After 1989, China's leadership began more radical market reforms, and Wang turned his attention firmly towards the question of development. Responding to the dominant ideas emanating from neoliberalism, Wang forcefully rejected its foundational assumptions and positive relevance for China's development. The dominant discourse of development in the 1990s lionised the market, with the state's role theorised primarily as serving it. Wang used *Du Shu* to initiate debates about the role of the state, the negative influence of capitalism and the increasing hegemony of Western power. He published widely, promoting social justice, calling for the protection of peoples deprived of their rights during the reform process, particularly the working class and peasants. This made him the intellectual leader of China's New Left movement, providing an important counterpoint to widely accepted economic orthodoxy. Wang challenged neoliberalism's pretension of science and its claim of objectivity. The relationship between democracy and development was another of his subjects, as science and democracy were central to the historic 4th May 1919 Movement of which Lu Xun was a leader, and thus re-entered the discourse in 1989. Since then, the debate in China on forms of democracy has continued, but below the radar given political sensitivities.

Wang is a social theorist who focuses on the intellectual history of modern China. He has been a leading figure in the intellectual debates in China since the 1990s on the how, where and why of development pathways. Wang builds on a deep historical and theoretical analysis of

China and modernity to address various problems of post-1978 reforms, including socio-economic inequality and its political and cultural consequences. His key intellectual contributions are, first, to problematise development discourse, rather than take for granted linear and traditional approaches and theories. Second is his comparative study on development in industrialised societies, China, and the other so-called developing countries. And thirdly, he not only contributed to but launched several key intellectual debates amongst Chinese scholars on China's place in the world, broadly conceived. This debate engages the ongoing question of how the world influences China's development, the emergent "China model" and China's changing global position.

Wang's current work on the 20th century argues that the world must be understood by integrating the roles of other parts of the world beyond the typical East/West dichotomy. He challenges standard views of modernity and social movements. His main concern is to rethink the role of modern China in the context of global history, during the interaction between the East and West. He coined the term "short 20th century" to relocate the historical position of 20th-century China, its revolution, reform logic and legacy. Wang argues that the first half of the 20th century was composed of a two-layered process: the great transformation from empire to modern nation-state and the anti-Western imperialism movement, which together resulted in socialist revolution. Because of perceived failures of socialist development, China adopted reform policies and returned to a "normal" capitalist track in the 1990s, marking the end of this short 20th century.

In his most recent work, *The Rise of Chinese Modern Thought: China from Empire to Modern State* (2015; forthcoming in English), Wang argues that the dichotomy between nation-state and empire cannot explain Chinese history. The history of the Qing dynasty (1636–1912) embodies both features. This dichotomy is a central tenet of modern development theory, which asserts that despotic empires are less economically advanced, while nation-states have liberal democracy and are more advanced and modern. Modernity is therefore associated with the nation-state, and tradition is associated with empire. Wang challenges this dichotomy. He simultaneously argues that the imagination of an imperial-despotic-traditional view concerning China's state is wrong, being both simplistic and Orientalist. The Qing had many aspects of the nation-state, for example boundary-drawing, well before that undertaken in Europe. And modern China assumed the territory of the Qing empire, so it is a multinational state. According to the dominant discourse of the nation-state, this would be exceptional, and therefore

China is bound to fall apart, keeping itself as a political body simply because despotic power controls the diverse regions. Wang rejects this Western idea of the nation-state, arguing that modern China is a product of a dual process of imperial and nation-state building. This somewhat conservative position makes traditional China much more relevant to a contemporary China built upon the unity of this large territorial empire. The radical side of this analysis embodies Wang's reflections on Western imperialism. According to Wang, the key international context of China's modern history is the 19th century's imperialist expansion. This included the very tangible and geographically significant element of the treaty port system and British control of Hong Kong and Portuguese control of Macau. In this respect, China's transformation and revolution are not an isolated phenomenon but rather an integral part of the anti-imperialist movements in global history.

According to Wang, empire is not only a term that explains the development of modern China, but also the integration of a regional order. Modern China can only be understood in the Asian context. His many articles on the Korean War, Taiwan and Tibet reflect this concern. Historically, he argues, these regions were an integral part of a traditional Chinese world until invasion by Western imperialist powers. Importantly, this Chinese world was not homogeneous, but composed of multiple systems of organisation and administration, and interwoven with multiple and varied societies (a '"trans-societal system"').

Wang was a postdoctoral researcher at the University of California at Los Angeles (UCLA) in the early 1990s. But it was *China's New Order* (2003), a decade later, that marked Wang's widespread introduction to English-speaking audiences. In the first essay on the "historical conditions of the 1989 social movement and the anti-historical explanation of 'neoliberalism'", Wang argues for the importance of emphasising the pursuit of social equality in this movement. Citizens and workers rose against corruption, inflation, nepotism and widening social disparities in the 1980s. This was improperly interpreted by neoliberal thinkers simply in terms of Western notions of liberal democracy associated with markets. Wang, in contrast, asserts that workers had different targets from those that were widely understood, attempting to halt the decline in workers' social and economic power fundamental to the rapid rise in societal inequality post-reform. However, this has not been fully reflected in analyses after the failure of the movement. In the 1990s, as China marched towards a market economy, a neoliberal interpretation took over the analysis of the social movement and China's development path.

Wang subsequently argued that Hayek's view of markets as a spontaneous invention of society is wrong and that, in fact, the market economy is created by the state from above, and markets have long histories in different societies. Drawing on and introducing to China **Polanyi**'s ideas concerning the relation between market and state, he argued that China was not neoliberal in this key sense because of the market's active creation by the state. Wang drew upon *Civilization and Capitalism* by Braudel (1981–1984) to confirm the role of smaller-scale transparent markets without accepting capitalism and free market economics *per se*.

The dominant position, challenged by Wang, was that the market would create a middle class, followed by democracy. In *The Rise of Modern Chinese Thought* (2015), Wang argues that democracy is not simply representative, but rather that social equality is the basis for democracy. Therefore, he asserts, the reform itself, centred on the market in the neoliberal discourse as preparation for democracy, actually destroyed China's existing social equality and undermined the capacity for real democracy.

Wang's influence today is based in part on his reputation in the West, and the publication of a number of important books in English, including *China's New Order* (2003), *The End of Revolution: China and the Limits of Modernity* (2010) and *The Politics of Imagining Asia* (2011). He won Italy's Luca Pacioli Prize in 2013, and Germany's Anneliese Maier Award in 2018, demonstrating his scholarly influence in the fields of social theory and humanities. As Ted Huters points out in his Introduction to the translated essays that compose the latter book, Wang's ambition is "to rethink world history as a whole as something other than a triumphant procession of ascendant Western social, economic, and political models" (2011: 4).

Wang's influence on development debates in China correlates with the rise of the New Left, heavily influenced by Wang's combination of writings and his editorial work for the journal *Du Shu*. The New Left's influence on development theory and practice in China encompasses critical approaches to neoliberal economics as well as other market-based answers to complex historical, social and political problems. China's New Left draws upon both critical ideas from the international Left and China's own critical historical discourse. This integration of modern Chinese intellectual history with Western critical Left ideas marks the emergence of a new school of thought in China closely associated with Wang. The revalorisation of direct democracy in the Jacobin and Maoist traditions is also representative of this synthesis.

Wang Hui has developed a broad intellectual network as one of China's leading public intellectuals. In response to Western ideas of the nation-state, he activates the concept of empire to denote traditional and modern China as a multi-ethnic territorial political entity. This connects him to postcolonial and cultural studies in the West. His broad vision of Asia allows him to find intellectual colleagues in Japan (沟口雄三, Mizoguchi Yuzo; 柄谷行人，Karatani Kojin), India (Partha Chatterjee), Egypt (**Samir Amin**), South Korea, Turkey and elsewhere. In response to Western ideas of capitalism and the nation-state, Wang engages and modernises the traditional Left discourse and reinvigorates it in the context of China's opening and reform process, placing 20th-century revolution at the core of his historiography.

Wang Hui's work is both sophisticated and controversial within China. Unwilling to accept the adequacy of standard orthodox notions, both Left and Right, of China's development processes, he has ultimately attempted a new fusion that is more than just the combining of separate components. It is simultaneously a new vision of China and of the world within which it is an integral part.

Major works

Wang has authored dozens of books, articles and public statements on scholarly and sociopolitical issues. A portion of his work has been translated into English and other languages. Key monographs in Chinese include:

Wang, H. (2015) *The Short 20th Century*, *短二十世纪*.

—— (2010) *From an Asian Perspective*: The Narrations of Chinese History, *亞洲視野: 中國歷史的敘述*.

——- (2015) *The Rise of Modern Chinese Thought, 現代中國思想的興起, 2004–2009*, 4 vols, 3rd edn, SDX Joint Publishing Company.

Works translated into English

Wang, H. (2003) *China's New Order: Society, Politics, and Economy in Transition*, trans. T. Huters and R. Karl, Cambridge, MA: Harvard University Press.

—— (2010) *The End of Revolution: China and the Limits of Modernity*, London: Verso.

—— (2011) *The Politics of Imagining Asia*, trans. T. Huters, Cambridge, MA: Harvard University Press.

—— (2014) *China from Empire to Nation-State*, Cambridge, MA: Harvard University Press.

—— (2016) *China's Twentieth Century: Revolution, Retreat and the Road to Equality*, London: Verso.

—— (forthcoming) *The Rise of Modern Chinese Thought*, Cambridge, MA: Harvard University Press.

Further reading

Braudel, F. (1981–1984) *Civilization and Capitalism, 15th–18th Century*, 3 vols, New York: Harper and Row. (Original editions in French, 1979.)

Gang, Yue (2005) '"Wang Hui"', in E.L. Davis (ed.), *Encyclopedia of Contemporary Chinese Culture*, London: Routledge.

Huters, T. (2011) '"Introduction"', in *The Politics of Imagining Asia*, trans. T. Huters, Cambridge, MA: Harvard University Press, 1–10.

Weber, R. (2011) '"On Wang Hui's Re-imagination of Asia and Europe"', *Europa Regional* 17: 221–228.

Zhang, Xiang, and He Jixian (eds) (2014) *In Search for the Modernity of China: Review Articles on Wang Hui's Works* (in Chinese), Oriental Press. *张翔，何吉贤编：《探寻中国的现代性：汪晖学术思想评论集（一）》，东方出版社2014年版.*

Zhang, Yongle (2008) '"No Forbidden Zone in Reading? Dushu and the Chinese Intelligentsia"', *New Left Review* 49: 5–26.

JOSHUA MULDAVIN WITH KONG YUAN

ERIC R. WOLF (1923–1999)

As an anthropologist, Eric Wolf's unique contribution to the theory of development stems from his critical approach, insisting on the inseparable linkages between the local and the global. This helps to overcome vital impasses and difficulties in development thinking.

Born in Vienna in 1923, Wolf was the son of a "multicultural" couple. His father, a secular Jew from Austria, had been in Siberia as a prisoner of war during World War I. There he met his wife, from a Jewish family that had been exiled for their participation in the 1905 revolution. His mother's tales of Siberia, as well as his father's accounts of travels in Latin America, enhanced the cosmopolitan bent of the family and may well have fostered Wolf's later interest in cultural anthropology and particularly in Latin America. In 1933, the year the Nazis came to power in Germany, Wolf moved hastily to the Sudetenland (then Czechoslovakia) with his family. In 1938, with Nazi expansion into Austria and Czechoslovakia, Eric Wolf was sent to school in England. Ironically confined to an internment camp for enemy aliens in early 1940, he met the sociologist Norbert Elias, who raised his interest in social science. The impact on Wolf of these traumatic early experiences and how they found resonance in some of his later professional work – especially his choice of the Nazis as one of the three extended case studies along with the Kwakiutl and Aztecs in his important book *Envisioning Power* (Wolf 1999) – are examined by Simon (2019).

After arriving in New York City, Wolf "discovered" cultural anthropology after stints in biology, political science, economics and sociology. After serving in the US Army, he gained his BA from Queens College in 1946 and became a PhD student at Columbia. There he belonged to a group of students sharing a similar background as veterans and Marxists who later went on to become quite eminent cultural anthropologists. Studying under Julian Steward, the "founding father" of "cultural ecology" or "cultural materialism" from 1947, Wolf commenced field research in Puerto Rico among poor coffee growers and gained first-hand knowledge of their risky and hard everyday life. After completing his PhD in 1951, Wolf held various teaching positions and conducted research in Mexico and South Tyrol. A professor at the University of Michigan from 1961, he became increasingly committed to political issues: in 1965, together with fellow-anthropologist Marshall Sahlins, he originated the idea of and participated in the first teach-in against the Vietnam War on the University of Michigan campus. Under the impact of this war, Wolf wrote his famous *Peasant Wars of the Twentieth Century* (1969). In this period, together with Joseph C. Jorgensen, he also led an investigation of the American Anthropological Association's Ethics Committee on the misuse of ethnographical research for counter-insurgency in Southeast Asia (cf. Jorgensen and Wolf 1971). A Distinguished Professor at the City University of New York (CUNY) from 1971, Wolf, encouraged by his second wife and fellow-anthropologist, Sydel Silverman, shifted his research towards world systems theory and the development of capitalism. This resulted in *Europe and the People Without History* (1982). Besides teaching at CUNY, he continued research into the worldwide spread of capitalism and its consequences for the rise of new forms of power and domination. These were also the themes of his last book (Wolf 1999). Eric Wolf died in the same year.

During the early years of Eric Wolf's academic career, development thinking was virtually possessed by various modernisation theories. For all their differences, these theories shared "dualistic" concepts: "tradition", seen as an obstacle to development, had to be modernised away to achieve "rational, progressive modernity". Arguably the "corporate community" was considered the most formidable of these traditional obstacles. Populated by a "backwards-oriented peasantry", such communities were pictured as regulated by immutable tradition that did not allow for individual development and success (see **Walt Rostow**).

One of Wolf's early contributions (1957) showed that this "ancient traditional" corporate community was a red herring. Referring to Mesoamerica and Central Java, Wolf unveiled such communities as, in fact, products of European expansion, thus anticipating critical debate by ten to twenty years. In *Peasants* (1966), he argued that the peasantry is not an amorphous mass beholden to tradition, but has its own structured forms of social, economic and political organisation, set between "tradition" and "modernity", and quite capable both of change and of stubborn persistence. Responding to US involvement in Vietnam, Wolf (1969) examined the reasons why peasants emerged as the most revolutionary force of the 20th century. He had a substantial impact on the debate on the role of the peasantry in history as well as the modern, especially the underdeveloped world (see Shanin 1971).

As an anthropologist, Wolf grappled with a problem of cultural-relativist scientific fiction that is still current today: the idea of bounded, more or less self-contained community. In contradistinction to this fiction, Wolf insisted on the intimate and "dialectical" interrelation between local communities or "small populations" and complex, or "total societies" (Cole and Wolf 1999: 3). The task was to unveil the close interconnections between processes evolving on a regional, national or even global level. It was from this vantage point that Wolf contributed decisively to social science development theory. Rather than attributing pristine, "traditional" conditions to the local, while "modern" would be the big world somewhere outside, for Wolf, peasant communities, as well as individual persons and households, constitute interfaces between local environments and overarching social connections, reaching from regions right up to the global nexus of the capitalist world market.

In *Europe and the People Without History*, his vital contribution to development theory, Wolf posits that "the world of humankind constitutes a manifold, a totality of interconnected processes" (1982: 3). He stresses that we all live in one huge and complex ecosystem, but also insists that throughout history, there have been "connections everywhere" (p. 3). Therefore, the constructions of modernisation theories are, in fact, mere mythology, based on the false assumption of discrete, solitary cultures or communities. US history, the penultimate model and goal of modernisation theory, also was "a complex orchestration of antagonistic forces" rather than "the unfolding of timeless essence" (p. 5). Therefore, history did not follow an inexorable trajectory. Rather, there had been alternatives. Any idea, then, of human history or social development as a triumphant march to progress was deeply misconceived. Yet, as Wolf noted, such conceptions underlie the splintered array of disciplines that since the mid-19th century deal with "inquiry into the nature and varieties of human kind" (p. 7).

Against widespread reification, Wolf reconstructed what happened with the *People Without History* in the course of European expansion since the 15th century. Above all he was concerned with the role that non-Europeans had played in the making of today's world and thus in shaping our ideas about development. In reconstructing historical process both at the global and at the micro-levels, he also hoped to overcome the blind spots of historical and developmental vision that he saw inherent in the relations of domination established on a world scale during the last 500 years. In this conception, "both the people who claim history as their own and the people to whom history has been denied emerge as participants in the same historical trajectory" (1982: 23). Accordingly, the idea of differential stages of development among societies is mistaken. Modernisation theory was fundamentally wrong precisely in its fundamental tenets that necessarily discriminate between different peoples and regions in the world: the fundamental changes of modernity

> affected not only people singled out as the carriers of "real" history but also the populations anthropologists have called "primitives" … The global processes set in motion by European expansion constitute *their* history as well. There are thus no "contemporary ancestors," no people without history.
>
> (Wolf 1982: 385)

The book title, *Europe and the People Without History*, is therefore deeply ironic. Wolf first sketches the situation of the world around 1400 CE as one of manifold interconnectedness, by trade, by relations between pastoralists and agriculturalists and by war. Western Europe developed from a marginal peninsula of the Eurasian land mass into a "hub of wealth and power" (1982: 123) by processes that set in around 800 CE, and from the sixteenth century onwards, England and the Netherlands were able to harness commerce and war to a grand strategy of expansion that was to transform the world. Wolf explores four exemplary cases: the implantation of Iberian colonies in the Americas, the North American fur trade, the transatlantic slave trade, and the establishment and expansion of European control in South and East Asia.

By giving centre stage to the intricate relationships of agency, Wolf manages to overcome the interrelated views of the colonised as basically backward and as mere victims and objects of colonial power. Rather, social structures pertaining in "Indian" communities in Spanish America or at the Guinea coast in West Africa during the age of the slave trade were shaped deeply by contemporary relationships of domination, exchange and exploitation in the context of an expanding world market. Moreover, Africans or American Indians were actively involved in shaping these conditions, even though the trading system was centred around Western Europe and precious metals flowed from America towards the Iberian peninsula and East Asia. While peasant communities in the Spanish colonies resulted from complex relationships and conflict between colonial administrators and peasants, the famous League of the Iroquois underwent fundamental changes in re-orienting towards the fur trade and warfare in an age of intense British–French rivalry around the Great Lakes; further, the lifestyles of Prairie Indians resembled today's standard image of "Indian" life only in the wake of deep transformations through the spread of trade, horses and firearms. The transatlantic slave trade that transported millions from West and Central Africa to the plantations of the "New World"[1] also built on previously existing institutions in African societies which were then fundamentally transformed by its consequences. In particular, centralised military and commercial powers were enormously enhanced in the Guinea Coast kingdoms that thrived on the slave trade.

To date, the historical process that has been common to all societies reviewed here has culminated in the advent and universal expansion of capitalism. In contradistinction to authors such as Wallerstein or **Andre Gunder Frank**, Wolf relates the capitalist mode of production strictly to the domination of capital at the point of production, not just to the spread of market relations around the world. His detailed account of the specific constellation that conditioned the industrial revolution and the advent of capitalism in England reveals an "unusual development" (1982: 268) with immediate global repercussions. Overseas regions were at various times converted into raw material bases for the burgeoning British cotton industry. Again, Wolf stresses the ambivalent dimension of agency: besides suffering oppression, the slave population in the Cotton South also developed "its own store of experience and modes of coping" which it "would pass ... on across the generations" (p. 281). During the 20th century, novel social relationships were spawned by capitalism's "tendency ... to expand in search of new raw materials ... and ... of cheap labor to process them" (p. 302). For example, the commercialisation of rubber caused vastly divergent transformations, from one stage of a successive chain of fundamental changes touched off by capitalist expansion in the Amazon rainforest, to the recruitment of Indian labourers to Malaya to work the rubber plantations. The worldwide recruitment of labour for widely varying concerns caused outcomes including migration, proletarianisation and ethnic segmentation. For Wolf, "working classes are not "made" in the place of work alone" (p. 360), underscoring the need for multidisciplinary approaches, integrating cultural factors. Here as in his other works, Wolf combines rich and diverse empirical data with a conceptual rigour to forge a vitally creative theoretical contribution.

Two decades after his death, Wolf's interventions remain just as relevant. This applies to a realistic, more long-term view of "globalisation" and its deep though differential effects across the world. It also holds true for a view of "labour" which has never been captured adequately by being confined to industrial work, but hinges on an adequate positioning of this form as well as of the host of other human productive activities which today remain subsumed under capital.

Note

1 Itself a very Eurocentric construct – *DS.*

Major works

Cole, J.W. and Wolf, E.R. (1999) *The Hidden Frontier: Ecology and Ethnicity in an Alpine Valley*, Berkeley: University of California Press. (Original publication New York: Academic Press 1974.)

Jorgensen, J. and Wolf, E. (1971) "Anthropology on the Warpath: An Exchange", *New York Review of Books*, 8 April and 22 July 1971.

Wolf, E.R. (1957) "Closed Corporate Peasant Communities in Mesoamerica and Central Java", *Southwestern Journal of Anthropology* 13 (Spring): 1–18.

—— (1966) *Peasants*, Englewood Cliffs, NJ: Prentice Hall.

—— (1969) *Peasant Wars of the Twentieth Century*, New York: Harper and Row.

—— (1982) *Europe and the People Without History*, Berkeley: University of California Press.

—— (1999) *Envisioning Power: Ideologies of Dominance and Crisis*, Berkeley: University of California Press.

—— with Silverman, S. (2001) *Pathways of Power: Building an Anthropology of the Modern World*, Berkeley: University of California Press.

Further reading

Schneider, J. (2002) "Eric R. Wolf", *American National Bibliography* online: www.anb.org/articles/14/14-01125.html.

—— and Rapp, R. (eds) (1995) *Articulating Hidden Histories: Exploring the Influence of Eric R. Wolf*, Berkeley: University of California Press.

Shanin, T. (ed.) (1971) *Peasants and Peasant Societies*, Harmondsworth: Pelican.

Simon, D. (2019) *Holocaust Escapees and Global Development: Hidden Histories*, London: Zed Books.

REINHART KÖßLER AND TILMAN SCHIEL

PETER WORSLEY (1924–2013)

Peter Worsley, the British social anthropologist, was born in Birkenhead (Cheshire) in 1924. He began to read English at Cambridge University but then joined the British Army during the Second World War, serving as an officer in Africa and India. Returning to Cambridge after the war, he switched his studies to the field of anthropology. After a short spell in East Africa he returned to Britain, taking a Master's degree at Manchester University where he focused on a critique of Meyer Fortes's work. He subsequently worked as a researcher with S.F. Nadel at the Australian National University, completing a dissertation on the Australian Aborigines in 1954. Back in Britain he returned to Manchester, joining Max Gluckman there and starting his famous work on cargo cults. His career, however, was to switch from social anthropology to sociology when he took up an early post in that discipline at Hull University. He then became the first Professor of Sociology at Manchester University, a post he held from 1964 to 1982 (see Worsley 2008 for his autobiography).

Worsley's intellectual career, however, cannot be understood in purely academic terms. At around the time of the Battle of Stalingrad he joined the British Communist Party, along with other intellectuals of his generation. Also in common with them, he left the CPGB in 1956 after the Soviet invasion of Hungary. This "political" side of Worsley's intellectual career is rarely made explicit but it helps to explain, for example, why the Preface to his book *The Third World* contains an acknowledgement to labour historians E.P. Thompson and John Saville. Their humanist, democratic and "English" vision of **Marx**ism was to inform all his subsequent work. It never became sectarian, pedantic or merely scholarly. The negative side of Worsley's political commitment was a succession of Cold War establishment vetoes on his work as an anthropologist, especially refusals of posts and access to fieldwork sites. This partly explains his conversion to sociology, although, as we shall see, he never stopped being a social anthropologist at heart.

The Trumpet Shall Sound (1957) is Worsley's first and still best-known work, virtually creating the area of "cargo cult" studies in social anthropology. He set these millenarian social movement cults in Melanesia in a broad historical and comparative context for the first time. Against the prevailing "nativist" interpretation that saw these movements as traditional and frankly irrational flights of fancy, Worsley stressed their proto-nationalist character and the protest they articulated against racial and colonial oppression. In a methodological appendix to the riveting story of the cargo cults, Worsley argues that:

> The concept of "nativism" … fails to account for many cult elements … Where old beliefs and practices are retained or revived, they have a new content … And there is little that is reversionary or perpetuative in the concept of expelling the Europeans … The movements are thus forward-looking, not regressive, the order of the future being the *reversion* of the present.
>
> (Worsley 1957: 275–276)

Worsley not only writes an extremely effective (and affecting) account of the Melanesian cargo cults but also sets them within their anticolonial context against the prevailing traditional anthropological focus of the day. Many generations of students were "converted" by reading *The Trumpet Shall Sound* to a passionate commitment to development and anticolonialism.

In 1964 Worsley published *The Third World*, which certainly popularised if not started the circulation of a term that had originated in French development thinking. It dealt with the colonial relationship of the majority world with the West, the rise of nationalism, the importance of populism, the structure of the new emerging postcolonial states and the "positive neutralism" of the Third World in the Cold War, then at its height. Peter Worsley, against the dominant modernisation perspective in social and political theory, did not see "democracy" versus "totalitarianism" as the main global divide but, rather, one "between those movements and organisations oriented to radical social change and those resisting it" (1964: 355), primarily located in the USA. In terms of the politics emerging in the post-colonial world he argued, against the orthodox Marxists of the day, for the importance of populism, a cross-class movement that "has been infinitely greater than has been recognised and the potential under-estimated" (p. 170). Despite his own recent communist past, Worsley argued that the Third World challenges both "late capitalism" and "late communism" (p. 271) and offers even to transcend itself. Against the dominant materialism of both "bourgeois" and "radical" development thinkers alike, Worsley declared: "People who consider ideas and ideals unimportant ... are ignorant of the real world" (p. 271).

Sometimes it is an author's lesser-known work that is more revealing in a way. Not having the cult status of *The Trumpet Shall Sound* or *The Third World*, these works have not received the attention they deserve, but they are perhaps even more relevant to Peter Worsley as development thinker. The first to be considered is *Two Blades of Grass* (Worsley 1971), an edited collection on the role of rural co-operatives in agricultural modernisation. Worsley sets the theme of co-operation in the broad context of the struggle against inequality and the possessive individualism of contemporary capitalism. He also develops a theme that is central in his later writings, namely the "populist" dimension of democracy in the Third World, an aspect "that so much structural political sociology ignores, concentrating as it does on the formal elements in democracy" (p. 10). He also rightly sees the state being more active than in the advanced industrialised countries "because capital accumulation will be inadequate if left to the free play of the market" (pp. 15–16). Yet the state most often intervenes to assist those who already have power and it is "traditional ties of kinship – be it those of neighbourhood, ethnicity or caste – that most often work against the requirements of strict economic rationality" (p. 23).

The mid-1970s were a period in which many Western intellectuals made pilgrimages to communist China and enthused on the Cultural Revolution and agrarian reform in that country. Not many people know Worsley's *Inside China* (1975), but it provides great insight into his own politics. As an ethnographic study, the book works well and is more balanced than some others appearing at the time. Worsley was sympathetic to the socialist transformation experiment going on in China at the time, but bemoaned "the virtual absence of 'development theory' and of social science as we know it" (p. 98) in China, and even acknowledged that "bourgeois scholars have often been right about very many things, big and small, and have contributed enormously in areas where the contribution of Marxism has been virtually invisible" (p. 199). While acknowledging the "pragmatic" Chinese role in world politics, Worsley still concluded that "there is no reason to assume that

China has abandoned her commitment to world socialist revolution" (p. 247). Even more jarring was his presumably naïve statement: "I have no idea how common the death penalty is but I should think it very rare. Nor do I believe that there are many concentration camps" (p. 215).

In the mid-1980s, Worsley came back into development theory with the magnificent monograph, *The Three Worlds* (1984), which focused on culture and world development. While in one way it was simply *The Third World* 20 years on, its focus was quite different. Reflecting his much earlier engagement with British labour historians like Edward Thompson and John Saville, Worsley had two of his three core chapters dealing with the "undoing of the peasantry" and the "making of the working class" respectively. Ethnicity and nationalism, traditional concerns of his, formed the third leg of this great piece of work. The rise of agribusiness and the development of a strong urban working class described in *The Three Worlds* reflected work he was then beginning on international labour studies. Worsley also, if belatedly, moved beyond his previous British Africa/Asia focus to engage with the rapidly transforming Latin America, with its stronger indigenous intellectual and analytical traditions. But the main point of this work was that Worsley identified "culture" as the main missing ingredient in the traditional sociology of development as well as in Marxist sociology. Taking his cue from the British cultural theorist Raymond Williams as much as from anthropology, Worsley re-introduced culture into the analysis of world development.

Worsley's last book, *Knowledges* (1997), returned to his early interest in language, thought and culture more generally. His basic argument was that knowledge takes many forms and is by nature plural. Extremely diverse in its topics – from Oceanic navigation techniques to Disneyland, and from aboriginal Australian botanical classification to the secularisation of Christmas – this was an ambitious and challenging book. Conventional Western distinctions between science and culture are, in the course of their discrete narratives, deconstructed and undermined. In the era of resurgent US imperialism, Worsley effectively shattered the premise of Western cultural superiority. Worsley was seemingly returning to his vocation as cultural anthropologist and revisiting the fieldwork he had carried out among the aboriginal people of Groote Eylandt in Australia's Gulf of Carpentaria in 1952–1953 (see http://pittrivers-photo.blogspot.co.uk/2013/11/peter-worsleys-groote-eylandt-fieldwork.html). In a worthy reprise of *The Trumpet Shall Sound*, Worsley engaged with contemporary debates on the science/culture and knowledge/belief distinctions with an original and radical intervention. As with all Worsley's work, it is supremely accessible.

Peter Worsley was a theorist of development, but in Britain he is probably best known for his edited sociology textbooks that served to educate a whole generation of students in the 1970s and 1980s (Worsley 1970). There was a breadth and accessibility to these texts that completely surpassed the competition for a whole generation. Then there was his popular text on *Marx and Marxism* (Worsley 2002) that systematised the open **Marx**ist method underpinning his own particular brand of social anthropology. For Worsley, Marxism was anything but an abstract philosophy, and his engagement with the influential contemporaneous "French school" of anthropology (Claude Meillassoux, Maurice Godelier and Emmanuel Terray) was tangential at best. Rather, and quite characteristically, Worsley portrayed Marxism as a political philosophy that had been put to use in various parts of the world with varying degrees of success. That open, flexible and pragmatic approach to social and political theory was reflected in his continuous commitment to interdisciplinarity, really a post-disciplinarity where terms like anthropology, sociology and social anthropology never took on a life of their own and whose disputes were never allowed to dominate the problems of development.

With any thinker there are, finally, invariably certain silences or lacunae. The strengths of an "understated" Marxism mean that Worsley did not engage in the important (if perhaps "theological") Marxist debates in the 1970s and 1980s on issues of development and underdevelopment. Likewise, he shows no interest whatsoever in the "post" debates – postcolonialism, post-structuralism and postmodernism – that were well under way in the mid-1980s when *The Three Worlds* appeared. Also one of his strengths – a groundedness in the British communist labour historian tradition – was his distinctly Anglocentric perspective in terms of sources and debates, however global his own view was. Thus while **Andre Gunder Frank**'s dependency approach is flagged up already in the 1960s work, *The Three Worlds* of 1984 fails even to mention the classic *Dependency and Development in Latin America* by **Fernando Henrique Cardoso** (and Enzo Faletto) which had already appeared in English in 1979. Notwithstanding these ethnocentric tendencies (even stronger in the earlier works), Worsley did "see" globalisation coming over the horizon in the early 1980s when he referred to "a new 'transcendental' global level" (1984: 317) and articulated the globalisation and culture debates before they even happened.

Major works

Worsley, P. (1957) *The Trumpet Shall Sound: A Study of "Cargo" Cults in Melanesia*, London: McGibbon and Kee. (2nd edn 1968.)

—— (1964) *The Third World*, London: Weidenfeld and Nicolson. (2nd edn 1967.)
—— (1975) *Inside China*, London: Allen Lane.
—— (1984) *The Three Worlds: Cultural and World Development*, London: Weidenfeld and Nicolson.
—— (1997) *Knowledges: Culture, Counter-culture, Subculture*, New York: New Press.
—— (ed.) with Ann Allen (1971) *Two Blades of Grass: Rural Co-operatives in Agricultural Modernization*, Manchester: Manchester University Press.

Other works

Worsley, P. (ed.) (1970) *Introducing Sociology*, Harmondsworth: Penguin.
—— Jefferys, M. and Brown, R.E. (1987) *The New Introducing Sociology*, Harmondsworth: Penguin.
—— (2002) *Marx and Marxism*, London and New York: Routledge.
—— (2008) *An Academic Skating on Thin Ice*, Oxford: Berghahn Books.

RONALDO MUNCK

INDEX